CHEMICAL ENGINEERING

VOLUME ONE
THIRD EDITION
(SI units)

Other Titles in the Series

CHEMICAL ENGINEERING

VOLUME ONE

FLUID FLOW, HEAT TRANSFER
AND MASS TRANSFER

J. M. COULSON

Emeritus Professor of Chemical Engineering
University of Newcastle-upon-Tyne

and

J. F. RICHARDSON

University College of Swansea

WITH EDITORIAL ASSISTANCE FROM
J. R. BACKHURST and J. H. HARKER

University of Newcastle-upon-Tyne

THIRD EDITION
(SI units)

PERGAMON PRESS

OXFORD · NEW YORK · TORONTO · SYDNEY · PARIS · FRANKFURT

U.K.	Pergamon Press Ltd., Headington Hill Hall, Oxford, OX3 0BW, England
U.S.A.	Pergamon Press Inc., Maxwell House, Fairview Park, Elmsford, New York 10523, U.S.A.
CANADA	Pergamon of Canada Ltd., 75, The East Mall, Toronto, Ontario M8Z 2L9, Canada
AUSTRALIA	Pergamon Press (Aust.) Pty. Ltd., 19a Boundary Street, Rushcutters Bay, N.S.W. 2011, Australia
FRANCE	Pergamon Press SARL, 24 rue des Ecoles, 75240 Paris, Cedex 05, France
WEST GERMANY	Pergamon Press GmbH, 6242 Kronberg/Taunus, Pferdstrasse 1, Frankfurt-am-Main, West Germany

Copyright © 1977 J. M. Coulson and J. F. Richardson

All Rights Reserved. No part of this publication may be reproduced, stored in a retrieval system, or transmitted, in any form or by any means, electronic, mechanical, photocopying, recording or otherwise, without the prior permission of the publishers.

First edition 1954
Reprinted with corrections 1955, 1956, 1957, 1959, 1960
Reprinted 1961, 1962
Revised Second Edition 1964
Reprinted with corrections 1966, 1970
Revised Third Edition (SI units) 1977

Library of Congress Cataloging in Publication Data

Coulson, John Metcalfe.
Chemical engineering.

Includes index.
CONTENTS: v. 1. Fluid flow, heat transfer, and mass transfer.
v. 4. Backhurst, J. R. and Harker, J. H. Solutions to the problems in Chemical engineering volume 1.
1. Chemical engineering. I. Richardson, John Francis joint author. II. Title.
TP155.C69 1976   660.2   75-42295
ISBN 0-08-020614-X (v. 1) Hard cover
      0-08-021015-5 Flexi cover

Printed in Great Britain by A. Wheaton & Co. Exeter

Contents

Preface to Third Edition

THE introduction of the SI system of units by the United Kingdom and many other countries has itself necessitated the revision of this engineering text. This clear implementation of a single system of units will be welcomed not only by those already in the engineering profession, but even more so by those who are about to join. The system which is based on the c.g.s. and m.k.s. systems using length (**L**), mass (**M**), and time (**T**) as the three basic dimensions, as is the practice in the physical sciences, has the very great advantage that it removes any possible confusion between mass and force which arises in the engineering system from the common use of the term *pound* for both quantities. We have therefore presented the text, problems, and examples in the SI system, but have arranged the tables of physical data in the Appendix to include both SI and other systems wherever possible. This we regard as important because so many of the physical data have been published in c.g.s. units. For similar reasons, engineering units have been retained as an alternative where appropriate.

In addition to the change to the SI system of units, we have taken the opportunity to update and to clarify the text. A new section on the flow of two-phase gas–liquid mixtures has been added to reflect the increased interest in the gas and petroleum industries and in its application to the boiling of liquids in vertical tubes.

The chapter on Mass Transfer, the subject which is so central and specific to chemical engineering, has been considerably extended and modernised. Here we have thought it important in presenting some of the theoretical work to stress its tentative nature and to show that, although some of the theories may often lack a full scientific basis, they provide the basis of a workable technique for solving problems. In the discussion on Fluid Flow reference has been made to American methods, and the emphasis on Flow Measurement has been slanted more to the use of instruments as part of a control system. We have emphasised the importance of pipe-flow networks which represent a substantial cost item in modern large-scale enterprises.

This text covers the physical basis of the three major transfer operations of fluid flow, heat transfer, and mass transfer. We feel that it is necessary to provide a thorough grounding in these operations before introducing techniques which have been developed to give workable solutions in the most convenient manner for practical application. At the same time, we have directed the attention of the reader to such invaluable design codes as TEMA and the British Standards for heat exchanger design and to other manuals for pipe-flow systems.

It is important for designers always to have in their minds the need for reliability and safety: this is likely to follow from an understanding of the basic principles involved, many of which are brought out in the text.

We would like to thank our many friends from several countries who have written with suggestions, and it is our hope that this edition will help in furthering growth and interest in the profession. We should also like to thank a number of industrialists who have made available much useful information for incorporation in this edition; this help is acknowledged at the appropriate point. Our particular thanks are due to Dr. B. Waldie for his contribution to the high temperature aspects of heat transfer and to the Kellogg International Corporation and Humphreys and Glasgow Limited for their help. In conclusion, we would like to thank Dr. J. R.

Backhurst and Dr. J. H. Harker for their editorial work and for recalculating the problems in SI units and converting the charts and tables.

Since the publication of the Second Edition of this Volume, Volume 3 of *Chemical Engineering* has been published in order to give a more complete coverage of those areas of chemical engineering which are of importance in both universities and industry in the 1970's.

January 1976

Preface to Second Edition

IN presenting this second edition, we should like to thank our many friends from various parts of the world who have so kindly made suggestions for clarifying parts of the text and for additions which they have felt to be important. During the last eight years there have been changes in the general approach to chemical engineering in the universities with a shift in emphasis towards the physical mechanisms of transport processes and with a greater interest in unsteady state conditions. We have taken this opportunity to strengthen those sections dealing with the mechanisms of processes, particularly in Chapter 7 on mass transfer and in the chapters on fluid mechanics where we have laid greater emphasis on the use of momentum exchange. Many chemical engineers are primarily concerned with the practical design of plant and we have tried to include a little more material of use in this field in Chapter 6 on heat transfer. An introductory section on dimensional analysis has been added but it has been possible to do no more than outline the possibilities opened up by the use of this technique. Small changes will be found throughout the text and we have tried to meet many readers' requests by adding some more worked examples and a further selection of problems for the student. The selection of material and its arrangement are becoming more difficult and must be to a great extent a matter of personal choice, but we hope that this new edition will provide a sound basis for the study of the fundamentals of the subject and will perhaps be of some value to practising engineers.

J. M. COULSON
J. F. RICHARDSON

Note: Chapter numbers given above have been altered in the current edition.

Preface to First Edition

THE idea of treating the various processes of the chemical industry as a series of unit operations was first brought out as a basis for a new technology by Walker, Lewis and McAdams in their book in 1923. Before this, the engineering of chemical plants had been regarded as individual to an industry and there was little common ground between one industry and another. Since the early 1920's chemical engineering as a separate subject has been introduced into the universities of both America and England and has expanded considerably in recent years so that there are now a number of university courses in both countries. During the past twenty years the subject matter has been extensively increased by various researches described in a number of technical journals to which frequent reference is made in the present work.

Despite the increased attention given to the subject there are few general books, although there have been a number of specialised books on certain sections such as distillation, heat transfer, etc. It is the purpose of the present work to present to the student an account of the fundamentals of the subject. The physical basis of the mechanisms of many of the chemical engineering operations forms a major feature of chemical engineering technology. Before tackling the individual operations it is important to stress the general mechanisms which are found in so many of the operations. We have therefore divided the subject matter into two volumes, the first of which contains an account of these fundamentals—diffusion, fluid flow and heat transfer. In Volume 2 we shall show how these theoretical foundations are applied in the design of individual units such as distillation columns, filters, crystallisers, evaporators, etc.

Volume 1 is divided into four sections, fluid flow, heat transfer, mass transfer and humidification. Since the chemical engineer must handle fluids of all kinds, including compressible gases at high pressures, we believe that it is a good plan to consider the problem from a thermodynamic aspect and to derive general equations for flow which can be used in a wide range of circumstances. We have paid special attention to showing how the boundary layer is developed over plane surfaces and in pipes, since it is so important in controlling heat and mass transfer. At the same time we have included a chapter on pumping since chemical engineering is an essentially practical subject, and the normal engineering texts do not cover the problem as experienced in the chemical and petroleum industries.

The chapter on heat transfer contains an account of the generally accepted techniques for calculation of film transfer coefficients for a wide range of conditions, and includes a section on the general construction of tubular exchangers which form a major feature of many works. The possibilities of the newer plate type units are indicated.

In section three, the chapter on mass transfer introduces the mechanism of diffusion and this is followed by an account of the common relationships between heat, mass and momentum transfer and the elementary boundary layer theory. The final section includes the practical problem of humidification where both heat and mass transfer are taking place simultaneously.

It will be seen that in all chapters there are sections in small print. In a subject such as this, which ranges from very theoretical and idealised systems to the practical problems with empirical or experimentally determined relations, there is much to be said for omitting the more theoretical features in a first reading, and in fact this is frequently done in the more practical courses. For this reason the more difficult theoretical sections have been put in small print and the whole of Chapter 9 may be omitted by those who are more concerned with the practical utility of the subject.

In many of the derivations we have given the mathematical analysis in more detail than is customary. It is our experience that the mathematical treatment should be given in full and that the student should then apply similar analysis to a variety of problems.

We have introduced into each chapter a number of worked examples which we believe are essential to a proper understanding of the methods of treatment given in the text. It is very desirable for a student to understand a worked example before tackling fresh practical problems himself. Chemical engineering problems require a numerical answer and it is essential to become familiar with the different techniques so that the answer is obtained by systematic methods rather than by intuition.

In preparing this text we have been guided by courses of lectures which we have given over a period of years and have presented an account of the subject with the major emphasis on the theoretical side. With a subject that has grown so rapidly, and which extends from the physical sciences to practical techniques, the choice of material must be a matter of personal selection. It is, however, more important to give the principles than the practice, which is best acquired in the factory. We hope that the text may also prove useful to those in industry who, whilst perhaps successfully employing empirical relationships, feel that they would like to find the extent to which the fundamentals are of help.

We should like to take this opportunity of thanking a number of friends who have helped by their criticism and suggestions, amongst whom we are particularly indebted to Mr. F. E. Warner, to Dr. M. Guter, to Dr. D. J. Rasbash and to Dr. L. L. Katan. We are also indebted to a number of companies who have kindly permitted us to use illustrations of their equipment. We have given a number of references to technical journals and we are grateful to the publishers for permission to use illustrations from their works. In particular we would thank the Institution of Chemical Engineers, the American Institute of Chemical Engineers, the American Chemical Society, the Oxford University Press and the McGraw-Hill Book Company.

South Kensington,
London S.W.7
1953

Acknowledgements

THE authors and publisher acknowledge the kind assistance of the following organisations in providing illustrative material and tables of properties and data.

Fig. 5.5.   Budenberg Gauge Co.
Fig. 5.6.   Foxboro-Yoxall Ltd.
Fig. 5.15, 5.16, 5.25.   Kent Meters Ltd.
Fig. 5.17.   GEC–Elliott Process Instruments Ltd.
Fig. 5.23.   Baird and Tatlock Ltd.
Fig. 6.1.   Worthington–Simpson Ltd.
Figs. 6.3, 6.10.   Metering Pumps Ltd.
Fig. 6.5.   Watson-Marlow Ltd.
Figs. 6.6, 6.7.   Sigmund Pulsometer Pumps Ltd., Reading.
Figs. 6.8, 6.9.   Mono Pumps Ltd.
Fig. 6.19.   Crane Packing Ltd.
Fig. 6.20.   Hayward Tyler and Co. Ltd.
Figs. 6.25, 6.26, 6.28.   Reavell and Co. Ltd.
Fig. 6.22.   International Combustion Ltd.
Fig. 6.27.   Nash Engineering Co. (GB) Ltd.
Fig. 6.29.   CompAir Industrial Ltd.
Figs. 6.30, 6.31.   Dresser Industries, Dresser Clark Div., New York, USA.
Figs. 6.32, 6.33, 6.34, 6.35, 6.36.   Hick Hargreaves and Co. Ltd.
Fig. 7.41.   American Chemical Society.
Fig. 7.49.   Shell Thornton Research Centre.
Fig. 7.59.   Brown Fintube Co.
Fig. 7.62.   The APV Co. Ltd.
Fig. 7.63.   Ashmore Benson, Pease and Co. Ltd.
Figs. 9.2, 9.3.   Professor F. N. M. Brown, University of Notre Dame.
Figs. 11.10, 11.11, 11.12.   C. F. Casella & Co. Ltd.
Fig. 11.14.   Film Cooling Towers (1925) Ltd.

CHAPTER 1

Units and Dimensions

1.1. INTRODUCTION

At an early stage the student of chemical engineering will discover that the data which he uses are expressed in a great variety of different units, so that he must convert his quantities into a common system before proceeding with his calculations. An attempt at standardisation has been made with the introduction of the *Système International d'Unités* (SI) which will be discussed later and is the main system used in this new edition. This system is widely used on the continent of Europe and is now accepted in the United Kingdom and is likely to spread internationally. Most of the physical properties determined in the laboratory will have been expressed in the c.g.s. system, whereas the dimensions of the full-scale plant, its throughput, design, and operating characteristics will have appeared either in some form of general engineering units or in special units which have their origin in the history of the particular industry. This inconsistency is quite unavoidable and is a reflection of the fact that chemical engineering has in many cases developed as a synthesis of scientific knowledge and practical experience. Familiarity with the various systems of units and an ability to convert from one to another are therefore essential. In this chapter the main systems of units will be discussed and the importance of understanding dimensions emphasised. It will then be shown how dimensions can be used to help very considerably in the formulation of relationships between large numbers of parameters.

The magnitude of any physical quantity is expressed as the product of two quantities; one is the magnitude of the unit and the other is the number of those units. Thus the distance between two points may be expressed as 1 m or as 100 cm or as 3·28 ft. The metre, centimetre, and foot are respectively the size of the units, and 1, 100, and 3·28 are the corresponding number of units.

Since the physical properties of a system are interconnected by a series of mechanical and physical laws, it is possible to regard certain quantities as basic and other quantities as derived. The choice of basic dimensions varies from one system to another but it is usual to take length and time as fundamental. These quantities will be denoted by L and T. The dimensions of velocity, which is a rate of increase of distance with time, can be written as LT^{-1}, and those of acceleration, the rate of increase of velocity, are LT^{-2}. An area has dimensions L^2 and a volume has the dimensions L^3.

The volume of a body does not completely define the amount of material which it contains, and therefore it is usual to define a third basic quantity, the amount of matter in the body, i.e. its mass M. Thus the density of the material, its mass per unit volume, has the dimensions ML^{-3}. Alternatively, the third basic quantity which is chosen may be force F; in some cases, both force and mass can be used simultaneously as basic quantities as will be shown later.

Physical and mechanical laws provide a further set of relations between dimensions. The most important of these is that the force required to produce a given acceleration of a body is proportional to its mass and, similarly, the acceleration imparted to a body is proportional to the applied force.

Thus force is proportional to the product of mass and acceleration,

i.e.

$$F = \text{const } M(LT^{-2})$$

1

The proportionality constant therefore has the dimensions:

$$F/M(LT^{-2}) = FM^{-1}L^{-1}T^2$$

In any set of consistent units it is convenient to put the proportionality constant arbitrarily equal to unity and to define unit force as that which will impart unit acceleration to unit mass. Provided that no other relationship between force and mass is used, the constant may be arbitrarily regarded as dimensionless and the dimensional relationship obtained:

$$F = MLT^{-2}$$

If, however, some other physical law were to be introduced so that, for instance, the attractive force between two bodies is proportional to the product of their masses, then this relation between **F** and **M** will no longer hold.

1.2.   SYSTEMS OF UNITS

Although in scientific work mass is taken as the third fundamental quantity and in engineering force is sometimes used as mentioned above, four fundamental quantities **L, M, F, T** may be used, the main attraction being that the four units are themselves commonly used to express quantities.

The various systems of units and the basic quantities associated with them will now be considered. A summary is given in Table 1.1.

1.2.1.   The Centimetre–Gram–Second (c.g.s.) System

In this system the basic units are of length **L**, mass **M**, and time **T** with the nomenclature:

Length:	Dimension **L**:	Unit 1 centimetre	(1 cm)
Mass:	Dimension **M**:	Unit 1 gram	(1 g)
Time:	Dimension **T**:	Unit 1 second	(1 s)

The unit of force is that which will give a mass of 1 g an acceleration of 1 cm/s^2 and is known as the dyne,

Force: Dimension $F = MLT^{-2}$:   Unit 1 dyne   (1 dyn)

1.2.2.   The Système International d'Unités (SI)

This system is a modification of the c.g.s. system but uses larger units. The basic dimensions are again of **L, M**, and **T**, but their values are different.

Length:	Dimension **L**:	Unit 1 metre	(1 m)
Mass:	Dimension **M**:	Unit 1 kilogram	(1 kg)
Time:	Dimension **T**:	Unit 1 second	(1 s)

The unit of force, known as the newton, is that force which will give an acceleration of 1 m/s^2 to a mass of one kilogram. Thus $1 \text{ N} = 1 \text{ kg m/s}^2$ with dimensions MLT^{-2}, and it will be seen that one newton equals 10^5 dynes. The energy unit, the newton metre, is 10^7 ergs and is called the joule; and the power unit, equal to one joule per second, is the watt.

Thus	Force:	Dimension MLT^{-2}:	Unit 1 newton	(1 N) or 1 kg m/s^2
	Energy:	Dimension ML^2T^{-2}:	Unit 1 joule	(1 J) or $1 \text{ kg m}^2/\text{s}^2$
	Power:	Dimension ML^2T^{-3}:	Unit 1 watt	(1 W) or $1 \text{ kg m}^2/\text{s}^3$

For many purposes, the chosen unit in the SI system will be either too large or too small for practical purposes. The following prefixes are now adopted as standard. Multiples or sub-multiples in powers of 10^3 are preferred; thus, for instance, millimetre should always be used in preference to centimetre.

10^{12}	tera	(T)
10^9	giga	(G)
10^6	mega	(M)
10^3	kilo	(k)
10^2	hecto	(h)
10^1	deca	(da)
10^{-1}	deci	(d)
10^{-2}	centi	(c)
10^{-3}	milli	(m)
10^{-6}	micro	(μ)
10^{-9}	nano	(n)
10^{-12}	pico	(p)

The prefixes should be used with great care and be written immediately adjacent to the unit to be qualified; furthermore only one prefix should be used at a time to precede a given unit. Thus, for example, 10^{-3} metre, which is one millimetre, is written 1 mm. 10^3 kg is written as 1 Mg, not as 1 kkg—this shows immediately that the name *kilogram* is an unsuitable one for the basic unit of mass and a new name may well be given to it in the near future.

1.2.3. The British Engineering System

This system uses the foot and second as the units of length and time but employs the pound force as the third fundamental unit. The pound force is defined as that force which gives a mass of one pound an acceleration of $32 \cdot 17$ ft/s². It is therefore a fixed quantity and must not be confused with the pound weight which is the force exerted by the earth's gravitational field on a mass of one pound and which varies from place to place.

The unit of mass in this system is known as the slug, and is the mass which is given an acceleration of 1 ft/s² by a one pound force:

$$1 \text{ slug} = 1 \text{ (pound force) (ft)}^{-1} \text{ (second)}^2$$

1.2.4. The Foot–Pound–Second (f.p.s.) System

Here the three basic units are length L, value 1 ft, mass M, value 1 lb, and time T, value 1 s.

The unit of force is the poundal which is the force required to give an acceleration of 1 ft/s² to a mass of 1 lb:

$$1 \text{ poundal} = 1 \text{ (lb mass) (ft) (s)}^{-2}$$

It will be noted on comparing the British engineering and the f.p.s. systems that:

$$1 \text{ slug} = 32 \cdot 17 \text{ lb mass} \quad \text{and} \quad 1 \text{ lb force} = 32 \cdot 17 \text{ poundals}$$

Misunderstanding often arises from the fact that the unit of mass in the one system has the same name as the unit of force in the other, viz. the pound. To avoid confusion the pound mass should be written as lb or even lb_m and the unit of force lb_f.

Many writers, particularly in America, use both the pound mass and pound force as basic units in the same equation because they are the units in common use. This is an essentially inconsistent system and requires great care in use. In this system a proportionality factor between force and mass is defined as g_c given by:

$$\text{Force (in pounds force)} = \text{(mass in pounds) (acceleration in ft/s²)}/g_c$$

Thus in terms of dimensions: $F = (M)(LT^{-2})/g_c$

and g_c has the dimensions $F^{-1}MLT^{-2}$.

1.2.5. Derived Units

The three fundamental units of the SI and of the c.g.s. systems are length, mass, and time. It has been shown that force can be regarded as having the dimensions of MLT^{-2}; the dimensions of many other parameters may be worked out in terms of the basic MLT system.
   For example:

   Energy is given by the product of force and distance with dimensions ML^2T^{-2}.
   Pressure is the force per unit area with dimensions $ML^{-1}T^{-2}$.
   Viscosity is defined as the shear stress per unit velocity gradient with dimensions $(MLT^{-2}/L^2)/(LT^{-1}/L) = ML^{-1}T^{-1}$.
   Kinematic viscosity is the viscosity divided by the density with dimensions $ML^{-1}T^{-1}/ML^{-3} = L^2T^{-1}$.

The units, dimensions, and normal method of expression for these quantities in the SI system may now be stated.

Parameter	Unit	Dimensions	In terms of kg, m, s
Force	newton	MLT^{-2}	1 kg m/s²
Energy or work	joule	ML^2T^{-2}	1 kg m²/s² (= 1 N m)
Power	watt	ML^2T^{-3}	1 kg m²/s³ (= 1 J/s)
Pressure	pascal	$ML^{-1}T^{-2}$	1 kg/m s² (= 1 N/m²)
Viscosity		$ML^{-1}T^{-1}$	1 kg/m s (= 1 N s/m²)
Frequency	hertz	T^{-1}	1 1/s

1.2.6. Thermal Units

Since heat is a form of energy, it can be expressed in units with dimensions ML^2T^{-2}, but it is not always convenient to express it in this way. If the concept of temperature is introduced with dimension θ then the thermal energy H of a body is conveniently expressed as $(H) \propto$ mass $(M) \times$ temperature (θ). The proportionality constant is the specific heat capacity which varies from material to material unlike the proportionality constant in Newton's law of motion. It is therefore necessary to define heat quantities in terms of a particular material, usually water at 15°C (298 K). If this is done using thermal units for heat energy the proportionality constant can be put equal to unity for water and, dimensionally:

$$H = M\theta$$

In the SI system the quantity heat is always expressed as joules, in the same way as other forms of energy. In the c.g.s. system the unit of heat is defined as that quantity which will raise the temperature of one gram of water through 1°C at atmospheric pressure.
   In the f.p.s. and British engineering systems the British thermal unit is the quantity of heat required to raise one pound of water through 1°F (60–61°F) and the pound calorie or centigrade heat unit is that quantity required to raise it through 1°C.
   The calorie and kilocalorie have now been redefined in terms of the joule as exactly 4·1868 J and 4186·8 J respectively.
   Expressing thermal energy in the same units as mechanical energy, specific heat has the dimensions $ML^2T^{-2}/M\theta = L^2T^{-2}\theta^{-1}$ and the specific heat of water becomes equal to the

mechanical equivalent of heat (see below). Thus according to the system adopted the specific heat of water will have the following values:

c.g.s. system   1 cal/g °C
Engineering units   1 Btu/lb °F
SI system   4186·8 J/kg K

It is necessary to be able to relate mechanical and thermal energy and this is conveniently done by introducing the term J as the mechanical equivalent of heat:

$$J = \frac{\text{Mechanical energy}}{\text{Thermal energy}} = \frac{\mathbf{LF}}{\mathbf{H}} = \frac{(\mathbf{L})(\mathbf{MLT}^{-2})}{\mathbf{M}\theta}$$

$$= \mathbf{L^2 T^{-2} \theta^{-1}}$$

Thus where mechanical and thermal energies appear in the same equation, care must be taken to ensure that both have been expressed in the same units. In the SI system, J is in fact unity since, in this system, the relation $\mathbf{H} = \mathbf{M}\theta$ is not valid.

1.2.7. Molar Units

Where chemical reactions are involved it is frequently useful to work in terms of molar units. Thus the mole (mol) is the quantity of substance whose mass in grams is numerically equal to its molecular weight, the kilogram mole (kmol) is the corresponding quantity in terms of kilograms and the pound mole (lb mol) in terms of pounds.

The mole is the molar unit adopted in the SI system but in this work the kilogram mole will be used in preference as it is the more consistent unit when mass is expressed in kilograms.

1.3. CONVERSION OF UNITS

Conversion of units from one system to another is simply carried out if the quantities are expressed in terms of the fundamental units of mass, length, time, temperature. The conversion factors for the British and metric systems are, then:

Mass   1 lb = 1/32·2 slug = 453·6 g = 0·4536 kg
Length   1 ft = 30·48 cm = 0·3048 m
Time   1 s = 1/3600 h
Temperature
  difference   1°F = (1/1·8)°C = (1/1·8) K
Force   1 pound force = 32·2 poundal = 4·44 × 10⁵ dyne = 4·44 N

If a viscosity is to be converted from poises (grams per centimetre-second) to a viscosity in British or in SI units, the procedure is as follows in Example 1.1.

Example 1.1. Convert 1 P to British engineering units and the SI unit.

Solution. $1 \text{ P} = 1 \text{ g/cm s} = \dfrac{1 \text{ g}}{1 \text{ cm} \times 1 \text{ s}}$

$= \dfrac{(1/453\cdot6) \text{ lb}}{(1/30\cdot48) \text{ ft} \times 1 \text{ s}}$

$= 0\cdot0672 \text{ lb/ft s}$

$= \underline{\underline{242 \text{ lb/ft h}}}$

TABLE 1.1
Units

Quantity	c.g.s.	SI	f.p.s.	Dimensions in M, L, T, θ	British/American engineering system	Dimensions in F, L, T, θ	Dimensions in F, M, L, T, θ
Mass	gram	kilogram	pound	M	slug	$FL^{-1}T^2$	M
Length	centimetre	metre	foot	L	foot	L	L
Time	second	second	second	T	second	T	T
Force	dyne	newton	poundal	MLT^{-2}	pound force	F	F
Energy	erg ($= 10^{-7}$ joules)	joule	foot-poundal	ML^2T^{-2}	foot-pound	FL	FL
Pressure	dynes/square centimetre	newtons/sq metre	poundals/square foot	$ML^{-1}T^{-2}$	pound force/square foot	FL^{-2}	FL^{-2}
Power	ergs/second	watt	foot-poundals/second	ML^2T^{-3}	foot-pounds/second	FLT^{-1}	FLT^{-1}
Entropy per unit mass	ergs/gram °C	joules/kilogram K	foot-poundals/pound °C	$L^2T^{-2}\theta^{-1}$	foot-pounds/slug °F	$L^2T^{-2}\theta^{-1}$	$FM^{-1}L\theta^{-1}$

Heat units

Quantity	c.g.s.	SI	British/American engineering system	Dimensions in M, L, T, θ	Dimensions in H, M, L, T, θ
Temperature	degree centigrade	degree Kelvin	degree Fahrenheit	θ	θ
Thermal energy or heat	gram-calorie	joule	British thermal unit (Btu)	$M\theta$	H
Entropy per unit mass, specific heat	gram-calories/gram-°C	joule/kilogram K	Btu/pound °F	—	$HM^{-1}\theta^{-1}$
Mechanical equivalent of heat, J	$4 \cdot 18 \times 10^7$ ergs/gram-°C	1 J (heat energy) = 1 J (mechanical energy)	$2 \cdot 50 \times 10^4$ foot-poundals/pound °F	$L^2T^{-2}\theta^{-1}$	$H^{-1}ML^2T^{-2}$

$$1 \text{ P} = 1 \text{ g/cm s} = \frac{1 \text{ g}}{1 \text{ cm} \times 1 \text{ s}}$$

$$= \frac{(1/1000) \text{ kg}}{(1/100) \text{ m} \times 1 \text{ s}}$$

$$= 0 \cdot 1 \text{ kg/m s}$$

$$= \underline{0 \cdot 1 \text{ N s/m}^2} \quad [(\text{kg m/s}^2) \text{ s/m}^2]$$

The conversion of power in SI units to horse power is illustrated in Example 1.2.

Example 1.2. Convert 1 kW to h.p.

Solution. $1 \text{ kW} = 10^3 \text{ W} = 10^3 \text{ J/s}$

$$= 10^3 \times \frac{1 \text{ kg} \times 1 \text{ m}^2}{1 \text{ s}^3}$$

$$= \frac{10^3 \times (1/0 \cdot 4536) \text{ lb} \times (1/0 \cdot 3048)^2 \text{ ft}^2}{1 \text{ s}^3}$$

$$= 23{,}730 \text{ lb ft}^2/\text{s}^3$$

$$= 23{,}730/32 \cdot 2 = 737 \text{ slug ft}^2/\text{s}^3$$

$$= 737 \text{ lb}_f \text{ ft/s}$$

$$= 737/550 = \underline{1 \cdot 34 \text{ h.p.}}$$

or $\qquad\qquad 1 \text{ h.p.} = 0 \cdot 746 \text{ kW}.$

Table 1.2 provides factors to enable conversions to be made to SI units.

1.4. DIMENSIONAL ANALYSIS

Dimensional analysis depends upon the fundamental principle that any equation or relation between variables must be *dimensionally consistent*; that is to say that each term in the relationship must have the same dimensions. Thus in the simple application of the principle, an equation may consist of a number of terms, each representing, and therefore having, the dimensions of length. It is not permissible to add, say, lengths and velocities in an algebraic equation because they are quantities of different characters. The corollary of this principle is that if the whole equation is divided through by any one of the terms, each remaining term in the equation must be dimensionless. The use of these *dimensionless groups*, or *dimensionless numbers* as they are called, is of considerable value in developing relationships in chemical engineering.

The requirement of dimensional consistency places a number of constraints on the form of the functional relation between variables in a problem and forms the basis of the technique of *dimensional analysis* which enables the variables in a problem to be grouped into the form of dimensionless groups. Since the dimensions of the physical quantities can be expressed in terms of a number of fundamentals, usually mass, length, and time, and sometimes temperature and thermal energy in addition, the requirement of dimensional consistency must be satisfied in respect of each of the fundamentals. Dimensional analysis will give no information about the form of the functions, nor will it provide any means of evaluating numerical proportionality constants.

The study of problems in fluid dynamics and in heat transfer is made difficult by the many parameters which appear to affect them. In most instances further study shows that the

TABLE 1.2. *Factors for conversion to SI units*

mass		*pressure*	
1 lb	0·454 kg	1 lbf/in²	6·895 kN/m²
1 ton	1016 kg	1 atm	101·3 kN/m²
		1 bar	100 kN/m²
length		1 ft water	2·99 kN/m²
1 in	25·4 mm	1 in water	249 N/m²
1 ft	0·305 m	1 in Hg	3·39 kN/m²
1 mile	1·609 km	1 mm Hg	133 N/m²
time		*viscosity*	
1 min	60 s	1 P	0·1 N s/m²
1 h	3·6 ks	1 lb/ft h	0·414 mN s/m²
1 day	86·4 ks	1 stoke	10⁻⁴ m²/s
1 year	31·5 Ms	1 ft²/h	0·258 cm²/s
area		*mass flow*	
1 in²	645·2 mm²	1 lb/h	0·126 g/s
1 ft²	0·093 m²	1 ton/h	0·282 kg/s
		1 lb/h ft²	1·356 g/s m²
volume			
1 in³	16,387·1 mm³	*thermal*	
1 ft³	0·0283 m³	1 Btu/h ft²	3·155 W/m²
1 UK gal	4546 cm³	1 Btu/h ft² °F	5·678 W/m² K
1 US gal	3786 cm³	1 Btu/lb	2·326 kJ/kg
		1 Btu/lb °F	4·187 kJ/kg K
		1 Btu/h ft °F	1·731 W/m K
force			
1 pdl	0·138 N		
1 lbf	4·45 N	*energy*	
1 dyne	10⁻⁵ N	1 kWh	3·6 MJ
		1 therm	105·5 MJ
energy		*calorific value*	
1 ft lbf	1·36 J	1 Btu/ft³	37·26 kJ/m³
1 cal	4·187 J	1 Btu/lb	2·326 kJ/kg
1 erg	10⁻⁷ J		
1 Btu	1·055 kJ		
		density	
		1 lb/ft³	16·02 kg/m³
power			
1 h.p.	745 W		
1 Btu/h	0·293 W		

variables may be grouped together in dimensionless groups, thus reducing the effective number of variables. It is rarely possible, and certainly time consuming, to try to vary these many variables separately, and the method of dimensional analysis by providing a smaller number of independent groups is most helpful to the experimenter.

The application of the principles of dimensional analysis may best be understood by considering the problem expressed as Example 1.3.

Example 1.3. It is found, as a result of experiment, that the pressure difference (ΔP) between two ends of a pipe in which a fluid is flowing is a function of the following variables: pipe diameter d, pipe length l, fluid velocity u, fluid density ρ, and fluid viscosity μ.

Show by dimensional analysis how these variables are related.

Solution. The relationship may be written as:

$$\Delta P = f_1(d, l, u, \rho, \mu) \tag{1.1}$$

The form of the function is unknown, but since any function can be expanded as a power series, the function can be regarded as the sum of a number of terms each consisting of products of powers of the variables. The simplest form of relation will be where the function consists simply of a single term, when:

$$\Delta P = \text{const } d^{n_1} l^{n_2} u^{n_3} \rho^{n_4} \mu^{n_5} \tag{1.2}$$

The requirement of dimensional consistency is that the combined term on the right-hand side will have the same dimensions as that on the left, i.e. it must have the dimensions of pressure.

Each of the variables in equation 1.2 can be expressed in terms of mass, length, and time. Thus, dimensionally:

$$\Delta P = \mathbf{ML^{-1}T^{-2}} \qquad u = \mathbf{LT^{-1}}$$
$$d = \mathbf{L} \qquad \rho = \mathbf{ML^{-3}}$$
$$l = \mathbf{L} \qquad \mu = \mathbf{ML^{-1}T^{-1}}$$

i.e. $$\mathbf{ML^{-1}T^{-2}} = \mathbf{L^{n_1}L^{n_2}(LT^{-1})^{n_3}(ML^{-3})^{n_4}(ML^{-1}T^{-1})^{n_5}}$$

The conditions of dimensional consistency must be met for each of the fundamentals of $\mathbf{M}$, $\mathbf{L}$, and $\mathbf{T}$ and the indices of each of these variables can be equated. Thus:

In $\qquad\qquad\qquad \mathbf{M} \qquad 1 = n_4 + n_5$

$\qquad\qquad\qquad\quad \mathbf{L} \qquad -1 = n_1 + n_2 + n_3 - 3n_4 - n_5$

$\qquad\qquad\qquad\quad \mathbf{T} \qquad -2 = -n_3 - n_5$

Thus three equations and five unknowns result and the equations may be solved in terms of any two unknowns. Solving in terms of n_2 and n_5:

$$n_4 = 1 - n_5 \quad \text{(from the equation in } \mathbf{M})$$
$$n_3 = 2 - n_5 \quad \text{(from the equation in } \mathbf{T})$$

Substituting in the $\mathbf{L}$ equation:

$$-1 = n_1 + n_2 + (2 - n_5) - 3(1 - n_5) - n_5$$

i.e. $$0 = n_1 + n_2 + n_5$$

∴ $$n_1 = -n_2 - n_5$$

Thus, substituting into equation 1.2:

$$\Delta P = \text{const } d^{-n_2 - n_5} l^{n_2} u^{2 - n_5} \rho^{1 - n_5} \mu^{n_5}$$

i.e. $$\frac{\Delta P}{\rho u^2} = \text{const } \left(\frac{l}{d}\right)^{n_2} \left(\frac{\mu}{du\rho}\right)^{n_5} \tag{1.3}$$

Since n_2 and n_5 are arbitrary constants, the above equation can only be satisfied if each of the terms $\Delta P / \rho u^2$, l/d, and $\mu / du\rho$ is dimensionless. As a check, if the dimension of each group is evaluated, it will be found to be dimensionless.

The group $ud\rho / \mu$, known as the *Reynolds number*, is one which will frequently arise in the study of fluid flow and affords a criterion by which the type of flow in a given geometry is characterised. Equation 1.3 involves the reciprocal of the Reynolds number, but this can be rewritten as

$$\frac{\Delta P}{\rho u^2} = \text{const } \left(\frac{l}{d}\right)^{n_2} \left(\frac{ud\rho}{\mu}\right)^{-n_5} \tag{1.4}$$

The right-hand side of equation 1.4 is a typical term in the function for $\Delta P / \rho u^2$. More generally:

$$\frac{\Delta P}{\rho u^2} = f\left(\frac{l}{d}, \frac{ud\rho}{\mu}\right) \tag{1.5}$$

Comparing equations 1.1 and 1.5, it is seen that a relationship between six variables has been reduced to a relationship between three dimensionless groups. Later, this statement will be generalised to show that the number of dimensionless groups is normally the number of variables less the number of fundamentals.

A number of important points emerge from a consideration of the preceding example:

(1) If the index of a particular variable is found to be zero, this indicates that this variable is not of significance in the problem.

(2) If two of the fundamentals always appear in the same combination, e.g. if L and T always occur as powers of LT^{-1}, then the same equation for the indices will be obtained for both L and T and the number of effective fundamentals is thus reduced by one.

(3) The form of the final solution will depend upon the method of solution of the simultaneous equations. If the equations had been solved, say, in terms of n_3 and n_4 instead of n_2 and n_5, different dimensionless groups would have been obtained. However, these groups would have been obtainable from those obtained in the above example, any number of new dimensionless groups being capable of formation by multiplying together products of powers of the existing groups.

Clearly the maximum degree of simplification of the problem is achieved by using the greatest possible number of fundamentals since each yields a simultaneous equation of its own. In certain problems, force may be used as a fundamental in addition to mass, length, and time, provided that at no stage in the problem is force defined in terms of mass and acceleration. In heat transfer problems, temperature is usually an additional fundamental, and heat can also be used as a fundamental provided it is not defined in terms of mass and temperature and provided that the relation between mechanical and thermal energy is not used. Considerable experience is needed in the proper use of dimensional analysis, and its application in a number of areas of fluid flow and heat transfer will be seen in the relevant chapters of this book.

The choice of physical variables to be included in the dimensional analysis must be based on an understanding of the nature of the phenomenon being studied. Nevertheless, on occasions there may be some doubt as to whether a particular quantity is relevant or not.

If a variable is included which does not exert a significant influence on the problem, the value of the dimensionless group in which it appears will have little effect on the final numerical solution of the problem, and therefore the exponent of that group must approach zero. This presupposes that the dimensionless groups are so constituted that the variable in question appears in only one of them. If an important variable is omitted, on the other hand, it will be found that there is no unique relationship between the dimensionless groups.

Chemical engineering analysis requires the formulation of relationships which will apply over a wide range of size of the individual items of a plant. This problem of scale up is vital and can be much helped by dimensional analysis.

Since linear size is included among the variables, the influence of scale, which may be regarded as the influence of linear size without change of shape or other variables, has been introduced. Thus in the viscous flow past an object, a change in linear dimension L will alter the Reynolds number and therefore the flow pattern around the solid, though if the change in scale is accompanied by a change in any other variable in such a way that the Reynolds number remains unchanged, then the flow pattern around the solid will not be altered. This ability to change scale but still maintain a design relationship is one of the many attractions of dimensional analysis.

1.5. BUCKINGHAM'S Π THEOREM

The need for dimensional consistency imposes a restraint in respect of each of the fundamentals involved in the dimensions of the variables. This was apparent during the previous discussion in which a series of simultaneous equations was solved, one equation for each of the fundamentals. A generalisation of this statement is provided in Buckingham's Π theorem which states that the number of dimensionless groups is equal to the number of variables minus the number of fundamental dimensions. In mathematical terms, this can be expressed as follows:

If there are n variables, $Q_1, Q_2, \ldots, Q_n$, the functional relationship between them can be written:

$$f_3(Q_1, Q_2, \ldots, Q_n) = 0 \tag{1.6}$$

If there are m fundamental dimensions, there will be $n - m$ dimensionless groups $(\Pi_1, \Pi_2, \ldots, \Pi_{n-m})$ and the functional relationship between them can be written:

$$f_4(\Pi_1, \Pi_2, \ldots, \Pi_{n-m}) = 0 \tag{1.7}$$

The groups Π_1, Π_2, etc., must be independent of one another, and no one group should be capable of being formed by multiplying together powers of the other groups.

By making use of the theorem it is possible to obtain the dimensionless groups more simply than by solving the simultaneous equations for the indices. Furthermore, the functional relationship can often be obtained in a form which is of more immediate use.

The method involves choosing m of the original variables to form what is called a *recurring set*. Any set m of the variables may be chosen with the following two provisions:

(a) Each of the fundamentals must appear in at least one of the m variables.
(b) It must not be possible to form a dimensionless group from some or all of the variables within the recurring set. If it were so possible, this dimensionless group would, of course, be one of the Π terms.

The procedure which is then followed is to take each of the remaining $n - m$ variables on its own and to form it into a dimensionless group by combining it with one or more members of the recurring set. In this way the $n - m$ Π groups are formed, the only variables appearing in more than one group being those that constitute the recurring set. Thus if it is desired to obtain an explicit functional relation for one particular variable, that variable should not be included in the recurring set.

In some cases, the number of dimensionless groups will be greater than predicted by the Π theorem. Thus, for instance, if two of the fundamentals always occur in the same combination—e.g. length and time always as LT^{-1}—they will constitute a single fundamental instead of two fundamentals. By referring back to the method of equating indices, it is seen that each of the two fundamentals gives the same equation, and therefore only a single constraint is placed on the relationship by considering the two variables.

The procedure is more readily understood by consideration of the example given before. The relationship between the variables affecting the pressure drop for flow of fluid in a pipe can be written:

$$f_5(\Delta P, d, l, \rho, \mu, u) = 0 \tag{1.8}$$

Equation 1.8 includes six variables, and three fundamental quantities (mass, length, and time) are involved.

Number of groups $= 6 - 3 = 3$

The recurring set must contain three variables that cannot themselves be formed into a dimensionless group. This imposes the following two restrictions:

(a) Both l and d cannot be chosen as they can be formed into the dimensionless group l/d.
(b) $\Delta P, \rho$ and u cannot be used since $\Delta P/\rho u^2$ is dimensionless.

Outside these constraints, any three variables can be chosen. However, it should be remembered that the variables forming the recurring set are liable to appear in all the dimensionless groups. As one is concerned in this problem with the effect of conditions on the pressure difference ΔP, it is convenient if ΔP appears in only one group and therefore it is preferable not to include it in the recurring set.

If the variables d, u, ρ are chosen as the recurring set, it fulfils all the above conditions.
Dimensionally:

$$d = \mathbf{L}$$
$$u = \mathbf{LT}^{-1}$$
$$\rho = \mathbf{ML}^{-3}$$

Each of the dimensions $\mathbf{M}, \mathbf{L}, \mathbf{T}$ can then be obtained explicitly in terms of the variables d, u, ρ. Thus:

$$\mathbf{L} = d$$
$$\mathbf{M} = \rho d^3$$
$$\mathbf{T} = du^{-1}$$

The three dimensionless groups are thus obtained by taking each of the remaining variables $\Delta P, l,$ and μ in turn.

ΔP has dimensions $\mathbf{ML}^{-1}\mathbf{T}^{-2}$.
$\Delta P \mathbf{M}^{-1}\mathbf{LT}^2$ is therefore dimensionless.
Group Π_1 is, therefore, $\Delta P(\rho d^3)^{-1}(d)(du^{-1})^2 = \Delta P/\rho u^2$

l has dimensions $\mathbf{L}$.
$l\mathbf{L}^{-1}$ is therefore dimensionless.
Group Π_2 is, therefore, $l(d^{-1}) = l/d$

μ has dimensions $\mathbf{ML}^{-1}\mathbf{T}^{-1}$.
$\mu \mathbf{M}^{-1}\mathbf{LT}$ is therefore dimensionless.
Group Π_3 is, therefore, $\mu(\rho d^3)^{-1}(d)(du^{-1}) = \mu/du\rho.$

Thus $$f_6\left(\frac{\Delta P}{\rho u^2}, \frac{l}{d}, \frac{\mu}{u d\rho}\right) = 0 \quad \text{or} \quad \frac{\Delta P}{\rho u^2} = f_7\left(\frac{l}{d}, \frac{u d\rho}{\mu}\right)$$

$\mu/ud\rho$ is arbitrarily inverted because the Reynolds number is usually expressed in the form $ud\rho/\mu$.

Dimensionless analysis is a useful tool in chemical engineering in that it shows the possible ways in which the variables involved can be grouped. It is a dangerous tool in that it can give incorrect results if the physical nature of the problem is not understood and can lead to false conclusions if a significant variable is omitted from the problem. It should therefore be used with caution.

1.6. FURTHER READING

BRIDGMAN, P. W.: *Dimensional Analysis* (Yale University Press, 1931).
FOCKEN, C. M.: *Dimensional Methods and their Applications* (Arnold, 1953).
IPSEN, D. C.: *Units, Dimensions, and Dimensionless Numbers* (McGraw-Hill, 1960).

JOHNSTONE, R. E. and THRING, M. W.: *Pilot Plants, Models and Scale-up in Chemical Engineering* (McGraw-Hill, 1957).

KLINKENBERG, A. and MOOY, H. H.: *Chem. Eng. Prog.* **44** (1948) 17. Dimensionless groups in fluid friction, heat and material transfer.

MASSEY, B. S.: *Units, Dimensional Analysis and Physical Similarity* (Van Nostrand Reinhold, 1971).

MULLIN, J. W.: *The Chemical Engineer*, No. 211 (Sept. 1967) 176. SI units in chemical engineering.

MULLIN, J. W.: *The Chemical Engineer*, No. 254 (Oct. 1971) 352. Recent developments in the change-over to the international system of units (SI).

British Standard BS 3763. *The International System (SI) Units.*

British Standard PD 5686. *The Use of SI Units.*

1.7.   NOMENCLATURE

		Units in SI system	Dimensions M, L, T, θ
d	Pipe diameter	m	L
f	A function of	—	—
g	Acceleration due to gravity	m/s^2	LT^{-2}
g_c	Numerical constant equal to "standard value" of g	—	—
J	Mechanical equivalent of heat	—	$L^2T^{-2}\theta^{-1}$ or —
l	Length	m	L
m	Number of fundamental dimensions	—	—
n	Number of variables	—	—
ΔP	Pressure difference	N/m^2	$ML^{-1}T^{-2}$
Q	Physical quantity	—	—
u	Mean velocity of flow	m/s	LT^{-1}
μ	Viscosity of fluid	N s/m^2	$ML^{-1}T^{-1}$
Π	Dimensionless group	—	—
ρ	Density of fluid	kg/m^3	ML^{-3}

Dimensions

F	Force
H	Heat
L	Length
M	Mass
T	Time
θ	Temperature

CHAPTER 2

Flow of Fluids —Energy and Momentum Relationships

2.1. INTRODUCTION

The chemical engineer is interested in many aspects of the problems involved in the flow of fluids. In the first place, in common with many other engineers, he is concerned with the transport of fluids from one location to another through pipes or open ducts; this requires determination of the pressure drops in the system, and hence of the power required for pumping, selection of the most suitable type of pump, and measurement of the flow rates. In many cases, the fluid contains solid particles in suspension and it is necessary to determine the effect of these particles on the flow characteristics of the fluid or, alternatively, the drag force exerted by the fluid on the particles. In some cases, such as filtration, the particles are in the form of a fairly stable bed and the fluid has to pass through the tortuous channels formed by the pore spaces. In other cases the shape of the boundary surfaces must be so arranged that a particular flow pattern is obtained: for example, when solids are maintained in suspension in a liquid by means of agitation, the desired effect can be obtained with the minimum expenditure of energy if the most suitable flow pattern is produced in the fluid. Further, in those processes where heat transfer or mass transfer to a flowing fluid occurs, the nature of the flow may have a profound effect on the transfer coefficient for the process.

It is necessary to be able to calculate the energy and momentum of a fluid at various positions in a flow system. It will be seen that energy occurs in a number of forms and that some of these are influenced by the motion of the fluid. In the first part of this chapter the thermodynamic properties of fluids will be discussed. It will then be seen how the thermodynamic relations are modified if the fluid is in motion. In the following chapters, the effects of frictional forces will be considered, and the principal methods of measuring flow will be described.

2.2. INTERNAL ENERGY

When a fluid flows from one location to another, energy will, in general, be converted from one form to another. The energy which is attributable to the physical state of the fluid is known as internal energy; it is arbitrarily taken as zero at some reference state, such as the absolute zero of temperature or the melting point of ice at atmospheric pressure. A change in the physical state of a fluid will, in general, cause an alteration in the internal energy. An elementary reversible change results from an infinitesimal change in one of the intensive factors acting on the system; the change proceeds at an infinitesimal rate and a small change in the intensive factor in the opposite direction would have caused the process to take place in the reverse direction. Truly reversible changes never occur in practice but they provide a useful standard with which actual processes can be compared. In an irreversible process, changes are caused by a finite difference in the intensive factor and take place at a finite rate. In general the process will be accompanied by the conversion of electrical or mechanical energy into heat, or the reduction of the temperature difference between different parts of the system.

For a stationary material the change in the internal energy is equal to the difference between the net amount of heat added to the system and the net amount of work done by the system on its surroundings. For an infinitesimal change:

$$dU = \delta q - \delta W \tag{2.1}$$

where dU is the small change in the internal energy, δq the small amount of heat added, and δW the net amount of work done on the surroundings.

In this expression consistent units must be used—either heat units or mechanical energy units. dU is a small change in the internal energy which is a property of the system; it is therefore a perfect differential. On the other hand, δq and δW are small quantities of heat and work; they are not properties of the system and their values depend on the manner in which the change is effected; they are, therefore, not perfect differentials. For a reversible process, however, both δq and δW can be expressed in terms of properties of the system. For convenience, reference will be made to systems of unit mass and the effects on the surroundings will be disregarded.

A property called entropy may be defined by the relation:

$$dS = \delta q / T \tag{2.2}$$

where dS is the small change in entropy resulting from the addition of a small quantity of heat δq, at a temperature T, under reversible conditions. From the definition of the thermodynamic scale of temperature, $\oint \dfrac{\delta q}{T} = 0$ for a reversible cyclic process, and the net change in the entropy is also zero. Thus, for a particular condition of the system, the entropy has a definite value and must be a property of the system; dS is, therefore, a perfect differential.

For an irreversible process:

$$(\delta q / T) < dS = (\delta q / T) + (\delta F / T) \,(\text{say}) \tag{2.3}$$

δF is then a measure of the degree of irreversibility of the process; it represents the amount of mechanical energy converted into heat or the conversion of heat energy at one temperature to heat energy at another temperature. For a finite process:

$$\int_1^2 T\, dS = \Sigma \delta q + \Sigma \delta F = q + F \,(\text{say}) \tag{2.4}$$

When a process is isentropic, $q = -F$; a reversible process is isentropic when $q = 0$, i.e. a reversible adiabatic process is isentropic.

The increase in the entropy of an irreversible process can be illustrated in the following manner. Consider the spontaneous transfer of a quantity of heat δq from one part of a system at a temperature T_1 to another part at a temperature T_2. The net change in the entropy of the system as a whole is then:

$$dS = (\delta q / T_2) - (\delta q / T_1)$$

T_1 must be greater than T_2 and dS is therefore positive. If the process had been carried out reversibly, there would have been an infinitesimal difference between T_1 and T_2 and the change in entropy would have been zero.

Now the change in the internal energy can be expressed in terms of properties of the system itself. For a reversible process:

$$\delta q = T\, dS \quad (\text{from equation 2.2}) \quad \text{and} \quad \delta W = P\, dv$$

if the only work done is that resulting from a change in volume, dv.

Thus, from equation 2.1:

$$dU = T \, dS - P \, dv \tag{2.5}$$

Since this relation is in terms of properties of the system, it must also apply to a system in motion and to irreversible changes where the only work done is the result of change of volume.

Thus, in an irreversible process, for a stationary system:

from equations 2.1 and 2.2: $dU = \delta q - \delta W = T \, dS - P \, dv$

and from equation 2.3: $\delta q + \delta F = T \, dS$

$\therefore$ $\delta W = P \, dv - \delta F \tag{2.6}$

i.e. the useful work performed by the system is less than $P \, dv$ by an amount δF, which represents the amount of mechanical energy converted into heat energy.

The relation between the internal energy and the temperature of a fluid will now be considered. In a system consisting of unit mass of material and where the only work done is that resulting from volume change, the change in internal energy after a reversible change is given by:

$$dU = \delta q - P \, dv \quad \text{(from equation 2.1)}$$

If there is no volume change:

$$dU = \delta q = C_v \, dT \tag{2.7}$$

where C_v is the specific heat at constant volume.

As this relation is in terms of properties of the system, it must be applicable to all changes at constant volume.

In an irreversible process:

$$dU = \delta q - (P \, dv - \delta F) \quad \text{(from equations 2.1 and 2.6)}$$
$$= \delta q + \delta F \qquad\qquad \text{(under conditions of constant volume)} \tag{2.8}$$

This quantity δF thus represents the mechanical energy which has been converted into heat and which is therefore available for increasing the temperature.

Thus $\delta q + \delta F = C_v \, dT = dU. \tag{2.9}$

For changes that take place under conditions of constant pressure, it is more satisfactory to consider variations in the enthalpy H. The enthalpy is defined by the relation:

$$H = U + Pv. \tag{2.10}$$

Thus $dH = dU + P \, dv + v \, dP$

$$= \delta q - P \, dv + \delta F + P \, dv + v \, dP \quad \text{(from equation 2.8)}$$

for an irreversible process: (For a reversible process $\delta F = 0$)

$\therefore$ $dH = \delta q + \delta F + v \, dP \tag{2.11}$

$$= \delta q + \delta F \qquad \text{(at constant pressure)}$$

$$= C_p \, dT \tag{2.12}$$

where C_p is the specific heat at constant pressure.

No assumptions have been made concerning the properties of the system and, therefore, the following relations apply to all fluids.

From equation 2.7: $(\partial U / \partial T)_v = C_v \tag{2.13}$

From equation 2.12: $(\partial H / \partial T)_P = C_p \tag{2.14}$

2.3. TYPES OF FLUID

Fluids may be classified in two different ways; either according to their behaviour under the action of externally applied pressure, or according to the effects produced by the action of a shear stress.

If the volume of an element of fluid is independent of its pressure and temperature, the fluid is said to be incompressible; if its volume changes it is said to be compressible. No real fluid is completely incompressible but liquids may generally be regarded as such when their flow is considered. Gases have a very much higher compressibility than liquids, and appreciable changes in volume may occur if the pressure or temperature is altered. However, if the percentage change in the pressure or in the absolute temperature is small, for practical purposes a gas may also be regarded as incompressible. Thus, in practice, volume changes are likely to be important only when the pressure or temperature of a gas changes by a large proportion. The relation between pressure, temperature, and volume of a real gas is generally complex, but except at very high pressures the behaviour of gases approximates to that of the ideal gas for which the volume of a given mass is inversely proportional to the pressure and directly proportional to the absolute temperature. At high pressures and when pressure changes are large, however, there may be appreciable deviations from this law and an approximate equation of state must then be used.

The behaviour of a fluid under the action of a shear stress is important in that it determines the way in which it will flow. The most important physical property affecting the stress distribution within the fluid is its viscosity. For a gas, the viscosity is low and even at high rates of shear, the viscous stresses are small. Under such conditions the gas approximates in its behaviour to an inviscid fluid. In many problems involving the flow of a gas or a liquid, the viscous stresses are important and give rise to appreciable velocity gradients within the fluid, and dissipation of energy occurs as a result of the frictional forces set up. In gases and in most pure liquids the ratio of the shear stress to the rate of shear is constant and equal to the viscosity of the fluid. These fluids are said to be *Newtonian* in their behaviour. However, in some liquids, particularly those containing a second phase in suspension, the ratio is not constant and the apparent viscosity of the fluid is a function of the rate of shear. The fluid is then said to be *non-Newtonian* and to exhibit rheological properties. The importance of the viscosity of the fluid in determining velocity profiles and friction losses will be discussed in Chapter 3.

The effect of pressure on the properties of an incompressible fluid, an ideal gas, and a non-ideal gas will now be considered.

2.3.1. The Incompressible Fluid (Liquid)

By definition, v is independent of P, so that $(\partial v/\partial P)_T = 0$. The internal energy will be a function of temperature but not a function of pressure.

2.3.2. The Ideal Gas

An ideal gas is defined as a gas whose properties obey the law:

$$PV = n\mathbf{R}T \tag{2.15}$$

where V is the volume occupied by n molar units of the gas, $\mathbf{R}$ the universal gas constant, and T the absolute temperature. In this text n will be expressed in kmol when using the SI system.

This law is closely obeyed by real gases under conditions where the actual volume of the molecules is small compared with the total volume, and where the molecules exert only a very small attractive force on one another. These conditions are met at very low pressures when the distance apart of the individual molecules is large. The value of **R** is then the same for all gases and in SI units has the value of $8 \cdot 314 \, \text{kJ/kmol K}$.

When the only external force on a gas is the fluid pressure, the equation of state is given by

$$f(P, V, T, n) = 0$$

Any property can be expressed in terms of any three other properties. Consider the dependence of the internal energy on temperature and volume:

$$U = f(T, V, n)$$

For unit mass of gas:

$$U = f(T, v)$$

where v is the volume per unit mass and

$$Pv = \mathbf{R}T/M \tag{2.16}$$

so that

$$dU = \left(\frac{\partial U}{\partial T}\right)_v dT + \left(\frac{\partial U}{\partial v}\right)_T dv \tag{2.17}$$

Now

$$T \, dS = dU + P \, dv \quad \text{(from equation 2.5)}$$

$$\therefore \qquad T \, dS = \left(\frac{\partial U}{\partial T}\right)_v dT + \left[P + \left(\frac{\partial U}{\partial v}\right)_T\right] dv$$

and

$$dS = \left(\frac{\partial U}{\partial T}\right)_v \frac{dT}{T} + \frac{1}{T}\left[P + \left(\frac{\partial U}{\partial v}\right)_T\right] dv \tag{2.18}$$

Thus

$$\left(\frac{\partial S}{\partial T}\right)_v = \frac{1}{T}\left(\frac{\partial U}{\partial T}\right)_v \tag{2.19}$$

and

$$\left(\frac{\partial S}{\partial v}\right)_T = \frac{1}{T}\left[P + \left(\frac{\partial U}{\partial v}\right)_T\right] \tag{2.20}$$

Then differentiating equation 2.19 by v and equation 2.20 by T and equating:

$$\frac{1}{T}\frac{\partial^2 U}{\partial T \partial v} = \frac{1}{T}\left[\left(\frac{\partial P}{\partial T}\right)_v + \frac{\partial^2 U}{\partial v \partial T}\right] - \frac{1}{T^2}\left[P + \left(\frac{\partial U}{\partial v}\right)_T\right]$$

i.e.

$$\left(\frac{\partial U}{\partial v}\right)_T = T\left(\frac{\partial P}{\partial T}\right)_v - P \tag{2.21}$$

This relation applies to any fluid. For the particular case of an ideal gas, since $Pv = \mathbf{R}T/M$ (equation 2.16):

$$T\left(\frac{\partial P}{\partial T}\right)_v = T\frac{\mathbf{R}}{Mv} = P$$

so that

$$(\partial U/\partial v)_T = 0 \tag{2.22}$$

and

$$\left(\frac{\partial U}{\partial P}\right)_T = \left(\frac{\partial U}{\partial v}\right)_T \left(\frac{\partial v}{\partial P}\right)_T = 0 \tag{2.23}$$

Thus the internal energy of an ideal gas is a function of temperature only. The variation of internal energy and enthalpy with temperature will now be calculated.

$$dU = \left(\frac{\partial U}{\partial T}\right)_v dT + \left(\frac{\partial U}{\partial v}\right)_T dv \tag{equation 2.17}$$

$$= C_v \, dT \quad \text{(from equations 2.13 and 2.22)} \tag{2.24}$$

Thus for an ideal gas under all conditions:

$$dU/dT = C_v \tag{2.25}$$

In general, this relation applies only to changes at constant volume. For the particular case of the ideal gas, however, it applies under all circumstances.

Again, since $H = f(T, P)$:

$$dH = \left(\frac{\partial H}{\partial T}\right)_P dT + \left(\frac{\partial H}{\partial P}\right)_T dP$$

$$= C_p\, dT + \left(\frac{\partial U}{\partial P}\right)_T dP + \left(\frac{\partial (Pv)}{\partial P}\right)_T dP \quad \text{(from equations 2.12 and 2.10)}$$

$$= C_p\, dT$$

since $(\partial U/\partial P)_T = 0$ and $[\partial(Pv)/\partial P]_T = 0$ for an ideal gas.

Thus under all conditions for an ideal gas:

$$dH/dT = C_p \tag{2.26}$$

$$\therefore \quad C_p - C_v = (dH/dT) - (dU/dT) = d(Pv)/dT = \mathbf{R}/M \tag{2.27}$$

Isothermal Processes

In fluid flow it is important to know how the volume of a gas will vary as the pressure changes. Two important idealised conditions which are rarely obtained in practice are changes at constant temperature and changes at constant entropy. Although not actually reached, these conditions are approached in many flow problems.

For an *isothermal* change in an ideal gas, the product of pressure and volume is a constant. For unit mass of gas:

$$Pv = \mathbf{R}T/M = \text{constant} \tag{equation 2.16}$$

Isentropic Processes

For an *isentropic* process the enthalpy can be expressed as a function of the pressure and volume:

$$H = f(P, v)$$

$$dH = \left(\frac{\partial H}{\partial P}\right)_v dP + \left(\frac{\partial H}{\partial v}\right)_P dv$$

$$= \left(\frac{\partial H}{\partial T}\right)_v \left(\frac{\partial T}{\partial P}\right)_v dP + \left(\frac{\partial H}{\partial T}\right)_P \left(\frac{\partial T}{\partial v}\right)_P dv$$

Now

$$\left(\frac{\partial H}{\partial T}\right)_P = C_p \tag{equation 2.14}$$

and

$$\left(\frac{\partial H}{\partial T}\right)_v = \left(\frac{\partial U}{\partial T}\right)_v + \left(\frac{\partial Pv}{\partial T}\right)_v$$

$$= C_v + v\left(\frac{\partial P}{\partial T}\right)_v \quad \text{(from equation 2.13)}$$

Further,

$$dH = dU + P\, dv + v\, dP \quad \text{(from equation 2.10)}$$

$$= T\, dS - P\, dv + P\, dv + v\, dP \quad \text{(from equation 2.5)}$$

$$= T\, dS + v\, dP \tag{2.28}$$

$$= v\, dP \quad \text{(for an isentropic process)} \tag{2.29}$$

Thus for an isentropic process:

$$v\, dP = \left[C_v + v\left(\frac{\partial P}{\partial T}\right)_v\right]\left(\frac{\partial T}{\partial P}\right)_v dP + C_p\left(\frac{\partial T}{\partial v}\right)_P dv$$

i.e.
$$\left(\frac{\partial T}{\partial P}\right)_v dP + \frac{C_p}{C_v}\left(\frac{\partial T}{\partial v}\right)_P dv = 0$$

From the equation of state for an ideal gas (2.15):

$$(\partial T/\partial P)_v = T/P \quad \text{and} \quad (\partial T/\partial v)_P = T/v$$

∴
$$(dP/P) + \gamma(dv/v) = 0$$

where $\gamma = C_p/C_v$.

Integration gives

$$\ln P + \gamma \ln v = \text{constant}$$

i.e.
$$Pv^\gamma = \text{constant} \tag{2.30}$$

This relation holds only approximately, even for an ideal gas, since γ has been taken as a constant in the integration, although it does vary somewhat with pressure.

2.3.3.   The Non-ideal Gas

For a non-ideal gas, equation 2.15 is modified by including a compressibility factor Z which is a function of both temperature and pressure:

$$PV = Zn\mathbf{R}T \tag{2.31}$$

At very low pressures, deviations from the ideal gas law are caused mainly by the attractive forces between the molecules and the compressibility factor has a value less than unity. At higher pressures, deviations are caused mainly by the fact that the volume of the molecules themselves, which can be regarded as incompressible, becomes significant compared with the total volume of the gas.

Many equations have been given to denote the approximate relation between the properties of a non-ideal gas. Of these the simplest, and probably the most commonly used, is van der Waals' equation:

$$[P + (an^2/V^2)](V - nb) = n\mathbf{R}T \tag{2.32}$$

where b is a quantity which is a function of the incompressible volume of the molecules themselves, and a/V^2 is a function of the attractive forces between the molecules. It is seen that as P approaches zero and V approaches infinity, this equation reduces to the equation of state for the ideal gas.

A chart which correlates experimental P–V–T data for all gases is included as Fig. 2.1 and is known as the generalised compressibility-factor chart.[1] Use is made of *reduced* coordinates where the *reduced temperature* T_R, the *reduced pressure* P_R, and the *reduced volume* V_R are defined as the ratio of the actual temperature, pressure, and volume of the gas to the corresponding values of these properties at the critical state. It is found that, at a given value of T_R and P_R, nearly all gases have the same molar volume, compressibility factor, and other thermodynamic properties. This empirical relationship applies to within about 2 per cent for most gases, and the most important exception to the rule is ammonia.

The generalised compressibility-factor chart is not to be regarded as a substitute for experimental P–V–T data. If accurate data are available, as they are for some of the more common gases,[2] they should be used.

It will be noted from Fig. 2.1 that Z approaches unity for all temperatures as the pressure approaches zero. This serves to confirm the statement made previously that all gases approach ideality as the pressure is reduced to zero. For most gases the critical pressure is 3 MN/m² or above. Thus at atmospheric pressure (101·3 kN/m²), P_R is 0·033 or less. At this pressure, for any temperature above the critical temperature ($T_R = 1$), it will be seen that Z

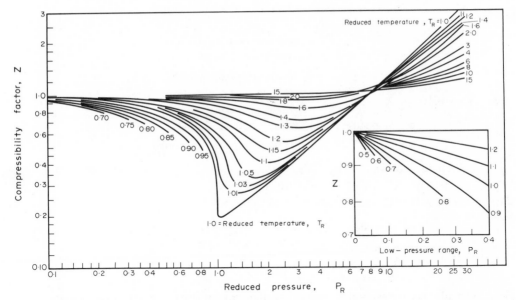

FIG. 2.1. Compressibility factors of gases and vapours.

deviates from unity by no more than 1 per cent. Thus at atmospheric pressure for temperatures greater than the critical temperature, the assumption that the ideal gas law is valid usually leads to errors of less than 1 per cent. It should also be noted that for reduced temperatures between 3 and 10 the compressibility factor is nearly unity for reduced pressures up to a value of 6. For very high temperatures the isotherms approach a horizontal line at $Z = 1$ for all pressures. Thus all gases tend towards ideality as the temperature approaches infinity.

Example 2.1. It is required to store 1 kmol of methane at 320 K and 60 MN/m². Using the following methods, estimate the volume of the vessel which must be provided:

(a) ideal gas law;
(b) van der Waals' equation;
(c) generalised compressibility-factor chart;
(d) experimental data from Perry.[2]

Solution. For 1 kmol of methane,
(a) $PV = \mathbf{R}T$, where $\mathbf{R} = 8 \cdot 314$ kJ/kmol K.

In this case: $$P = 60 \times 10^3 \text{ kN/m}^2; \quad T = 320 \text{ K}$$

∴ $$V = 8 \cdot 314 \times 320/(60 \times 10^3) = \underline{0 \cdot 0443 \text{ m}^3}$$

(b) In van der Waals' equation (2.32), the constants may be taken as:

$$a = 27\mathbf{R}^2 T_c^2/64 P_c; \quad b = \mathbf{R}T_c/8P_c$$

where the critical temperature $T_c = 191$ K and the critical pressure $P_c = 4640$ kN/m² for methane from the Appendix tables.

∴ $$a = 27 \times 8 \cdot 314^2 \times 191^2/(64 \times 4640) = 229 \cdot 3 \text{ (kN/m}^2)(\text{m}^3)^2/(\text{kmol})^2$$

and $$b = 8 \cdot 314 \times 191/(8 \times 4640) = 0 \cdot 0427 \text{ m}^3/\text{kmol}$$

Thus in equation 2.32:

$$(60 \times 10^3 + 229 \cdot 3 \times 1/V^2)(V - 1 \times 0 \cdot 0427) = 1 \times 8 \cdot 314 \times 320$$

or $$60{,}000\,V^3 - 5233\,V^2 + 229{\cdot}3\,V = 9{\cdot}79$$

Solving by trial and error: $$\underline{V = 0{\cdot}066\text{ m}^3}$$

(c) $$T_r = T/T_c = 320/191 = 1{\cdot}68$$
$$P_r = P/P_c = 60 \times 10^3/4640 = 12{\cdot}93$$

Thus from Fig. 2.1, $Z = 1{\cdot}33$

and $$V = Zn\mathbf{R}T/P \quad \text{(from equation 2.31)}$$
$$= 1{\cdot}33 \times 1{\cdot}0 \times 8{\cdot}314 \times 320/60 \times 10 = \underline{0{\cdot}0589\text{ m}^3}$$

(d) Perry[2] lists experimental values of A' for various gases, where $A' = PV/(PV)_0$, where $(PV)_0$ is the PV product at 273 K and 101·3 kN/m².
For methane at 320 K and 60 MN/m²: $A' = 1{\cdot}565$

$\therefore$ $$(PV)_0 = 101{\cdot}3 \times 10^3 \times 22{\cdot}4 = 2{\cdot}27 \times 10^6 \ (\text{N/m}^2)\,(\text{m}^3/\text{kmol})$$
$$V = A'(PV)_0/P = 1{\cdot}565 \times 2{\cdot}27 \times 10^6/(60 \times 10^6)$$
$$= \underline{0{\cdot}0592\text{ m}^3}$$

Compared with the experimental value, the percentage errors in the other values are:

(a) ideal gas law $-25{\cdot}2\%$
(b) van der Waals' equation $+11{\cdot}5\%$
(c) compressibility factor $-0{\cdot}5\%$

Joule–Thomson Effect

It has already been shown that the change of internal energy of unit mass of fluid with volume at constant temperature is given by the relation:

$$\left(\frac{\partial U}{\partial v}\right)_T = T\left(\frac{\partial P}{\partial T}\right)_v - P \qquad\qquad \text{(equation 2.21)}$$

for a non-ideal gas: $$T\left(\frac{\partial P}{\partial T}\right)_v \neq P$$

and therefore $(\partial U/\partial v)_T$ and $(\partial U/\partial P)_T$ are not equal to zero.

Thus the internal energy of the non-ideal gas is a function of pressure as well as temperature. As the gas is expanded, the molecules are separated from each other against the action of the attractive forces between them. Energy is therefore stored in the gas; this is released when the gas is compressed and the molecules are allowed to approach one another again.

A characteristic of the non-ideal gas is that it has a finite Joule–Thomson effect. This relates to the amount of heat which must be added during an expansion of a gas from a pressure P_1 to a pressure P_2 in order to maintain isothermal conditions. Imagine a gas flowing from a cylinder, fitted with a piston, at a pressure P_1 to a second cylinder at a pressure P_2 (Fig. 2.2).

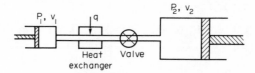

FIG. 2.2. Joule–Thomson effect.

The net work done by unit mass of gas on the surroundings in expanding from P_1 to P_2 is given by:

$$W = P_1 v_1 - P_2 v_2 \qquad (2.33)$$

A quantity of heat (q, say) is added during the expansion so as to maintain isothermal conditions. The change in the internal energy is therefore given by:

$$\Delta U = q - W \quad \text{(from equation 2.1)}$$

$$\therefore \qquad q = \Delta U - P_1 v_1 + P_2 v_2 \qquad (2.34)$$

For an ideal gas, under isothermal conditions, $\Delta U = 0$ and $P_2 v_2 = P_1 v_1$. Thus $q = 0$ and the ideal gas is said to have a zero Joule–Thomson effect. A non-ideal gas has a Joule–Thomson effect which may be either positive or negative.

2.4. THE FLUID IN MOTION

When a fluid flows through a duct or over a surface, the velocity over a plane at right angles to the stream is not normally uniform. The variation of velocity can be shown by the use of streamlines which are lines so drawn that the velocity vector is always tangential to them. The flowrate between any two streamlines is always the same. Constant velocity over a cross-section is shown by equidistant streamlines and an increase in velocity by closer spacing of the streamlines. There are two principal types of flow which will be discussed in detail later, namely streamline and turbulent flow. In streamline flow, movement across streamlines occurs solely as the result of diffusion on a molecular scale and the flowrate is steady. In turbulent flow the presence of circulating currents results in transference of fluid on a larger scale, and cyclic fluctuations occur in the flowrate, though the time-average rate remains constant.

A group of streamlines can be taken together to form a streamtube, and thus the whole area for flow can be regarded as being composed of bundles of streamtubes.

Figures 2.3, 2.4, and 2.5 show the flow patterns in a straight tube, through a constriction and past an immersed object. In the first case, the streamlines are all parallel to one another, whereas in the other two cases the streamlines approach one another as the passage becomes constricted, indicating that the velocity is increasing.

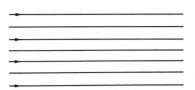

FIG. 2.3. Streamlines in a straight tube.

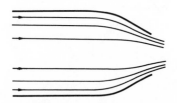

FIG. 2.4. Streamlines in a constriction.

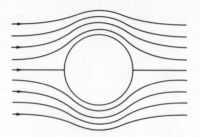

FIG. 2.5. Streamlines for flow past immersed object.

2.4.1.   Continuity

Consider the flow of a fluid through a streamtube, as shown in Fig. 2.6. Then equating the mass rates of flow at sections 1 and 2:

$$dG = \rho_1 \dot{u}_1 \, dA_1 = \rho_2 \dot{u}_2 \, dA_2 \tag{2.35}$$

where ρ_1, ρ_2 are the densities; $\dot{u}_1$, $\dot{u}_2$ the velocities in the streamtube; and dA_1, dA_2 the flow areas at sections 1 and 2 respectively.

On integration:

$$G = \int \rho_1 \dot{u}_1 \, dA_1 = \int \rho_2 \dot{u}_2 \, dA_2 = \rho_1 u_1 A_1 = \rho_2 u_2 A_2 \tag{2.36}$$

where u_1, u_2 are the average velocities (defined by the above equations) at the two sections. In many problems, the mass flowrate per unit area G' is the important quantity.

$$G' = G/A = \rho u \tag{2.37}$$

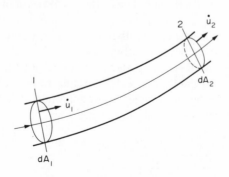

FIG. 2.6. Flow through a streamtube.

For an incompressible fluid (a liquid or a gas where the pressure changes are small)

$$u_1 A_1 = u_2 A_2 \tag{2.38}$$

It will be seen that it is important to be able to determine the velocity profile so that the flowrate can be calculated. For streamline flow in a pipe the mean velocity is 0·5 times the maximum stream velocity which occurs at the axis. For turbulent flow, the profile is flatter and the ratio of the mean velocity to the maximum velocity is 0·82.

2.4.2.   Momentum Changes in a Fluid

As a fluid flows through a duct its momentum and pressure may change. The magnitude of the changes can be considered by applying the momentum equation (force equals rate of change of momentum) to the fluid in a streamtube and then integrating over the cross-section of the duct. The effect of frictional forces will be neglected at first and the relations thus obtained will strictly apply only to an inviscid (frictionless) fluid.

Consider an element of length dl of a streamtube of cross-sectional area dA, increasing to $dA + \dfrac{d(dA)}{dl} dl$, as shown in Fig. 2.7.

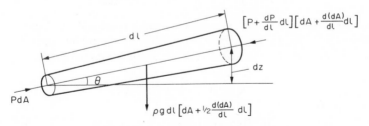

FIG. 2.7. Forces on fluid in a streamtube.

Then the upstream pressure $= P$ and force attributable to upstream pressure $= P\, dA$.

The downstream pressure $= P + (dP/dl)\, dl$

This pressure acts over an area $dA + \dfrac{d(dA)}{dl} dl$ and gives rise to a total force of $\left(P + \dfrac{dP}{dl} dl\right)\left(dA + \dfrac{d(dA)}{dl} dl\right)$.

In addition, the mean pressure of $P + \dfrac{1}{2}\dfrac{dP}{dl} dl$ acting on the sides of the streamtube will give rise to a force having a component $\left(P + \dfrac{1}{2}\dfrac{dP}{dl} dl\right)\dfrac{d(dA)}{dl} dl$ along the streamtube. Thus, the net force along the streamtube due to the pressure gradient is:

$$P\, dA - \left(P + \frac{dP}{dl} dl\right)\left(dA + \frac{d(dA)}{dl} dl\right) + \left(P + \frac{1}{2}\frac{dP}{dl} dl\right)\frac{d(dA)}{dl} dl \approx -\frac{dP}{dl} dl\, dA$$

The other force acting is the weight of the fluid

$$= \rho g\, dl \left(dA + \frac{1}{2}\frac{d(dA)}{dl} dl\right)$$

The component of this force along the streamtube is

$$-\rho g\, dl \left(dA + \frac{1}{2}\frac{d(dA)}{dl} dl\right) \sin\theta$$

Neglecting second order terms and noting that $\sin\theta = dz/dl$:

total force on fluid $\qquad = -\dfrac{dP}{dl} dl\, dA - \dfrac{dz}{dl} \rho g\, dl\, dA$ \hfill (2.39)

The rate of change of momentum of the fluid along the streamtube

$$= (\rho \dot{u}\, dA)\left[\left(\dot{u} + \frac{d\dot{u}}{dl} dl\right) - \dot{u}\right]$$

$$= \rho \dot{u}\, \frac{d\dot{u}}{dl} dl\, dA \hfill (2.40)$$

Equating equations 2.39 and 2.40:

$$\rho \dot{u}\, dl\, dA\, (d\dot{u}/dl) = -dl\, dA\, (dP/dl) - \rho g\, dl\, dA\, (dz/dl)$$

$\therefore \qquad\qquad \dot{u}\, d\dot{u} + (dP/\rho) + g\, dz = 0$ \hfill (2.41)

On integration:

$$(\dot{u}^2/2) + \int (dP/\rho) + gz = \text{constant} \qquad (2.42)$$

For the simple case of the incompressible fluid, ρ is independent of pressure, and

$$(\dot{u}^2/2) + (P/\rho) + gz = \text{constant} \qquad (2.43)$$

Equation 2.43 is known as Bernoulli's equation, which relates the pressure at a point in the fluid to its position and velocity. Each term in equation 2.43 represents energy per unit mass of fluid. Thus, if all the fluid is moving with a velocity u, the total energy per unit mass ψ is given by:

$$\psi = (\dot{u}^2/2) + (P/\rho) + gz \qquad (2.44)$$

Dividing equation 2.43 by g:

$$(\dot{u}^2/2g) + (P/\rho g) + z = \text{constant} \qquad (2.45)$$

In equation 2.45 each term represents energy per unit weight of fluid and has the dimensions of length and can be regarded as representing a contribution to the total fluid head.

Thus $\dot{u}^2/2g$ is the velocity head

$P/\rho g$ is the pressure head

and z is the potential head

Equation 2.42 can also be obtained from consideration of the energy changes in the fluid.

2.4.3.   Energy of a Fluid in Motion

The total energy of a fluid in motion is made up of a number of components. Reference will be made to unit mass of fluid and neglect changes in magnetic and electrical energy, etc.

Internal Energy U

This has already been discussed in Section 2.2.

Pressure Energy

This represents the work which must be done in order to introduce the fluid, without change in volume, into the system. It is therefore given by the product Pv, where P is the pressure of the system and v is the volume of unit mass of fluid.

Potential Energy

The potential energy of the fluid, due to its position in the earth's gravitational field, is equal to the work which must be done on it in order to raise it to that position from some arbitrarily chosen datum level at which the potential energy is taken as zero. Thus, if the fluid is situated at a height z above the datum level, the potential energy is zg, where g is the acceleration due to gravity which is taken as constant unless otherwise stated.

Kinetic Energy

The fluid possesses kinetic energy by virtue of its motion with reference to some arbitrarily fixed body (normally taken as the earth). If the fluid is moving with a velocity u, the kinetic energy is $u^2/2$.

The total energy of unit mass of fluid is, therefore:

$$U + Pv + gz + u^2/2 \qquad (2.46)$$

If the fluid flows from section 1 to section 2 (where the values of the various quantities are denoted by suffixes 1 and 2 respectively) and q is the net heat absorbed from the surroundings and W_s is the net work done by the fluid on the surroundings, other than that done by the fluid in entering or leaving the section under consideration:

$$U_2 + P_2v_2 + gz_2 + (u_2^2/2) = U_1 + P_1v_1 + gz_1 + (u_1^2/2) + q - W_s \qquad (2.47)$$

$$q - W_s = \Delta U + \Delta(Pv) + g\,\Delta z + \Delta(u^2/2) \qquad (2.48)$$

where Δ denotes a finite change in the quantities.

It should be noted that the shaft work W_s is related to the total work W by the relation:

$$W = W_s + \Delta(Pv) \qquad (2.49)$$

Thus
$$q - W_s = \Delta H + g\,\Delta z + (\Delta u^2/2) \qquad (2.50)$$

For a small change in the system:

$$\delta q - \delta W_s = dH + g\,dz + u\,du \qquad (2.51)$$

For many purposes it is convenient to eliminate H by using equation 2.11:

$$dH = \delta q + \delta F + v\,dP \qquad \text{(equation 2.11)}$$

Here δF represents the amount of mechanical energy irreversibly converted into heat.

Thus
$$u\,du + g\,dz + v\,dP + \delta W_s + \delta F = 0 \qquad (2.52)$$

When no work is done by the fluid on the surroundings and when friction can be neglected, it will be noted that equation 2.52 is identical to equation 2.41 derived from consideration of momentum, since:

$$v = 1/\rho$$

Integrating the above equation for flow from section 1 to section 2 and summing the terms δW_s and δF:

$$(\Delta u^2/2) + g\,\Delta z + \int_1^2 v\,dP + W_s + F = 0 \qquad (2.53)$$

Equations 2.41 to 2.53 are quite general and apply therefore to any type of fluid.

With incompressible fluids the energy F is either lost to the surroundings or causes a very small rise in temperature. If the fluid is compressible, however, the rise in temperature may result in an increase in the pressure energy and part of it may be available for doing useful work.

If the fluid is flowing through a channel or pipe a frictional drag arises in the region of the boundaries and gives rise to a velocity distribution across any section perpendicular to the direction of flow. For the unidirectional flow of fluid, the mean velocity of flow has been defined by equation 2.36 as the ratio of the volumetric flowrate to the cross-sectional area of the channel. When equation 2.53 is applied over the whole cross-section, therefore, allowance must be made for the fact that the mean square velocity is not equal to the square of the mean velocity, and a correction factor α must therefore be introduced into the kinetic energy term. Thus, considering the fluid over the whole cross-section, for small changes:

$$(u\,du/\alpha) + g\,dz + v\,dP + \delta W_s + \delta F = 0 \qquad (2.54)$$

and for finite changes:

$$(\Delta u^2/2\alpha) + g\,\Delta z + \int_{P_1}^{P_2} v\,dP + W_s + F = 0 \qquad (2.55)$$

Equation 2.50 becomes:

$$(\Delta u^2/2\alpha) + g\,\Delta z + \Delta H = q - W_s \qquad (2.56)$$

For flow in a pipe of circular cross-section α will be shown to be exactly 0·5 for streamline flow and to approximate to unity for turbulent flow.

For turbulent flow, and where no external work is done, equation 2.54 becomes:

$$u \, du + g \, dz + v \, dP = 0 \qquad (2.57)$$

if frictional effects can be neglected.

For horizontal flow, or where the effects of change of height can be neglected, as normally with gases, equation 2.57 simplifies to:

$$u \, du + v \, dP = 0 \qquad (2.58)$$

2.4.4. Pressure and Fluid Head

In equation 2.54 each term represents energy per unit mass of fluid. If it is multiplied throughout by density ρ, each term has the dimensions of pressure and represents energy per unit volume of fluid:

$$(\rho u \, du / \alpha) + \rho g \, dz + dP + \rho \delta W_s + \rho \delta F = 0 \qquad (2.59)$$

If equation 2.52 is divided throughout by g, each term has the dimensions of length, and, as already mentioned, can be regarded as a component of the total head of the fluid and represents energy per unit weight:

$$(u \, du / \alpha g) + dz + (v \, dP / g) + (\delta W_s / g) + (\delta F / g) = 0 \qquad (2.60)$$

For an incompressible fluid flowing in a horizontal pipe of constant cross-section, in the absence of work being done by the fluid on the surroundings, the pressure change due to frictional effects is given by:

$$(v \, dP_f / g) + (\delta F / g) = 0$$

i.e.
$$-dP_f = \delta F / v = \rho g \, dh_f \qquad (2.61)$$

2.4.5. Constant Flow per Unit Area

When the flowrate of the fluid per unit area G' is constant, equation 2.37 can be written:

$$G / A = G' = u_1 / v_1 = u_2 / v_2 = u / v \qquad (2.62)$$

or
$$G' = u_1 \rho_1 = u_2 \rho_2 = u\rho \qquad (2.63)$$

Equation 2.58, which is the momentum balance for horizontal turbulent flow, is:

$$u \, du + v \, dP = 0 \qquad \text{(equation 2.58)}$$

i.e.
$$(u \, du / v) + dP = 0$$

Because u/v is constant, this gives on integration:

$$[u_1(u_2 - u_1)/v_1] + P_2 - P_1 = 0$$

i.e.
$$(u_1^2 / v_1) + P_1 = (u_2^2 / v_2) + P_2 \qquad (2.64)$$

Before equation 2.55 can be applied to any particular flow problem, the term $\int_1^2 v \, dP$ must be evaluated.

2.4.6. Separation

It should be noted that the energy and mass balance equations assume that the fluid is continuous. This is so, in the case of a liquid, provided that the pressure does not fall to such a low value that boiling, or the evolution of dissolved gases, takes place. For water at normal temperatures the pressure should not be allowed to fall below the equivalent of a head of 1·2 m of liquid. With gases, there is no lower limit to the pressures at which the fluid remains continuous, but the various equations which are derived need modification if the pressures are so low that the linear dimensions of the channels become comparable with the mean free path of the molecules, i.e. when the so-called molecular flow sets in.

2.5. PRESSURE–VOLUME RELATIONSHIPS

2.5.1. Incompressible Fluids

Here v is independent of pressure so that

$$\int_1^2 v \; dP = (P_2 - P_1)v \tag{2.65}$$

Therefore equation 2.55 becomes:

$$(u_1^2/2\alpha_1) + gz_1 + P_1v = (u_2^2/2\alpha_2) + gz_2 + P_2v + W_s + F \tag{2.66}$$

or $$(\Delta u^2/2\alpha) + g\,\Delta z + v\,\Delta P + W_s + F = 0 \tag{2.67}$$

In a frictionless system in which the fluid does not work on the surroundings and α_1 and α_2 are taken as unity (turbulent flow)

$$(u_1^2/2) + gz_1 + P_1v = (u_2^2/2) + gz_2 + P_2v \tag{2.68}$$

Example 2.2. Water flows from a tap at a pressure of 350 kN/m². What is the velocity of the jet if frictional effects are neglected?

Solution. From equation 2.68:

$$0\cdot5(u_2^2 - u_1^2) = g(z_1 - z_2) + (P_1 - P_2)/\rho$$

Using suffix 1 to denote conditions in the pipe and suffix 2 to denote conditions in the jet and neglecting the velocity of approach in the pipe

$$0\cdot5(u_2^2 - 0) = g \times 0 + (350 - 101\cdot3) \times 10^3/1000$$

$$\underline{\underline{u_2 = 22\cdot3 \text{ m/s}}}$$

2.5.2. Compressible Fluids

For a gas, the mean value of the specific volume can be used unless the pressure or temperature change is very large, when the relation between pressure and volume must be taken into account.

The term $\int_1^2 v \; dP$ will now be evaluated for the ideal gas under various conditions. In most cases the results so obtained can be applied to the non-ideal gas without introducing an error greater than is involved in estimating the other factors concerned in the process. The only common exception to this occurs in the flow of gases at very high pressures, when it is necessary to employ one of the approximate equations for the state of a non-ideal gas, in place of the equation for the ideal gas. Alternatively, one may use equation 2.50 and work in terms of changes in enthalpy. For a gas, the potential energy term is usually small compared with the other energy terms.

The relation between the pressure and the volume of an ideal gas depends on the rate of transfer of heat to the surroundings and the degree of irreversibility of the process. The following conditions will be considered:

    (a) an isothermal process;
    (b) an isentropic process;
    (c) a reversible process which is neither isothermal nor adiabatic;
    (d) an irreversible process which is not isothermal.

Isothermal Process

For an isothermal process $Pv = \mathbf{R}T/M = P_1v_1$, where the subscript 1 denotes the initial values and M is the molecular weight.

Thus
$$\int_1^2 v\,dP = P_1v_1 \int_1^2 (1/P)dP = P_1v_1 \ln(P_2/P_1) \tag{2.69}$$

Isentropic Process

From equation 2.30, for an isentropic process:
$$Pv^{\gamma} = P_1v_1^{\gamma} = \text{constant}$$

$$\therefore \quad \int_1^2 v\,dP = \int_1^2 \left(\frac{P_1v_1^{\gamma}}{P}\right)^{1/\gamma} dP$$

$$= P_1^{1/\gamma} v_1 \int_1^2 P^{-1/\gamma}\,dP$$

$$= P_1^{1/\gamma} v_1 \frac{1}{1-(1/\gamma)} (P_2^{1-(1/\gamma)} - P_1^{1-(1/\gamma)})$$

$$= \frac{\gamma}{\gamma-1} P_1v_1 \left[\left(\frac{P_2}{P_1}\right)^{(\gamma-1)/\gamma} - 1\right] \tag{2.70}$$

$$= \frac{\gamma}{\gamma-1} \left[P_1 \left(\frac{P_2}{P_1}\right)^{(\gamma-1)/\gamma} \left(\frac{P_2}{P_1}\right)^{1/\gamma} v_2 - P_1v_1\right]$$

$$= \frac{\gamma}{\gamma-1} (P_2v_2 - P_1v_1) \tag{2.71}$$

Further, from equations 2.29 and 2.26:
$$\int_1^2 v\,dP = \int_1^2 dH = C_p\Delta T \tag{2.72}$$

(taking C_p as constant).

The above relations apply for an ideal gas to a reversible adiabatic process which, as already shown, is isentropic.

Reversible Process—Neither Isothermal Nor Adiabatic

In general the conditions under which a change in state of a gas takes place are neither isothermal nor adiabatic and the relation between pressure and volume is approximately of the form $Pv^k = $ constant for a reversible process, where k is a numerical quantity whose value depends on the heat transfer between the gas and its surroundings. k usually lies between 1 and γ but may, under certain circumstances, lie outside these limits; it will have the same value for a reversible compression as for a reversible expansion under similar conditions. Under these conditions therefore, equation 2.70 becomes:

$$\int_1^2 v \, dP = \frac{k}{k-1} P_1 v_1 \left[\left(\frac{P_2}{P_1}\right)^{(k-1)/k} - 1 \right] \tag{2.73}$$

Irreversible Process

For an irreversible process it may not be possible to express the relation between pressure and volume as a continuous mathematical function, but by choosing a suitable value for the constant k an equation of the form $Pv^k = $ constant may be used over a limited range of conditions. The above equation can then be used for the evaluation of $\int_1^2 v \, dP$. It should be noted that, for an irreversible process, k will have different values for compression and expansion under otherwise similar conditions. Thus, for the irreversible adiabatic compression of a gas, k will be greater than γ: for the corresponding expansion k will be less than γ. This means that more energy has to be put into an irreversible compression than will be received back when the gas expands to its original condition.

2.6. ROTATIONAL OR VORTEX MOTION IN A FLUID

In many chemical engineering problems a liquid is in rotational motion—for instance, in the casing of a centrifugal pump, in a stirred vessel, or in the basket of a centrifuge. Usually, fluid friction will have a comparatively minor effect and energy dissipation can be disregarded. It is necessary to consider the effect of the forces on the fluid in determining the pressure gradients in the vertical and horizontal directions. It will be assumed that rotation is taking place about a vertical axis.

Vertically, the only force acting is the gravitational force and therefore:

$$(\partial P/\partial z) = -\rho g = -(g/v) \tag{2.74}$$

In the horizontal direction, a centrifugal force is acting and its magnitude is a function of the velocity of the fluid and the radius of rotation r. If the angular velocity at radius r is ω, a force balance in a radial direction can be taken on an element of liquid of inner radius r, of depth dz, and subtending an angle $d\theta$ at the centre of rotation, as shown in Fig. 2.8. Thus:

$$\left(P + \frac{\partial P}{\partial r} dr\right)(r + dr)d\theta \, dz - Pr \, d\theta \, dz - 2\left(P + \frac{1}{2}\frac{\partial P}{\partial r} dr\right)dr \, dz \frac{d\theta}{2} - r \, d\theta \, dr \, dz \, \rho \omega^2 = 0$$

Neglecting small quantities of a second order:

$$(\partial P/\partial r) = r\rho\omega^2 = (\rho u^2/r) \tag{2.75}$$

The variation of pressure on the liquid is then given by:

$$P = \int \frac{\partial P}{\partial z} dz + \int \frac{\partial P}{\partial r} dr \tag{2.76}$$

Equation 2.76 can be integrated only when the variation of ω with r is specified. Two important cases will be considered—the forced vortex and the free vortex.

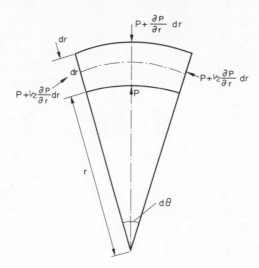

FIG. 2.8. Forces acting on element of fluid in a vortex.

2.6.1. The Forced Vortex

A forced vortex is defined as one in which the angular velocity ω is independent of the radius r. Substitution from equations 2.74 and 2.75 into equation 2.76 and integrating gives:

$$P - P_0 = (z_0 - z)\rho g + (\rho\omega^2 r^2/2) \tag{2.77}$$

where P_0 and z_0 denote the values of P and z at $r = 0$.

Now the total energy ψ per unit mass of liquid is given by equation 2.44:

$$\psi = (u^2/2) + (P/\rho) + gz \tag{equation 2.44}$$

Substituting $u = r\omega$ and for (P/ρ) from equation 2.77:

$$\psi = r^2\omega^2 + (P_0/\rho) + gz_0 \tag{2.78}$$

Thus the energy per unit mass increases with radius, and a forced vortex is therefore not stable because the shear stresses within the liquid would tend to make ψ constant throughout. A forced vortex is maintained only in the presence of an external influence. The most common cases of liquid in a forced vortex are:

(a) the liquid within the impeller of a centrifugal pump when the delivery valve is closed;
(b) the liquid within the confines of a stirrer in an agitated tank;
(c) liquid in the rotating basket of a centrifuge.

From equation 2.77 it is seen that lines of constant pressure are paraboloids of revolution (Fig. 2.9).

2.6.2. The Free Vortex

In a free vortex the energy per unit mass of fluid is constant, and thus a free vortex is inherently stable. The variation of pressure with radius is obtained by differentiating equation 2.44 with respect to radius at constant depth z:

i.e.

$$u\frac{\partial u}{\partial r} + \frac{1}{\rho}\frac{\partial P}{\partial r} = 0 \tag{2.79}$$

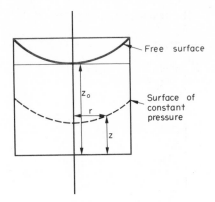

FIG. 2.9. Forced vortex.

But $$(\partial P/\partial r) = (\rho u^2/r) \qquad \text{(equation 2.75)}$$

$\therefore$ $$(\partial u/\partial r) + (u/r) = 0$$

i.e. $$ur = \text{constant} = \kappa \qquad (2.80)$$

Hence the angular momentum of the liquid is everywhere constant.

Thus $$(\partial P/\partial r) = (\rho\kappa^2/r^3) \qquad (2.81)$$

Substituting from equations 2.74 and 2.81 into equation 2.76 and integrating:

$$P - P_\infty = (z_\infty - z)\rho g - (\rho\kappa^2/2r^2) \qquad (2.82)$$

where P_∞ and z_∞ are the values of P and z at $r = \infty$.
Putting $P = P_\infty =$ atmospheric pressure:

$$z = z_\infty - (\kappa^2/2r^2 g) \qquad (2.83)$$

Substituting into equation 2.44 gives:

$$\psi = (P_\infty/\rho) + gz_\infty \qquad (2.84)$$

ψ is constant by definition and equal to the value at $r = \infty$ where $u = 0$.
A free vortex (Fig. 2.10) exists:

(a) outside the impeller of a centrifugal pump;
(b) outside the region of the agitator in a stirred tank.

In both cases the free vortex may be modified by the frictional effect exerted by the external walls.

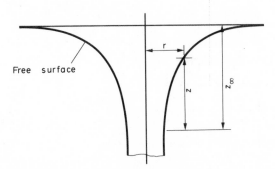

FIG. 2.10. Free vortex.

2.7. FURTHER READING

SMITH, J. M. and VAN NESS, H. C.: *Introduction to Chemical Engineering Thermodynamics*, 2nd edn. (McGraw-Hill, New York, 1959).
DODGE, B. F.: *Chemical Engineering Thermodynamics* (McGraw-Hill, New York, 1944).
SCHLICTING, H.: *Boundary Layer Theory*, 4th edn. (McGraw-Hill, New York, 1960).

2.8. REFERENCES

1. HOUGEN, O. A. and WATSON, K. M.: *Chemical Process Principles* (Wiley, New York, 1964).
2. PERRY, J. H. (ed.): *Chemical Engineers Handbook*, 5th edn. (McGraw-Hill, New York, 1974).

2.9. NOMENCLATURE

		Units in SI system	Dimensions in M, L, T, θ
A	Area perpendicular to direction of flow	m²	$\mathbf{L^2}$
a	Coefficient in van der Waals' equation	(kN/m²)(m³)²/(kmol)²	$\mathbf{M^{-1}L^5T^{-2}}$
b	Coefficient in van der Waals' equation	m³/kmol	$\mathbf{M^{-1}L^3}$
C_p	Specific heat at constant pressure per unit mass	J/kg K	$\mathbf{L^2T^{-2}\theta^{-1}}$ or —
C_v	Specific heat at constant volume per unit mass	J/kg K	$\mathbf{L^2T^{-2}\theta^{-1}}$ or —
F	Energy per unit mass degraded because of irreversibility of process	J/kg	$\mathbf{L^2T^{-2}}$
G	Mass rate of flow	kg/s	$\mathbf{MT^{-1}}$
G'	Mass rate of flow per unit area	kg/m² s	$\mathbf{ML^{-2}T^{-1}}$
g	Acceleration due to gravity	m/s²	$\mathbf{LT^{-2}}$
H	Enthalpy per unit mass	J/kg	$\mathbf{L^2T^{-2}}$
h_f	Head lost due to friction	m	$\mathbf{L}$
k	Numerical constant used as index for compression	—	—
l	Length of streamtube	m	$\mathbf{L}$
M	Molecular weight	kg/kmol	—
n	Number of molar units of fluid	kmol	$\mathbf{M}$
P	Pressure	N/m²	$\mathbf{ML^{-1}T^{-2}}$
P_R	Reduced pressure	—	—
q	Net heat flow into system	J/kg	$\mathbf{L^2T^{-2}}$
$\mathbf{R}$	Universal gas constant	8·314 kJ/kmol K	$\mathbf{L^2T^{-2}\theta^{-1}}$
r	Radius	m	$\mathbf{L}$
S	Entropy per unit mass	J/kg K	$\mathbf{L^2T^{-2}\theta^{-1}}$
T	Absolute temperature	K	$\boldsymbol{\theta}$
T_R	Reduced temperature	—	—
t	Time	s	$\mathbf{T}$
U	Internal energy per unit mass	J/kg	$\mathbf{L^2T^{-2}}$
u	Mean velocity	m/s	$\mathbf{LT^{-1}}$
$\dot{u}$	Velocity in streamtube	m/s	$\mathbf{LT^{-1}}$
V	Volume of fluid	m³	$\mathbf{L^3}$
V_R	Reduced volume	—	—
v	Volume per unit mass of fluid	m³/kg	$\mathbf{M^{-1}L^3}$
W	Net work per unit mass done by system on surroundings	J/kg	$\mathbf{L^2T^{-2}}$
W_s	Shaft work per unit mass	J/kg	$\mathbf{L^2T^{-2}}$
Z	Compressibility factor for non-ideal gas	—	—
z	Distance in vertical direction	m	$\mathbf{L}$
α	Constant in expression for kinetic energy of fluid	—	—
γ	Ratio of specific heats C_p/C_v	—	—
ψ	Total mechanical energy per unit mass of fluid	J/kg	$\mathbf{L^2T^{-2}}$
ρ	Density of fluid	kg/m³	$\mathbf{ML^{-3}}$
θ	Angle	—	—
ω	Angular velocity of rotation	rad/s	$\mathbf{T^{-1}}$

CHAPTER 3

Friction in Pipes and Channels

3.1. INTRODUCTION

In the process industries it is often necessary to pump fluids over long distances from storage to reactor units, and there may be a substantial drop in pressure in both the pipeline and in the individual units themselves. Many intermediate products are pumped from one factory site to another, and raw materials such as natural gas and petroleum products may be pumped very long distances to domestic or industrial consumers. It is necessary, therefore, to consider the problems concerned with calculating the power requirements for pumping, with designing the most suitable flow system, with estimating the most economical sizes of pipes, with measuring the rate of flow, and frequently with controlling this flow at a steady rate. The flow may take place at high pressures, for instance when nitrogen and hydrogen flow to an ammonia synthesis plant, or at low pressures when, for example, vapour leaves the top of a vacuum distillation plant. The fluid may consist of one or more phases, it may contain suspended solids, it may be near its boiling point, and it may have non-Newtonian properties, all of which complicate the analysis.

The design and layout of pipe systems is an important factor in the planning of modern plants and may represent a significant part of the total cost. In this chapter, the problem of determining the drop in pressure during flow will be discussed; in Chapter 5 the techniques of flow measurement, and in Chapter 6 the problem of selecting the most suitable pump for a given application will be considered.

Chemical engineering design frequently concerns equipment for the transfer of material or heat from one phase to another, and, in order to understand the mechanism of the transport process, the flow pattern of the fluid, and particularly the distribution of velocity near a surface, must be studied. It must be realised that when a fluid is flowing over a surface or through a pipe, the velocity at various points in a plane at right angles to the stream velocity is rarely uniform, and the rate of change of velocity with distance from the surface will exert a vital influence on the resistance to flow and on the rate of mass or heat transfer.

3.2. THE NATURE OF FLUID FLOW

When a fluid is flowing through a tube or over a surface, the pattern of the flow will vary with the velocity, the physical properties of the fluid, and the geometry of the surface. This problem was first examined by Reynolds[1] in 1883 using an apparatus shown in Fig. 3.1. A glass tube with a flared entrance was immersed in a glass tank fed with water and, by means of the valve, the rate of flow from the tank through the glass tube could be controlled. By introducing a fine filament of coloured water from a small reservoir centrally into the flared entrance of the glass tube, the nature of the flow could be observed. At low rates of flow the coloured filament remained at the axis of the tube indicating that the flow was in the form of parallel streams which did not interfere with each other. Such flow is called laminar or streamline and is characterised by the absence of bulk movement at right angles to the main stream direction, though a small amount of radial dispersion will occur as a result of diffusion. As the flowrate was increased, a condition was reached in which oscillations appeared in the coloured filament which broke up into eddies causing dispersion across the

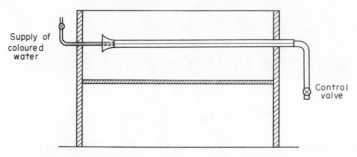

FIG. 3.1. Reynolds' apparatus for tracing flow patterns.

tube section. This type of flow is known as turbulent flow and is characterised by the rapid movement of fluid as eddies in random directions across the tube. The general pattern is as shown in Fig. 3.2. These experiments clearly showed the nature of the transition from streamline to turbulent flow. Below the critical velocity, oscillations in the flow were unstable and any disturbance would quickly disappear. At higher velocities, however, the oscillations were stable and increased in amplitude, causing a high degree of radial mixing. However, it was found that even when the main flow was turbulent there was a region near the wall (the *laminar sub-layer*) in which streamline flow persisted.

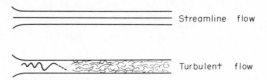

FIG. 3.2. Break-up of laminar thread in Reynolds' experiment.

Only the problem of steady flow will be considered in which the time average velocity in the main stream direction X is constant and equal to u_x. In laminar flow, the instantaneous velocity at any point then has a steady value of u_x and does not fluctuate. In turbulent flow the instantaneous velocity at a point will vary about the mean value of u_x. It is convenient to consider the components of the eddy velocities in two directions—one along the main stream direction X and the other at right angles to the stream flow Y. Since the net flow in the X-direction is steady, the instantaneous velocity u_i can be imagined as being made up of a steady velocity u_x and a fluctuating velocity u_{Ex}, so that:

$$u_i = u_x + u_{Ex} \tag{3.1}$$

Since the average value of the main stream velocity is u_x, the average value of u_{Ex}, $\bar{u}_{Ex}$, will be zero, although the fluctuating component may at any instant amount to several percent of the stream velocity. The fluctuating velocity in the Y-direction will also vary but, again, must have an average value of zero since there is no net flow at right angles to the stream flow. Turbulent flow is of great importance in chemical engineering because it causes rapid mixing of the fluid elements and is therefore responsible for effecting high rates of heat and mass transfer.

3.2.1. Flow over a surface

When a fluid flows over a surface the elements in contact with the surface will be brought to rest and the adjacent layers retarded by the viscous drag of the fluid. Thus the velocity in the neighbourhood of the surface will change in a direction at right angles to the stream flow. It is important to realise that this change in velocity originates at the walls or surface.

Suppose a fluid flowing with uniform velocity approaches a surface in the form of a plane as in Fig. 3.3. When the fluid reaches the surface, a velocity gradient is set up at right angles to the surface because of the viscous forces acting within the fluid. The fluid in contact with the surface must be brought to rest as otherwise there would be an infinite velocity gradient at the wall, and an infinite stress to correspond. If u_x is the velocity in the X-direction at distance y from the surface, u_x will increase from zero at the surface ($y = 0$) and will gradually approach the stream velocity u_s at some distance from the surface. Thus, if the values of u_x are measured, the velocity profile will be as shown in Fig. 3.3. The velocity

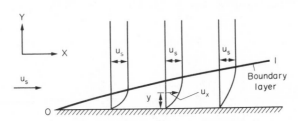

FIG. 3.3. Development of boundary layer.

distributions are shown for three different distances downstream, and it is seen that in each case there is a rapid change in velocity near the wall and that the thickness of the layer in which the fluid is retarded becomes greater with distance in the direction of flow. The line 01 divides the stream into two sections; in the lower part the velocity is increasing away from the surface, whilst in the upper portion the velocity is approximately equal to u_s. This line indicates the thickness of the zone of retarded fluid which was termed the *boundary layer* by Prandtl.[2] As will be shown in Chapter 9, the main stream velocity is approached asymptotically, and therefore the boundary layer strictly has no precise outer limit. However, it is convenient to define the boundary layer thickness such that the velocity at its outer edge equals 99 per cent of the stream velocity. Other definitions will be given later. Thus, by making certain assumptions covering the velocity profile, it will be shown in Chapter 9 that the boundary layer thickness δ at a distance x from the leading edge of a surface is dependent on the Reynolds number of flow.

Towards the leading edge of the surface, the flow in the boundary layer is laminar, but at a critical distance eddies start to form giving a turbulent boundary layer. In the turbulent layer there is a small region near the surface where the flow remains laminar; this is known as the laminar sub-layer. The change from laminar to turbulent flow in the boundary layer will occur at different distances downstream depending on the roughness of the surface and the physical properties of the fluid. This is discussed at length in Chapter 9.

3.2.2. Flow in a Pipe

When a fluid with uniform flow over the cross-section enters a pipe, the layers of fluid adjacent to the walls are slowed down as on a plane surface and a boundary layer forms at the entrance. This builds up in thickness as the fluid passes into the pipe. At some distance downstream from the mouth, the boundary layers reach a thickness equal to the pipe radius and join at the axis, after which conditions remain constant and *fully developed flow* exists. If the flow in the boundary layers is streamline when they meet, laminar flow exists in the pipe. If the transition has already taken place before they meet, turbulent flow will persist in the region of fully developed flow. The region before the boundary layers join is known as the entry length and is discussed in greater detail in Chapter 9.

3.3. SHEARING CHARACTERISTICS OF A FLUID

In the laminar flow of most simple fluids, such as gases and many low molecular weight liquids, the shear rate increases linearly with shear stress over a wide range of shear rate, provided that temperature and pressure remain constant. The viscosity μ, which is the ratio of shear stress R_y to shear rate, is therefore constant, and the behaviour is termed *Newtonian* (Fig. 3.4, curve A).

Then:
$$R_y = -\mu(du_x/dy) \tag{3.2}$$

It will be noted that the rate of shear has been expressed as the velocity gradient du_x/dy in the fluid.

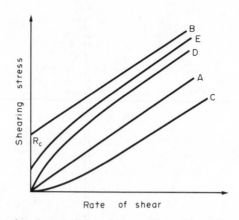

FIG. 3.4. Shearing characteristics of Newtonian and non-Newtonian fluids.

However, a wide range of industrially important liquids, such as solutions of high molecular weight, colloids, suspensions, and emulsions, exhibit more complex behaviour, which is termed *non-Newtonian*. Some of the most common types of non-Newtonian behaviour are described below. The terminology used is taken where possible from the *British Standard Glossary of Rheological Terms*.

When the viscosity of a fluid decreases with increasing shear rate under steady shear flow, the fluid is said to exhibit *shear-thinning* or *pseudoplasticity* (curve D); if it increases, the fluid is said to show *shear-thickening* or *dilatancy* (curve C). It is frequently possible to represent *shear thinning* or *shear thickening* behaviour by a power law equation:

$$R_y = -k(du_x/dy)^n \tag{3.3}$$

where k is a measure of consistency and the difference between n and unity indicates the degree of departure from Newtonian behaviour. For shear thinning materials, $n < 1$; for shear thickening materials, $n > 1$.

Viscosity can also show complex time-dependent changes: it may decrease with time during shear but recover when the shear stress is removed, in which case the substance is said to be *thixotropic*; the opposite behaviour is termed *negative thixotropy*. Recovery or solidification following thixotropic behaviour may be accelerated by gentle and regular motion; this phenomenon is known as *rheopexy*. Sometimes the decrease in viscosity is irreversible, and the fluid is then said to have suffered *shear breakdown* or *rheodestruction*.

Fluids may exhibit some of the properties of a solid. Those that will not flow until a critical *yield stress* is exceeded are termed *viscoplastic*. Others which show a finite elasticity are termed *viscoelastic*.

Viscoplastic behaviour is represented by curves B and E. For curve B:

$$R_y - R_c = -\mu(du_x/dy) \tag{3.4}$$

Further discussion on non-Newtonian behaviour is included in Volume 3.[3]

3.4. THE DROP IN PRESSURE FOR FLOW THROUGH A TUBE

Experimental work by Reynolds,[1] Stanton and Pannell,[4] Moody,[5] and others for the drop in pressure for flow through a pipe is most conveniently represented by plotting the head loss per unit length against the average velocity through the tube. The graph shown in Fig. 3.5 is obtained in which $i = h_f/l$ is known as the hydraulic gradient. At low velocities the curve is a straight line showing that i is directly proportional to the velocity, but at higher velocities the pressure drop increases more rapidly. If a logarithmic plot is used, as in Fig. 3.6, the results fall into three sections. Over the region of low velocity the line PB has a slope of unity but beyond this region over section BC there is instability with poorly defined points. At higher velocities, the line CQ has a slope of about 1·8. If QC is produced

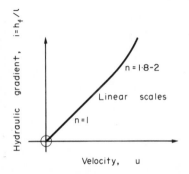

FIG. 3.5. Hydraulic gradient versus velocity.

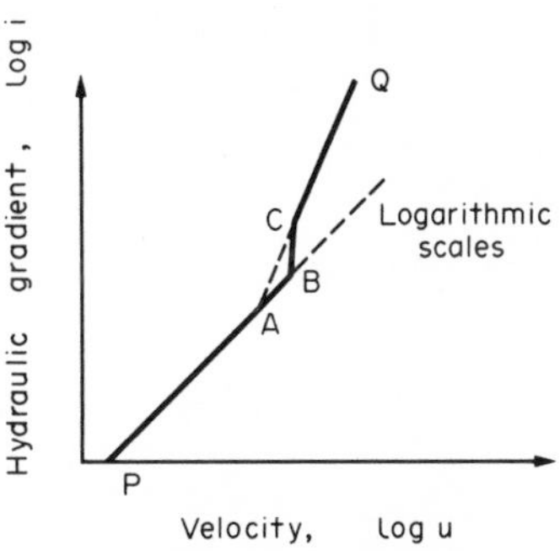

FIG. 3.6. Critical velocities.

backwards, it cuts PB at the point A, corresponding in Reynolds' earlier experiments to the change from laminar to turbulent flow and representing the critical velocity. Thus for streamline flow the pressure gradient is directly proportional to the velocity, and for turbulent flow it is proportional to the velocity raised to the 1·8 power.

The velocity corresponding to point A is taken as the lower critical velocity and to B as the higher critical velocity. Experiments with pipes of various sizes showed that the critical velocity was inversely proportional to the diameter, and that it was less at higher temperatures where the viscosity was lower. This led Reynolds to develop his own criterion based on the velocity of the fluid, the diameter of the tube, and the viscosity and density of the fluid. The dimensionless group $du\rho/\mu$ is termed the Reynolds number (Re) and is of vital importance in the study of fluid flow. It has been found that for values of Re less than about 2000 the flow is usually laminar and for values above 4000 the flow is usually turbulent. There is, however, no sharp change from laminar to turbulent flow but the transition occurs gradually as the velocity is increased in a given system. The precise velocity will depend on the geometry and on the pipe roughness. It is important to realise that there is no such thing as stable transitional flow. When the Reynolds number lies between 2000 and 4000 the flow is changing rapidly from laminar to turbulent and the portion of any cycle time for turbulent flow increases as Re approaches 4000.

If a turbulent fluid passes into a pipe so that the Reynolds number there is less than 2000, the flow pattern will change and the fluid will become streamline at some distance from the point of entry. On the other hand, if the fluid is initially streamline ($Re < 2000$), the diameter of the pipe can be gradually increased so that the Reynolds number exceeds 2000 and yet streamline flow will persist in the absence of any disturbance. Unstable streamline flow has been obtained in this manner at Reynolds numbers as high as 40,000. The setting up of turbulence requires a small force at right angles to the flow to initiate eddies.

3.4.1. Shear Stress in Fluid

It is convenient to relate the pressure drop $-\Delta P_f$ to the shear stress R_0, at the walls of a pipe. If R_y is the shear stress at a distance y from the wall of the pipe, the corresponding value at the wall R_0 is given by:

$$R_0 = -\mu (du_x/dy)_{y=0} \quad \text{(from equation 3.2)}$$

In this equation the negative sign is introduced in order to maintain a consistency of sign convention when shear stress is related to momentum transfer (see Chapter 9). Since $(du_x/dy)_{y=0}$ must be positive (velocity increases towards the pipe centre), R_0 is negative. It is therefore more convenient to work in terms of R ($= -R_0$) when calculating friction data.

Suppose a fluid is flowing through a length l of pipe of radius r (diameter d) over which the change in pressure due to friction is $-\Delta P_f$. Then a force balance on the fluid in the pipe gives:

$$-\Delta P_f \pi r^2 = 2\pi r l (-R_0)$$

i.e.
$$R_0 = \Delta P_f (r/2l) \tag{3.5}$$

If a force balance is now taken over the central core of fluid of radius s:

$$-\Delta P_f \pi s^2 = 2\pi s l (-R_y)$$

i.e.
$$R_y = \Delta P_f s /2l \tag{3.6}$$

Thus from equations 3.5 and 3.6:

$$R_y/R_0 = s/r = 1 - (y/r) \tag{3.7}$$

Thus the shear stress increases linearly from zero at the centre of the pipe to a maximum at the walls.

3.4.2. Resistance to Flow in Pipes

Stanton and Pannell[4] measured the drop in pressure due to friction for a number of fluids flowing in pipes of various diameters and surface roughnesses. They expressed their results by using the concept of a friction factor, defined as the dimensionless group $R/\rho u^2$, which is plotted as a function of the Reynolds number. $R(= -R_0)$ represents the resistance to flow per unit area of pipe surface. For a given surface a single curve was found to express the results for all fluids, pipe diameters, and velocities. As with the results of Reynolds the curve was in three parts (Fig. 3.7). At low values of Reynolds number ($Re < 2000$), $R/\rho u^2$ was independent of the surface roughness, but at high values ($Re > 2500$), $R/\rho u^2$ varied with the surface roughness. At very high Reynolds numbers the friction factor became independent of Re and a function of the surface roughness only. Over the transition region of Re, from 2000 to 2500, $R/\rho u^2$ increased very rapidly, showing the great increase in friction as soon as turbulent motion commenced. This general relationship is one of the most widely used in all problems associated with fluid motion, heat transfer, and mass transfer. Moody[5] worked in terms of a friction factor (here denoted by f') equal to $8R/\rho u^2$ and expressed this factor as a function of two dimensionless terms Re and e/d where e is a length representing the magnitude of the surface roughness. This relationship can be seen from dimensional analysis.

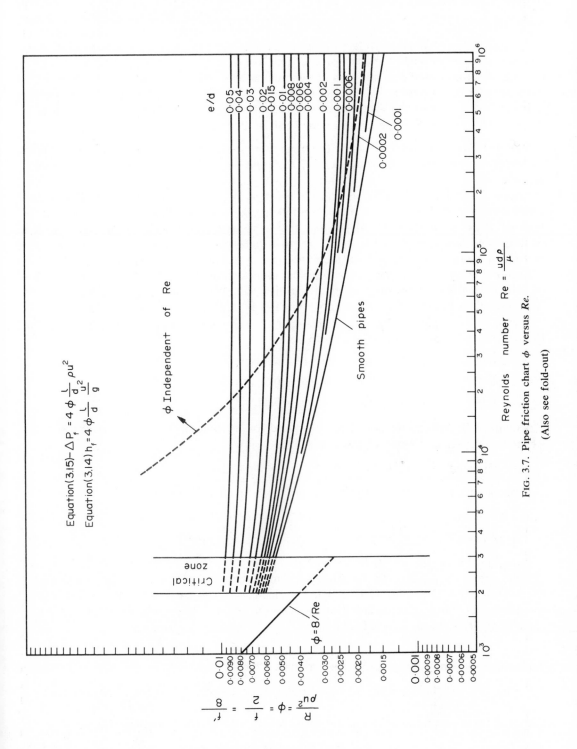

FIG. 3.7. Pipe friction chart ϕ versus Re.

(Also see fold-out)

Thus if R is function of u, d, ρ, μ, e

the analysis gives $R/\rho u^2 = $ function of $(ud\rho/\mu)$ and (e/d)

Thus a single curve will correlate the friction factor with the Reynolds group for all pipes with the same degree of roughness of e/d. This curve is of very great importance since it not only determines the pressure loss in the flow but can often be related to heat transfer or mass transfer (see Chapter 10). Such a series of curves for varying values of e/d is given in Fig. 3.7 which shows the values of $R/\rho u^2$ and the values of the Moody factor f' related to the Reynolds group. Four separate regions can be distinguished:

Region 1 $(Re < 2000)$ corresponds to streamline motion and a single curve represents all the data, irrespective of the roughness of the pipe surface. The equation of the curve is $R/\rho u^2 = 8/Re$.

Region 2 $(2000 < Re < 3000)$ is a transition region between streamline and turbulent flow conditions. Reproducible values of pressure drop cannot be obtained in this region, but the value of $R/\rho u^2$ is considerably higher than that in the streamline region. If an unstable form of streamline flow does persist at Re greater than 2000, the frictional force will correspond to that given by the curve $R/\rho u^2 = 8/Re$, extrapolated to values of Re greater than 2000.

Region 3 $(Re > 3000)$ corresponds to turbulent motion of the fluid and $R/\rho u^2$ is a function of both Re and e/d, with rough pipes giving high values of $R/\rho u^2$. For smooth pipes there is a lower limit below which $R/\rho u^2$ does not fall for any particular value of Re.

Region 4. Rough pipes at high Re. In this region the friction factor becomes independent of Re and depends only on (e/d) as shown below:

$e/d = 0.05$ $Re > 1 \times 10^5$ $R/\rho u^2 = 0.0087$

$e/d = 0.0075$ $Re > 1 \times 10^5$ $R/\rho u^2 = 0.0042$

$e/d = 0.001$ $Re > 1 \times 10^6$ $R/\rho u^2 = 0.0024$

A number of expressions have been proposed for calculating $R/\rho u^2 (= \phi)$ in terms of the Reynolds number and some of these are given below.

Smooth pipes: $2.5 \times 10^3 < Re < 10^5$ $\phi = 0.0396 Re^{-0.25}$ (3.8)

Smooth pipes: $2.5 \times 10^3 < Re < 10^7$ $\phi^{-0.5} = 2.5 \ln(Re\phi^{0.5}) + 0.3$ (3.9)

Rough pipes: $\phi^{-0.5} = -2.5 \ln(0.27 e/d + 0.885 Re^{-1}\phi^{-0.5})$ (3.10)

Rough pipes: $(e/d)Re\phi^{0.5} \gg 3.3$ $\phi^{-0.5} = 3.2 - 2.5 \ln(e/d)$ (3.11)

Equation 3.8 is due to Blasius[6] and the others are derived from considerations of velocity profile. In addition to the Moody friction factor $f' = 8R/\rho u^2$, the Fanning or Darcy friction factor $f = 2R/\rho u^2$ is often used in American texts. It is extremely important therefore to ensure that the precise meaning of the friction factor is clear when using this term in calculating head losses due to friction.

3.4.3. Calculation of Drop in Pressure Along a Pipe

For the flow of a fluid in a short length of pipe dl of diameter d, the total frictional force at the walls is the product of the shear stress R and the surface area of the pipe $(R\pi d\, dl)$. This frictional force results in a change in pressure dP_f so that for a horizontal pipe:

$$R\pi d\, dl = -dP_f\pi(d^2/4) \tag{3.12}$$

i.e. $-dP_f = 4R(dl/d) = 4(R/\rho u^2)(dl/d)(\rho u^2)$ (3.13)

The head lost due to friction:

$$dh_f = -(dP_f/\rho g) = 4(R/\rho u^2)(dl/d)(u^2/g) \tag{3.14}$$

For an incompressible fluid flowing in a pipe of constant cross-section, u is not a function of pressure or length, and the above equations can be integrated, over a length l to give:

$$-\Delta P_f = 4(R/\rho u^2)(l/d)(\rho u^2) \tag{3.15}$$

The energy lost per unit mass F is then given by equation 3.16:

$$F = -\Delta P_f/\rho = 4(R/\rho u^2)(l/d)(u^2) = 4\phi(l/d)u^2 \tag{3.16}$$

To obtain ΔP_f is is therefore necessary to evaluate e/d and then from a knowledge of the value of Re the corresponding value of $R/\rho u^2$ is obtained. This is then used in equation 3.15 to give ΔP_f or the head loss due to friction h_f as:

$$h_f = -\Delta P_f/\rho g = 8\phi(l/d)(u^2/2g) \tag{3.17}$$

Using the friction factors of Moody and Fanning, f' and f respectively, the head loss due to friction is obtained from the following equations:

Moody: $\qquad\qquad\qquad\qquad h_f = f'(l/d)(u^2/2g) \tag{3.18}$

Fanning: $\qquad\qquad\qquad\qquad h_f = 4f(l/d)(u^2/2g) \tag{3.19}$

The energy loss per unit mass because of the irreversibility of the process is given by $\delta F = -dP_f/\rho$ or, over a length of pipe l, the energy loss is $F = 4\phi(l/d)u^2$, i.e. equation 3.16.

If it is required to calculate the flow in a pipe where the maximum pressure drop is specified, the velocity u is required but the Reynolds number is unknown, the above method cannot be used directly to give $R/\rho u^2$. One method of solution is to estimate the value of $R/\rho u^2$ and calculate the velocity and the value of Re. The value of $R/\rho u^2$ is then found and if different from the assumed value, a further trial becomes necessary.

An alternative solution to this problem arises from consideration of equation 3.20:

$$\frac{R}{\rho u^2}\left(\frac{\rho u d}{\mu}\right)^2 = \phi\, Re^2 = \frac{Rd^2\rho}{\mu^2} = \frac{-\Delta P_f d^3\rho}{4l\mu^2} \tag{3.20}$$

If Re is plotted as a function of $\phi\, Re^2$ and e/d as shown in Fig. 3.8, the group $(-\Delta P_f d^3\rho/4l\mu^2) = \phi\, Re^2$ may be evaluated directly as it is independent of velocity. Hence Re may be found from the graph and the required velocity obtained. Similarly, if the diameter of pipe is required to transport fluid at a mass rate of flow G with a given fall in pressure, the following group, which is independent of d, is useful.

$$\left(\frac{R}{\rho u^2}\right)\left(\frac{u d\rho}{\mu}\right)^{-1} = \phi\, Re^{-1} = \frac{-\Delta P_f \mu}{4\rho^2 u^3 l} \tag{3.21}$$

3.4.4. Roughness of Pipe Surfaces

The estimation of the roughness of the surface of the pipe normally presents considerable difficulty. The use of an incorrect value, however, will not be serious for turbulent flow at low Reynolds numbers because the pressure drop is not critically dependent on the roughness in this region. However, at high values of Reynolds number, the effect of pipe roughness is considerable, as can be seen from the graph of $R/\rho u^2$ against the Reynolds number shown in Fig. 3.7. The values of the absolute roughness have been measured for a number of materials and examples are quoted below. Where the value for the pipe surface in question is not given, it is necessary to take an approximate value estimated from the known figures. Where pipes have become corroded, the value of the roughness will commonly be increased, up to tenfold.

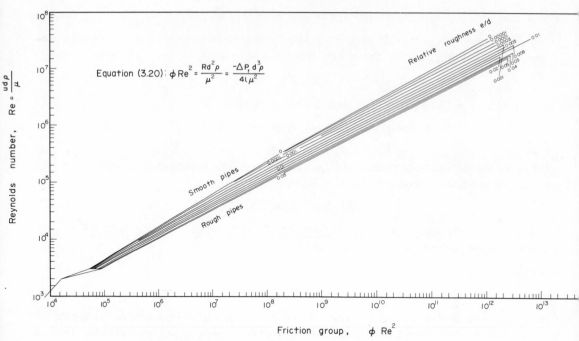

FIG. 3.8. Pipe friction chart ϕRe^2 versus Re for various values of e/d.

TABLE 3.1. *Values of absolute roughness e*

	(ft)	(mm)
Drawn tubing	0·000005	0·0015
Commercial steel and wrought-iron	0·00015	0·046
Asphalted cast-iron	0·0004	0·122
Galvanised iron	0·0005	0·152
Cast-iron	0·00085	0·260
Wood stave	0·0006–0·003	0·18–0·91
Concrete	0·001–0·01	0·3–3·0
Riveted steel	0·003–0·03	0·9–9·0

Further values of roughness applicable to materials used in the construction of open channels may be found in Table 3.3.

For practical installations it must be remembered that the frictional losses cannot be estimated with very great accuracy both because the roughness will change with use and the pumping unit must always have ample excess capacity.

Example 3.1. Ninety-eight per cent sulphuric acid is pumped at 1·25 kg/s through a 25 mm diameter pipe, 30 m long, to a reservoir 12 m higher than the feed point. Calculate the pressure drop in the pipeline.

$$\text{Viscosity of acid} = 25 \text{ mN s/m}^2 \text{ or } 25 \times 10^{-3} \text{ N s/m}^2$$

$$\text{Density of acid} = 1840 \text{ kg/m}^3$$

Solution. Reynolds number:
$$Re = ud\rho/\mu = 4G/\pi\mu d$$
$$= 4 \times 1·25/(\pi \times 25 \times 10^{-3} \times 25 \times 10^{-3})$$
$$= 2545$$

For a mild steel pipe, suitable for conveying the acid, the roughness e will be between 0·00005 and 0·0005 m.

Relative roughness: $e/d = 0·002$ to $0·02$

From Fig. 3.7: $R/\rho u^2 = 0·006$ over this range of e/d

Velocity: $u = G/\rho A = 1·25/[1840 \times (\pi/4)(0·025)^2]$

$$= 1·38 \text{ m/s}$$

The kinetic energy attributable to this velocity will be dissipated when the liquid enters the reservoir. The pressure drop can now be calculated from the energy balance equation and equation 3.16. For turbulent flow of an incompressible fluid:

$$\Delta(u^2/2) + g\,\Delta z + v(P_2 - P_1) + 4(R/\rho u^2)(l/d)u^2 = 0 \quad \text{(from equation 2.55)}$$

$\therefore$

$$(P_1 - P_2) = \rho[(0·5 + 4(R/\rho u^2)(l/d)]u^2 + g\,\Delta z)$$

$$= 1840\{[0·5 + 4 \times 0·006(30/0·025)]1·38^2 + 9·81 \times 12\}$$

$$= 3·19 \times 10^5 \text{ N/m}^2$$

or

$$\underline{\underline{319 \text{ kN/m}^2}}$$

Example 3.2. Water flows in a 50 mm pipe, 100 m long, whose roughness e is equal to 0·013 mm. If the pressure drop across this length of pipe is not to exceed 50 kN/m², what is the maximum allowable water velocity? The density and viscosity of water may be taken as 1000 kg/m³ and 1·0 mN s/m² respectively.

Solution. From equation 3.20:

$$(R/\rho u^2)Re^2 = -\Delta P_f d^3 \rho/(4l\mu^2) = \phi\,Re^2$$

$$-\Delta P_f d^3 \rho/(4l\mu^2) = 50\,000(0·05)^3 1000/(4 \times 1000(1 \times 10^{-3})^2)$$

$$= 1·56 \times 10^7$$

$$e/d = 0·013/50 = 0·00026$$

From Fig. 3.8, when $\phi\,Re^2 = 1·56 \times 10^7$ and $(e/d) = 0·00026$:

$$Re = \rho ud/\mu = 7·9 \times 10^4$$

Hence

$$u = 7·9 \times 10^4(1 \times 10^{-3})/(1000 \times 0·05)$$

$$= \underline{\underline{1·58 \text{ m/s}}}$$

Example 3.3. A cylindrical tank, 5 m in diameter, discharges through a horizontal mild steel pipe 100 m long and 225 mm in diameter connected to the base. Find the time taken for the water level in the tank to drop from 3 m to 0·3 m above the bottom. The viscosity may be taken as 1 mN s/m².

Solution. At time t let the liquid level be D m above the bottom of the tank. Designating point 1 as the liquid level and point 2 as the pipe outlet, then applying the energy balance equation for turbulent flow:

$$\Delta(u^2/2) + g\,\Delta z + v(P_2 - P_1) + F = 0$$

Now

$$P_2 = P_1 = 101·3 \text{ kN/m}^2$$

$$u_1/u_2 = (0·225/5)^2 = 0·0020, \quad \text{i.e. } u_1 \text{ can be neglected}$$

and

$$\Delta z = -D$$

Thus
$$u_2^2/2 - Dg + 4(R/\rho u^2)(l/d)u_2^2 = 0$$
or
$$u_2^2 - 19\cdot62D + 8(R/\rho u^2)(444)u_2^2 = 0$$
$$\therefore \qquad u_2 = 4\cdot43\sqrt{D}/\sqrt{[1 + 3552(R/\rho u^2)]}$$

As the level of liquid in the tank changes from D to $(D + dD)$, the quantity of water discharged $= (\pi/4)5^2(-dD) = -19\cdot63\,dD$ m³.

Time taken for the level to change by an amount dD is given by:

$$dt = -19\cdot63\,dD/\{(\pi/4)0\cdot225^2 \times 4\cdot43\sqrt{D}/\sqrt{[1 + 3555(R/\rho u^2)]}\}$$
$$= -111\cdot5\sqrt{[1 + 3555(R/\rho u^2)]}D^{-0\cdot5}\,dD$$

Total time: $$t = -\int_3^{0\cdot3} 111\cdot5\sqrt{[1 + 3555(R/\rho u^2)]}D^{-0\cdot5}\,dD$$

Assuming $R/\rho u^2$ is constant over the range of flow rates considered:

$$t = 264\sqrt{[1 + 3555(R/\rho u^2)]}\ \text{s}$$

Assuming that the kinetic energy of the liquid is small compared with the frictional losses, an approximate value of $R/\rho u^2$ may be calculated.

Pressure drop along the pipe $= D\rho g = 4Rl/d$:

$$(R/\rho u^2)Re^2 = Rd^2\rho/\mu^2 = Dg\rho^2 d^3/4l\mu^2$$
$$= (D \times 9\cdot81 \times 1000^2 \times 0\cdot225^3)/(4 \times 100 \times 0\cdot001^2)$$
$$= 2\cdot79 \times 10^8 D$$

As D varies from 3 m to 0·3 m, $(R/\rho u^2)Re^2$ varies from $8\cdot38 \times 10^8$ to $0\cdot838 \times 10^8$. It is of interest to consider whether the difference in roughness of a new or old pipe has a significant effect at this stage.

For a mild steel pipe:

New: $e = 0\cdot00005$ m, $e/d = 0\cdot00022$, $Re = 7\cdot0$ to $2\cdot2 \times 10^5$ (from Fig. 3.7)

Old: $e = 0\cdot0005$ m, $e/d = 0\cdot0022$, $Re = 6\cdot0$ to $2\cdot2 \times 10^5$ (from Fig. 3.7)

For a new pipe $R/\rho u^2$ therefore varies from 0·0019 to 0·0020 and for an old pipe $R/\rho u^2 = 0\cdot0029$ (from Fig. 3.7).

Taking a value of 0·002 for a new pipe, assumed constant:

$$t = 264\sqrt{(1 + 3555 \times 0\cdot002)} = 264\sqrt{8\cdot1} = 750\ \text{s}$$

Pressure drop due to friction is approximately $(7\cdot1/8\cdot1) = 0\cdot88$ of the total pressure drop, i.e. due to friction plus the change in kinetic energy.

Thus $(R/\rho u^2)Re^2$ varies from about $7\cdot4 \times 10^8$ to $0\cdot74 \times 10^8$

Re varies from about $6\cdot2 \times 10^5$ to $1\cdot9 \times 10^5$

and $R/\rho u^2$ varies from about 0·0019 to 0·0020

which is sufficiently close to the assumed value of $R/\rho u^2 = 0\cdot002$.

The time taken for the level to fall is therefore about <u>750 s</u>.

Example 3.4. Two storage tanks, A and B, containing a petroleum product, discharge through pipes each 0·3 m in diameter and 1·5 km long to a junction at D. From D the liquid is passed through a 0·5 m diameter pipe to a third storage tank C, 0·75 km away as shown in Fig. 3.9. The surface of the liquid in A is initially 10 m above that in C and the liquid level in B is 6 m higher than that in A. Calculate the initial rate of discharge of liquid into tank C assuming

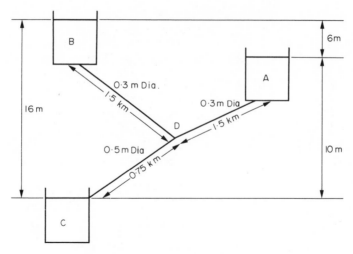

FIG. 3.9. Tank layout for Example 3.4.

the pipes are mild steel. The density and viscosity of the liquid are 870 kg/m^3 and 0·7 mN s/m^2 respectively.

Solution. Because the pipes are long, the kinetic energy of the fluid and minor losses at the entry to the pipes may be neglected.

Assume, as a first approximation, that $R/\rho u^2$ is the same in each pipe and that the velocities in pipes AD, BD, and DC are u_1, u_2, and u_3 respectively. Let the pressure at D be P_D and let point D be z_d m above the datum for the calculation of potential energy, the liquid level in C.

Applying the energy balance equation between D and the liquid level in each of the tanks:

A–D $\qquad\qquad\qquad (z_d - 10)g + vP_D + 4(R/\rho u^2)(1500/0\cdot3)u_1^2 = 0$ (1)

B–D $\qquad\qquad\qquad (z_d - 16)g + vP_D + 4(R/\rho u^2)(1500/0\cdot3)u_2^2 = 0$ (2)

D–C $\qquad\qquad\qquad\quad - z_d g - vP_D + 4(R/\rho u^2)(750/0\cdot5)u_3^2 = 0$ (3)

From equations 1 and 2: $\qquad\quad 6g + 20\,000(R/\rho u^2)(u_1^2 - u_2^2) = 0$ (4)

From equations 2 and 3: $\quad -16g + 20\,000(R/\rho u^2)(u_2^2 + 0\cdot30u_3^2) = 0$ (5)

Taking the roughness of mild steel pipe e as 0·00005 m, therefore e/d varies from 0·00007 to 0·00017.

As a first approximation, take $R/\rho u^2 = 0\cdot002$ in each pipe. Substituting this value in equations 4 and 5,

$$58\cdot9 + 40(u_1^2 - u_2^2) = 0 \qquad (6)$$

$$-156\cdot0 + 40(u_2^2 + 0\cdot30\ u_3^2) = 0 \qquad (7)$$

The flowrate in DC is equal to the sum of the flowrates in AD and BD.

$\therefore \qquad\qquad (\pi/4)0\cdot3^2u_1 + (\pi/4)0\cdot3^2u_2 = (\pi/4)0\cdot5^2u_3$

or $\qquad\qquad\qquad\qquad u_1 + u_2 = 2\cdot78u_3$ (8)

From equation 6: $\qquad\qquad\qquad u_1^2 = u_2^2 - 1\cdot47$ (9)

From equations 7, 8, and 9:

$$-156\cdot0 + 40\{u_2^2 + 0\cdot3 \times (1/2\cdot78)^2[u_2^2 + u_3^2 - 1\cdot47 + 2u_2\sqrt{(u_2^2 - 1\cdot47)]}\} = 0$$

$$\therefore \qquad u_2\sqrt{(u_2^2 - 1\cdot47)} = 50\cdot7 - 13\cdot8u_2^2$$

$$\therefore \qquad u_2^4 - 7\cdot38\,u_2^2 + 13\cdot57 = 0$$

$$\therefore \qquad u_2^2 = 0\cdot5[7\cdot38 \pm \sqrt{(54\cdot46 - 54\cdot28)}] = 3\cdot90 \text{ or } 3\cdot48$$

$$\therefore \qquad u_2 = 1\cdot975 \text{ or } 1\cdot87 \text{ m/s}$$

Substituting in equation 9: $\qquad u_1 = 1\cdot56$ or $1\cdot42$ m/s

Substituting in equation 8: $\qquad u_3 = 1\cdot30$ or $1\cdot18$ m/s

When these values of u_1, u_2, and u_3 are substituted back in equation 7, the lower set of values satisfies the equation and the higher set, introduced as false roots during squaring do not.

Thus $\qquad\qquad\qquad\qquad\qquad u_1 = 1\cdot42$ m/s

$$u_2 = 1\cdot87 \text{ m/s}$$

$$u_3 = 1\cdot18 \text{ m/s}$$

The assumed value of $0\cdot002$ for $R/\rho u^2$ can now be checked:

For pipe AD: $\qquad Re = 0\cdot3 \times 1\cdot42 \times 870/(0\cdot7 \times 10^{-3}) = 5\cdot3 \times 10^5$

For pipe BD: $\qquad Re = 0\cdot3 \times 1\cdot87 \times 870/(0\cdot7 \times 10^{-3}) = 6\cdot9 \times 10^5$

For pipe DC: $\qquad Re = 0\cdot5 \times 1\cdot18 \times 870/(0\cdot7 \times 10^{-3}) = 7\cdot3 \times 10^5$

For $e/d = 0\cdot00007$ to $0\cdot00017$ and this range of Re, $R/\rho u^2$ varies from $0\cdot0019$ to $0\cdot0017$. The assumed value of $0\cdot002$ is therefore sufficiently close.

The volumetric flowrate is, therefore:

$$(\pi/4) \times 0\cdot5^2 \times 1\cdot18 = \underline{\underline{0\cdot23 \text{ m}^3/\text{s}}}$$

3.5.   TYPES OF FLOW

3.5.1.   Reynolds Number, Shear Stress, and Momentum Transfer

For a fluid flowing through a pipe the momentum per unit cross-sectional area is given by ρu^2. This quantity may be regarded as proportional to the inertia force per unit area, or alternatively as the force required to destroy the momentum.

The ratio u/d represents the velocity gradient in the fluid, and thus the group $\mu u/d$ is proportional to the shear stress in the fluid, so that $\rho u^2/\mu u d = du\rho/\mu = Re$ is proportional to ratio of the inertia forces to the viscous forces. This is an important physical interpretation of the Reynolds number.

In turbulent flow with high values of Re, the inertia forces become predominant and the viscous shear stress becomes correspondingly less important.

Shear Stress and Viscosity

In steady streamline flow the direction and velocity of flow at any point remain constant and the shear stress R_y at a point where the velocity gradient at right angles to the direction of flow is given for a Newtonian flow by the relation:

$$R_y = -\mu\frac{\partial u_x}{\partial y} = -\frac{\mu}{\rho}\frac{\partial(\rho u_x)}{\partial y} \tag{3.22}$$

showing the relation between shear stress and momentum for unit volume (ρu_x). (See equation 3.2.)

In turbulent motion, the presence of circulating or eddy currents brings about the much-increased exchange of momentum in all three directions of the stream flow, and these eddies are responsible for the random fluctuations in velocity u_E. The high rate of transfer in turbulent flow is accompanied by a much higher shear stress for a given velocity gradient.

Thus
$$R_y = - \left(\frac{\mu}{\rho} + E \right) \frac{\partial (\rho u_x)}{\partial y} \tag{3.23}$$

where E is known as the *eddy kinematic viscosity* of the fluid, which will depend upon the degree of turbulence in the fluid, is not a physical property of the fluid and varies with position.

In streamline flow, E is very small and approaches zero, so that μ/ρ determines the shear stress. In turbulent flow, E is negligible at the wall and increases very rapidly with distance from the wall. Laufer,[7] using very small hot-wire anemometers, measured the velocity fluctuations and gave a valuable account of the structure of turbulent flow. In typical chemical engineering transport operations of mass, heat, and momentum transfer, the transfer has to be effected through the laminar layer near the wall, and it is here that there is the greatest resistance, again determined by μ/ρ. For this reason it is very important to note that when evaluating the Reynolds number, the kinematic viscosity μ/ρ is used irrespective of the type of flow.

The Reynolds group will often be used where a moving fluid is concerned. Thus the drag produced as a fluid flows past a particle is related to the Reynolds number in which the diameter of the particle is used in place of the diameter of the pipe. Under these conditions the transition from streamline to turbulent flow occurs at a very much lower value. Again for the flow of fluid through a bed composed of granular particles, a mean dimension of the particles is used, and the velocity is usually calculated by dividing the flowrate by the total area of the bed. In this case there is no sharp transition from streamline to turbulent flow because the sizes of the individual flow passages vary.

If the surface over which the fluid is flowing contains a series of relatively large projections, turbulence may arise at a very low Reynolds number. Under these conditions, the frictional force will be increased but so will the coefficients for heat transfer and mass transfer, and therefore turbulence is often purposely induced by this method.

Analysis of Inertia and Viscous Forces

Consider an element of fluid of dimensions dx by dy by dz moving with a velocity u_x in the X-direction (Fig. 3.10). In all cases where there is a solid boundary within a finite distance of the element there will be relative motion in the fluid because of the existence, either of a stable velocity distribution, or of eddy currents. Thus the velocity of the element in the X-direction will be changing as it moves from one point to another.

The net force acting on the element is equal to its rate of change of momentum. Thus the

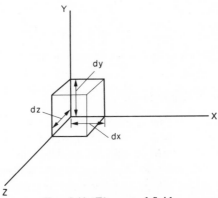

FIG. 3.10. Element of fluid.

net force in the X-direction

$$= \rho \, dx \, dy \, dz (du_x/dt)$$

$$= \rho \, dx \, dy \, dz \, u_x (\partial u_x / \partial x) \tag{3.24}$$

where ρ is the density of the fluid.

The shear stresses acting on the two faces of the element which lie in $X-Z$ planes are R_y (say) and $R_y + (\partial R_y/\partial y) dy$.

Thus the net shear force acting on these faces

$$= -(\partial R_y/\partial y) dy (dx \, dz)$$

$$= \mu (\partial^2 u_x/\partial y^2) dx \, dy \, dz \quad \text{(from equation 3.22)} \tag{3.25}$$

Similarly, the net shear force acting on the two faces in the $X-Y$ planes

$$= \mu (\partial^2 u_x/\partial z^2) dx \, dy \, dz \tag{3.26}$$

These two forces are referred to as the viscous forces in the fluid. The resultant force will produce a pressure gradient in the X-direction.

The force acting on the element because of the pressure gradient

$$= \left(-\frac{\partial P}{\partial x} dx\right) dy \, dz \tag{3.27}$$

Then equating the net force acting in the X-direction to the rate of change of momentum in that direction:

$$\rho u_x \frac{\partial u_x}{\partial x} dx \, dy \, dz = \mu \frac{\partial^2 u_x}{\partial y^2} dx \, dy \, dz + \mu \frac{\partial^2 u_x}{\partial z^2} dx \, dy \, dz + \left(-\frac{\partial P}{\partial x}\right) dx \, dy \, dz$$

i.e.

$$\rho u_x \frac{\partial u_x}{\partial x} = \mu \frac{\partial^2 u_x}{\partial y^2} + \mu \frac{\partial^2 u_x}{\partial z^2} - \frac{\partial P}{\partial x} \tag{3.28}$$

It is now convenient to work in terms of dimensionless derivatives of the length, pressure, and velocity. These are obtained by dividing the actual length, pressure, and velocity by a reference length, pressure, and velocity respectively.

Let the reference length $= L$, some characteristic dimension of the system, the reference velocity $= u_0$, the velocity at some particular point, and the reference pressure $= \rho u_0^2$, which is proportional to the kinetic energy per unit volume of fluid at that point.

Then the dimensionless velocity, $u' = u/u_0$, the dimensionless lengths x', y', and z', $= x/L$, y/L, and z/L respectively, and the dimensionless pressure $P' = P/\rho u_0^2$. The equation can therefore be written:

$$\rho \frac{u_0^2}{L} u' \frac{\partial u'}{\partial x'} = \mu \frac{u_0}{L^2} \left(\frac{\partial^2 u'}{\partial y'^2} + \frac{\partial^2 u'}{\partial z'^2}\right) - \frac{1}{L} \rho u_0^2 \frac{\partial P'}{\partial x'}$$

i.e.

$$\underbrace{\frac{\partial P'}{\partial x'} + u' \frac{\partial u'}{\partial x'}}_{\text{(inertia forces)}} = \underbrace{\frac{\mu}{u_0 L \rho} \left(\frac{\partial^2 u'}{\partial y'^2} + \frac{\partial^2 u'}{\partial z'^2}\right)}_{\text{(viscous forces)}} \tag{3.29}$$

3.5.2. Velocity Distribution, Streamline Flow

The distribution of velocity (in the stream direction) over the cross-section of a fluid flowing in a pipe is not uniform. Whilst this distribution in velocity over a diameter can be calculated for streamline flow this is not possible in the same basic manner for turbulent flow.

The pressure drop due to friction and the velocity distribution resulting from the shear stresses within a fluid in streamline Newtonian flow will be considered for three cases: (a) the flow through a pipe of circular cross-section, (b) the flow between two parallel plates, and (c) the flow through an annulus. The velocity at any distance from the boundary surfaces will be calculated and the mean velocity of the fluid will be related to the pressure gradient in the system. For flow through a circular pipe, the kinetic energy of the fluid will be calculated in terms of the mean velocity of flow.

Pipe of Circular Cross-section

Figure 3.11 shows a horizontal pipe with a concentric element marked *ABCD*. Since the flow is steady, the net force on this element must be zero. The forces acting are the normal pressures over the ends and shear forces over the curved sides.

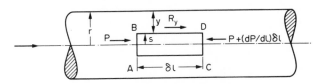

FIG. 3.11. Flow through pipe.

Force over $AB = P\pi s^2$

Force over $CD = -\left(P + \dfrac{dP}{dl}\,\delta l\right)\pi s^2$

Force over curved surface $= 2\pi\,s\,\delta l\,R_y$

where R_y is the shear stress and equals $\mu\dfrac{du_x}{ds}\left(=-\mu\dfrac{du_x}{dy}\right)$

Taking a force balance:

$$P\pi s^2 - \left(P + \dfrac{dP}{dl}\,\delta l\right)\pi s^2 + 2\pi s\delta l\mu\dfrac{du_x}{ds} = 0$$

i.e.

$$-\dfrac{dP}{dl}\,s + 2\mu\dfrac{du_x}{ds} = 0$$

The velocity at any distance s from the axis of the pipe can now be found by integrating this expression.

Thus

$$u_x = \dfrac{1}{2\mu}\dfrac{dP}{dl}\dfrac{s^2}{2} + \text{constant}$$

At the walls of the pipe (i.e. where $s = d/2$) the velocity u_x must be zero; otherwise there would be an infinite velocity gradient and consequently an infinite shear stress at the walls of the pipe. Substituting the value $u_x = 0$, when $s = d/2$:

$$\text{constant} = -\dfrac{1}{2\mu}\dfrac{dP}{dl}\dfrac{d^2}{8}$$

and, therefore:

$$u_x = -\dfrac{1}{4\mu}\dfrac{dP}{dl}(d^2/4 - s^2) \tag{3.30}$$

Thus the velocity over the cross-section varies in a parabolic manner with the distance from the axis of the pipe. The velocity of flow is seen to be a maximum when $s = 0$, i.e. at the pipe axis.

Thus the maximum velocity, at the pipe axis, is given by u_{CL} where:

$$u_{max} = u_{CL} = -\dfrac{1}{4\mu}\dfrac{dP}{dl}\left(\dfrac{1}{4}d^2\right) = -\dfrac{dP}{dl}\dfrac{d^2}{16\mu} \tag{3.31}$$

Hence
$$u_x/u_{CL} = 1 - (4s^2/d^2) \qquad (3.32)$$
$$= 1 - (s^2/r^2) \qquad (3.33)$$

The velocity is thus seen to vary in a parabolic manner over the cross-section of the pipe. This agrees well with experimental measurements.

Volumetric Rate of Flow and Average Velocity

If the velocity is taken as constant over an annulus of radii s and $(s + ds)$, the volumetric rate of flow dQ through the annulus is given by:

$$dQ = 2\pi s \, ds \, u_x$$
$$= 2\pi u_{CL}s \left(1 - \frac{4s^2}{d^2}\right) ds$$

Then the total flow over the cross-section is given by:

$$Q = 2\pi u_{CL} \int_0^{d/2} s \left(1 - \frac{4s^2}{d^2}\right) ds$$

$$= 2\pi u_{CL} \left[\frac{s^2}{2} - \frac{s^4}{d^2}\right]_0^{d/2}$$

$$= \pi d^2 u_{CL}/8 \qquad (3.34)$$

Thus the average velocity u is given by:

$$u = Q/(\pi d^2/4) \qquad (3.35)$$

$$= -\frac{dP}{dl}\frac{d^2}{32\mu} = u_{CL}/2 = u_{max}/2 \qquad (3.36)$$

This relation was derived by Hagen[8] in 1839 and independently by Poiseuille[9] in 1840.

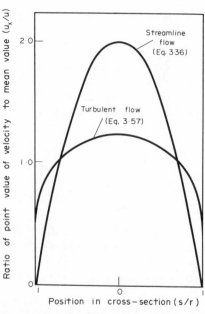

FIG. 3.12. Shape of velocity profiles for streamline and turbulent flow.

Now $\qquad$ $32\mu u \; dl/d^2 = -dP = 4(R/\rho u^2)(dl/d)(\rho u^2)$ (from equations 3.13 and 3.36)

so that $\qquad\qquad\qquad\qquad$ $R/\rho u^2 = 8\mu/ud\rho = 8Re^{-1}$ $\qquad\qquad$ (3.37)

as already indicated earlier in this chapter.
From equations 3.31 and 3.34:

$$\frac{u_x}{u} = \frac{2(d^2 - 4s^2)}{d^2} = 2\left(1 - \frac{s^2}{r^2}\right) \qquad\qquad (3.38)$$

Equation 3.36 is plotted in Fig. 3.12 which shows the shape of the velocity profile for streamline flow.

Kinetic Energy of Fluid

In order to obtain the kinetic energy term for insertion in the energy balance equation, it is necessary to obtain the average kinetic energy per unit mass in terms of the mean velocity.

The kinetic energy of the fluid flowing per unit time in the annulus between s and $(s + ds)$

$$= (\rho 2\pi s \; ds \; u_x)u_x^2/2$$
$$= \pi \rho s u_x^3 \; ds$$
$$= \pi \rho u^3 \, 8s \, [1 - (s^2/r^2)]^3 ds$$

The total kinetic energy per unit time of the fluid flowing in the pipe

$$= -4\pi\rho u^3 r^2 \int_{s=0}^{s=r} \left(1 - \frac{s^2}{r^2}\right)^3 d\left(1 - \frac{s^2}{r^2}\right)$$

$$= \rho \pi r^2 u^3 \qquad\qquad (3.39)$$

Thus the kinetic energy per unit mass:

$$\rho \pi r^2 u^3 / \rho \pi r^2 u = u^2 \qquad\qquad (3.40)$$

In the energy balance equation, the kinetic energy per unit mass was expressed as $u^2/2\alpha$ and thus $\alpha = 0.5$ for the streamline flow of a fluid in a round pipe.

Flow Between Two Parallel Plates

Consider the flow of fluid between two plates of unit width, a distance d_a apart (Fig. 3.13). Then, considering the equilibrium of an element $ABCD$, a force balance can be set up in a similar manner to that used for flow through pipes.

$$P2s - \left(P + \frac{dP}{dl} \, \delta l\right) 2s + 2 \, \delta l \, \mu \, \frac{du_x}{ds} = 0$$

i.e. $\qquad\qquad\qquad\qquad$ $-\frac{dP}{dl} s + \mu \frac{du_x}{ds} = 0$

and $\qquad\qquad\qquad\qquad$ $u_x = \frac{1}{\mu} \frac{dP}{dl} \frac{s^2}{2} + \text{constant}$

When $s = d_a/2$, $u_x = 0$.

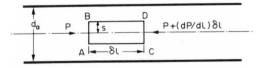

FIG. 3.13. Streamline flow between parallel plates.

$\therefore$

$$\text{constant} = -\frac{d_a^2}{8} \frac{1}{\mu} \frac{dP}{dl}$$

i.e.

$$u_x = -\frac{1}{2\mu} \frac{dP}{dl} \left(\frac{d_a^2}{4} - s^2 \right)$$

The total rate of flow of fluid between the plates is obtained by calculating the flow through two laminae of thickness ds and situated at a distance s from the centre plane and then integrating. Flow through laminae:

$$dQ' = -\frac{1}{2\mu} \frac{dP}{dl} (d_a^2/4 - s^2) 2 \, ds$$

Total rate of flow:

$$Q' = -\frac{1}{\mu} \frac{dP}{dl} (d_a^3/8 - d_a^3/24)$$

$$= -\frac{dP}{dl} \frac{d_a^2}{18\mu} \tag{3.42}$$

The average velocity of the fluid: $u = Q'/d_a$

$$= -\frac{dP}{dl} \frac{d_a^2}{12\mu} \tag{3.43}$$

The maximum velocity occurs at the centre plane and its value is obtained by putting $s = 0$ in equation 3.41:

$$\text{Maximum velocity} = u_{max} = -\frac{dP}{dl} \frac{d_a^2}{8\mu} = 1 \cdot 5u \tag{3.44}$$

It has been assumed that the width of the plates is large compared with the distance between them so that the flow can be considered as unidirectional.

Flow through an Annulus

The velocity distribution and the mean velocity of a fluid flowing through an annulus of outer radius r and inner radius r_i is more complex (Fig. 3.14). If the pressure changes by an amount dP as a result of friction in a length dl of annulus, the resulting force can be equated to the shearing force acting on the fluid. Consider the flow of the fluid situated at a distance not greater than s from the centre line of the pipes. The shear force acting on this fluid consists of two parts; one is the drag on its outer surface; this can be expressed in terms of the viscosity of the fluid and the velocity gradient at that radius: the other is the drag occurring at the inner boundary of the annulus; this cannot be estimated at present and will be denoted by the symbol λ for unit length of pipe.

Then

$$(dP/dl) dl \pi (s^2 - r_i^2) = \mu 2\pi s \, dl (du_x/ds) + \lambda \, dl$$

where u_x is the velocity of the fluid at radius s.

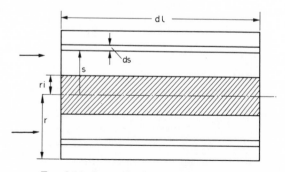

FIG. 3.14. Streamline flow through annulus.

$$du_x = \frac{s^2 - r_i^2}{2\mu s} \frac{dP}{dl} ds - \lambda \frac{ds}{2\pi\mu s} \tag{3.45}$$

Integrating:
$$u_x = \frac{1}{2\mu} \frac{dP}{dl} (s^2/2 - r_i^2 \ln s) - \frac{\lambda}{2\pi\mu} \ln s + u_c \tag{3.46}$$

where u_c is an integration constant with dimensions of velocity.

Substituting the boundary conditions $s = r_i$, $u_x = 0$, and $s = r$, $u_x = 0$ in equation 3.46 and solving for λ and u_c:

$$\lambda = \pi \frac{dP}{dl} \left(\frac{r^2 - r_i^2}{2 \ln (r/r_i)} - r_i^2 \right) \tag{3.47}$$

and
$$u_c = -\frac{1}{2\mu} \frac{dP}{dl} \left(r^2/2 - \frac{r^2 - r_i^2}{2 \ln(r/r_i)} \ln r \right) \tag{3.48}$$

Now substituting these values of λ and u_c in equation 3.46, and simplifying:

$$u_x = -\frac{1}{4\mu} \frac{dP}{dl} \left(r^2 - s^2 + \frac{r^2 - r_i^2}{\ln(r/r_i)} \ln \frac{s}{r} \right) \tag{3.49}$$

The rate of flow of fluid through a small annulus of inner radius s and outer radius $(s + ds)$, is given by:

$$dQ = 2\pi s \, ds \, u_x$$
$$= -\frac{\pi}{2\mu} \frac{dP}{dl} \left(r^2 s - s^3 + \frac{r^2 - r_i^2}{\ln(r/r_i)} s \ln \frac{s}{r} \right) ds \tag{3.50}$$

Integrating between the limits $s = r_i$ and $s = r$:

$$Q = -\frac{\pi}{8\mu} \frac{dP}{dl} \left(r^2 + r_i^2 - \frac{r^2 - r_i^2}{\ln(r/r_i)} \right) (r^2 - r_i^2) \tag{3.51}$$

The average velocity:
$$u = \frac{Q}{\pi(r^2 - r_i^2)}$$
$$= -\frac{1}{8\mu} \frac{dP}{dl} \left(r^2 + r_i^2 - \frac{r^2 - r_i^2}{\ln(r/r_i)} \right) \tag{3.52}$$

3.5.3. Velocity Distribution, Turbulent Flow

No exact mathematical analysis of the conditions within a turbulent fluid has yet been developed, though a number of semi-theoretical expressions for the shear stress at the walls of a pipe of circular cross-section have been suggested.[6]

The shear stresses within the fluid are responsible for the frictional force at the walls and the velocity distribution over the cross-section. A given assumption for the shear stress at the walls therefore implies some particular velocity distribution. It will be shown in Chapter 9 that the velocity at any point in the cross-section will be proportional to the one-seventh power of the distance from the walls if the shear stress is given by the Blasius equation (equation 3.8). This may be expressed as follows:

$$(u_x/u_{CL}) = (y/r)^{1/7} \tag{3.53}$$

where u_x is the velocity at a distance y from the walls, u_{CL} the velocity at the axis of the pipe, and r the radius of the pipe.

This equation is sometimes referred to as the Prandtl one-seventh power law.

Mean Velocity

In a thin annulus of inner radius s and outer radius $s + ds$, the velocity u_x may be taken as constant (see Fig. 3.11).

$$\therefore \quad dQ = 2\pi s \ ds \ u_x$$
$$= -2\pi(r - y)dy \ u_x \quad \text{(since } s + y = r\text{)} \tag{3.54}$$
$$= -2\pi(r - y)dy \ u_{CL}(y/r)^{1/7}$$

$$\therefore \quad Q = \int_{y=r}^{y=0} -2\pi r^2 u_{CL}\left(1 - \frac{y}{r}\right)\left(\frac{y}{r}\right)^{1/7} d\left(\frac{y}{r}\right)$$

Integrating gives: $Q = 2\pi r^2 u_{CL}[(7/8) - (7/15)]$
$$= 49/60 \pi r^2 u_{CL} = 0.82\pi r^2 u_{CL} \tag{3.55}$$

The mean velocity of flow: $u = Q/\pi r^2 = 0.82 u_{CL} \tag{3.56}$

This relation holds provided that the one-seventh power law can be assumed to apply over the whole of the cross-section of the pipe. This is strictly so only at high Reynolds numbers when the thickness of the laminar sub-layer is small. By combining equations 3.53 and 3.56 it is seen that the velocity profile is given by:

$$\frac{u_x}{u} = \frac{1}{0.82}\left(\frac{y}{r}\right)^{1/7}$$

$$= 1.22\left(1 - \frac{s}{r}\right)^{1/7} \tag{3.57}$$

Equation 3.57 is plotted in Fig. 3.12, and it will be noted that the velocity profile is very much flatter than for streamline flow.

The variation of (u/u_{max}) with the Reynolds number is shown in Fig. 3.15. The sharp change at a Reynolds number between 2000 and 3000 will be noted.

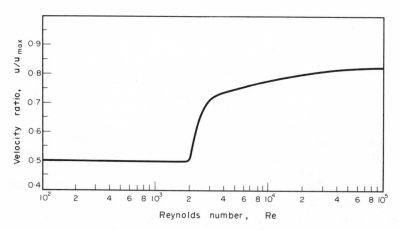

FIG. 3.15. Variation of (u/u_{max}) with Reynolds number in a pipe.

Kinetic Energy

Since $dQ = -2\pi(r - y)dy \ u_x \qquad$ (equation 3.54)

the kinetic energy per unit time of the fluid passing through the annulus

$$= -2\pi(r - y)\rho \ dy \ u_x(u_x^2/2)$$

$$= -\pi u_{CL}^3 \rho r^2\left(1 - \frac{y}{r}\right)\left(\frac{y}{r}\right)^{3/7} d\left(\frac{y}{r}\right)$$

Integration gives total kinetic energy per unit time

$$= \pi u_{CL}^3 \rho r^2 [(7/10) - (7/17)]$$

$$= (49/170) \pi r^2 \rho u_{CL}^3 \qquad (3.58)$$

Mean kinetic energy per unit mass of fluid using equation 3.56

$$= ((49/170) \pi r^2 \rho u_{CL}^3)/((49/60) \pi r^2 \rho u_{CL})$$

$$= (6/17) u_{CL}^2$$

$$= (6/17)(60 u/49)^2 \quad \text{(from equation 3.56)}$$

$$= 0 \cdot 53 u^2 \approx u^2/2 \qquad (3 \cdot 59)$$

$$= u^2/2\alpha \quad \text{(by definition of } \alpha)$$

Thus for turbulent flow at high Reynolds numbers, where the thickness of the laminar sub-layer can be neglected, $\alpha \approx 1$.

When the thickness of the laminar sub-layer cannot be neglected, α will be slightly less than 1.

Flow in Non-circular Ducts

For turbulent flow in a duct of non-circular cross-section, the hydraulic mean diameter may be used in place of the pipe diameter and the formulae for circular pipes can then be applied without introducing a large error. This method of approach is entirely empirical.

The hydraulic mean diameter d_m is defined as four times the cross-sectional area divided by the wetted perimeter: some examples are given. For a circular pipe:

$$d_m = 4(\pi d^2/4)/\pi d = d \qquad (3.60)$$

For an annulus of outer radius r and inner radius r_i:

$$d_m = 4\pi (r^2 - r_i^2)/2\pi (r + r_i) = 2(r - r_i) \qquad (3.61)$$

For a duct of rectangular cross-section d_a by d_b:

$$d_m = 4 d_a d_b / 2(d_a + d_b)$$

$$= 2 d_a d_b / (d_a + d_b) \qquad (3.62)$$

For streamline flow this method is not applicable, and exact expressions relating the pressure drop to the velocity can be obtained for ducts of certain shapes only. Two examples of this have already been given.

Flow Through Curved Pipes

If the pipe is not straight, the velocity distribution over the section is altered and the direction of flow of the fluid is continuously changing. The frictional losses are therefore somewhat greater than for a straight pipe of the same length. If the radius of the pipe divided by the radius of the bend is less than about 0·002, however, the effects of the curvature are negligible.

It has been found[10] that stable streamline flow persists at higher values of the Reynolds number in coiled pipes. Thus, for instance, when the ratio of the diameter of the pipe to the diameter of the coil is 1 to 15, the transition occurs at a Reynolds number of about 8000.

3.5.4. Miscellaneous Friction Losses for Incompressible Fluids

The friction losses occurring as a result of a sudden enlargement or contraction in the cross-section of the pipe, and the resistance of various standard pipe fittings, will now be considered.

Sudden Enlargement

If the diameter of the pipe suddenly increases, as shown in Fig. 3.16, the effective area available for flow will gradually increase from that of the smaller pipe to that of the larger one and the velocity of flow will progressively decrease. Thus fluid with a relatively high velocity will be injected into relatively slow moving fluid; turbulence will be set up and much of the excess kinetic energy will be converted into heat and therefore wasted. If the change of cross-section is gradual, the kinetic energy can be recovered as pressure energy.

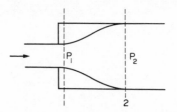

FIG. 3.16. Sudden enlargement.

For the fluid flowing as shown in Fig. 3.16, from section 1 (the pressure just inside the enlargement is found to be equal to that at the end of the smaller pipe) to section 2, the net force = the rate of change of momentum, i.e.

$$P_1 A_2 - P_2 A_2 = \rho_2 A_2 u_2 (u_2 - u_1)$$

i.e.
$$(P_1 - P_2) v_2 = u_2^2 - u_1 u_2$$

Now
$$(P_1 - P_2) v_2 = - \int_1^2 v \, dP \quad \text{for an incompressible fluid} \tag{3.63}$$

Applying the energy equation between the two sections:

$$\frac{u_1^2}{2\alpha_1} = \frac{u_2^2}{2\alpha_2} + \int_1^2 v \, dP + F$$

i.e.
$$F = \frac{u_1^2}{2\alpha_1} - \frac{u_2^2}{2\alpha_2} + u_2^2 - u_1 u_2 \tag{3.64}$$

For fully turbulent flow:
$$\alpha_1 = \alpha_2 = 1 \text{ and}$$
$$F = (u_1 - u_2)^2 / 2 \tag{3.65}$$

The change in pressure $-\Delta P_f$ is therefore given by:

$$-\Delta P_f = \rho (u_1 - u_2)^2 / 2 \tag{3.66}$$

The loss of head h_f is given by:

$$h_f = (u_1 - u_2)^2 / 2g \tag{3.67}$$

Substituting into equation 3.66, from the relation $u_1 A_1 = u_2 A_2$:

$$-\Delta P_f = \frac{\rho u_1^2}{2} \left[1 - \left(\frac{A_1}{A_2} \right) \right]^2 \tag{3.68}$$

The loss can be substantially reduced if a tapering enlarging section is used. For a circular pipe, the optimum angle is about 7° and for a rectangular duct is about 11°.

If A_2 is very large compared with A_1, as for instance at a pipe exit:

$$-\Delta P_f = \rho u_1^2 / 2 \tag{3.69}$$

The use of a small angle enlarging section will be seen in the venturi meter (Chapter 5).

Example 3.5. Water flows at 2000 cm³/s through a sudden enlargement from a 40 mm to a 50 mm diameter pipe. What is the loss in head?

Solution. Velocity in 50 mm pipe $= 2000 \times 10^{-6}/(\pi/4)(50 \times 10^{-3})^2 = 1 \cdot 02$ m/s

Velocity in 40 mm pipe $= 2000 \times 10^{-6}/(\pi/4)(40 \times 10^{-3})^2 = 1 \cdot 59$ m/s

Head lost is given by equation 3.67 as:

$$h_f = (u_1 - u_2)^2/2g$$
$$= 0 \cdot 57^2/(2 \times 9 \cdot 81) = 0 \cdot 0165 \text{ m}$$

or
$$\underline{16 \cdot 5 \text{ mm water}}$$

Sudden Contraction

The effective area for flow gradually decreases as the sudden contraction is approached and then continues to decrease, for a short distance, to what is known as the *vena contracta*. After the vena contracta the flow area gradually approaches that of the smaller pipe. As the fluid moves towards the vena contracta it is accelerated and pressure energy is converted into kinetic energy; this process does not give rise to eddy formation and losses are very small. However, beyond the vena contracta the velocity falls as the flow area increases and conditions are equivalent to those for a sudden enlargement. The expression for the loss at a sudden enlargement can, therefore, be applied for the fluid flowing from the vena contracta to some section a small distance away, where the whole of the cross-section of the pipe is available for flow.

Applying equation 3.65 between sections C and 2, as shown in Fig. 3.17 the frictional loss per unit mass of fluid is given by:

$$F = (u_c - u_2)^2/2$$
$$= (u_2^2/2)[(u_c/u_2) - 1]^2 \tag{3.70}$$

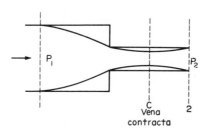

C
Vena
contracta

2

FIG. 3.17. Sudden contraction.

Denoting the ratio of the area at section C to that at section 2 by a coefficient of contraction C_c:

$$F = (u_2^2/2)[(1/C_c) - 1]^2 \tag{3.71}$$

Thus the change in pressure ΔP_f is $-(\rho u^2/2)[(1/C_c) - 1]^2$ and the head lost is

$$(u_2^2/2g)[(1/C_c) - 1]^2$$

C_c varies from about $0 \cdot 6$ to $1 \cdot 0$ as the ratio of the pipe diameters varies from 0 to 1. For a common value of C_c of $0 \cdot 67$,

$$F = u_2^2/8 \tag{3.72}$$

It should be noted that the maximum possible frictional loss which can occur at a change in cross-section is the whole of the kinetic energy of the fluid.

Pipe Fittings

Most pipes are made from steel with or without small alloying ingredients, and they are made by welding or drawing to give a seamless pipe. Tubes from 6 to 50 mm are frequently made from non-ferrous metals such as copper, brass, or aluminium, and these are very widely used in heat exchangers. For special purposes there is a very wide range of materials including industrial glass, many varieties of plastics, rubber, stoneware, and ceramic materials. The normal metal piping is supplied in standard lengths of about 6 m and these are joined to give longer lengths as required. Such jointing is by screw flanging or welding, and the small diameter copper or brass tubes are often brazed or soldered or jointed by compression fittings.

A very large range of pipe fittings is available to enable branching and changes in size to be arranged for industrial pipe layouts. For the control of the flow of a fluid, valves of various kinds are used. The most important of these are known as gate, globe, and needle valves. In addition, check valves are supplied for relieving the pressure in pipelines, and reducing valves are available for controlling the pressure on the downstream side of the valve. Gate and globe valves are supplied in all sizes and may be controlled by motor units operated from an automatic control system. Hand wheels should be fitted for emergency use.

On the whole, gate valves give coarse control, globe valves finer and needle valves the finest control of the rate of flow. Diaphragm valves are also very much in use for the handling of corrosive fluids since the diaphragm can be made of corrosion resistant materials.

Some average figures are given in Table 3.2 for the friction losses in various pipe fittings for the turbulent flow of fluid. They are expressed in terms of the equivalent length of straight pipe with the same resistance and as the number of velocity heads ($u^2/2g$) lost. Considerable variation will occur according to the exact construction of the fittings.

Typical fittings are shown in Figs. 3.18 and 3.19.

TABLE 3.2. *Friction losses in pipe fittings*

	Number of pipe diameters	Number of velocity heads ($u^2/2g$)
45° elbows (a)*	15	0·3
90° elbows (standard radius) (b)	30–40	0·6–0·8
90° square elbows (c)	60	1·2
Entry from leg of T-piece (d)	60	1·2
Entry into leg of T-piece (d)	90	1·8
Unions and couplings (e)	Very small	Very small
Globe valves fully open	60–300	1·2–6·0
Gate valves: fully open	7	0·15
$\frac{3}{4}$ open	40	1
$\frac{1}{2}$ open	200	4
$\frac{1}{4}$ open	800	16

* See Fig. 3.18.

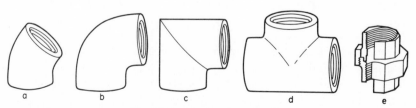

FIG. 3.18. Standard pipe fittings.

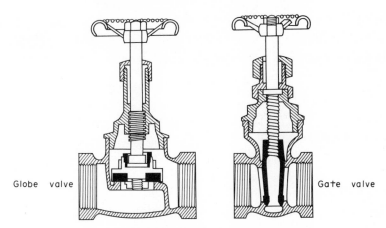

Globe valve   Gate valve

FIG. 3.19. Standard valves.

Example 3.6. 630 cm³/s water at 320 K is pumped in a 40 mm i.d. pipe through a length of 150 m in a horizontal direction and up through a vertical height of 10 m. In the pipe there is a control valve which may be taken as equivalent to 200 pipe diameters and also other pipe fittings equivalent to 60 pipe diameters. Also in the line is a heat exchanger across which there is a loss in head of 1·5 m of water. If the main pipe has a roughness of 0·0002 m, what power must be supplied to the pump if it is 60% efficient?

Solution.

Relative roughness: $e/d = 0.0002/0.04 = 0.005$

Viscosity at 320 K: $\mu = 0.65$ mN s/m² or 0.65×10^{-3} N s/m²

Flowrate = 630 cm³/s or 6.3×10^{-4} m³/s

Area for flow = $(\pi/4)(40 \times 10^{-3})^2 = 1.26 \times 10^{-3}$ m²

Velocity = $6.3 \times 10^{-4}/(1.26 \times 10^{-3}) = 0.50$ m/s

$\therefore$ $Re = 40 \times 10^{-3} \times 0.50 \times 1000/(0.65 \times 10^{-3}) = 30{,}770$

$\therefore$ $R/\rho u^2 = 0.004$

Equivalent length of pipe $= 150 + 10 + (260 \times 40 \times 10^{-3}) = 170.4$ m

$$h_f = 4(R/\rho u^2)(l/d)(u^2/g)$$
$$= 4 \times 0.004(170.4/40 \times 10^{-3})(0.5^2/9.81)$$
$$= 1.74 \text{ m}$$

Total head to be developed $= 1.74 + 1.5 + 10 = 13.24$ m

Mass throughput = $6.3 \times 10^{-4} \times 1000 = 0.63$ kg/s

Power required = $0.63 \times 13.24 \times 9.81 = 81.8$ W

Since the pump efficiency is 60%, power taken $= 81.8/0.60 = 136.4$ W or $\underline{0.136 \text{ kW}}$

The kinetic energy head, $u^2/2g$ has been omitted as this amounts to $0.5^2/(2 \times 9.81) = 0.013$ m, which may be neglected.

Example 3.7. Water in a tank flows through an outlet 25 m below the water level into a 0·15 m diameter horizontal pipe 30 m long, with a 90° elbow at the end leading to a vertical pipe of the same diameter 15 m long. This is connected to a second 90° elbow which leads to a horizontal pipe of the same diameter, 60 m long, containing a fully open globe valve and discharging to atmosphere 10 m below the level of the water in the tank. Taking $e/d = 0.01$ and the viscosity of water as $1\,\text{mN s/m}^2$, what is the initial rate of discharge?

Solution. From equation 3.14 head lost due to friction:

$$h_f = 4(R/\rho u^2)(l/d)(u^2/g)\ \text{m water}$$

Using Bernoulli's equation 2.43:

total head lost: $h = (u^2/2g) + hf + \text{losses in fittings}$

Losses in fittings (from Table 3.2)

$$= 2 \times 0.8\ u^2/2g\ \text{(for the elbows)} + 5.0\ u^2/2g\ \text{(for the valve)}$$
$$= 6.6\ u^2/2g\ \text{m water}$$

Taking $R/\rho u^2 = 0.0045$, then

$$10 = (6.6 + 1)u^2/2g + 4 \times 0.0045[(30 + 15 + 60)/0.15]u^2/2g$$
$$= (7.6 + 12.6)u^2/2g$$
$$u^2 = 9.71$$

and $u = 3.12\ \text{m/s}$

The assumed value of $R/\rho u^2$ must now be checked.

$$Re = du\rho/\mu = 0.15 \times 3.12 \times 1000/1 \times 10^{-3} = 4.68 \times 10^5$$

For $Re = 4.68 \times 10^5$ and $e/d = 0.01$, from Fig. 3.7, $R/\rho u^2 = 0.0046$ which agrees with the assumed value.

$\therefore$ rate of discharge $= 3.12 \times (\pi/4)0.15^2 = 0.055\ \text{m}^3/\text{s}$ or $0.055 \times 1000 = \underline{\underline{55\ \text{kg/s}}}$.

Flow over Banks of Tubes

The frictional loss for a fluid flowing parallel to the axes of the tubes can be calculated in the normal manner by considering the hydraulic mean diameter of the system. This is applicable only to turbulent flow.

For flow at right angles to the axes of the tubes, the cross-sectional area is continually changing, and the problem can be treated as one involving a series of sudden enlargements and sudden contractions. Thus the friction loss would be expected to be directly proportional to the number of banks of pipes j in the direction of flow and to the kinetic energy of the fluid. Thus the pressure drop $-\Delta P_f$ can be written:

$$-\Delta P_f = C_f j \rho u_t^2/6 \tag{3.73}$$

where C_f is a coefficient dependent on the arrangement of the tubes and the Reynolds number. Values of C_f are given in Chapter 7. u_t is the velocity of flow at the narrowest cross-section.

3.6. FLOW IN OPEN CHANNELS

The chief characteristic of flow in an open channel is that the pressure at the surface is everywhere equal to atmospheric pressure and the channel may be either partly or completely full. The flow of liquid may be either streamline or turbulent, but streamline flow occurs, in practice, only when the liquid is present as a thin film, and will be considered in Chapter 7; therefore attention will be confined only to turbulent flow. The transition from streamline to turbulent flow occurs over the range of Reynolds numbers, $u\rho d_m/\mu = 4000-11,000$, where d_m is the hydraulic mean diameter discussed earlier under *Flow in Non-circular Ducts* in Section 3.5.3.

Three different types of turbulent flow can be obtained in open channels. They are *tranquil flow, rapid flow*, and *critical flow*. In tranquil flow the velocity is less than that at which some disturbance, such as a surge wave, will be transmitted. Therefore the flow is influenced by conditions at both the upstream and the downstream end of the channel. In rapid flow, the velocity of the fluid is greater than the velocity of a surge wave and therefore the conditions at the downstream end do not influence the flow. Critical flow occurs when the velocity is exactly equal to the velocity of a surge wave.

3.6.1. Uniform Flow

For a channel of constant cross-section and slope, the flow is said to be uniform when the depth of the liquid D is constant throughout the length of the channel. For these conditions, as indicated in Fig. 3.20, for a length l of channel the accelerating force acting on liquid

$$= lA\rho g \sin \theta.$$

The force resisting the motion $\qquad = R_m pl$

where R_m is the mean value of the shear stress at the solid surface of the channel, p the wetted perimeter and A the cross-sectional area of the flowing liquid.

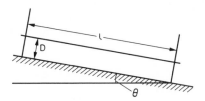

FIG. 3.20. Uniform flow in open channel.

For uniform motion,

$$lA\rho g \sin \theta = R_m pl$$

$\therefore$
$$R_m = (A/p)\rho g \sin \theta$$
$$= (d_m/4)\rho g \sin \theta \qquad (3.74)$$

where d_m is the hydraulic mean diameter $= 4A/p$ (see page 57).
Dividing both sides of the equation by ρu^2, where u is the mean velocity in the channel:

$$\frac{R_m}{\rho u^2} = \frac{d_m}{4} \frac{g \sin \theta}{u^2} \qquad (3.75)$$

i.e.
$$u^2 = \frac{d_m}{4(R_m/\rho u^2)} g \sin \theta \qquad (3.76)$$

For turbulent flow, $R_m/\rho u^2$ is almost independent of velocity but is a function of the surface roughness of the channel. Thus the resistance force is proportional to the square of the velocity. $R_m/\rho u^2$ is found experimentally to be proportional to the one-third power of the relative roughness of the channel surface and can conveniently be written:

$$\frac{R_m}{\rho u^2} = \frac{1}{16} \left(\frac{e}{d_m}\right)^{1/3} \qquad (3.77)$$

for the values of the roughness e given in Table 3.3.

$$u^2 = 4d_m (d_m/e)^{1/3} g \sin \theta \qquad (3.78)$$

$$= 4d_m^{4/3} e^{-1/3} g \sin \theta \qquad (3.79)$$

TABLE 3.3. *Values of the roughness e, for use in equation 3.77*

	ft	mm
Planed wood or finished concrete	0·00015	0·046
Unplaned wood	0·00024	0·073
Unfinished concrete	0·00037	0·113
Cast iron	0·00056	0·171
Brick	0·00082	0·250
Riveted steel	0·0017	0·518
Corrugated metal	0·0055	1·68
Rubble	0·012	3·66

The volumetric rate of flow Q is then given by:

$$Q = uA$$

$$= 2Ad_m^{2/3}e^{-1/6}\sqrt{g \sin \theta} \tag{3.80}$$

The loss of energy due to friction for unit mass of fluid flowing isothermally through a length l of channel is equal to its loss of potential energy because the other forms of energy remain unchanged

i.e. $$F = gl \sin \theta = 4(R_m/\rho u^2)(l/d_m)u^2 \tag{3.81}$$

An empirical equation for the calculation of the velocity of flow in an open channel is the Chezy equation, which can be expressed as follows:

$$u = C\sqrt{(d_m/4) \sin \theta} \tag{3.82}$$

where the value of the coefficient C is a function of the units of the other quantities in the equation. This expression takes no account of the effect of surface roughness on the velocity of flow.

The velocity of the liquid varies over the cross-section and is usually a maximum at a depth of between $0·05D$ and $0·25D$ below the surface, at the centre line of the channel. The velocity distribution can be measured by means of a pitot tube as described in Chapter 5.

The shape and the proportions of the channel can be chosen for any given flow rate so that the perimeter, and hence the cost of the channel, is a minimum. From equation 3.77,

$$Q = 2Ad_m^{2/3}e^{-1/6}\sqrt{g \sin \theta}$$

Assuming that slope and the roughness of the channel are fixed then, for a given flowrate Q from equation 3.80,

$$Ad_m^{2/3} = \text{constant}$$

$$\therefore \qquad A^{5/3}p^{-2/3} = \text{constant} \tag{3.83}$$

The perimeter is therefore a minimum when the cross-section for flow is a minimum.

For a rectangular channel, of depth D and width B:

$$A = DB = D(p - 2D)$$

$$\therefore \qquad D^{5/3}(p - 2D)^{5/3}p^{-2/3} = \text{constant}$$

$$\therefore \qquad Dp^{3/5} - 2D^2p^{-2/5} = \text{constant}$$

p is a minimum when $dp/dD = 0$. Differentiating with respect to D and putting dp/dD equal to 0:

$$p^{3/5} - p^{-2/5}4D = 0$$

$$\therefore \qquad 4D = p = 2D + B$$

i.e. $$B = 2D \tag{3.84}$$

Thus the most economical rectangular section is such that the width is equal to twice the depth of liquid flowing. The most economical proportions for other shapes can be determined in a similar manner.

3.6.2. Specific Energy of Liquid

Consider a liquid flowing in a channel inclined at an angle θ to the horizontal (Fig. 3.21). The various energies possessed by unit mass of fluid at a depth h below the liquid surface

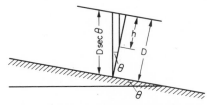

FIG. 3.21. Energy of fluid in open channel.

(measured at right angles to the bottom of the channel) are as follows:

Internal energy:
$$U$$

Pressure energy:
$$(P_a + h\rho g \sec \theta)v = P_a v + hg \sec \theta$$

where P_a is atmospheric pressure.

Potential energy:
$$zg + (D - h)g \sec \theta$$

where z is the height of the bottom of the channel above the datum level at which the potential energy is reckoned as zero.

Kinetic energy:
$$u^2/2.$$

The total energy per unit mass is, therefore:

$$U + P_a v + zg + Dg \sec \theta + u^2/2 \qquad (3.85)$$

It will be seen that at any cross-section the total energy is independent of the depth h below the liquid surface. As the depth is increased, the pressure energy increases at the same rate as the potential energy decreases. If the fluid flows through a length dl of channel, the net change in energy per unit mass is given by:

$$\delta q - \delta W_s = dU + g\,dz + g \sec \theta\,dD + u\,du \qquad (3.86)$$

For an irreversible process, from Chapter 2:

$$dU = T\,dS - P\,dv \qquad \text{(equation 2.5)}$$
$$= \delta q + \delta F \qquad \text{(equation 2.8)}$$

if the fluid is incompressible, since $dv = 0$.
If no work is done on the surroundings, $\delta W_s = 0$ and

$$g\,dz + g \sec \theta\,dD + u\,du + \delta F = 0 \qquad (3.87)$$

For a fluid at a constant temperature, the first three terms in equation 3.85 are independent of the flow conditions within the channel. On the other hand, the last two terms are functions of the velocity and the depth of liquid. The *specific energy* of the fluid is defined by the relation:

$$J = Dg \sec \theta + u^2/2 \qquad (3.88)$$

For a horizontal channel, rectangular in section:

$$J = Dg + u^2/2 = Dg + Q^2/2B^2D^2$$

The specific energy will vary with the velocity of the liquid and will be a minimum for some critical value of D; for a given rate of flow Q the minimum will occur when $dJ/dD = 0$

i.e. when $$g + (-2Q^2)/2B^2D^3 = 0$$

i.e. when $$u^2/D = g$$

i.e. when $$u = \sqrt{gD} \qquad (3.89)$$

This value of u is known as the *critical velocity*.

The corresponding values of D the *critical depth* and J are given by:

$$D = (Q^2/B^2g)^{1/3} \qquad (3.90)$$

and $$J = Dg + Dg/2 = 3Dg/2 \qquad (3.91)$$

It can similarly be shown that, at the critical conditions, the flow rate is a maximum for a given value of the specific energy J. At the critical velocity, $u/\sqrt{(gD)}$ is equal to unity. This dimensionless group is known as the Froude number. For velocities greater than the critical velocity it is greater than unity, and vice versa. It will now be shown that the velocity with which a small disturbance is transmitted through a liquid in an open channel is equal to the critical velocity. Thus the Froude number is the criterion by which the type of flow, tranquil or rapid, is determined. Tranquil flow occurs when the Froude number is less than unity and rapid flow when it is greater than unity.

3.6.3. Velocity of Transmission of a Wave

Consider a liquid which is flowing with a velocity u in a rectangular channel of width B. Initially, the depth of liquid is D_1 but, as a result of a change in conditions at the downstream end of the channel, the level there is suddenly increased to some value D_2. A wave therefore tends to move upstream against the motion of the oncoming fluid. Consider two sections, 1 and 2, one on each side of the wave at any instant (Fig. 3.22).

The rate of accumulation of fluid between the two sections is:

$$u_1D_1B - u_2D_2B$$

This accumulation of fluid results from the propagation of the wave and is therefore equal to $u_wB(D_2 - D_1)$.

Thus $$u_1D_1 - u_2D_2 = u_w(D_2 - D_1)$$

$\therefore$ $$u_2 = [u_w(D_1 - D_2) + u_1D_1]/D_2 \qquad (3.92)$$

The velocity of the fluid is changed from u_1 to u_2 by the passage of the wave. The rate of travel of the wave relative to the upstream liquid is $(u_1 + u_w)$ and therefore the mass of fluid whose velocity is changed in unit time

$$= (u_1 + u_w)BD_1\rho \qquad (3.93)$$

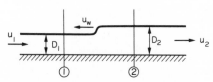

FIG. 3.22. Transmission of a wave.

The force acting on the fluid at any section where the liquid depth is D is equal to:

$$\int_0^D (h\rho g)B \, \mathrm{d}h = B\rho gD^2/2 \tag{3.94}$$

where h is any depth below the liquid surface.

The net force acting on the fluid between sections 1 and 2 is:

$$B\rho g(D_1^2 - D_2^2)/2 \text{ in the direction of flow}$$

Then, neglecting the frictional drag of the walls of the channel between sections 1 and 2, the net force can be equated to the rate of increase of momentum:

i.e.
$$B\rho g(D_1^2 - D_2^2)/2 = (u_1 + u_w)BD_1\rho(u_2 - u_1)$$

$\therefore$
$$(u_1 + u_w)D_1\left\{\frac{1}{D_2}[u_w(D_1 - D_2) + u_1D_1] - u_1\right\} = g(D_1^2 - D_2^2)/2$$

$\therefore$
$$(u_1 + u_w)^2 = \frac{D_2}{D_1}(D_1 + D_2)\frac{g}{2}$$

$$= \frac{gD_2}{2}\left(1 + \frac{D_2}{D_1}\right) \tag{3.95}$$

where $u_1 + u_w$ is the velocity of the wave relative to the oncoming fluid. For a very small wave, $D_1 \rightarrow D_2$ and:

$$u_1 + u_w = \sqrt{(gD_2)} \tag{3.96}$$

It is thus seen that the velocity of an elementary wave is equal to the critical velocity, at which the specific energy of the fluid is a minimum for a given flow rate. The criterion for critical conditions is therefore that the Froude number, $u/\sqrt{(gD)}$, is equal to unity.

3.6.4. Hydraulic Jump

If a liquid enters a channel under a gate, it will flow at a high velocity through and just beyond the gate and the depth will be correspondingly low. This is an unstable condition, and at some point the depth of the liquid may suddenly increase and the velocity fall. This change is known as the *hydraulic jump*, and is accompanied by a reduction of the specific energy of the liquid as the flow changes from rapid to tranquil; the excess energy is dissipated as a result of turbulence.

Consider a liquid flowing in a rectangular channel in which a hydraulic jump occurs between sections 1 and 2 (Fig. 3.23). Then the conditions after the jump can be determined by equating the net force acting on the liquid between the sections to the rate of change of momentum; the frictional forces at the walls of the channel are neglected.

Net force acting on the fluid

$$= B\rho g(D_1^2 - D_2^2)/2$$

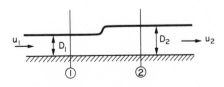

FIG. 3.23. Hydraulic jump.

Rate of change of momentum of fluid

$$= u_1 BD_1 \rho (u_2 - u_1)$$

$$\therefore \qquad g(D_1^2 - D_2^2)/2 = u_1 D_1(u_2 - u_1)$$

The volumetric rate of flow of the fluid is the same at sections 1 and 2.

$$\therefore \qquad Q = u_1 BD_1 = u_2 BD_2$$

$$\therefore \qquad g(D_1^2 - D_2^2)/2 = u_1^2 D[(D_1/D_2) - 1]$$

If $D_1 \neq D_2$,

$$g(D_1 + D_2)/2 = u_1^2 D_1/D_2$$

$$\therefore \qquad D_2^2 + D_1 D_2 - 2u_1^2 D_1/g = 0 \qquad\qquad (3.97)$$

$$\therefore \qquad D_2 = \frac{1}{2}\left(-D_1 \genfrac{}{}{0pt}{}{+}{[\neq]} \sqrt{D_1^2 + \frac{8u_1^2 D_1}{g}}\right) \qquad\qquad (3.98)$$

This expression gives D_2 as a function of the conditions at the upstream side of the hydraulic jump. The corresponding velocity u_2 is obtained by substituting in the equation:

$$u_1 D_1 = u_2 D_2$$

Corresponding values of D_1 and D_2 are referred to as *conjugate depths*.

The minimum depth at which a hydraulic jump can occur is found by putting $D_1 = D_2 = D$ in equation 3.97:

i.e.
$$2D^2 = 2u_1^2 D/g$$

$$\therefore \qquad D = u_1^2/g \qquad\qquad (3.99)$$

This value of D corresponds to the critical depth for flow in a channel.

Thus a hydraulic jump can occur provided that the depth of the liquid is less than the critical depth. After the jump it will be greater than the critical depth, the flow having changed from rapid to tranquil.

The energy loss associated with the hydraulic jump will now be calculated. For a small change in the flow of a fluid in an open channel:

$$u\,du + g\,dz + g\,\sec\theta\,dD + \delta F = 0 \qquad\qquad \text{(equation 3.87)}$$

For a horizontal channel, therefore:

$$F = g(D_1 - D_2) + (u_1^2 - u_2^2)/2 \qquad\qquad (3.100)$$

From equation 3.97: $\qquad u_1^2 = gD_2(D_1 + D_2)/2D_1$

Similarly: $\qquad u_2^2 = gD_1(D_1 + D_2)/2D_2$

$$\therefore \qquad F = g(D_1 - D_2) + \frac{1}{4}g(D_1 + D_2)\left(\frac{D_2}{D_1} - \frac{D_1}{D_2}\right)$$

$$= \frac{1}{4}g(D_1 - D_2)\left[4 + (D_1 + D_2)\frac{-(D_1 + D_2)}{D_1 D_2}\right]$$

$$= (D_2 - D_1)^3 g/4D_1 D_2 \qquad\qquad (3.101)$$

The hydraulic jump may be compared with the shock wave for the flow of a compressible fluid which will be discussed in Chapter 4.

3.7. FURTHER READING

STREETER, V. L.: *Fluid Mechanics*, 5th edn. (McGraw-Hill, New York, 1971).

FRANCIS, J. R. D.: *A Textbook of Fluid Mechanics*, 3rd edn. (Edward Arnold, London, 1971).

DUNCAN, W. J., THOM, A. S., and YOUNG, A. D.: *Mechanics of Fluids*, 2nd edn. (Edward Arnold, London, 1972).

SABERSKY, R. H., ACOSTA, A. J., and HAUPTMANN, E. G.: *Fluid Flow*, 2nd edn. (Macmillan, New York, 1971).

CRANE: Technical Paper No. 410: *Flow Fluids Through Valves, Fittings and Pipe*, 14th printing (Crane Co., New York, 1974).

KNUDSEN, J. G. and KATZ, D. L.: *Fluid Dynamics and Heat Transfer* (McGraw-Hill, New York, 1958).

3.8. REFERENCES

1. REYNOLDS, O.: *Papers on Mechanical and Physical Subjects* **2** (1881–1901) 51. An experimental investigation of the circumstances which determine whether the motion of water shall be direct or sinuous and the law of resistance in parallel channels. 535. On the dynamical theory of incompressible viscous fluids and the determination of the criterion.
2. PRANDTL, L.: *Z. Ver. deut. Ing.* **77** (1933) 105. Neuere Ergebnisse der Turbulenzforschung.
3. COULSON, J. M. and RICHARDSON, J. F.: *Chemical Engineering*, vol. 3 (Pergamon Press, Oxford, 1971).
4. STANTON, T. and PANNELL, J.: *Phil. Trans. R. Soc.* **214** (1914) 199. Similarity of motion in relation to the surface friction of fluids.
5. MOODY, L. F.: *Trans. Am. Soc. Mech. Engrs.* **66** (1944) 671. Friction factors for pipe flow.
6. BLASIUS, H.: *Forschft. Ver. deut. Ing.* **131** (1913). Das Ähnlichkeitsgesetz bei Reibungsvorgangen in Flüssigkeiten.
7. LAUFER, J.: United States National Advisory Committee for Aeronautics. Report No. 1174 (1954).
8. HAGEN, G.: *Ann Phys. (Pogg. Ann.)* **46** (1839) 423. Ueber die Bewegung des Wassers in engen cylindrischen Röhren.
9. POISEUILLE, J.: *Inst. de France Acad. des Sci. Memoires presente par divers savantes* **9** (1846) 433. Recherches experimentales sur le mouvement des liquides dans les tubes de très petit diametre.
10. WHITE, C. M.: *Proc. R. Soc.* A, **123** (1929) 645. Streamline flow through curved pipes.

3.9. NOMENCLATURE

		Units in SI System	Dimensions in M, L, T, θ
A	Area perpendicular to direction of flow	m^2	L^2
B	Width of rectangular channel or notch	m	L
C_c	Coefficient of contraction	—	—
C_f	Coefficient for flow over a bank of tubes	—	—
D	Depth of liquid in channel	m	L
d	Diameter of pipe	m	L
d_a	Dimension of rectangular duct, or distance apart of parallel plates	m	L
d_b	Dimension of rectangular duct	m	L
d_m	Hydraulic mean diameter ($=4A/p$)	m	L
E	Eddy kinematic viscosity	m^2/s	L^2T^{-1}
e	Surface roughness	m	L
F	Energy per unit mass degraded because of irreversibility of process	J/kg	L^2T^{-2}
f	Fanning friction factor ($=2R/\rho u^2$)	—	—
f'	Moody friction factor ($8R/\rho u^2$)	—	—
G	Mass rate of flow	kg/s	MT^{-1}
g	Acceleration due to gravity	m/s^2	LT^{-2}
h	Depth below surface measured perpendicular to bottom of channel or notch	m	L
h_f	Head lost due to friction	m	L
i	Hydraulic gradient (h_f/l)	—	—
J	Specific energy of fluid in open channel	J/kg	L^2T^{-2}
j	Number of banks of pipes in direction of flow	—	—
k	Numerical constant used as index in rheological equation	—	—
L	Characteristic linear dimension	m	L
l	Length of pipe or channel	m	L
P	Pressure	N/m^2	$ML^{-1}T^{-2}$
P_f	Pressure due to friction	N/m^2	$ML^{-1}T^{-2}$
ΔP	Pressure difference	N/m^2	$ML^{-1}T^{-2}$
P'	Dimensionless derivative of P	—	—
p	Wetted perimeter	m	L

		Units in SI System	Dimensions in M, L, T, θ
Q	Volumetric rate of flow	m³/s	L^3T^{-1}
q	Net heat flow into system	J/kg	L^2T^{-2}
R	Shear stress on surface	N/m²	$ML^{-1}T^{-2}$
R_c	Yield stress	N/m²	$ML^{-1}T^{-2}$
R_m	Mean value of shear stress at surface	N/m²	$ML^{-1}T^{-2}$
R_y	Shear stress at some point in fluid	N/m²	$ML^{-1}T^{-2}$
R_0	Shear stress in fluid at boundary surface	N/m²	$ML^{-1}T^{-2}$
r	Radius of pipe, or outer pipe in case of annulus	m	L
r_i	Radius of inner pipe of annulus	m	L
S	Entropy per unit mass	J/kg K	$L^2T^{-2}\theta^{-1}$
s	Distance from axis of pipe or from centre plane	m	L
T	Absolute temperature	K	θ
U	Internal energy per unit mass	J/kg	L^2T^{-2}
u	Mean velocity	m/s	LT^{-1}
u_{CL}	Velocity in pipe at centre line	m/s	LT^{-1}
u_{Ex}	Fluctuating velocity component	m/s	LT^{-1}
u_i	Instantaneous value of velocity	m/s	LT^{-1}
u_{max}	Maximum velocity	m/s	LT^{-1}
u_t	Velocity at narrowest cross section of bank of tubes	m/s	LT^{-1}
u_x	Velocity in X-direction at a point	m/s	LT^{-1}
u_w	Velocity of propagation wave	m/s	LT^{-1}
u_0	Reference velocity	m/s	LT^{-1}
u'	Dimensionless derivative of velocity	—	—
v	Volume per unit mass of fluid	m³/kg	$M^{-1}L^3$
W_s	Shaft work per unit mass	J/kg	L^2T^{-2}
x	Distance in X-direction or in direction of motion	m	L
x'	Dimensionless derivative of x	—	—
y	Distance in Y-direction or distance from surface	m	L
y'	Dimensionless derivative of y	—	—
z	Distance in vertical direction	m	L
z'	Dimensionless derivative of z	—	—
α	Constant in expression for kinetic energy of fluid	—	—
δ	Boundary layer thickness	m	L
λ	Shear force acting on unit length of inner surface of annulus	N/m	MT^{-2}
μ	Viscosity of fluid	N s/m²	$ML^{-1}T^{-1}$
ϕ	Friction factor $(= R/\rho u^2)$	—	—
ρ	Density of fluid	kg/m³	ML^{-3}
θ	Angle	—	—
Re	Reynolds number with respect to pipe diameter	—	—
Re_x	Reynolds number with respect to distance x	—	—

CHAPTER 4

Flow of Compressible Fluids and Two-Phase Flow

4.1. INTRODUCTION

Single-phase Flow

In this chapter the problems will be considered which arise when a gas flows through a tube under conditions in which the pressure or temperature changes by an appreciable amount. The density of the gas can no longer be taken as constant but will vary along the tube. It is important to note that it is the ratio of the difference in pressure $P_1 - P_2$ along the tube to the inlet pressure P_1 that is important. Thus if $P_1 - P_2$ is a significant fraction (say 10 per cent) of P_1, then the change in density must be taken into consideration in design: similarly, it is the temperature change that is important—not necessarily the temperature level. The flow of gas at high pressure may not mean significant change in density, and can frequently be treated as the flow of an incompressible fluid.

The flow of a gas through pipes or through nozzles presents a new feature in that the velocity has a maximum possible value, and it will be necessary to consider this maximum value and show that it corresponds to the velocity of sound under those conditions.

Two-phase Flow

With the flow of a two-phase fluid formed from a liquid and a vapour or gas, it is the presence and behaviour of the vapour or gas which complicates the analysis. With such a flow, the gas velocity is usually substantially greater than that of the liquid, and this leads to energy losses. The change in condition of the gas phase along the pipe results in further energy losses. Where heat is added to the two-phase system, as in an evaporator, the problems of design will be even more complex.

4.2. SINGLE-PHASE GAS FLOW

It has been seen in Chapter 2 that the general energy equation for the flow of a fluid through a pipe is of the form:

$$(u \ du/\alpha) + g \ dz + v \ dP + \delta W_s + \delta F = 0 \qquad \text{(equation 2.54)}$$

For a fluid flowing through a length dl of pipe of constant cross-sectional area A:

$$d(u^2/2\alpha) + g \ dz + v \ dP + 4(R/\rho u^2)(dl/d)u^2 = 0 \qquad (4.1)$$

This equation cannot be integrated directly because the velocity u increases as the pressure falls and is, therefore, a function of l (Fig. 4.1). It is, therefore, convenient to use the mass flow G which remains constant throughout the length of pipe.

Since $\qquad G = uA/v \quad \text{(from equation 2.37)} \qquad (4.2)$

$$u = Gv/A = 4Gv/\pi d^2 \qquad (4.3)$$

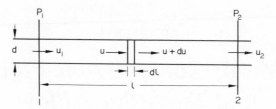

FIG. 4.1. Flow of compressible fluid.

The friction factor $R/\rho u^2$ is a function of the Reynolds number and of the relative roughness e/d which will be taken as constant throughout the pipe.

Thus $\qquad Re = ud\rho/\mu = 4G/\pi d\mu \quad$ (from equation 4.3)

and will be affected only by changes in the viscosity of the fluid. Unless the temperature variation is large, μ can be taken as constant and hence Re will be constant.

The Reynolds number for the flow of gases is usually high because the viscosities are low, and therefore small changes in the value of Re will result in negligible changes in $R/\rho u^2$. By substitution from equation 4.2 in equation 4.1:

$$\left(\frac{G}{A}\right)^2 v \, dv + g \, dz + v \, dP + 4\left(\frac{R}{\rho u^2}\right)\frac{dl}{d}\left(\frac{G}{A}\right)^2 v^2 = 0 \tag{4.4}$$

where α has been taken as unity for turbulent flow.

For a horizontal pipe $dz = 0$ and:

$$\left(\frac{G}{A}\right)^2 v \, dv + v \, dP + 4\left(\frac{R}{\rho u^2}\right)\frac{dl}{d}\left(\frac{G}{A}\right)^2 v^2 = 0 \tag{4.5}$$

Dividing through by v^2 and integrating over a length l of pipe:

$$\left(\frac{G}{A}\right)^2 \ln\frac{v_2}{v_1} + \int_1^2 \frac{dP}{v} + 4\left(\frac{R}{\rho u^2}\right)\frac{l}{d}\left(\frac{G}{A}\right)^2 = 0 \tag{4.6}$$

The relation between v and P must be known before $\int_1^2 dP/v$ can be evaluated. This integral will depend on the nature of the flow, and the two most important conditions are for isothermal and for adiabatic flow in pipes and in nozzles.

4.2.1.   Isothermal Flow of an Ideal Gas in a Horizontal Pipe

$$\int_1^2 \frac{dP}{v} = \frac{1}{P_1 v_1}\int_1^2 P \, dP = \frac{P_2^2 - P_1^2}{2P_1 v_1} \tag{4.7}$$

and, therefore, substituting in equation 4.6:

$$\left(\frac{G}{A}\right)^2 \ln\frac{P_1}{P_2} + \frac{P_2^2 - P_1^2}{2P_1 v_1} + 4\left(\frac{R}{\rho u^2}\right)\frac{l}{d}\left(\frac{G}{A}\right)^2 = 0 \tag{4.8}$$

Since v_m, the specific volume at the mean pressure in the pipe, is given by:

$$v_m(P_1 + P_2)/2 = P_1 v_1 \tag{4.9}$$

$$\left(\frac{G}{A}\right)^2 \ln\frac{P_1}{P_2} + (P_2 - P_1)\frac{1}{v_m} + 4\left(\frac{R}{\rho u^2}\right)\frac{l}{d}\left(\frac{G}{A}\right)^2 = 0 \tag{4.10}$$

If the pressure drop in the pipe is a small proportion of the inlet pressure, the first term is negligible and the fluid can be treated as an incompressible fluid at the mean pressure in the pipe.

It is sometimes convenient to substitute $\mathbf{R}T/M$ in place of P_1v_1 in equation 4.8 to give:

$$\left(\frac{G}{A}\right)^2 \ln\frac{P_1}{P_2} + \frac{P_2^2 - P_1^2}{2\mathbf{R}T/M} + 4\left(\frac{R}{\rho u^2}\right)\frac{l}{d}\left(\frac{G}{A}\right)^2 = 0 \tag{4.11}$$

Equations 4.8 and 4.11 are the most convenient for calculation of gas flow at moderate velocities under isothermal conditions. Some additional refinement can be added if a compressibility factor is introduced as defined by the relation $Pv = ZRT/M$.

Example 4.1. Over a 30 m length of a 150 mm vacuum line carrying air at 295 K, the pressure falls from 1·3 kN/m² to 0·13 kN/m². If the relative roughness e/d is 0·002, what is the approximate flowrate? (The flow of gases in vacuum systems is discussed fully by Griffiths.[1])

Solution.

Fall in pressure $= (1·30 - 0·13) = 1·17 \text{ kN/m}^2 \text{ or } 1170 \text{ N/m}^2$

Upstream specific volume

$$= (22·4/29)(295/273)(101·3/1·3) = 65·0 \text{ m}^3/\text{kg}$$

For $e/d = 0·002$ and an assumed Reynolds number of $1·5 \times 10^4$, $(R/\rho u^2) = 0·004$ from Fig. 3.7.

For isothermal flow, equation 4.8 applies:

$$(G/A)^2 \ln (P_1/P_2) + (P_2^2 - P_1^2)/2P_1v_1 + 4(R/\rho u^2)(l/d)(G/A)^2 = 0$$

Thus $(G/A)^2[\ln (1·3/0·13) + 4 \times 0·004(30/0·150)] = (P_1 - P_2)(P_1 + P_2)/2P_1v_1$

$$(G/A)^2(2·30 + 3·20) = (1170 \times 1430)(2 \times 1300 \times 65.0)$$

∴ $(G/A)^2 = 1·80$

∴ $G/A = \underline{1·34 \text{ kg/m}^2 \text{ s}}$

Viscosity of air $= 0·018 \text{ mN s/m}^2 = 1·8 \times 10^{-5} \text{ Ns/m}^2$

Reynolds number $= (1·34 \times 0·150)/(1·8 \times 10^{-5}) = 1·12 \times 10^4,$

which is near enough to the assumed value.

An example of compressible flow, where the mass flowrate is specified and the downstream is to be calculated, is included here as Example 4.2.

Example 4.2. Air is flowing at 30 kg/m² s through a smooth pipe 50 mm in diameter and 300 m long. If the upstream pressure is 800 kN/m², what will the downstream pressure be if the flow is isothermal at 273 K? The viscosity of air is 0·015 mN s/m² and the kilogram molecular volume is 22·4 m³. What is the significance of the change in the kinetic energy of the fluid?

Solution. An approximate value of P_2 will be obtained by neglecting the kinetic energy of the fluid and using equation 4.10, omitting the first term:

$$(P_2 - P_1)/v_m + 4(R/\rho u^2)(l/d)(G/A)^2 = 0$$

Assume a mean pressure of, say, 795 kN/m².

∴   specific volume: $v_m = (22·4/29·0)(101·3/795)(273/273) = 0·098 \text{ m}^3/\text{kg}$

Reynolds number: $Re = d(G/A)/\mu$

$$= 0{\cdot}05 \times 30/0{\cdot}015 \times 10^{-3} = 100{,}000$$

Taking $e/d = 0.001$, from Fig. 3.7:

$$(R/\rho u^2) = 0{\cdot}0027$$

∴ $$(P_2 - 800{,}000)/0{\cdot}098 + 4 \times 0{\cdot}0027(300/0{\cdot}05)30^2 = 0$$

∴ $$P_2 = 794{,}300 \text{ N/m}^2 \quad \text{or} \quad \underline{794{\cdot}3 \text{ kN/m}^2}$$

Taking the kinetic energy into account, in equation 4.10

$$(G/A)^2 \ln (800/794{\cdot}3) + (794{\cdot}3 - 800) \times 10^3/0{\cdot}098 + 4 \times 0{\cdot}0027(300/0{\cdot}05)(G/A)^2 = 0$$

$$0{\cdot}066(G/A)^2 - 58163 + 64{\cdot}8(G/A)^2 = 0$$

$$(G/A)^2 = 897$$

and $$(G/A) = 29{\cdot}95 \text{ kg/m}^2 \text{ s}$$

and hence the kinetic energy term is negligible.

Example 4.3. A flow of 50 m³/s methane, measured at 288 K and 101·3 kN/m², has to be delivered along a 0·6 m diameter line, 3·0 km long, linking a compressor and a processing unit. The methane is to be discharged at the plant at 288 K and 170 kN/m² and it leaves the compressor at 297 K. What pressure must be developed at the compressor in order to achieve this flowrate?

Solution. Molecular weight of methane $= 16$ kg/kmol

Density of methane at 288 K and 101·3 kN/m²

$$= (16/22{\cdot}4)(273/288) = 0{\cdot}677 \text{ kg/m}^3$$

Mass flow: $$G = (50 \times 0{\cdot}677) = 33{\cdot}9 \text{ kg/s}$$

Area for flow: $$A = (\pi/4)0{\cdot}60^2 = 0{\cdot}283 \text{ m}^2$$

∴ $$(G/A) = 33{\cdot}9/0{\cdot}283 = 119{\cdot}6 \text{ kg/m}^2 \text{ s}$$

Viscosity of methane at the mean temperature of 293 K,

$$\mu = 0{\cdot}011 \times 10^{-3} \text{ N s/m}^2$$

∴   Reynolds number: $Re = d(G/A)/\mu = 0{\cdot}60 \times 119{\cdot}6/(0{\cdot}011 \times 10^{-3}) = 6{\cdot}53 \times 10^6$

Taking $e/d = 0{\cdot}001$, then, from Fig. 3.7,

$$(R/\rho u^2) = 0{\cdot}0014$$

Neglecting the kinetic energy of the fluid, from equation 4.11:

$$(P_1 - P_2)\rho_m = 4(R/\rho u^2)(l/d)(G/A)^2$$

$$(P_1 - 170{,}000)0{\cdot}677 = 4 \times 0{\cdot}0014(3 \times 10^3/0{\cdot}60)(119{\cdot}6)^2$$

$$0{\cdot}677 P_1 - 115{,}090 = 400{,}516$$

∴ $$P_1 = 761{,}605 \text{ N/m}^2 \quad \text{or} \quad \underline{762 \text{ kN/m}^2}$$

The kinetic energy term

$$(G/A)^2 \ln P_1/P_2 = 11{\cdot}96^2 \ln (762/170) = 214{,}420 \text{ kg}^2/\text{m}^4 \text{ s}^2$$

In this case, this is comparable to the other terms in equation 4.11 and may not be neglected. Equation 4.10 becomes

$$(G/A)^2 \ln P_1/P_2 + (P_2 - P_1)\rho_m + 4(R/\rho u^2)(l/d)(G/A)^2 = 0$$
$$119{\cdot}6^2 \ln (P_1/170{,}000) + (170{,}000 - P_1)0{\cdot}677 + 4 \times 0{\cdot}0014(3 \times 10^3/0{\cdot}60)119{\cdot}6^2 = 0$$
$$114{,}300 \ln (P_1/170{,}000) + 115{,}090 - 0{\cdot}677P_1 + 400{,}576 = 0$$
$$\ln (P_1/170{,}000) - 0{\cdot}473 \times 10^{-4}P_1 + 36{\cdot}05 = 0$$

Solving by trial and error: $P_1 = 795{,}000$ N/m² or <u>795 kN/m²</u>

Conditions for Maximum Velocity and Critical Pressure Ratio

For a constant upstream pressure P_1, the rate of flow G changes as the downstream pressure is varied. From equation 4.8, when $P_1 = P_2$, $G = 0$; also when $P_2 = 0$, $G = 0$. At some intermediate value of P_2, the flow must therefore be a maximum.

Multiplying equation 4.8 by $(A/G)^2$:

$$-\ln \left(\frac{P_2}{P_1}\right) + \left(\frac{A}{G}\right)^2 \frac{(P_2^2 - P_1^2)}{2P_1 v_1} + 4\left(\frac{R}{\rho u^2}\right)\frac{l}{d} = 0 \tag{4.12}$$

Differentiating with respect to P_2, for a constant value of P_1:

$$-\frac{P_1}{P_2}\frac{1}{P_1} + \left(\frac{A}{G}\right)^2 \frac{2P_2}{2P_1 v_1} + \frac{A^2}{2P_1 v_1}(P_2^2 - P_1^2)\frac{-2}{G^3}\frac{dG}{dP_2} = 0$$

The rate of flow is a maximum when $dG/dP_2 = 0$. Denoting conditions at the downstream end of the pipe by suffix w, when the flow is a maximum:

$$1/P_w = (A/G)^2(P_w/P_1 v_1) \tag{4.13}$$

i.e. $$(G/A)^2 = P_w^2/P_1 v_1 = P_w^2/P_w v_w = P_w/v_w$$

i.e. $$u_w = \sqrt{P_w v_w} \tag{4.14}$$

It will be shown later (equation 4.33) that the velocity u_w is equal to the velocity of transmission of a small pressure wave in the fluid at the pressure P_w if heat could be transferred sufficiently rapidly to maintain isothermal conditions. If the pressure at the downstream end of the pipe were P_w, the fluid there would then be moving with the velocity of a pressure wave, and therefore a wave could not be transmitted through the fluid in the opposite direction because its velocity relative to the pipe would be zero. If, at the downstream end, the pipe were connected to a reservoir in which the pressure was reduced below P_w, the flow conditions within the pipe would be unaffected and the pressure at the exit of the pipe would remain at the value P_w (Fig. 4.2). The drop in pressure from P_w to P_2 would then take place by virtue of lateral expansion of the gas beyond the end of the pipe. If the pressure P_2 in the reservoir at the downstream end were gradually reduced from P_1, the rate of flow would increase until the pressure reached P_w; it would then remain constant at this maximum value as the pressure was further reduced.

Thus with compressible flow there is a maximum mass flow rate G_w which can be attained by

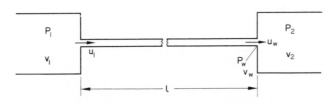

FIG. 4.2. Maximum flow conditions.

the gas for a given upstream pressure P_1 and further reduction in downstream pressure below P_w will not give any further increase. This maximum G_w is given by equation 4.15 and the corresponding pressure P_w from equation 4.16:

$$G_w = AP_w\sqrt{1/(P_1v_1)} \qquad (4.15)$$

P_w is given by substitution in equation 4.12

i.e.
$$\ln\left(\frac{P_1}{P_w}\right) + \frac{v_w}{P_w}\frac{1}{P_wv_w}\frac{P_w^2 - P_1^2}{2} + 4\left(\frac{R}{\rho u^2}\right)\frac{l}{d} = 0$$

or
$$\ln\left(\frac{P_1}{P_w}\right)^2 + 1 - \left(\frac{P_1}{P_w}\right)^2 + 8\left(\frac{R}{\rho u^2}\right)\frac{l}{d} = 0 \qquad (4.16)$$

Heat Flow Required to Maintain Isothermal Conditions

As the pressure in the pipe falls, the kinetic energy of the fluid increases at the expense of the internal energy and the temperature tends to fall. The maintenance of isothermal conditions therefore depends on the transfer of an adequate amount of heat from the surroundings. For a small change in the system, the energy balance is given in Chapter 2:

$$\delta q - \delta W_s = dH + g\,dz + u\,du \quad \text{(equation 2.51)}$$

For a horizontal pipe, $dz = 0$, and for isothermal expansion of an ideal gas $dH = 0$. Thus if the system does no work on the surroundings:

$$\delta q = u\,du \qquad (4.17)$$

and the required transfer of heat (in mechanical energy units) per unit mass is $\Delta u^2/2$. Thus the amount of heat required is equivalent to the increase in the kinetic energy of the fluid. If the mass rate of flow is G, the total heat to be transferred per unit time is $G\,\Delta u^2/2$. In cases where the change in the kinetic energy is small, the required flow of heat is correspondingly small, and conditions are almost adiabatic.

4.2.2. Non-isothermal Flow of an Ideal Gas in a Horizontal Pipe

In general, where an ideal gas expands or is compressed, the relation between the pressure P and the specific volume v can be represented approximately by the expression:

$$Pv^k = a \text{ constant} = P_1v_1{}^k$$

where k will depend on the heat transfer to the surroundings.

Thus
$$\int_1^2 \frac{dP}{v} = \frac{k}{k+1}\frac{P_1}{v_1}\left[\left(\frac{P_2}{P_1}\right)^{(k+1)/k} - 1\right]$$

Inserting this value in equation 4.6:

$$\left(\frac{G}{A}\right)^2\frac{1}{k}\ln\left(\frac{P_1}{P_2}\right) + \frac{k}{k+1}\frac{P_1}{v_1}\left[\left(\frac{P_2}{P_1}\right)^{(k+1)/k} - 1\right] + 4\left(\frac{R}{\rho u^2}\right)\frac{l}{d}\left(\frac{G}{A}\right)^2 = 0 \qquad (4.18)$$

For a given upstream pressure P_1 the maximum flow rate occurs when $u_2 = \sqrt{kP_2v_2}$, the velocity of transmission of a pressure wave under these conditions (equation 4.36). Flow under adiabatic conditions is considered in detail below, but an approximate result is obtained by putting k equal to γ in equation 4.18; this is only approximate because equating k to γ implies reversibility.

4.2.3. Adiabatic Flow of an Ideal Gas in a Horizontal Pipe[2, 3]

The conditions existing during the adiabatic flow in a pipe can be calculated using the approximate expression Pv^k = a constant to give the relation between the pressure and the specific volume of the fluid. In general, however, the value of the index k may not be known for an irreversible adiabatic process. An alternative approach to the problem is therefore desirable.

For a fluid flowing under turbulent conditions in a pipe, $\delta W_s = 0$ and:

$$\delta q = dH + g\ dz + u\ du \quad \text{(from equation 2.51)}$$

In an adiabatic process, $\delta q = 0$, and the equation can then be written as follows for the flow in a pipe of constant cross-sectional area A:

$$(G/A)^2 v\ dv + g\ dz + dH = 0 \quad \text{(from equation 4.3)} \tag{4.19}$$

Now $\qquad dH = dU + d(Pv)$

$$= C_v\ dT + d(Pv) \quad \text{for an ideal gas (from equation 2.24)}$$

Further: $C_p\ dT = C_v\ dT + d(Pv) \quad$ for an ideal gas (from equation 2.26)

$$\therefore \qquad dT = d(Pv)/(C_p - C_v) \tag{4.20}$$

so that $\qquad\qquad dH = d(Pv)\left(\dfrac{C_v}{C_p - C_v} + 1\right) \tag{4.21}$

$$= \frac{\gamma}{\gamma - 1} d(Pv) \tag{4.22}$$

Substituting this value of dH in equation 4.19 and writing $g\ dz = 0$ for a horizontal pipe:

$$\left(\frac{G}{A}\right)^2 v\ dv + \frac{\gamma}{\gamma - 1} d(Pv) = 0 \tag{4.23}$$

Integrating, a relation between P and v for adiabatic flow in a horizontal pipe results:

$$\frac{1}{2}\left(\frac{G}{A}\right)^2 v^2 + \frac{\gamma}{\gamma - 1} Pv = \frac{1}{2}\left(\frac{G}{A}\right)^2 v_1^2 + \frac{\gamma}{\gamma - 1} P_1 v_1 = \text{constant}, K \text{ (say)} \tag{4.24}$$

From equation 4.24:

$$P = \frac{\gamma - 1}{\gamma}\left[\frac{K}{v} - \frac{1}{2}\left(\frac{G}{A}\right)^2 v\right] \tag{4.25}$$

$$dP = \frac{\gamma - 1}{\gamma}\left[-\frac{K}{v^2} - \frac{1}{2}\left(\frac{G}{A}\right)^2\right] dv$$

$$\frac{dP}{v} = \frac{\gamma - 1}{\gamma}\left[-\frac{K}{v^3} - \frac{1}{2}\left(\frac{G}{A}\right)^2 \frac{1}{v}\right] dv \tag{4.26}$$

$$\int_1^2 \frac{dP}{v} = \frac{\gamma - 1}{\gamma}\left[\frac{K}{2}\left(\frac{1}{v_2^2} - \frac{1}{v_1^2}\right) - \frac{1}{2}\left(\frac{G}{A}\right)^2 \ln\frac{v_2}{v_1}\right] \tag{4.27}$$

Substituting for K from equation 4.24:

$$\int_1^2 \frac{dP}{v} = \frac{\gamma - 1}{\gamma}\left[\left(\frac{G}{A}\right)^2 \frac{v_1^2}{4}\left(\frac{1}{v_2^2} - \frac{1}{v_1^2}\right) + \frac{\gamma P_1 v_1}{2(\gamma - 1)}\left(\frac{1}{v_2^2} - \frac{1}{v_1^2}\right) - \frac{1}{2}\left(\frac{G}{A}\right)^2 \ln\frac{v_2}{v_1}\right]$$

$$= \frac{\gamma - 1}{4\gamma}\left(\frac{G}{A}\right)^2\left(\frac{v_1^2}{v_2^2} - 1 - 2\ln\frac{v_2}{v_1}\right) + \frac{P_1 v_1}{2}\left(\frac{1}{v_2^2} - \frac{1}{v_1^2}\right) \tag{4.28}$$

Inserting the value of $\int_1^2 \dfrac{dP}{v}$ from equation 4.28 into equation 4.6:

$$\left(\frac{G}{A}\right)^2 \ln \frac{v_2}{v_1} + \frac{\gamma - 1}{4\gamma} \left(\frac{G}{A}\right)^2 \left(\frac{v_1^2}{v_2^2} - 1 - 2\ln \frac{v_2}{v_1}\right) + \frac{P_1 v_1}{2}\left(\frac{1}{v_2^2} - \frac{1}{v_1^2}\right) + 4\left(\frac{R}{\rho u^2}\right)\frac{l}{d}\left(\frac{G}{A}\right)^2 = 0$$

Simplifying:

$$8\left(\frac{R}{\rho u^2}\right)\frac{l}{d} = \left[\frac{\gamma - 1}{2\gamma} + \frac{P_1}{v_1}\left(\frac{A}{G}\right)^2\right]\left[1 - \left(\frac{v_1}{v_2}\right)^2\right] - \frac{\gamma + 1}{\gamma}\ln \frac{v_2}{v_1} \qquad (4.29)$$

This expression enables v_2, the specific volume at the downstream end of the pipe, to be calculated for the fluid flowing at a mass rate G from an upstream pressure P_1. Alternatively, the mass rate of flow G can be calculated in terms of the specific volume of the fluid at the two pressures P_1 and P_2.

The pressure P_2 at the downstream end of the pipe is obtained by substituting the value of v_2 in equation 4.24.

For constant upstream conditions, the maximum flow through the pipe is found by differentiating with respect to v_2 and putting (dG/dv_2) equal to zero. The maximum flow is thus shown to occur when the velocity at the downstream end of the pipe is the sonic velocity $\sqrt{\gamma P_2 v_2}$ (equation 4.35).

The rate of flow of gas under adiabatic conditions is never more than 20 per cent greater than that obtained for the same pressure difference with isothermal conditions. For pipes of length at least 1000 diameters, the difference does not exceed about 5 per cent. In practice the rate of flow may be limited, not by the conditions in the pipe itself but by the development of sonic velocity at some valve or other constriction in the pipe. Care should, therefore, be taken in the selection of fittings for pipes conveying gases at high velocities.

4.2.4.   Flow of Non-ideal Gases

Methods have been given for the calculation of the pressure drop for the flow of an incompressible fluid and for a compressible fluid which behaves as an ideal gas. If the fluid is compressible and deviations from the ideal gas law are appreciable, one of the approximate equations of state, such as van der Waals' equation, can be used in place of the law $PV = n\mathbf{R}T$ to give the relation between temperature, pressure, and volume. Alternatively, if the enthalpy of the gas is known over a range of temperature and pressure, the energy balance equation, 2.56, which involves a term representing the change in the enthalpy, can be employed:

$$\Delta u^2/2\alpha + g\,\Delta z + \Delta H = q - W_s \qquad \text{(equation 2.56)}$$

This method of approach is useful in considering the flow of steam at high pressures.

4.2.5.   Velocity of Propagation of a Pressure Wave

When the pressure at some point in a fluid is changed, the new condition takes a finite time to be transmitted to some other point in the fluid because the state of each intervening element of fluid has to be changed. The velocity of propagation is a function of the bulk modulus of elasticity ε, where ε is defined by the relation:

$$\varepsilon = \frac{\text{increase of stress within the fluid}}{\text{resulting volumetric strain}} = \frac{dP}{-(dv/v)} = -v(dP/dv) \qquad (4.30)$$

Suppose a pressure wave to be transmitted at a velocity u_w over a distance dl in a fluid of cross-sectional area A, from section A to section B (Fig. 4.3).

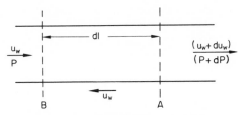

FIG. 4.3. Propagation of pressure wave.

Now imagine the pressure wave to be brought to rest by causing the fluid to flow at a velocity u_w in the opposite direction. Let the pressure and specific volume at B be P and v, and at $A(P + dP)$ and $(v + dv)$, respectively. As a result of the change in pressure, the velocity of the fluid changes from u_w at B to $(u_w + du_w)$ at A.

The mass rate of flow of fluid:

$$G = u_w A/v = (u_w + du_w)A/(v + dv)$$

The net force acting on the fluid between sections A and B is equal to the rate of change of momentum of the fluid:

i.e.
$$PA - (P + dP)A = G\, du_w$$

i.e.
$$- A\, dP = G(G/A)\, dv$$

$$-(dP/dv) = G^2/A^2$$

$$= \varepsilon/v \quad \text{(from equation 4.30)}$$

$\therefore$
$$u_w = \sqrt{\varepsilon v} \tag{4.31}$$

For an ideal gas, ε can be calculated from the equation of state. Under isothermal conditions:

$$Pv = \text{constant.}$$

$\therefore$
$$-(dP/dv) = P/v$$

$\therefore$
$$\varepsilon = P \tag{4.32}$$

and
$$u_w = \sqrt{Pv} \tag{4.33}$$

Under isentropic conditions:

$$Pv^\gamma = \text{constant}$$

$$-(dP/dv) = \gamma P/v$$

$\therefore$
$$\varepsilon = \gamma P \tag{4.34}$$

and
$$u_w = \sqrt{\gamma Pv} \tag{4.35}$$

This value of u_w is found to correspond closely to the velocity of sound in the fluid. That is, for normal conditions of transmission of a small pressure wave, the process is almost isentropic. When the relation between pressure and volume is $Pv^k = \text{constant}$:

$$u_w = \sqrt{kPv} \tag{4.36}$$

It will be noted that these values of u_w correspond to the velocity of the fluid at the downstream end of a pipe under conditions of maximum flow.

It will now be shown from purely thermodynamic considerations that for adiabatic conditions supersonic flow cannot be obtained in a pipe of constant cross-sectional area because the fluid is in a condition of maximum entropy when flowing at the sonic velocity.

The condition of the gas at any point in the pipe where the pressure is P is given by the equations:

$$Pv = \mathbf{R}T/M \qquad \text{(equation 2.16)}$$

and

$$\frac{\gamma}{\gamma - 1} Pv + \frac{1}{2} \left(\frac{G}{A}\right)^2 v^2 = K \qquad \text{(equation 4.22)}$$

It will be noted that if the changes in the kinetic energy of the fluid are small, the process is almost isothermal.

Eliminating v, an expression for T results:

$$\frac{\gamma}{\gamma - 1} \frac{\mathbf{R}T}{M} + \frac{1}{2} \left(\frac{G}{A}\right)^2 \frac{\mathbf{R}^2 T^2}{P^2 M^2} = K \qquad (4.37)$$

The corresponding value of the entropy is obtained as follows:

$$dH = T \, dS + v \, dP = C_p \, dT \quad \text{for an ideal gas (from equations 2.28 and 2.26)}$$

$$\therefore \qquad dS = C_p \frac{dT}{T} - \frac{\mathbf{R}}{MP} \, dP$$

$$\therefore \qquad S = C_p \ln \frac{T}{T_0} - \frac{\mathbf{R}}{M} \ln \frac{P}{P_0} \quad \text{(if } C_p \text{ is constant)} \qquad (4.38)$$

where T_0, P_0 represents the condition of the gas at which the entropy is arbitrarily taken as zero.

The temperature or enthalpy of the gas can then be plotted to a base of entropy to give a *Fanno line*.[4] This line shows the condition of the fluid as it flows along the pipe. If the velocity at entrance is subsonic (normal condition), then the enthalpy will decrease along the pipe and the velocity will increase until sonic velocity is reached. If the flow is supersonic at the entrance, the velocity will decrease along the duct until it becomes sonic. The entropy has a maximum value corresponding to sonic velocity as shown in Fig. 4.4.

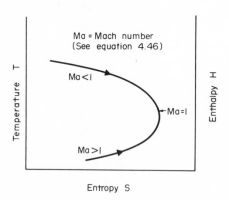

FIG. 4.4. The Fanno line.

Fanno lines are useful in presenting conditions in nozzles, turbines, and other units where supersonic flow arises.[15]

For small changes in pressure and entropy, the kinetic energy of the gas increases only very slowly, and therefore the temperature remains almost constant. As the pressure is further reduced, the kinetic energy changes become important and the rate of fall of temperature increases and eventually dT/dS becomes infinite. Any further reduction of the pressure would cause a decrease in the entropy of the fluid and is, therefore, impossible.

The condition of maximum entropy occurs when $dS/dT = 0$.

$$\frac{dS}{dT} = \frac{C_p}{T} - \frac{R}{MP}\frac{dP}{dT} \quad \text{(from equation 4.38)}$$

The entropy is, therefore, a maximum when:

$$dP/dT = MPC_P/RT \tag{4.39}$$

Now, for an ideal gas:

$$C_p - C_v = R/M \tag{equation 2.27}$$

Substituting in equation 4.39

$$\frac{dP}{dT} = \frac{PC_p}{(C_p - C_v)T} = \frac{P}{T}\frac{\gamma}{\gamma - 1} \tag{4.40}$$

The general value of dP/dT can be obtained by differentiating equation 4.37 with respect to T.

Then

$$\frac{R}{M} + \frac{\gamma - 1}{2\gamma}\left(\frac{G}{A}\right)^2 \frac{R^2}{M^2}\left(\frac{P^2 2T - T^2 2P(dP/dT)}{P^4}\right) = 0$$

$$\therefore \quad 1 + \frac{\gamma - 1}{\gamma}\left(\frac{G}{A}\right)^2 \frac{R}{M}\left(\frac{T}{P^2} - \frac{T^2}{P^3}\frac{dP}{dT}\right) = 0$$

$$\therefore \quad \frac{dP}{dT} = \frac{P}{T} + \frac{\gamma}{\gamma - 1}\left(\frac{A}{G}\right)^2 \frac{M}{R}\frac{P^3}{T^2} \tag{4.41}$$

The maximum value of the entropy occurs when the values of dP/dT given by equations 4.40 and 4.41 are the same:

i.e. when

$$\frac{\gamma}{\gamma - 1}\frac{P}{T} = \frac{P}{T} + \frac{\gamma}{\gamma - 1}\left(\frac{A}{G}\right)^2 \frac{M}{R}\frac{P^3}{T^2}$$

i.e. when

$$\left(\frac{G}{A}\right)^2 = \frac{\gamma}{\gamma - 1}\frac{M}{R}\frac{P^2}{T}(\gamma - 1)$$

$$= \gamma\frac{P^2}{T}\frac{T}{Pv} - \gamma\frac{P}{v}$$

i.e. when

$$u = \sqrt{\gamma Pv} = u_w \quad \text{(from equation 4.35)}$$

which has been shown to be the velocity of propagation of a pressure wave.

Thus the gas can expand until its velocity is equal to the velocity of sound.

Shock Waves. It has been seen in deriving equations 4.31 to 4.36 that for a small disturbance the velocity of propagation of the pressure wave is equal to the velocity of sound. If the changes are much larger and the process is not isentropic, the wave developed is known as a shock wave, and the velocity may be much greater than the velocity of sound. Material and momentum balances must be maintained and the appropriate equation of state for the fluid must be followed. Furthermore, any change which takes place must be associated with an increase, never a decrease, in entropy. For an ideal gas in a uniform pipe under adiabatic conditions a material balance gives:

$$u_1/v_1 = u_2/v_2 \tag{equation 2.62}$$

Momentum balance:

$$(u_1^2/v_1) + P_1 = (u_2^2/v_2) + P_2 \tag{equation 2.64}$$

Equation of state:

$$u_1^2/2 + \frac{\gamma}{\gamma - 1}P_1 v_1 = \tfrac{1}{2}u_2^2 + \frac{\gamma}{\gamma - 1}P_2 v_2 \tag{equation 4.23}$$

Substituting from equation 2.62 into 2.64:

$$v_1 = (u_1^2 - u_1 u_2)/(P_2 - P_1) \tag{4.42}$$

$$v_2 = (u_2^2 - u_1 u_2)/(P_1 - P_2) \tag{4.43}$$

Then, from equation 4.23:

$$\tfrac{1}{2}(u_1^2 - u_2^2) + \frac{\gamma}{\gamma - 1} \frac{P_1}{P_2 - P_1} u_1(u_1 - u_2) - \frac{\gamma}{\gamma - 1} \frac{P_1}{P_1 - P_2} u_2(u_2 - u_1) = 0$$

i.e. $(u_1 - u_2) = 0$, representing no change in conditions, or:

$$\tfrac{1}{2}(u_1 + u_2) + \frac{\gamma}{\gamma - 1} \frac{1}{P_2 - P_1} (u_1 P_1 - u_2 P_2) = 0$$

i.e.
$$\frac{u_2}{u_1} = \frac{(\gamma - 1)(P_2/P_1) + (\gamma + 1)}{(\gamma + 1)(P_2/P_1) + (\gamma - 1)} \tag{4.44}$$

and
$$\frac{P_2}{P_1} = \frac{(\gamma + 1) - (\gamma - 1)(u_2/u_1)}{(\gamma + 1)(u_2/u_1) - (\gamma - 1)} \tag{4.45}$$

Equation 4.45 gives the pressure changes associated with a sudden change of velocity. In order to understand the nature of a possible velocity change, it is convenient to work in terms of Mach numbers. The Mach number (Ma) is defined as the ratio of the velocity at a point to the corresponding velocity of sound:

i.e.
$$Ma_1 = u_1/\sqrt{\gamma P_1 v_1} \tag{4.46}$$

$$Ma_2 = u_2/\sqrt{\gamma P_2 v_2} \tag{4.47}$$

From equations 4.42, 4.43, 4.46, and 4.47:

$$v_1 = \frac{u_1^2}{Ma_1^2} \frac{1}{\gamma P_1} = \frac{u_1(u_1 - u_2)}{P_2 - P_1} \tag{4.48}$$

$$v_2 = \frac{u_2^2}{Ma_2^2} \frac{1}{\gamma P_2} = \frac{u_2(u_2 - u_1)}{P_1 - P_2} \tag{4.49}$$

Giving:
$$\frac{Ma_2^2}{Ma_1^2} = \frac{u_2}{u_1} \frac{P_1}{P_2} \tag{4.50}$$

From equation 4.48:

$$\frac{1}{\gamma Ma_1^2} = \frac{1 - (u_2/u_1)}{(P_2/P_1) - 1} \tag{4.51}$$

$$= \frac{2}{(\gamma + 1)(P_2/P_1) + (\gamma - 1)} \quad \text{(from equation 4.44)} \tag{4.52}$$

Thus
$$\frac{P_2}{P_1} = \frac{2\gamma Ma_1^2 - (\gamma - 1)}{\gamma + 1} \tag{4.53}$$

and
$$\frac{u_2}{u_1} = \frac{(\gamma - 1)Ma_1^2 + 2}{Ma_1^2(\gamma + 1)} \tag{4.54}$$

From equation 4.50:

$$\frac{Ma_2^2}{Ma_1^2} = \frac{(\gamma - 1)Ma_1^2 + 2}{Ma_1^2(\gamma + 1)} \frac{(\gamma + 1)}{2\gamma Ma_1^2 - (\gamma - 1)}$$

i.e.
$$Ma_2^2 = \frac{(\gamma - 1)Ma_1^2 + 2}{2\gamma Ma_1^2 - (\gamma - 1)} \tag{4.55}$$

For a sudden change or normal shock wave to occur, the entropy change per unit mass of fluid must be positive.

From equation 4.38, the change in entropy is given by:

$$S_2 - S_1 = C_p \ln\frac{T_2}{T_1} - \frac{R}{M}\ln\frac{P_2}{P_1}$$

$$= C_p \ln\frac{P_2}{P_1} + C_p \ln\frac{v_2}{v_1} - \frac{R}{M}\ln\frac{P_2}{P_1}$$

$$= C_v \ln\frac{P_2}{P_1} + C_p \ln\frac{u_2}{u_1} \quad \text{(from equations 2.62 and 2.27)}$$

$$= C_v \ln\frac{2\gamma Ma_1{}^2 - (\gamma - 1)}{\gamma + 1} - C_p \ln\frac{Ma_1{}^2(\gamma + 1)}{(\gamma - 1)Ma_1{}^2 + 2} \quad (4.56)$$

$S_2 - S_1$ is positive when $Ma_1 > 1$. Thus a normal shock wave can occur only when the flow is supersonic. From equation 4.55, if $Ma_1 > 1$, then $Ma_2 < 1$, and therefore the flow necessarily changes from supersonic to subsonic. If $Ma_1 = 1$, $Ma_2 = 1$ also, from equation 4.55, and no change therefore takes place. It should be noted that there is no change in the energy of the fluid as it passes through a shock wave, but the entropy increases and therefore the change is irreversible.

4.2.6. Converging–Diverging Nozzles for Gas Flow

Converging–diverging nozzles (Fig. 4.5), sometimes known as Laval nozzles, are used for the expansion of gases where the pressure drop is large. If the nozzle is carefully designed so that the contours closely follow the lines of flow, the resulting expansion of the gas is almost reversible. Because the flow rate is large for high-pressure differentials, there is little time for heat transfer to take place between the gas and surroundings and the expansion is effectively isentropic. In the analysis of the nozzle, the change in flow will be examined for various pressure differentials across the nozzle.

The specific volume v_2 at a downstream pressure P_2, is given by:

$$v_2 = v_1(P_1/P_2)^{1/\gamma} = v_1(P_2/P_1)^{-1/\gamma} \quad (4.57)$$

If gas flows under turbulent conditions from a reservoir at a pressure P_1, through a horizontal nozzle, the velocity of flow u_2, at the pressure P_2 is given by:

$$(u_2{}^2/2) + \int_1^2 v \, dP = 0 \quad \text{(from equation 2.42)}$$

Thus

$$u_2{}^2 = \frac{2\gamma}{\gamma - 1} P_1 v_1 \left[1 - \left(\frac{P_2}{P_1}\right)^{(\gamma-1)/\gamma}\right] \quad (4.58)$$

Since

$$A_2 = Gv_2/u_2 \quad \text{(from equation 2.36)} \quad (4.59)$$

the required cross-sectional area for flow when the pressure has fallen to P_2 can be found.

Maximum Velocity and Critical Pressure Ratio

In the flow of a gas through a nozzle, the pressure falls from its initial value P_1 to a value P_2 at some point along the nozzle; at first the velocity rises more rapidly than the specific volume and therefore the area required for flow decreases. For low values of the pressure ratio P_2/P_1, however, the velocity changes much less rapidly than the specific volume so that the area for flow must increase again. The effective area for flow presented by the nozzle must therefore pass through a minimum. It will be shown that this occurs if the pressure ratio

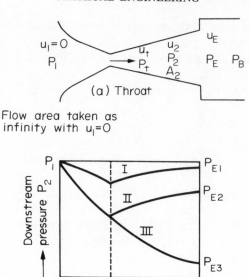

(a) Throat

Flow area taken as
infinity with $u_1=0$

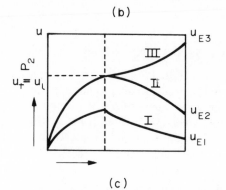

(b)

(c)

FIG. 4.5. Flow through converging–diverging nozzles.

P_2/P_1 is less than the critical pressure ratio (usually approximately 0·5) and that the velocity at the throat is then equal to the velocity of sound. For expansion to pressures below the critical value the flowing stream must be diverging. Thus in a converging nozzle the velocity of the gas stream will never exceed the sonic velocity though in a converging–diverging nozzle supersonic velocities may be obtained in the diverging section. The flow changes which occur as the back pressure P_B is steadily reduced are shown in Fig. 4.5. The pressure at the exit of the nozzle is denoted by P_E, which is not necessarily the same as the back pressure P_B. In the three cases given below, the exit and back pressures correspond and the flow is approximately isentropic. The effect of a mismatch in the two pressures will be considered later.

Case I. Back-pressure P_B quite high. Curves I show how pressure and velocity change along the nozzle. The pressure falls to a minimum at the throat and then rises to a value $P_{E1} = P_B$. The velocity increases to maximum at the throat (less than sonic velocity) and then decreases to a value of u_{E1} at the exit of the nozzle. This situation corresponds to conditions in a venturi operating entirely at subsonic velocities.

Case II. Back-pressure reduced (curves II). The pressure falls to the critical value at the throat where the velocity is sonic. The pressure then rises to $P_{E2} = P_B$ at the exit. The velocity rises to the sonic value at the throat and then falls to u_{E2} at the outlet.

Case III. Back-pressure low, with pressure less than critical value at the exit. The pressure falls to the critical value at the throat and continues to fall to give an exit pressure $P_{E3} = P_B$. The velocity increases to sonic at the throat and continues to increase to supersonic in the diverging cone to a value u_{E3}.

With a converging–diverging nozzle, the velocity increases beyond the sonic velocity only if the velocity at the throat is sonic and the pressure at the outlet is lower than the throat pressure. For a converging nozzle the rate of flow is independent of the downstream pressure, provided the critical pressure ratio is reached and the throat velocity is sonic.

The Pressure and Area for Flow

As indicated in the previous section, the area required at any point will depend upon the ratio of the downstream to the upstream pressure P_2/P_1 and it is helpful to establish the minimum value of A_2. A_2 may be expressed in terms of P_2 and $w = (P_2/P_1)$ using equations 4.57 and 4.58.

Thus
$$A_2{}^2 = G^2 \frac{v_1{}^2 (P_2/P_1)^{-2/\gamma}}{\dfrac{2\gamma}{\gamma - 1} P_1 v_1 [1 - (P_2/P_1)^{(\gamma-1)/\gamma}]}$$

$$= \frac{G^2 v_1 (\gamma - 1)}{2 P_1 \gamma} \frac{w^{-2/\gamma}}{1 - w^{(\gamma-1)/\gamma}} \tag{4.60}$$

For a given rate of flow G, A_2 decreases from an effectively infinite value at pressure P_1 at the inlet to a minimum value given by:

$$dA_2{}^2/dw = 0$$

i.e. when
$$(1 - w^{(\gamma-1)/\gamma}) \frac{-2}{\gamma} w^{-1-2/\gamma} - w^{-2/\gamma} \frac{1 - \gamma}{\gamma} w^{-1/\gamma} = 0$$

i.e. when
$$w = \left(\frac{2}{\gamma + 1} \right)^{\gamma/(\gamma-1)} \tag{4.61}$$

The value of w given by equation 4.61 will be shown to equal the critical pressure ratio w_c in Chapter 5. Thus the velocity at the throat is equal to the sonic velocity. Alternatively, equation 4.60 can be put in terms of the flowrate (G/A_2) as:

$$\left(\frac{G}{A_2} \right)^2 = \frac{2\gamma}{\gamma - 1} \left(\frac{P_2}{P_1} \right)^{2/\gamma} \frac{P_1}{v_1} \left[1 - \left(\frac{P_2}{P_1} \right)^{(\gamma-1)/\gamma} \right] \tag{4.62}$$

and the flowrate G/A_2 can then be shown to have a maximum value of $G_{max}/A_2 = \sqrt{\gamma P_2/v_2}$.

The nozzle is correctly designed for any outlet pressure between P_1 and P_{E1} in Fig. 4.5; under these conditions the velocity will not exceed the sonic velocity at any point, and the flowrate will be independent of the exit pressure $P_E = P_B$. It is also correctly designed for supersonic flow in the diverging cone for an exit pressure of P_{E3}.

It has been shown above that when the pressure in the diverging section is greater than the throat pressure, subsonic flow occurs. Conversely, if the pressure in the diverging section is less than the throat pressure the flow will be supersonic beyond the throat. Thus at a given point in the diverging cone where the area is equal to A_2 the pressure may have one of two values for isentropic flow.

As an example, when γ is 1·4:

$$v_2 = v_1 w^{-0.71} \tag{4.63}$$

and
$$u_2{}^2 = 7 P_1 v_1 (1 - w^{0.29}) \tag{4.64}$$

Thus
$$A_2{}^2 = \frac{v_1 w^{-1.42}}{7 P_1 (1 - w^{0.29})} G^2 \tag{4.65}$$

In Fig. 4.6 values of v_2/v_1, $u_2/\sqrt{P_1v_1}$, and $(A_2/G)\sqrt{P_1/v_1}$ which are proportional to v_2, u_2, and A_2 respectively are plotted as abscissae against P_2/P_1. It is seen that the area A_2 decreases to a minimum and then increases again. At the minimum cross-section the velocity is equal to the sonic velocity and P_2/P_1 is the critical ratio. On the same plot is given the velocity of sound at the pressure P_2 as $\sqrt{\gamma P_2 v_2}$.

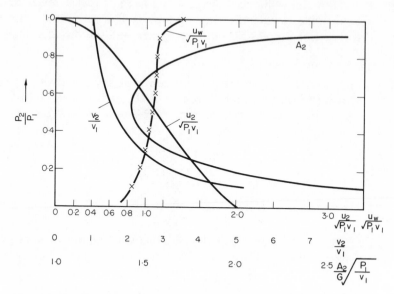

FIG. 4.6. Specific volume, velocity, and nozzle area as a function of pressure.

Effect of Back Pressure P_B on Flow in Nozzle

It is of interest to study how the flow in the nozzle will vary with the back pressure P_B.

(1) $P_B > P_{E2}$. The flow is subsonic throughout the nozzle and the rate of flow G is determined by the value of the back pressure P_B. Under these conditions $P_E = P_B$. It is shown as P_{E1} in Fig. 4.5.

(2) $P_B = P_{E2}$. The flow is subsonic throughout the nozzle, except at the throat where the velocity is sonic. Again $P_E = P_B$.

(3) $P_{E3} < P_B < P_{E2}$. The flow in the converging section and at the throat is exactly as for Case 2. For this range of exit pressures the flow is not isentropic through the whole nozzle and the area at the exit section is greater than the design value for a back pressure P_B. One of two things may occur. Either the flow will be supersonic but not occupy the whole area of the nozzle, or a shock wave will occur at some intermediate point in the diverging cone, giving rise to an increase in pressure and a change from supersonic to subsonic flow.

(4) $P_B = P_{E3}$. The flow in the converging section and at the throat is again the same as for Cases 2, 3, and 4. The flow in the diverging section is supersonic throughout and the pressure of P_B is reached, without the occurrence of a shock wave at the exit. P_{E3} is therefore the design pressure for supersonic flow at the exit.

(5) $P_B < P_{E3}$. In this case the flow throughout the nozzle is exactly as for Case 4. The pressure at the exit will again be P_{E3} and the pressure will fall beyond the end of the nozzle from P_{E3} to P_B by continued expansion of the gas.

It should be noted that the flowrate through the nozzle is a function of back pressure only for Case 1.

Example 4.4. A reaction vessel in a building is protected by means of a bursting disc and the gases are vented to atmosphere through a stack pipe having a cross-sectional area of 0·07 m². The ruptured disc has a flow area of 4000 mm² and the gases expand to the full area of the stack pipe in a divergent section. If the gas in the vessel is at a pressure of 10 MN/m² and a temperature of 500 K, calculate: (a) the initial rate of discharge of gas, (b) the pressure and Mach number immediately upstream of the shock wave, and (c) the pressure of the gas immediately downstream of the shock wave.

Assume that isentropic conditions exist on either side of the shock wave and that the gas has a mean molecular weight of 40 kg/kmol, a ratio of specific heats of 1·4, and obeys the ideal gas law.

Solution. The pressure ratio w_c at the throat is given by equation 4.61:

$$w_c = (P_c/P_1) = [2/(\gamma + 1)]^{\gamma/(\gamma-1)} = (2/2\cdot4)^{1\cdot4/0\cdot4} = 0\cdot53$$

Thus, the throat pressure $= (10 \times 0\cdot53) = 5\cdot3$ MN/m².

Specific volume of gas in reactor:

$$v_1 = (22\cdot4/40)(500/273)(101\cdot3/10,000) = 0\cdot0103 \text{ m}^3/\text{kg}$$

Specific volume of gas at the throat $= (0\cdot0103)(1/0\cdot53)^{1/1\cdot4} = (0\cdot0103 \times 1\cdot575) = 0\cdot0162$ m³/kg

∴ velocity at the throat = sonic velocity (from equation 4.35)

$$= \sqrt{\gamma P v}$$

$$= \sqrt{1\cdot4 \times 5\cdot3 \times 10^6 \times 0\cdot0162} = 347 \text{ m/s}$$

Initial rate of discharge of gas; $G = uA/v$ (at throat)

$$= 347 \times 4000 \times 10^{-6}/0\cdot0162$$

$$= \underline{\underline{85\cdot7 \text{ kg/s}}}$$

The gas continues to expand isentropically and the pressure ratio w is related to the flow area by equation 4.65. If the cross-sectional area of the exit to the divergent section is such that $w^{-1} = (10,000/101\cdot3) = 98\cdot7$, the pressure here will be atmospheric and the expansion will be entirely isentropic. The duct area, however, has nearly twice this value, and the flow is *over-expanded*, atmospheric pressure being reached within the divergent section. In order to satisfy the boundary conditions, a shock wave occurs further along the divergent section across which the pressure increases. The gas then expands isentropically to atmospheric pressure.

Suppose the shock wave occurs when the flow area is A, then the flow conditions at this point can be calculated by solution of the equations for:

(1) *The isentropic expansion from conditions at the vent.* The pressure ratio w (pressure/pressure in the reactor) is given by equation 4.65:

$$w^{-1\cdot42}/(1 - w^{0\cdot29}) = (A/G)^2(7P_1/v_1)$$
$$= (A/85\cdot7)^2(7 \times 10,000 \times 10^3/0\cdot0103) = 9\cdot25 \times 10^5 A^2 \qquad (1)$$

The pressure at this point is $10 \times 10^6 w$ N/m².
Specific volume of gas at this point is given by equation 4.63:

Thus $v = v_1 w^{-0\cdot71} = 0\cdot0103 w^{-0\cdot71}$

The velocity is given by equation 4.64:

Thus
$$u^2 = 7P_1 v_1(1 - w^{0.29})$$
$$= 7 \times 10 \times 10^6 \times 0.0103(1 - w^{0.29})$$

$\therefore$
$$u = 0.849 \times 10^3 (1 - w^{0.29})^{0.5} \text{ m/s}$$

Velocity of sound at a pressure of $10 \times 10^6 w$ N/m^2

$$= \sqrt{1.4 \times 10 \times 10^6 w \times 0.0103 w^{-0.71}}$$
$$= 380 w^{0.145} \text{ m/s}$$

Mach number $= 0.849 \times 10^3(1 - w^{0.29})^{0.5}/(380 w^{0.145}) = 2.23(w^{-0.29} - 1)^{0.5}$ (2)

(2) *The non-isentropic compression across the shock wave.* The velocity downstream from the shock wave (suffix s) is given by equation 4.54:

$$u_s = u_1[(\gamma - 1)Ma_1^2 + 2]/[Ma_1^2(\gamma + 1)]$$
$$= 0.849 \times 10^3(1 - w^{0.29})^{0.5}(0.4 \times 4.97(w^{-0.29} - 1) + 2)/[4.97(w^{-0.29} - 1) \times 2.4]$$
$$= 141(1 - w^{0.29})^{-0.5} \text{ m/s} \tag{3}$$

The pressure downstream from the shock wave P_s is given by equation 4.53:

$$P_s/10 \times 10^6 w = [2\gamma Ma_1^2 - (\gamma - 1)/(\gamma + 1)]$$

Substituting from equation 2:

$$P_s = 56.3 w(w^{-0.29} - 1) \times 10^6 \text{ N/m}^2 \tag{4}$$

(3) *The isentropic expansion of the gas to atmospheric pressure.* The gas now expands isentropically from P_s to P_a (= 101.3 kN/m^2) and the flow area increases from A to the full bore of 0.07 m^2. Denoting conditions at the outlet by suffix a, from equation 4.64:

$$u_a^2 - u_s^2 = 7P_s v_s[1 - (P_a/P_s)^{0.25}] \tag{5}$$
$$u_a/v_a = 85.7/0.07 = 1224 \text{ kg/m}^2 \text{ s} \tag{6}$$
$$u_s/v_s = 85.7/A \text{ kg/m}^2 \text{ s} \tag{7}$$
$$v_a/v_s = (P_a/P_s)^{-0.71} \tag{8}$$

Equations 1, 3 to 8, involving seven unknowns, can be solved by trial and error methods to give $w = 0.0057$.

Thus the pressure upstream from the shock wave is:

$$10 \times 10^6 \times 0.0057 = 0.057 \times 10^6 \text{ N/m}^2$$

or
$$\underline{\underline{57 \text{ kN/m}^2}}$$

Mach Number, from equation 2: $= \underline{\underline{4.15}}$

Pressure downstream from shock wave P_s, from equation 4:

$$= \underline{\underline{1165 \text{ kN/m}^2}}$$

4.3. TWO-PHASE FLOW

In this section some features of the flow of a two-phase fluid formed from a gas or vapour and a liquid will be discussed. Steady flow of such fluids occurs in oil and gas lines and in many situations in the process industries. The flow of water and steam in boilers and evaporators and the flow of partially condensed vapour–liquid mixtures are further illustrations.

Because of the presence of the two phases there are considerable complications in describing and quantifying the nature of the flow compared with conditions with a single phase. The lack of knowledge of the velocities at a point in the individual phases makes it impossible to give any real picture of the velocity distribution. In most cases the gas phase, which is flowing with a much greater velocity than the liquid, continuously accelerates the liquid, thus involving further loss of energy. Either phase may be in streamline or in turbulent condition, but the most important case is that in which both phases are turbulent. This variation in velocity may be expressed in terms of a velocity ratio u_G/u_L.

The more important factors in two-phase flow which a designer must seek to understand and be able to evaluate are the stability of the flow, the likelihood of the flow causing erosion problems, and the drop in pressure caused by the presence of an additional phase.

4.3.1.   Flow Regime

It is first necessary to realise the wide differences in the flow regimes that may occur. Baker,[6] Hoogendoorn,[7] Griffith and Wallis,[8] and others have suggested approximate methods for determining the flow regime, and the diagram first produced by Baker is shown in Fig. 4.7. Here the mass velocity G' (mass flow rate per unit area) of the gas phase is plotted against the ratio of the mass of the two phases L'/G' with the additional parameters λ and ψ defined in the following way:

$$\lambda = \left[\left(\frac{\rho_G}{\rho_a} \right) \left(\frac{\rho_L}{\rho_w} \right) \right]^{0.5}$$

$$\psi = \frac{\sigma_w}{\sigma_L} \left[\frac{\mu_L}{\mu_w} \left(\frac{\rho_w}{\rho_L} \right)^2 \right]^{0.33}$$

where ρ_G, ρ_L are the densities of the gas and liquid, σ_L is the surface tension of the liquid, μ_L the viscosity of the liquid, L', G' are the mass velocities of the liquid and gas phases

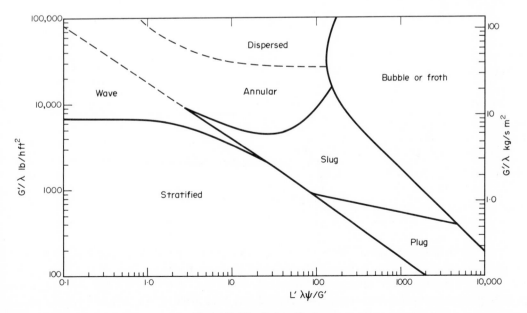

FIG. 4.7. The Baker diagram.

respectively, and the subscripts G, L, a, w refer to the properties of gas, liquid, air and water respectively.

Much of Baker's work was carried out with the air–water system at atmospheric pressure in horizontal pipes of up to 100 mm diameter. Some indication of the appearance of the flow patterns is shown in Fig. 4.8, and an indication of the velocities that produce these regimes is given in Table 4.1.

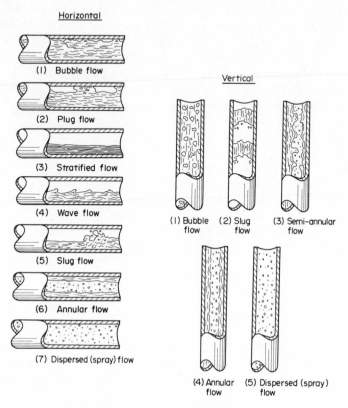

FIG. 4.8. Flow patterns in two-phase flow.

TABLE 4.1. *Flow regimes in horizontal two-phase flow*

Regime	Description	Typical velocities (m/s)	
		Liquid	Vapour
Bubble flow	Bubbles of gas dispersed throughout the liquid	1·5–5	0·3–3
Plug flow	Plugs of gas in liquid phase	0·6	<1·0
Stratified flow	Layer of liquid with a layer of gas above	<0·15	0·6–3
Wavy flow	As stratified but with a wavy interface due to higher velocities	<0·3	>5
Slug flow	Slugs of gas in liquid phase	Occurs over a wide range of velocities	
Annular flow	Liquid film on inside walls with gas in centre		>6
Spray flow	Liquid droplets dispersed in gas		>60

Slug flow should be avoided since it gives rise to unsteady conditions, and it is desirable to design so that annular flow still persists at loadings down to 50 per cent of the normal flow rates. Whilst both phases should be turbulent, excessive gas velocity will lead to a high pressure drop, particularly in small pipes. Instability may arise at any point where the phases separate, and in upward vertical flow a high vapour velocity may give rise to instability. An empirical relation for the avoidance of the slug flow condition is that the minimum two-phase velocity, u (m/s), should exceed $(3 \cdot 05 + 0 \cdot 024D)$ where D mm is the pipe diameter.

Example 4.5. Water and steam flow in a 50 mm pipe at flow rates of 0·1 and 0·03 kg/s respectively. Under the conditions of temperature and pressure existing in the pipe, the water and steam have the properties listed below. Using the Baker diagram, determine the two-phase flow regime.

Solution.

	Water	Steam	Air (STP)	Water (STP)
Viscosity (mN s/m²)	0·52	—	—	1·0
Density (kg/m³)	1009	0.79	1.20	998
Surface tension (J/m²)	0·067	—	—	0·073

The parameters on the Baker diagram are:

$$G'/\lambda \quad \text{and} \quad L'\lambda\psi/G'$$

where

$$\lambda = [(\rho_G/\rho_a)(\rho_L/\rho_w)]^{0 \cdot 5}$$
$$= [(0 \cdot 79/1 \cdot 20)(1009/998)]^{0 \cdot 5}$$
$$= 0 \cdot 815$$

and

$$\psi = [(\sigma_w/\sigma_L)(\mu_L/\mu_w)(\rho_w/\rho_L)^2]^{0 \cdot 33}$$
$$= (0.073/0.067)[(0.52/1.0)(998/1009)^2]^{0.33}$$
$$= 0 \cdot 870$$

Cross-sectional area of pipe $= (\pi/4)(0 \cdot 050)^2 \times 0 \cdot 00196 \text{ m}^2$

So that

$$L' = 0 \cdot 1/0 \cdot 00196 = 51 \cdot 0 \text{ kg/s m}^2$$

and

$$G' = 0 \cdot 03/0 \cdot 00196 = 15 \cdot 30 \text{ kg/s m}^2$$

Hence

$$G'/\lambda = 15 \cdot 30/0 \cdot 815 = 18 \cdot 77 \text{ kg/s m}^2$$

and

$$L'\lambda\psi/G' = 51 \cdot 0 \times 0 \cdot 815 \times 0 \cdot 870/15 \cdot 30 = 2 \cdot 36$$

From the Baker diagram (Fig. 4.7) the flow regime is seen to be _annular_

4.3.2. Erosion

Two-phase systems are often accompanied by erosion, and many empirical relationships have been suggested to avoid the condition. Since high velocities are desirable to avoid the instability associated with slug flow, there is a danger that any increase in throughput above the normal operating condition will lead to a situation where erosion may become a serious possibility.

An indication of the velocity at which erosion becomes significant may be obtained from

$$\rho_m u_m^2 = 15,000 \tag{4.66}$$

where ρ_m is the mean density of the two-phase mixture (kg/m³) and u_m the mean velocity of the two-phase mixture (m/s),

$$\rho_m = (L' + G')/[(L'/\rho_L) + (G'/\rho_G)] \tag{4.67}$$

and

$$u_m = u_L + u_G \tag{4.68}$$

It is apparent that some compromise may be essential between the avoidance of both a slug-flow condition and the velocities which are likely to cause erosion conditions.

Example 4.6. Using the data of Example 4.5, are the flows likely to cause erosion?

Solution. In this case:

$$\rho_m = (51 \cdot 0 + 15 \cdot 30)/[(51 \cdot 0/1009) + (15 \cdot 30/0 \cdot 79)] = 3 \cdot 41 \text{ kg/m}^3$$

$$u_m = u_L + u_G = (51 \cdot 0/1009) + (51 \cdot 30/0 \cdot 79) = 19.42 \text{ m/s}$$

Hence $\rho_m u_m^2 = 3 \cdot 41(19 \cdot 42)^2 = 1286$

This value is considerably less than 15,000 and erosion will *not* be a problem.

4.3.3. Pressure, Momentum, and Energy Relations

Methods for determining the drop in pressure start with a proposal for the physical model of the two-phase system, and the analysis is developed as an extension of that used for single-phase flow. In the *separated flow* model the phases are first considered to flow separately, and their combined effect is then examined.

The total pressure gradient, $(-dP/dl)$, consists of two components which represent the frictional and the acceleration pressure gradients respectively.

i.e. $$-(dP/dl) = -(dP_f/dl) - (dP_a/dl) \tag{4.69}$$

A momentum balance for the flow of a two-phase fluid through a horizontal pipe and an energy balance may be written in an expanded form of that applicable to single-phase fluid flow. These equations for two-phase flow cannot be used in practice since the individual phase velocities and local densities are not known. Some simplification is possible if it is assumed that the two phases flow separately in the channel occupying fixed fractions of the total area, but even with this assumption of separated flow regimes, progress is difficult. It is important to note that, as in the case of single phase flow of a compressible fluid, it is no longer possible to relate the shear stress to the pressure drop in a simple form since the pressure drop now covers both frictional and acceleration losses. The shear at the wall is proportional to the total rate of momentum transfer (arising from friction, acceleration, and potential effects), so that the total drop in pressure ΔP_{TPF} is given by:

$$\Delta P_{TPF} = \Delta P_{\text{friction}} + \Delta P_{\text{acceleration}} \tag{4.70}$$

The pressure drop due to acceleration is important in two-phase flow because the gas is normally flowing much faster than the liquid, and therefore accelerates the liquid phase with consequent transfer of energy. For flow in a vertical direction, an additional term $\Delta P_{\text{gravity}}$ must be added to the right hand side of equation 4.70.

Analytical solutions for the equations of motion are not possible because of the difficulty of specifying the flow pattern and of defining the precise nature of the interaction between the phases. Rapid fluctuations in flow frequently occur and these cannot readily be taken into account. For these reasons, it is necessary for design purposes to use correlations which have been obtained using experimental data. Great care should be taken, however, if these are used outside the limits used in the experimental work.

Practical Methods for Evaluating Drop in Pressure

Probably the most widely used method for estimating the drop in pressure due to friction is that proposed by Lockhart and Martinelli[9] and later improved by Chisholm.[10] This is based on the physical model of separated flow in which each phase is considered separately and then a combined effect formulated. The two-phase pressure drop due to friction ΔP_{TPF} is taken as the pressure drop ΔP_L or ΔP_G that would arise for either phase flowing alone in the pipe at the stated rate, multiplied by some factor Φ_L or Φ_G. This factor is presented as a function of the ratio of the individual single-phase pressure drops.

$$\Delta P_{TPF}/\Delta P_G = \Phi_G^2 \tag{4.71}$$

$$\Delta P_{TPF}/\Delta P_L = \Phi_L^2 \tag{4.72}$$

and

$$\Delta P_L/\Delta P_G = X^2 \tag{4.73}$$

The relation between X, Φ_G, and Φ_L is shown in Fig. 4.9, where it is seen that separate curves are given according to the nature of the flow of the two phases. This relation was developed from studies on the flow in small tubes of up to 25 mm diameter with water, oils, and hydrocarbons using air at a pressure of up to 400 kN/m². Although $Re_L\left(\dfrac{L'd}{\mu}\right)$ and $Re_G\left(\dfrac{G'd}{\mu}\right)$ are used as criteria for defining the flow regime, values less than 1000 do not necessarily imply that the fluid is in truly viscous flow. Later experimental work showed that the total pressure has an influence and data presented by Griffith[11] should be consulted where pressures are in excess of 3 MN/m². Chisholm[10] has developed a relation between Φ_L and X which he puts in the form:

$$\Phi_L^2 = 1 + (C/X) + (1/X^2) \tag{4.74}$$

where C has a value of 20 for turbulent–turbulent flow and lower values for both phases in viscous flow, but the graphs for Fig. 4.9 are much to be preferred.

Chenoweth and Martin[12, 13] have presented an alternative method for calculating the drop in pressure. Their method is empirical and based on experiments with pipes of 75 mm and pressures of up to 0·7 MN/m². They have plotted the volume fraction of the inlet stream that is liquid as abscissa against the ratio of the two-phase pressure drop to that for liquid flowing at the same volumetric rate as the mixture. An alternative technique has been described by Baroczy.[14] If heat transfer gives rise to evaporation then reference should be made to work by Dukler.[15] An illustration of the method of calculation of two-phase pressure drop is included here as Example 4.7.

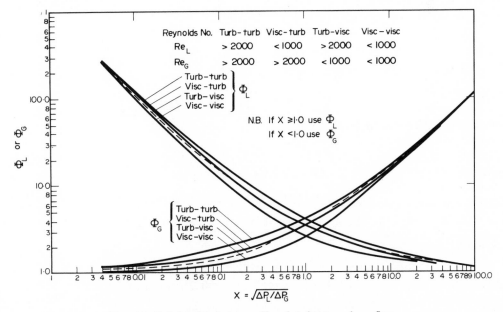

FIG. 4.9. Relationship between X and ϕ for two-phase flow.

Example 4.7. Steam and water flow through a 75 mm i.d. pipe at flowrates of 0·05 and 1·5 kg/s respectively. If the mean temperature and pressure are 330 K and 120 kN/m², what is the pressure drop per unit length of pipe assuming adiabatic conditions?

Solution. Cross area for flow $= \pi(0·075)^2/4 = 0·00442 \text{ m}^2$

Flow of water $= 1·5/1000 = 0·0015 \text{ m}^3/\text{s}$

equivalent to a velocity of $0·0015/0·00442 = 0.339 \text{ m/s}$

Density of steam at 330 K and 120 kN/m²

$$= (18/22·4)(273/330)(120/101.3) = 0.788 \text{ kg/m}^3$$

Flow of steam $= 0·05/0·788 = 0·0635 \text{ m}^3/\text{s}$

equivalent to a velocity of $0·0635/0·00442 = 14·37 \text{ m/s}$

Viscosities at 330 K and 120 kN/m²:

$$\text{steam} = 0·0113 \times 10^{-3} \text{ N s/m}^2; \quad \text{water} = 0.52 \times 10^{-3} \text{ N s/m}^2$$

Therefore $\quad Re_L = 0·075 \times 0·339 \times 1000/(0·52 \times 10^{-3}) = 4·89 \times 10^4$

$$Re_G = 0·075 \times 14·37 \times 0·788/(0·0113 \times 10^{-3}) = 7·52 \times 10^4$$

That is, both the gas and liquid are in turbulent flow.
From the friction chart (Fig. 3.7), assuming $e/d = 0.00015$:

$$(R/\rho u^2)_L = 0·0025 \quad \text{and} \quad (R/\rho u^2)_G = 0·0022$$

$\therefore$ From equation 3.15:

$$\Delta P_L = 4f_L(1/d)(\rho u^2)$$
$$= 4 \times 0·0025(1/0·075)(1000 \times 0·339^2)$$
$$= 15·32 \text{ (N/m}^2)/\text{m}$$

or $\quad (15·32/1000 \times 9·81) = 0·00156 \text{ m/m}$

$$\Delta P_G = 4 \times 0·0022(1/0·075)(0·778 \times 14·37^2)$$
$$= 18·85 \text{ (N/m}^2)/\text{m}$$

or $\quad (18·85/0·778 \times 9·81) = 2·47 \text{ m/m}$

$\therefore \quad \Delta P_L/\Delta P_G = 15·32/18·85 = 0·812$

that is $\quad X^2 = 0·812 \quad \text{and} \quad X = 0·901$

From Fig. 4.9, for turbulent–turbulent flow,

$$\Phi_L = 4·35 \quad \text{and} \quad \Phi_G = 3·95$$

Therefore $\quad \Delta P_{TPF}/\Delta P_G = 3·95^2 = 15·60$

and $\quad \Delta P_{TPF} = 15·60 \times 18·85 = 294 \text{ (N/m}^2)/\text{m}$

or $\quad \underline{\underline{\Delta P_{TPF} = 0·29 \text{ (kN/m}^2)/\text{m}}}$

Similarly, as a check:

$$\Delta P_{TPF}/\Delta P_L = 4·35^2 = 18·92$$

and $\quad \Delta P_{TPF} = 18·92 \times 15·32 = 290 \text{ (N/m}^2)/\text{m}$

or $\quad \underline{\underline{\Delta P_{TPF} = 0·29 \text{ (kN/m}^2)/\text{m}}}$

Critical Flow

For the flow of a compressible fluid, conditions of sonic velocity may be reached, and impose limitations on the maximum flow rate for a given upstream pressure. This situation can also occur with two-phase flow, and such critical velocities may sometimes be reached with a drop in pressure not much above 30 per cent of the inlet pressure.

4.4. FURTHER READING

LEE, J. F. and SEARS, F. W.: *Thermodynamics*, 2nd edn. (Addison–Wesley, 1962).
MAYHEW, Y. R. and ROGERS, G. F. C.: *Thermodynamics and Transport Properties of Fluids*, 2nd edn. (Blackwell, Oxford, 1971).
DODGE, B. F.: *Chemical Engineering Thermodynamics* (McGraw-Hill, New York, 1944).
GRIFFITH, P.: Two-phase Flow, sect. 14 in *Handbook of Heat Transfer*, ed. by Rohsenow, W. M. and Hartnett, J. P. (McGraw-Hill, New York, 1973).
COLLIER, J. G.: *Convective Boiling and Condensation* (McGraw Hill, New York, 1972).

4.5. REFERENCES

1. GRIFFITHS, H.: *Trans. Inst. Chem. Eng.* **23** (1945) 113. Some problems of vacuum technique from a chemical engineering standpoint.
2. LAPPLE, C. E.: *Trans. Am. Inst. Chem. Eng.* **39** (1948) 385. Isothermal and adiabatic flow of compressible fluids.
3. WOOLLATT, E.: *Trans. Inst. Chem. Eng.* **24** (1946) 17. Some aspects of chemical engineering thermodynamics with particular reference to the development and use of the steady flow energy balance equations.
4. STODOLA, A. and LOWENSTEIN, L. C.: *Steam and Gas Turbines* (McGraw-Hill, New York, 1945).
5. LEE, J. F. and SEARS, F. W.: *Thermodynamics*, 2nd edn. (Addison-Wesley, 1962).
6. BAKER, O.: *Oil & Gas J.* **53**, 26 July, (1954) 185; **57**, 10 Nov., (1958) 156. Two-phase flow regime correlation; Two-phase flow pattern sketches.
7. HOOGENDOORN, C. J.: *Chem. Eng. Sci.* **9** (1959) 205. Gas–liquid flow in horizontal pipes.
8. GRIFFITH, P. and WALLIS, G. B.: *Trans. Am. Inst. Mech. Eng.* **8**, C83 (1961) 307. Two-phase slug flow.
9. LOCKHART, R. N. and MARTINELLI, R. C.: *Chem. Eng. Prog.* **45** (1949) 39. Proposed correlation of data for isothermal two-phase, two-component flow in pipes.
10. CHISHOLM, D.: Min of Tech. NEL Report No. 310, August 1967. A theoretical basis for the Lockhart–Martinelli correlation for two-phase flow.
11. ROHSENOW, W. M. and HARTNETT, J. P. (eds.): *Handbook of Heat Transfer*, sect. 14, by P. Griffith (McGraw-Hill, New York, 1973).
12. CHENOWETH, J. M. and MARTIN, M. W.: *Pat. Reg.* **34** (10) (1955) 151.
13. CHENOWETH, J. M. and MARTIN, M. W.: *Trans. Am. Soc. Mech. Eng.*, Paper 55 (1955) 9. A pressure drop correlation for turbulent two-phase flow of gas–liquid mixtures in horizontal pipes.
14. BAROCZY, C. J.: *Chem. Eng. Prog. Symp. Ser.* **62** (44) (1966) 232. A systematic correlation for two-phase pressure drop.
15. DUKLER, A. E. MOYE WICKS III, and CLEVELAND, R. G.: *A.I.Ch.E.Jl* **10** (1964) 38. Frictional pressure drop in two-phase flow.

4.6. NOMENCLATURE

		Units in SI System	Dimensions in M, L, T, θ
A	area perpendicular to flow	m²	L^2
C_p	specific heat at constant pressure	J/kg K	$L^2T^{-2}\theta^{-1}$ or –
C_v	specific heat at constant volume	J/kg K	$L^2T^{-2}\theta^{-1}$ or –
d	diameter	m	L
g	acceleration due to gravity	m/s²	LT^{-2}
G	mass flowrate	kg/s	MT^{-1}
G'	mass flowrate per unit area	kg/s m²	$ML^{-2}T^{-1}$
H	enthalpy per unit mass	J/kg	L^2T^{-2}
K	numerical constant used as index	—	—
k	energy per unit mass	J/kg	L^2T^{-2}
L	liquid flowrate	kg/s	MT^{-1},

		Units in SI System	Dimensions in $\mathbf{M}, \mathbf{L}, \mathbf{T}, \theta$
L'	liquid flowrate per unit area	kg/s m²	$\mathbf{ML^{-2}T^{-1}}$
l	length	m	$\mathbf{L}$
M	molecular weight	kg/kmol	—
Ma	Mach number	—	—
P	pressure	N/m²	$\mathbf{ML^{-1}T^{-2}}$
ΔP	pressure drop	N/m²	$\mathbf{ML^{-1}T^{-2}}$
q	net heat flow into system from surroundings	J/kg	$\mathbf{L^2T^{-2}}$
R	shear stress	N	$\mathbf{ML^{-1}T^{-2}}$
$\mathbf{R}$	universal gas constant	8·314 kJ/kmol K	$\mathbf{L^2T^{-2}\theta^{-1}}$
Re	Reynolds number	—	—
S	entropy per unit mass	J/kg K	$\mathbf{L^2T^{-2}\theta^{-1}}$
T	temperature	K	$\boldsymbol{\theta}$
u	velocity	m/s	$\mathbf{LT^{-1}}$
V	volume	m³	$\mathbf{L^3}$
v	specific volume	m³/kg	$\mathbf{M^{-1}L^3}$
W_s	shaft work per unit mass	J/kg	$\mathbf{L^2T^{-2}}$
X	ratio of liquid phase to gas phase pressure drop	—	—
z	height	m	$\mathbf{L}$
α	constant in expression for kinetic energy of fluid	—	—
ρ	density	kg/m³	$\mathbf{ML^{-3}}$
μ	viscosity	Ns/m²	$\mathbf{ML^{-1}T^{-1}}$
γ	ratio of specific heats at constant pressure and constant volume	—	—
σ	surface tension	J/m²	$\mathbf{MT^{-2}}$
λ	correction factor for two-phase flow	—	—
ψ	correction factor for two-phase flow	—	—
Φ_G, Φ_L	ratio of two-phase to gas and liquid pressure drop	—	—

CHAPTER 5

Flow and Pressure Measurement

5.1. INTRODUCTION

The most important parameters that are measured to provide information on the operating conditions in a plant are flowrates, pressures, and temperatures. The instruments used may give either an instantaneous reading or, in the case of flow, may be arranged to give a cumulative flow over any given period. In either case the instrument may be required to give a signal to some control unit which will then govern one or more parameters on the plant. It should be noted that on industrial plants it is usually more important to have information on the change in the value of a given parameter than to use meters that give particular absolute accuracy. To maintain the value of a parameter at a desired value a control loop is used as outlined in Fig. 5.1.

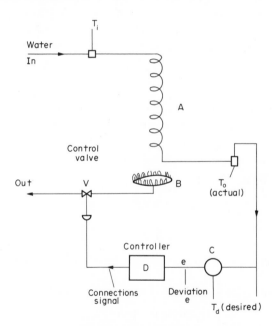

Fig. 5.1. A control loop.

Water enters coil A at a temperature T_i is heated by the burner B, and leaves at some higher temperature T_o. The temperature reading T_o is passed to the comparator C which records the error e $(T_d - T_o)$ between the value of T_o and the desired value T_d. The unit C passes on this error value to the controller D which produces a control signal proportional to e. The control valve V is actuated by the controller signal output, opening if e is negative and closing if e is positive. As a result of the controller D the final valve movement is proportional to the error e.

It is very important to note that in this loop system the parameter (T_o), which must be kept constant, is measured, though all subsequent action is concerned with the magnitude of the error and not with the actual value of T_o. This simple loop will frequently be complicated by

97

there being several parameters to control, which may necessitate considerable instrumental analysis and the control action will involve operation of several control valves.

This represents a simple form of control for a single variable, though in a modern plant many parameters are controlled at the same time from various measuring instruments, and the variables on a plant such as a distillation unit are frequently linked together, thus increasing the complexity of control that is required.

On industrial plants the instruments are therefore required not only to act as indicators but also to provide some link which can be used to help in the control of the plant. In this chapter, instruments for the measurement of flow will be discussed, and the method of transmitting the signal will also be briefly considered though the more general question of control is discussed in Volume 3.[1]

5.2. FLUID PRESSURE

In a stationary fluid the pressure is exerted equally in all directions and is referred to as the static pressure. In a moving fluid, the static pressure is exerted on any plane parallel to the direction of motion. The pressure exerted on a plane at right angles to the direction of flow is greater than the static pressure because the surface has, in addition, to exert sufficient force to bring the fluid to rest. This additional pressure is proportional to the kinetic energy of the fluid; it cannot be measured independently of the static pressure.

5.2.1. Static Pressure

The energy balance equation can be applied between any two sections in a continuous fluid. If the fluid is not moving, the kinetic energy and the frictional loss are both zero, and therefore:

$$v \, dP + g \, dz = 0$$

For an incompressible fluid:

$$v(P_2 - P_1) + g(z_2 - z_1) = 0$$

i.e.
$$(P_2 - P_1) = -\rho g(z_2 - z_1) \tag{5.1}$$

Thus the pressure difference can be expressed in terms of the height of a vertical column of fluid.

If the fluid is incompressible and behaves as an ideal gas, for isothermal conditions:

$$P_1 v_1 \ln(P_2/P_1) + g(z_2 - z_1) = 0$$

$$\therefore \quad P_2/P_1 = \exp[-(gM/\mathbf{R}T)(z_2 - z_1)] \tag{5.2}$$

This expression enables the pressure distribution within an ideal gas to be calculated for isothermal conditions.

When the static pressure in a moving fluid is to be determined, the measuring surface must be parallel to the direction of flow so that no kinetic energy is converted into pressure energy at the surface. If the fluid is flowing in a circular pipe the measuring surface must be perpendicular to the radial direction at any point. The pressure connection, which is known as a *piezometer tube*, should terminate flush with the wall of the pipe so that the flow is not disturbed: the pressure is then measured near the walls where the velocity is a minimum and the reading would be subject only to a small error if the surface were not quite parallel to the direction of flow. A piezometer tube of narrow diameter is used for accurate measurements.

The static pressure should always be measured at a distance of not less than 50 diameters from bends or other obstructions, so that the flow lines are almost parallel to the walls of the tube. If there are likely to be large cross-currents or eddies, a piezometer ring should be used. This consists of four pressure tappings equally spaced at 90° intervals round the circumference of the tube; they are joined by a circular tube which is connected to the pressure measuring device. By this means false readings due to irregular flow are avoided. If the pressure on one side of the tube is relatively high, the pressure on the opposite side is generally correspondingly low; with the piezometer ring a mean value is obtained. The cross-section of the piezometer tubes and ring should be small to prevent any appreciable circulation of the fluid.

5.2.2. Pressure Measuring Devices

(a) *The simple manometer* is shown in Fig. 5.2. It consists of a transparent U-tube containing the fluid **A** of density ρ whose pressure is to be measured and an immiscible fluid **B** of higher density ρ_m. The limbs are connected to the two points between which the pressure difference $P_2 - P_1$ is required; the connecting leads should be completely full of fluid **A**. If P_2 is greater than P_1, the interface between the two liquids in limb 2 will be depressed a distance h_m (say) below that in limb 1. The pressure at the level $a-a$ must be the same in each of the limbs and, therefore:

$$P_2 + z_m \rho g = P_1 + (z_m - h_m)\rho g + h_m \rho_m g$$

$\therefore$
$$\Delta P = P_2 - P_1 = h_m (\rho_m - \rho)g \tag{5.3}$$

If fluid **A** is a gas, ρ will normally be small compared with ρ_m so that:

$$\Delta P = h_m \rho_m g \tag{5.4}$$

(b) *The inverted manometer* (Fig. 5.3) is used for measuring pressure differences in liquids. The space above the liquid in the manometer is filled with air which can be admitted or expelled through the tap A in order to adjust the level of the liquid in the manometer.

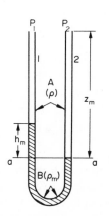

FIG. 5.2. Simple manometer.

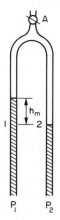

FIG. 5.3. Inverted manometer.

(c) *The inclined manometer.* For the measurement of small pressure differences the manometer can be inclined at an angle to the vertical in order to obtain a larger reading.

(d) *The two-liquid manometer.* Small differences in pressure in gases are often measured with a manometer of the form shown in Fig. 5.4. The reservoir at the top of each limb is of a sufficiently large cross-section for the liquid level to remain approximately the same on each

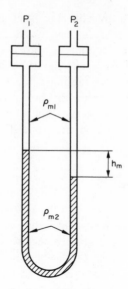

FIG. 5.4. Two-liquid manometers.

side of the manometer. The difference in pressure is then given by:

$$\Delta P = P_2 - P_1 = h_m(\rho_{m1} - \rho_{m2})g \qquad (5.5)$$

where ρ_{m1} and ρ_{m2} are the densities of the two manometer liquids. The sensitivity of the instrument is very high if the densities of the two liquids are nearly the same. To obtain accurate readings it is necessary to choose liquids which give sharp interfaces: paraffin oil and industrial alcohol are commonly used. According to Ower,[2] benzyl alcohol (specific gravity 1·048) and calcium chloride solutions give the most satisfactory results. The difference in density can be varied by altering the concentration of the calcium chloride solution.

(e) *The Bourdon gauge* (Fig. 5.5). The pressure to be measured is applied to a curved tube, oval in cross-section, and the deflection of the end of the tube is communicated through a system of levers to a recording needle. This gauge is widely used for steam and compressed gases, and frequently forms the indicating element on flow controllers.

The simple form of the gauge is shown in Fig. 5.5a. Figure 5.5b shows a Bourdon type gauge with the sensing element in the form of a helix. This instrument has a very much greater sensitivity and is suitable for very high pressures.

5.2.3. Pressure Signal Transmission—The Differential Pressure Cell

The meters described earlier provide a measurement usually in the form of a pressure differential, but in most modern plants these readings must be transmitted to a central control room where they form the basis either for recording or for automatic control. The transmission is most conveniently effected by pneumatic or electrical methods, and one typical arrangement known as a differential pressure (d.p.) cell operating on the force balance system is illustrated in Figs. 5.6 and 5.7.

In Fig. 5.7 increase in the pressure on the "in" bellows causes the force-bar to turn clockwise, thus reducing the separation in the flapper-nozzle. This nozzle is fed with air, and the increase in back pressure from the nozzle in turn increases the force developed by the negative feedback bellows producing a counterclockwise movement which restores the flapper-nozzle separation. Balance is achieved when the feedback pressure is proportional to the applied pressure.

(a)

(b)

FIG. 5.5. Bourdon gauge.

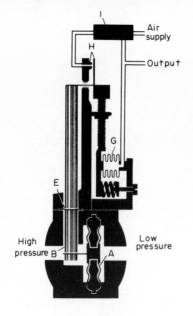

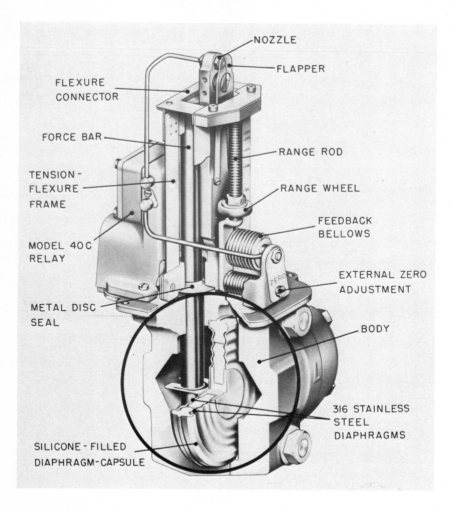

FIG. 5.6. Differential pressure cell.

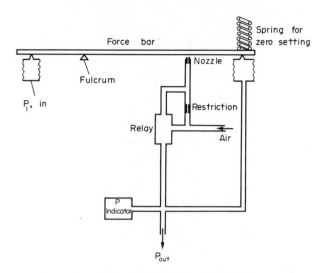

FIG. 5.7. Force balance system.

In this way the signal pressure is made to respond to the input pressure, and by adjusting the distance of the bellows from the fulcrum any range of force may be made to give an output pressure in the range 120–200 kN/m². It is important to note that since the flapper-nozzle unit has a high gain (large pressure change for small change in separation), the actual movements are very small, usually within the range 0·004–0·025 mm.

An instrument in which this feedback arrangement is used is shown in Fig. 5.6. The differential pressure from the meter is applied to the two sides of the diaphragm A giving a force on the bar B. The closure E in the force bar B acts as the fulcrum, so that the flapper-nozzle separation at H responds to the initial difference in pressure. This gives the change in air pressure or signal which can be transmitted through a considerable distance. A unit of this kind is thus an essential addition to all pressure meters which are to give a signal in a control room.

5.2.4. Impact Pressure

The pressure exerted on a plane at right angles to the direction of flow of the fluid consists of two components:

(a) static pressure;
(b) the additional pressure required to bring the fluid to rest at the point.

Consider a fluid flowing between two sections, 1 and 2 (Fig. 5.8), which are sufficiently close for friction losses to be negligible between the two sections; they are a sufficient distance apart, however, for the presence of a small surface at right angles to the direction of flow at section 2 to have negligible effect on the pressure at section 1. These conditions are normally met if the distance between the sections is one pipe diameter.

Now consider a small filament of liquid which is brought to rest at section 2.

Applying the energy balance equation between the two sections, since $g\,\Delta z$, W_s, and F are all zero:

$$\frac{\dot{u}_1^2}{2} = \frac{\dot{u}_2^2}{2} + \int_1^2 v\,dP \qquad \text{(from equation 2.55)}$$

where $\dot{u}_1$ and $\dot{u}_2$ are velocities at 1 and 2.

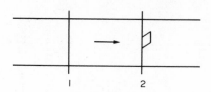

FIG. 5.8. Impact pressure.

If the fluid is incompressible or if the change in the density of the fluid between the sections is negligible, then (since $\dot{u}_2 = 0$):

$$\dot{u}_1{}^2 = v(P_2 - P_1) = h_i g$$

i.e.

$$\dot{u}_1 = \sqrt{2v(P_2 - P_1)} = \sqrt{2gh_i} \qquad (5.6)$$

where h_i is the difference between the impact pressure head at section 2 and the static pressure head at section 1.

Little error is introduced if this expression is applied to the flow of a compressible fluid provided that the velocity is not greater than about 60 m/s. When the velocity is high, the equation of state must be used to give the relation between the pressure and the volume of the gas. For non-isothermal flow, $Pv^k = $ a constant,

and

$$\int_1^2 v \, dP = \frac{k}{k-1} P_1 v_1 \left[\left(\frac{P_2}{P_1} \right)^{(k-1)/k} - 1 \right] \qquad \text{(equation 2.73)}$$

so that

$$\frac{\dot{u}_1{}^2}{2} = \frac{k}{k-1} P_1 v_1 \left[\left(\frac{P_2}{P_1} \right)^{(k-1)/k} - 1 \right] \qquad (5.7)$$

and

$$P_2 = P_1 \left(1 + \frac{\dot{u}_1{}^2}{2} \frac{k-1}{k} \frac{1}{P_1 v_1} \right)^{k/(k-1)} \qquad (5.8)$$

For isothermal flow:

$$(\dot{u}_1{}^2/2) = P_1 v_1 \ln(P_2/P_1)$$

∴

$$P_2 = P_1 \exp(\dot{u}_1{}^2/2P_1 v_1) \qquad (5.9)$$

or

$$P_2 = P_1 \exp(\dot{u}_1{}^2 M / 2RT) \qquad (5.10)$$

Equations 5.8 and 5.9 can be used for the calculation of the fluid velocity and the impact pressure in terms of the static pressure a short distance upstream. The two sections are chosen so that they are sufficiently close together for frictional losses to be negligible. Thus P_1 will be approximately equal to the static pressure at both sections and the equations give the relation between the static and impact pressures—and the velocity—at any point in the fluid.

5.3. MEASUREMENT OF FLUID FLOW

The most important class of flowmeter is that in which the fluid is either accelerated or retarded at the measuring section by reducing the flow area, and the change in the kinetic energy is measured by recording the pressure difference produced.

This class includes:

The pitot tube, in which a small element of fluid is brought to rest at an orifice situated at right angles to the direction of flow. The flowrate is then obtained from the difference between the impact and the static pressure. With this instrument the velocity measured is that of a small filament of fluid.

The orifice meter, in which the fluid is accelerated at a sudden constriction (the orifice) and the pressure developed is then measured. This is a relatively cheap and reliable instrument though the overall pressure drop is high because most of the kinetic energy of the fluid at the orifice is wasted.

The Venturi meter, in which the fluid is gradually accelerated to a throat and gradually retarded as the flow channel is expanded up to the pipe size. A high proportion of the kinetic energy is thus recovered but the instrument is expensive and bulky.

The nozzle, in which the fluid is gradually accelerated up to the throat of the instrument but expansion to pipe diameter is sudden as with an orifice. This instrument is again expensive because of the accuracy required over the inlet section.

The notch or weir, in which the fluid flows over the weir so that its kinetic energy is measured by determining the head of the fluid flowing above the weir. This instrument is used in open-channel flow and extensively in tray towers[3] where the height of the weir is adjusted to provide the necessary liquid depth for a given flow.

Each of the above devices will now be considered in more detail together with some less-common and special purpose meters.

5.3.1. The Pitot Tube

The pitot tube is used to measure the difference between the impact and static pressures in a fluid. It normally consists of two concentric tubes arranged parallel to the direction of flow; the impact pressure is measured on the open end of the inner tube. The end of the outer concentric tube is sealed and a series of orifices on the curved surface give an accurate indication of the static pressure. The position of these orifices must be carefully chosen because there are two disturbances which may cause an incorrect reading of the static pressure. These are due to:

(1) the head of the instrument;
(2) the portion of the stem which is at right angles to the direction of flow of the fluid.

These two disturbances cause errors in opposite directions, and the static pressure should therefore be measured at the point where the effects are equal and opposite.

If the head and stem are situated at a distance of 14 diameters from each other as on the standard instrument,[4] the two disturbances are equal and opposite at a section 6 diameters from the head and 8 from the stem. This is, therefore, the position at which the static pressure orifices should be located. If the distance between the head and the stem is too great, the instrument will be unwieldy; if it is too short, the magnitude of each of the disturbances will be relatively great, and a small error in the location of the static pressure orifices will appreciably affect the reading.

The two standard instruments are shown; the one with the rounded nose is preferred, since it is less subject to damage (Fig. 5.9a and b).

For Reynolds numbers, of 500–300,000, based on the external diameter of the pitot tube, an error of not more than 1 per cent is obtained with this instrument. A Reynolds number of 500 with the standard 7·94 mm pitot tube corresponds to a water velocity of 0·070 m/s or an air velocity of 0·91 m/s. Sinusoidal fluctuations in the flowrate up to 20 per cent also do not affect the accuracy by more than 1 per cent. Calibration of the instrument is not necessary.

A very small pressure difference is obtained for low rates of flow of gases, and the lower limit of velocity that can be measured is usually set by the minimum difference in pressure that can be measured. This limitation is serious, and various methods have been adopted for increasing the reading of the instrument although they involve the need for calibration. Correct alignment of the instrument with respect to the direction of flow is important; this is attained when the differential reading is a maximum.

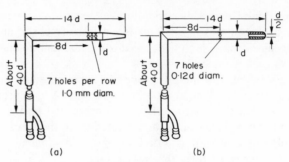

FIG. 5.9. Pitot tubes.

For the flow not to be appreciably disturbed, the diameter of the instrument must not exceed about one-fiftieth of the diameter of the pipe; the standard instrument (diameter 7·94 mm) should therefore not be used in pipes of less than 0·4 m diameter. An accurate measurement of the impact pressure can be obtained using a tube of very small diameter with its open end at right angles to the direction of flow; hypodermic tubing is convenient for this purpose. The static pressure is measured using a single piezometer tube or a piezometer ring upstream at a distance equal approximately to the diameter of the pipe: measurement should be made at least 50 diameters from any bend or obstruction.

The pitot tube measures the velocity of only a filament of fluid, and hence it can be used for exploring the velocity distribution across the pipe section. If, however, it is desired to measure the total flow of fluid through the pipe, the velocity must be measured at various distances from the walls and the results integrated. The total flowrate can be calculated from a single reading only if the velocity distribution across the section is already known.

5.3.2. The Orifice Meter, Nozzle, and Venturi Meter

In each of these three measuring devices the fluid is accelerated by causing it to flow through a constriction; the kinetic energy is thereby increased and the pressure energy therefore decreases. The flowrate is obtained by measuring the pressure difference between the inlet of the meter and a point of reduced pressure, as shown for each meter in Fig. 5.10. If the pressure is measured a short distance upstream where the flow is undisturbed (section 1) and at the position where the area of flow is a minimum (section 2), application of the energy and material balance equations gives:

$$(u_2^2/2\alpha_2) - (u_1^2/2\alpha_1) + g(z_2 - z_1) + \int_1^2 v \, dP + W_s + F = 0 \quad \text{(from equation 2.55)}$$

and the mass flow $\quad G = u_1 A_1/v_1 = u_2 A_2/v_2 \quad$ (from equation 2.37)

If the frictional losses are neglected, and the fluid does no work on the surroundings, i.e. putting W_s and F equal to zero:

$$(u_2^2/2\alpha_2) - (u_1^2/2\alpha_1) = g(z_1 - z_2) - \int_1^2 v \, dP \tag{5.11}$$

Inserting the value of u_1 in terms of u_2 in equation 5.11 enables u_2 and G to be obtained.
For an incompressible fluid:

$$\int_1^2 v \, dP = v(P_2 - P_1) \tag{equation 2.65}$$

and

$$u_1 = u_2(A_2/A_1)$$

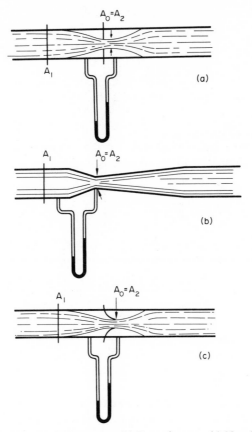

FIG. 5.10. (a) Orifice meter. (b) Venturi meter. (c) Nozzle.

Substituting these values in equation 5.11:

$$\frac{u_2^{\,2}}{2\alpha_2}\left(1-\frac{\alpha_2}{\alpha_1}\frac{A_2^{\,2}}{A_1^{\,2}}\right) = g(z_1 - z_2) + v(P_1 - P_2)$$

Thus

$$u_2^{\,2} = \frac{2\alpha_2[g(z_1 - z_2) + v(P_1 - P_2)]}{1 - \frac{\alpha_2}{\alpha_1}\left(\frac{A_2}{A_1}\right)^2} \tag{5.12}$$

For a horizontal meter $z = 0$:

$$u_2 = \sqrt{\frac{2\alpha_2 v(P_1 - P_2)}{1 - \frac{\alpha_2}{\alpha_1}\left(\frac{A_2}{A_1}\right)^2}}$$

and

$$G = u_2 A_2/v_2 = \frac{A_2}{v}\sqrt{\frac{2\alpha_2 v(P_1 - P_2)}{1 - \frac{\alpha_2}{\alpha_1}\left(\frac{A_2}{A_1}\right)^2}} \tag{5.13}$$

For an ideal gas in isothermal flow:

$$\int_1^2 v\,dP = P_1 v_1 \ln(P_2/P_1).$$

and

$$u_1 = u_2\frac{A_2 v_1}{A_1 v_2}$$

And again neglecting terms in z from equation 5.11:

$$\frac{u_2^2}{2\alpha_2}\left[1 - \frac{\alpha_2}{\alpha_1}\left(\frac{v_1 A_2}{v_2 A_1}\right)^2\right] = P_1 v_1 \ln(P_1/P_2)$$

i.e.

$$u_2^2 = \frac{2\alpha_2 P_1 v_1 \ln(P_1/P_2)}{1 - \frac{\alpha_2}{\alpha_1}\left(\frac{v_1 A_2}{v_2 A_1}\right)^2} \tag{5.14}$$

and the mass flow G is again $u_2 A_2/v_2$.

For an ideal gas in non-isothermal flow. If the pressure and volume are related by $Pv^k = $ constant, then a similar analysis gives:

$$\int_1^2 v \, dP = P_1 v_1 \frac{k}{k-1}\left[\left(\frac{P_2}{P_1}\right)^{(k-1)/k} - 1\right] \tag{equation 2.73}$$

and hence

$$\frac{u_2^2}{2\alpha_2}\left[1 - \frac{\alpha_2}{\alpha_1}\left(\frac{v_1 A_2}{v_2 A_1}\right)^2\right] = -P_1 v_1 \frac{k}{k-1}\left[\left(\frac{P_2}{P_1}\right)^{(k-1)/k} - 1\right]$$

i.e.

$$u_2^2 = \frac{2\alpha_2 P_1 v_1 [k/(k-1)][1 - (P_2/P_1)^{(k-1)/k}]}{1 - \frac{\alpha_2}{\alpha_1}\left(\frac{v_1 A_2}{v_2 A_1}\right)^2} \tag{5.15}$$

and the mass flow G is again $u_2 A_2/v_2$.

5.3.3. The Orifice Meter (Fig. 5.11)

The most important factors influencing the reading of an orifice meter are the size of the orifice and the diameter of the pipe in which it is fitted, though a number of other factors do affect the reading to some extent. Thus the exact position and the method of fixing the pressure tappings are important because the area of flow, and hence the velocity, gradually changes in the region of the orifice. The meter should be located not less than 50 pipe diameters from any pipe fittings. Details of the exact shape of the orifice, the orifice thickness, and other details are given in BS 1042[(4)] and the details must be followed if a

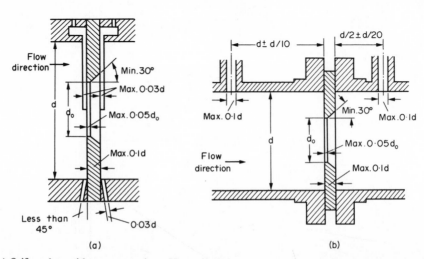

FIG. 5.11. (a) Orifice plate with corner tappings. Upper half shows construction with piezometer ring. Lower half shows construction with tappings into pipe flange. (b) Orifice plate with d and $d/2$ tappings. Nipples must finish flush with wall of pipe without burrs.

standard orifice is to be used without calibration—otherwise the meter must be calibrated. It should be noted that the standard applies only for pipes of at least 150 mm diameter.

A simple instrument can be made by inserting a drilled orifice plate between two pipe flanges and arranging suitable pressure connections. The orifice must be drilled with sharp edges and is best made from stainless steel to resist corrosion and abrasion. The size of the orifice should be chosen to give a convenient pressure drop. Since the flowrate is proportional to the square root of the pressure drop, accurate readings are given by this kind of meter, but it is difficult to cover a wide range in flow with any given size of orifice. Unlike the pitot tube, the orifice meter gives the average flowrate from a single reading.

The most serious disadvantage of the meter is that most of the pressure drop is not recoverable (i.e. it is inefficient). The velocity of the fluid is increased at the throat without much loss of energy. The fluid is subsequently retarded as it mixes with the relatively slow-moving fluid downstream from the orifice. A high degree of turbulence is set up and most of the excess kinetic energy is dissipated as heat. Usually only about 5 or 10 per cent of the excess kinetic energy can be recovered as pressure energy. The pressure drop over the orifice meter is therefore high, and this may preclude it from being used in a particular instance.

The area of flow decreases from A_1 at section 1 to A_0 at the orifice and then to A_2 at the vena contracta (Fig. 5.10). The area at the vena contracta can be conveniently related to the area of the orifice by the coefficient of contraction C_c, defined by the relation:

$$C_c = A_2/A_0$$

Inserting the value $A_2 = C_c A_0$ in equation 5.13 then for an *incompressible fluid* in a horizontal meter:

$$G = \frac{C_c A_0}{v} \sqrt{\frac{2\alpha_2 v(P_1 - P_2)}{1 - \frac{\alpha_2}{\alpha_1}\left(C_c \frac{A_0}{A_1}\right)^2}} \qquad (5.16)$$

Using a coefficient of discharge C_D to take account of the frictional losses in the meter and of the parameters C_c, α_1, and α_2:

$$G = \frac{C_D A_0}{v} \sqrt{\frac{2\rho(P_1 - P_2)}{1 - \left(\frac{A_0}{A_1}\right)^2}} \qquad (5.17)$$

For a meter in which the area of the orifice is small compared with that of the pipe:

$$[1 - (A_0/A_1)^2]^{1/2} \to 1$$

and

$$G = \frac{C_D A_0}{v} \sqrt{2v(P_1 - P_2)}$$

$$= C_D A_0 \sqrt{2\rho(P_1 - P_2)} \qquad (5.18)$$

$$= C_D A_0 \rho \sqrt{2gh_0} \qquad (5.19)$$

where h_0 is the difference in head across the orifice expressed in terms of the fluid in question.

This gives a simple working equation for evaluating G though the coefficient C_D is not a simple function and depends on the values of the Reynolds number in the orifice and the form of the pressure tappings. A value of 0·61 may be taken for the standard meter for Reynolds numbers in excess of 10^4, though the value changes noticeably at lower values of Reynolds number as shown in Fig. 5.12.

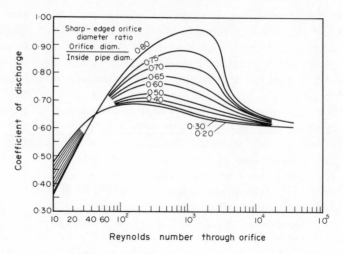

FIG. 5.12. Coefficient for orifice meter.

For the isothermal flow of an ideal gas, from equation 5.14 and using C_D as above:

$$G = \frac{C_D A_0}{v_2} \sqrt{\frac{2 P_1 v_1 \ln(P_1/P_2)}{1 - \left(\dfrac{v_1 A_0}{v_2 A_1}\right)^2}} \tag{5.20}$$

For a meter in which the area of the orifice is small compared with that of the pipe:

$$G = \frac{C_D A_0}{v_2} \sqrt{2 P_1 v_1 \ln (P_1/P_2)} \tag{5.21}$$

$$= C_D A_0 \sqrt{2 \frac{P_2}{v_2} \ln \frac{P_1}{P_2}}$$

$$= C_D A_0 \sqrt{2 \frac{P_2}{v_2} \ln\left(1 - \frac{\Delta P}{P_2}\right)} \quad \text{(where } \Delta P = P_2 - P_1)$$

$$= C_D A_0 \sqrt{2(-\Delta P / v_2)} \quad \text{if } \Delta P \text{ is small compared with } P_2 \tag{5.22}$$

For non-isothermal flow of an ideal gas, from equation 5.15:

$$G = \frac{C_D A_0}{v_2} \sqrt{\frac{2 P_1 v_1 [k/(k-1)][1 - (P_2/P_1)^{(k-1)/k}]}{1 - \left(\dfrac{v_1 A_0}{v_2 A_1}\right)^2}} \tag{5.23}$$

For a horizontal orifice in which (A_0/A_1) is small:

$$G = \frac{C_D A_0}{v_2} \sqrt{2 P_1 v_1 \frac{k}{k-1}\left[1 - \left(\frac{P_2}{P_1}\right)^{(k-1)/k}\right]} \tag{5.24}$$

$$= \frac{C_D A_0}{v_2} \sqrt{2 P_1 v_1 \frac{k}{k-1}\left[1 - \left(1 + \frac{\Delta P}{P_1}\right)^{(k-1)/k}\right]}$$

$$= C_D A_0 \sqrt{2(-\Delta P / v_2)} \quad \text{if } \Delta P \text{ is small compared with } P_2 \tag{5.25}$$

From equation 5.24:

$$G = A_0 C_D \left(\frac{P_2}{P_1}\right)^{1/k} \sqrt{2P_1 \frac{1}{v_1} \frac{k}{k-1} \left[1 - \left(\frac{P_2}{P_1}\right)^{(k-1)/k}\right]} \tag{5.26}$$

$$\therefore \quad G^2 = A_0^2 C_D^2 2P_1 \frac{1}{v_1} \frac{k}{k-1} (w^{2/k} - w^{(k+1)/k})$$

$$= k'(w^{2/k} - w^{(k+1)/k}) \tag{5.27}$$

where $w = (P_2/P_1)$ and k' is independent of P_2.

High Pressure Ratios

It is clear that, for a given upstream pressure P_1, the flow is zero when $w = 0$ or when $w = 1$. At some intermediate value of w the discharge is a maximum; this occurs when $(dG/dw) = 0$.

Differentiating equation 5.27 with respect to w:

$$2G \frac{dG}{dw} = k' \left(\frac{2}{k} w^{(2/k)-1} - \frac{k+1}{k} w^{1/k}\right)$$

When $(dG/dw) = 0$:

$$2w^{(2/k)-1} = (k+1)w^{1/k},$$

i.e.

$$w = \left(\frac{2}{k+1}\right)^{k/(k-1)} \tag{5.28}$$

$$= w_c \text{ (say)}$$

w_c is known as the *critical pressure ratio* (see Chapter 4).

From equation 5.15, the theoretical velocity u_2 at the vena contracta for turbulent flow ($\alpha = 1$) in a horizontal orifice in which A_2 is small compared with A_1 is given by:

$$u_2^2 = 2P_1 v_1 \frac{k}{k-1} \left[1 - \left(\frac{P_2}{P_1}\right)^{(k-1)/k}\right] \tag{5.29}$$

Inserting the critical value of (P_2/P_1) from equation 5.28:

$$u_2^2 = 2P_1 v_1 \frac{k}{k-1} \left(1 - \frac{2}{k+1}\right)$$

$$= \frac{2k}{k+1} P_1 v_1 \tag{5.30}$$

$$= \frac{2k}{k+1} \frac{P_2}{w_c} v_2 w_c^{1/k}$$

i.e.

$$u_2 = \sqrt{kP_2 v_2}$$

This has already been shown in Chapter 4 to be the velocity of propagation of a pressure wave (equation 4.36).

The maximum discharge is given by:

$$G = C_D A_0 u_2 / v_2$$

$$= C_D A_0 \sqrt{\frac{2k}{k+1} P_1 v_1 \frac{1}{v_1^2} \left(\frac{2}{k+1}\right)^{2/(k-1)}} \quad \text{(from equations 5.28 and 5.30)}$$

$$= C_D A_0 \sqrt{\frac{kP_1}{v_1} \left(\frac{2}{k+1}\right)^{(k+1)/(k-1)}} \,. \tag{5.31}$$

It should be noted that frictional losses have been neglected in the calculation of the theoretical velocity at the vena contracta. For an isentropic process, k is equal to γ in equation 5.28 and for air ($\gamma = 1\cdot4$), the critical ratio is then about $0\cdot5$. Thus, if the downstream pressure is reduced below about one-half of the upstream pressure, the rate of flow is given by equation 5.31 and is independent of the downstream pressure P_2. The pressure never falls below $w_c P_1$ at the vena contracta; it may fall from $w_c P_1$ to P_2 beyond the vena contracta.

Example 5.1. Water flows through an orifice of 25 mm diameter situated in a 75 mm diameter pipe, at a rate of 300 cm³/s. What will be the difference in level on a water manometer connected across the meter? (Take the viscosity of water as 1 mN s/m².)

Solution.

$$\text{Area of orifice} = (\pi/4) \times 25^2 = 491 \text{ mm}^2 \text{ or } 4\cdot91 \times 10^{-4} \text{ m}^2$$

$$\text{Flow of water} = 300 \text{ cm}^3/\text{s or } 3\cdot0 \times 10^{-4} \text{ m}^3/\text{s}$$

$\therefore$ Velocity of water through the orifice $= 3\cdot0 \times 10^{-4}/(4\cdot91 \times 10^{-4}) = 0\cdot61$ m/s

$$Re \text{ at orifice} = (25 \times 10^{-3} \times 0\cdot61 \times 1000)/1 \times 10^{-3} = 15250$$

From Fig. 5.12, the corresponding value of $C_D = 0\cdot67$ (diameter ratio $= 0\cdot33$):

$$[1 - (A_0/A_1)^2]^{0\cdot5} = [1 - (25^2/75^2)^2]^{0\cdot5} = 0\cdot994 \approx 1$$

Equation 5.19 can therefore be applied:

$$G = 3\cdot0 \times 10^{-4} \times 10^3 = 0\cdot30 \text{ kg/s}.$$

$\therefore$

$$0\cdot30 = 0\cdot67 \times 4\cdot91 \times 10^{-4} \times 10^3 \sqrt{(2 \times 9\cdot81 \times h_0)}.$$

Hence
$$\sqrt{h_0} = 0\cdot206$$

and
$$h_0 = 0\cdot042 \text{ m of water}$$

$$= \underline{\underline{42 \text{ mm of water}}}$$

Example 5.2. A cylinder contains air at a pressure of $6\cdot0$ MN/m² and discharges to atmosphere through a valve which may be taken as equivalent to a sharp-edged orifice of 6 mm diameter (coefficient of discharge $= 0\cdot6$). Plot the rate of discharge of air from the cylinder against cylinder pressure. Assume that the expansion at the valve is approximately isentropic and that the temperature of the air in the cylinder remains unchanged at 273 K.

Solution. Critical pressure ratio for discharge through the valve

$$= [2/(\gamma + 1)]^{[\gamma/(\gamma-1)]}$$

γ varies from $1\cdot40$ to $1\cdot54$ over the pressure range $0\cdot1$ to $6\cdot0$ MN/m². Critical conditions occur at a relatively low cylinder pressure. Taking $\gamma = 1\cdot40$:

$$\text{critical ratio} = (2/2\cdot40)^{1\cdot4/0\cdot4} = 0\cdot53$$

Sonic velocity will occur until the pressure in the cylinder falls to $101\cdot3/0\cdot53 = 191\cdot1$ kN/m².

Rate of discharge for cylinder pressures greater than $191\cdot1$ kN/m² is given by equation 5.31:

$$G = C_D A_0 \sqrt{(kP_1/v_1)[2/(k + 1)]^{(k+1)/(k-1)}}$$

For isentropic conditions, taking a mean value of γ as $1\cdot47$:

$$G = 0\cdot6(\pi/4)(6\times10^{-3})^2\sqrt{[(1\cdot47P_1/v_1)0\cdot810^{5\cdot26}]}$$

$$= 1\cdot697\times10^{-5}\times0\cdot693\sqrt{(P_1/v_1)}$$

$$= 1\cdot176\times10^{-5}(P_1/v_1)^{0\cdot5}$$

Now if P_a and v_a are atmospheric pressure and the specific volume of air at atmospheric pressure:

$$1/v_1 = P_1/(v_aP_a)$$

But $v_a = 22\cdot4/28\cdot9 = 0\cdot775\ \text{m}^3/\text{kg}, \quad P_a = 101\cdot3\ \text{kN/m}^2 \quad \text{or}\quad 101\,300\ \text{N/m}^2$

$\therefore$ $v_1 = 0\cdot775\times101\,300/P_1 = 78\,508/P_1\ \text{m}^3/\text{kg}$

$$G = 1\cdot176\times10^{-5}P_1/280\cdot2 = 4\cdot197\times10^{-8}P_1\ \text{kg/s}$$

or $4\cdot197\times10^{-2}P_{1a}\ \text{kg/s}$

where P_{1a} is pressure P_1, expressed in MN/m^2.

   When the pressure in the cylinder is less than $191\cdot1\ \text{kN/m}^2$, the rate of flow is found from equation 5.24.

Hence $$G^2 = (A_0C_D/v_2)^2 2P_1v_1[k/(k-1)][1-(P_2/P_1)^{(k-1)/k}]$$

For isentropic flow through the value, $k = \gamma$, and putting $\gamma = 1\cdot40$:

$$(P_1v_1/v_2^2) = (P_1v_1/v_1^2)(P_2/P_1)^{2/\gamma} = (P_1^2/v_aP_a)(P_a/P_1)^{2/\gamma} = P_1^{2-(2/\gamma)}P_a^{(2/\gamma)-1}v_a^{-1}$$

since $P_2 = P_a = 101\,300\ \text{N/m}^2$

Thus $G^2 = (A_0C_D)^2(1/v_a)P_a^{(2/\gamma)-1}[2\gamma/(\gamma-1)][1-(P_a/P_1)^{(\gamma-1)/\gamma}]P_1^{2-(2/\gamma)}$

$$= (1\cdot697\times10^{-5})^2(1/0\cdot775)(101\,300)^{0\cdot43}(2\cdot80/0\cdot4)(1-(P_{1a}/0\cdot101)^{-0\cdot29})P_1^{0\cdot58}$$

$$G = 0\cdot0335P_{1a}^{0\cdot29}(1-0\cdot675P_{1a}^{-0\cdot29})^{0\cdot5}\ \text{kg/s}$$

The discharge rate is plotted as a function of pressure in Fig. 5.13.

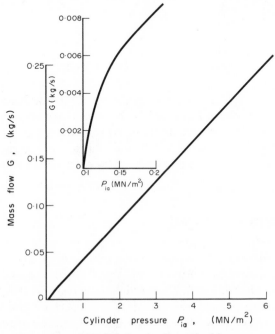

FIG. 5.13. Solution for Example 5.2.

Solution to G vs P_{1a} data

(i) Above $P_{1a} = 0.19 \, MN/m^2$		(ii) Below $P_{1a} = 0.19 \, MN/m^2$	
P_{1a} (MN/m²)	G (kg/s)	P_{1a} (MN/m²)	G (kg/s)
0·2	0·0084	0·10	0
0·5	0·021	0·125	0·0044
1·0	0·042	0·15	0·0064
2·0	0·084	0·17	0·0074
3·0	0·126	0·19	0·0085
4·0	0·168		
5·0	0·210		
6·0	0·252		

5.3.4. The Nozzle

This is similar to the orifice meter but has a converging tube in place of the orifice plate, as shown in Figs. 5.10c and 5.14. The velocity of the fluid is gradually increased and the contours are so designed that almost frictionless flow takes place in the converging portion; the outlet corresponds to the vena contracta on the orifice meter. The nozzle has a constant high coefficient of discharge (c. 0·99) over a wide range of conditions because the coefficient

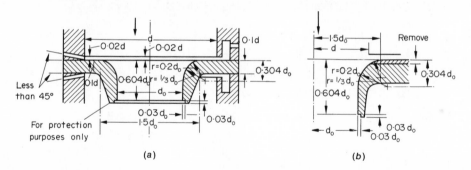

FIG. 5.14. (a) Standard nozzle (A_0/A_1) is less than 0.45. Left half shows construction for corner tappings. Right half shows construction for piezometer ring. (b) Standard nozzle where A_0/A_1 is greater than 0.45.

of contraction is unity, though because the simple nozzle is not fitted with a diverging cone, the head lost is very nearly the same as with an orifice. Although much more costly than the orifice meter, it is extensively used for metering steam. When the ratio of the pressure at the nozzle exit to the upstream pressure is less than the critical pressure ratio w_c, the flowrate is independent of the downstream pressure and can be calculated from the upstream pressure alone.

5.3.5. The Venturi Meter (Fig. 5.10b)

In this meter the fluid is accelerated by its passage through a converging cone of angle 15–20°. The pressure difference between the upstream end of the cone (section 1) and the throat (section 2) is measured and provides the signal for the rate of flow. The fluid is then retarded in a cone of smaller angle (5–7°) in which a large proportion of the kinetic energy is converted back to pressure energy. Because of the gradual reduction in the area of flow there is no vena contracta and the flow area is a minimum at the throat so that the coefficient of contraction is unity. The attraction of this meter lies in its high energy recovery so that it may be used where only a small pressure head is available, though its construction[4] is expensive. The flow relationship is given by a similar equation to that for the orifice.

For an incompressible fluid in horizontal meter:

$$G = \frac{C_D A_2}{v} \sqrt{\frac{2v(P_1 - P_2)}{1 - (A_2/A_1)^2}} \qquad \text{(equation 5.17)}$$

$$= C_D \rho \frac{A_1 A_2}{\sqrt{A_1^2 - A_2^2}} \sqrt{2v(P_1 - P_2)} \qquad (5.32)$$

$$= C_D \rho C' \sqrt{2gh_v} \qquad (5.33)$$

where C' is a constant for the meter and h_v is the loss in head over the converging cone expressed as height of fluid. The coefficient C_D is high, varying from 0·98 to 0·99. The meter is equally suitable for compressible and incompressible fluids.

Example 5.3. The rate of flow of water in a 150 mm diameter pipe is measured with a venturi meter with a 50 mm diameter throat. When the pressure drop over the converging section is 121 mm of water, the flowrate is 3·0 kg/s. What is the coefficient for the converging cone of the meter at this flowrate?

Solution. Mass rate of flow:

$$G = C_D \rho A_1 A_2 \sqrt{(2gh_v)} / \sqrt{(A_1^2 - A_2^2)} \qquad \text{(from equation 5.32)}$$

The coefficient for the meter is therefore given by:

$$3 \cdot 0 = C_D \times 1000 (\pi/4)^2 (121 \times 10^{-3})^2 (50 \times 10^{-3})^2$$
$$\times \sqrt{(2 \times 9 \cdot 81 \times 121 \times 10^{-3})} / \{(\pi/4) \sqrt{[(150 \times 10^{-3})^4 - (50 \times 10^{-3})^4]}\}$$
$$= C_D \times 1000 \times 0 \cdot 00204 \times 1 \cdot 63$$

$$\therefore \qquad C_D = \underline{\underline{0 \cdot 985}}$$

5.3.6. Pressure Recovery in Orifice-type Meters

The orifice meter has a low discharge coefficient of about 0·6 which means that there is a substantial loss in pressure over the instrument. This is shown in Fig. 5.15 where the pressure is shown in relation to the position along the tube. There is a steep drop in pressure of about 10 per cent as the vena contracta is reached and the subsequent pressure recovery is poor, amounting to about 50 per cent of the loss. The *Dall Tube* (Fig. 5.16) has been developed to provide a low head loss over the meter while still giving a reasonably high value of pressure head over the orifice and thus offering great advantages. A short length of parallel lead-in tube is followed by the converging upstream cone and then a diverging downstream cone. This recovery cone is formed by a liner which fits into the meter, the *throat* being formed by a circular slot located between the two cones. One pressure connection is taken to the throat through the annular chamber shown, and the second tapping is on the upstream side. The flow leaves the throat as a diverging jet which follows the walls of the downstream cone so that eddy losses are almost eliminated. These instruments are made for pipe sizes greater than 150 mm and the pressure loss is only 2–8 percent of the differential head. They are cheaper than a venturi, much shorter in length, and correspondingly lighter.

It should be noted that all these restricted flow type meters are primarily intended for pipe sizes greater than 50 mm, and when used on smaller tubes they must be individually calibrated.

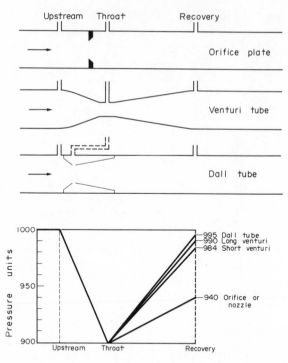

FIG. 5.15. Pressure distribution using orifice plate, venturi meter, and Dall tube. Pressure falls by 10% from upper pressure tapping to throat in each case.

Figure 5.15 shows the pressure changes over the four instruments considered. If the upstream pressure is P_1, the throat pressure P_2, and the final recovery pressure P_3, then the pressure recovery is conveniently expressed as $(P_1 - P_3)/(P_1 - P_2)$, and for these meters the values are given in Table 5.1.

TABLE 5.1. *Typical meter pressure recovery*

Meter type	Value of C_D	Value of $(P_1 - P_3)/(P_1 - P_2)$ (%)
Orifice	0·6	40–50
Nozzle	0·7	40–50
Venturi	0·9	80–90
Dall	0·9	92–98

Because of its simplicity the orifice meter is commonly used for process measurements, and this instrument is suitable for providing a signal of the pressure to some comparator as indicated in Fig. 5.1.

5.3.7. Variable Area Meters—Rotameters

In the meters so far described the area of the constriction or orifice is constant and the drop in pressure is dependent on the rate of flow. In the variable area meter the drop in pressure is constant and the flowrate is a function of the area of the constriction.

A typical meter of this kind, which is commonly known as a *rotameter* (Fig. 5.17), consists of a tapered tube with the smallest diameter at the bottom. The tube contains a freely moving float which rests on a stop at the base of the tube. When the fluid is flowing the float rises until

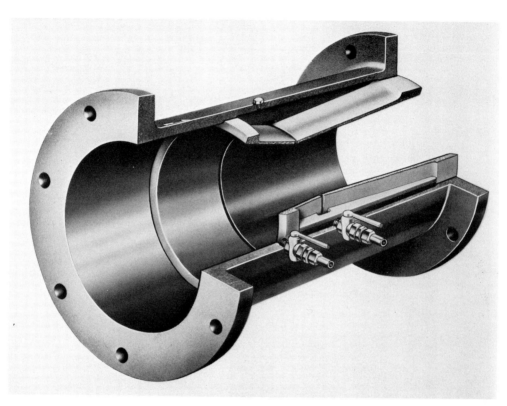
FIG. 5.16. Dall tube.

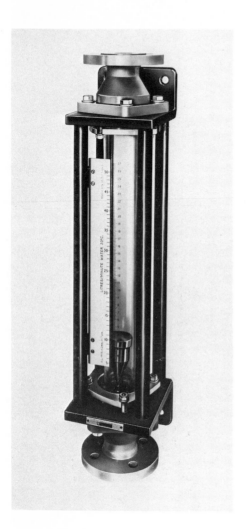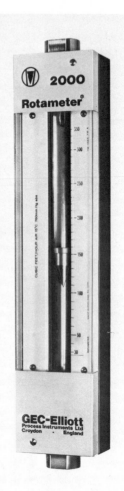

FIG. 5.17. Rotameters.

its weight is balanced by the upthrust of the fluid, its position then indicating the rate of flow. The pressure difference across the float is equal to its weight divided by its maximum cross-sectional area in a horizontal plane. The area for flow is the annulus formed between the float and the wall of the tube.

This meter can thus be considered as an orifice meter with a variable aperture, and the formulae already derived are therefore applicable with only minor changes. Both in the orifice-type meter and in the rotameter the pressure drop arises from the conversion of pressure energy to kinetic energy and from frictional losses which are accounted for in the coefficient of discharge. The pressure difference over the float $-\Delta P$, is given by:

$$-\Delta P = V_f(\rho_f - \rho)g/A_f \tag{5.34}$$

where V_f is the volume of the float, ρ_f the density of the material of the float, and A_f is the maximum cross-sectional area of the float in a horizontal plane.

If the area of the annulus between the float and tube is A_2 and the cross-sectional area of the tube is A_1, then from equation 5.17:

$$G = C_D A_2 \sqrt{\frac{2\rho(-\Delta P)}{1-(A_2/A_1)^2}} \tag{5.35}$$

Substituting for $-\Delta P$ from equation 5.34:

$$G = C_D A_2 \sqrt{\frac{2gV_f(\rho_f - \rho)\rho}{A_f[1-(A_2/A_1)^2]}} \tag{5.36}$$

The coefficient C_D depends on the shape of the float and the Reynolds number (based on the velocity in the annulus and the mean hydraulic diameter of the annulus) for the flow through the annular space of area A_2. In general, floats which give the most nearly constant coefficient are of such a shape that they set up eddy currents and give low values of C_D. The variation in C_D largely arises from variation in viscous drag of fluid on the float, and if turbulence is artificially increased the drag force rises quickly to a limiting but high value. As seen from Fig. 5.18, float A does not promote turbulence and the coefficient rises slowly to a high value of 0·98. Float C promotes turbulence and C_D rises quickly but only to a low value of 0·60.

The constant coefficient for float C arises from turbulence promotion, and for this reason the coefficient is also substantially independent of the fluid viscosity. The meter can be made relatively insensitive to changes in the density of the fluid by selection of ρ_f. Thus the flowrate for a given meter will be independent of ρ when $dG/d\rho = 0$.

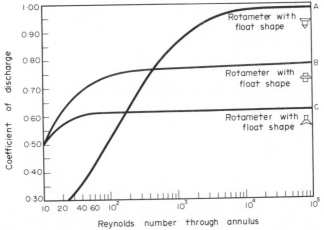

FIG. 5.18. Coefficient for rotameter.

From equation 5.36:

$$\frac{dG}{d\rho} = \frac{C_D A_2}{2} \sqrt{\frac{2gV_f}{A_f[1-(A_2/A_1)^2]}} \frac{(\rho_f - 2\rho)}{\sqrt{(\rho_f - \rho)\rho}} \tag{5.37}$$

When $dG/d\rho = 0,$

i.e. $\rho_f = 2\rho \tag{5.38}$

Thus if the density of the float is twice that of the fluid, then the position of the float for a given flow is independent of the fluid density.

The range of a meter can be increased by the use of floats of different densities, a given float covering a flowrate range of about 10:1. For high pressure work the glass tube is replaced by a metal tube. When a metal tube is used or when the liquid is very dark or dirty an external indicator is required.

Example 5.4. A rotameter tube is 0·3 m long with an internal diameter of 25 mm at the top and 20 mm at the bottom. The diameter of the float is 20 mm, its specific gravity is 4·80 and its volume is 6·0 cm³. If the coefficient of discharge is 0·7, what will be the flowrate of water when the float is halfway up the tube?

$$\rho_{\text{water}} = 1000 \text{ kg/m}^3$$

Solution. From equation 5.36:

$$G = C_D A_2 \sqrt{\frac{2gV_f(\rho_f - \rho)\rho}{A_f[1-(A_2/A_1)^2]}}$$

Cross-sectional area at top of tube $= (\pi/4)(25^2) = 491 \text{ mm}^2$ or $4\cdot91 \times 10^{-4} \text{ m}^2$

Cross-sectional area at bottom of tube $= (\pi/4)(20^2) = 314 \text{ mm}^2$ or $3\cdot14 \times 10^{-4} \text{ m}^2$

Area of float: $A_f = 3\cdot14 \times 10^{-4} \text{ m}^2$

Volume of float: $V_f = 6\cdot0 \times 10^{-6} \text{ m}^3$

When the float is halfway up the tube, the area at the height of the float A_1 is given by:

$$0\cdot5 = (A_1 - 0\cdot000314)/(0\cdot000506 - 0\cdot000314)$$

or $A_1 = 4\cdot10 \times 10^{-4} \text{ m}^2$

The area of the annulus A_2 is given by:

$$A_2 = A_1 - A_f = 0\cdot96 \times 10^{-4} \text{ m}^2$$

$\therefore$ $A_2/A_1 = 0\cdot234$ and $[1-(A_2/A_1)^2] = 0\cdot945$

Substituting into equation 5.36:

$$G = 0\cdot70 \times 0\cdot96 \times 10^{-4} \sqrt{\{[2 \times 9\cdot81 \times 6\cdot0 \times 10^{-6}(4800 - 1000)1000]/(3\cdot14 \times 10^{-4} \times 0\cdot945)\}}$$
$$= \underline{\underline{0\cdot082 \text{ kg/s}}}$$

5.3.8.   The Notch or Weir

The flow of a liquid presenting a free surface can be measured by means of a weir. The pressure energy of the liquid is converted into kinetic energy as it flows over the weir which may or may not cover the full width of the stream, and a calming screen may be fitted before the weir. Then the height of the weir crest gives a measure of the rate of flow. The velocity with which the liquid leaves depends on its initial depth below the surface. For unit mass of liquid, initially at a depth h below the free surface, discharging through a notch, the energy balance equation (2.67) gives:

$$\Delta u^2/2 + v\Delta P = 0 \qquad (5.39)$$

for a frictionless flow under turbulent conditions. If the velocity upstream from the notch is small and the flow upstream is assumed to be unaffected by the change of area at the notch:

$$u_2^2/2 = -v\Delta P = gh$$

i.e. $$u_2 = \sqrt{2gh} \qquad (5.40)$$

Rectangular Notch

For a rectangular notch (Fig. 5.19) the rate of discharge of fluid at a depth h through an element of cross-section of depth dh will be given by:

$$dQ = C_D B\, dh\sqrt{2gh} \qquad (5.41)$$

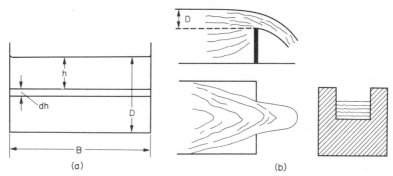

(a)   (b)

FIG. 5.19. Rectangular notch.

where C_D is the coefficient of discharge (usually about 0·6) and B is the width of the notch.
Thus the total rate of flow:

$$Q = C_D B\sqrt{2g}\,(2/3)D^{3/2} \qquad (5.42)$$

where D is the total depth of liquid above the bottom of the notch.
The empirical Francis formula:

$$Q = 1\cdot84(B - 0\cdot1nD)D^{1\cdot5} \qquad (5.43)$$

gives the flowrate in m³/s if the dimensions of the notch are expressed in metres with $C_D = 0\cdot62$; $n = 0$ if the notch is the full width of the channel; $n = 1$ if the notch is narrower than the channel but is arranged with one edge coincident with the edge of the channel; $n = 2$ if the notch is narrower than the channel and is situated symmetrically (see Fig. 5.19b): n is known as the number of end contractions.

Example 5.5. Water flows in an open channel across a weir which occupies the full width of the channel. The length of the weir is 0·5 m and the height of water over the weir is 100 mm. What is the volumetric flowrate of water?

Solution. Use is made of the Francis formula (equation 5.43):

$$Q = 1\cdot84(L - 0\cdot1nD)D^{1\cdot5} \quad \text{(m}^3\text{/s)} \qquad (5.43)$$

where L is the length of the weir (m), n the number of end contractions (in this case $n = 0$), and D the height of liquid above the weir (m).

Thus
$$Q = 1 \cdot 84 (0 \cdot 5) 0 \cdot 100^{1 \cdot 5}$$
$$= \underline{0 \cdot 030 \text{ m}^3/\text{s}}$$

Example 5.6. An organic liquid flows across a distillation tray and over a weir at the rate of 15 kg/s. The weir is 2 m long and the liquid density is 650 kg/m³. What is the height of liquid flowing over the weir?

Solution. Use is made of the Francis formula (equation 5.43), where, as in the previous example, $n = 0$. In the context of this example the height of liquid flowing over the weir is usually designated h_{ow} and the volumetric liquid flow by Q. Rearrangement of equation 5.43 gives:
$$h_{ow} = 0 \cdot 666 (Q/L_w)^{0 \cdot 67} \quad \text{(m)}$$

where Q is the liquid flowrate (m³/s) and L_w the weir length (m).

In this case:
$$Q = (15/650) = 0 \cdot 0230 \text{ m}^3/\text{s and } L_w = 2 \cdot 0 \text{ m}$$
$$\therefore \qquad h_{ow} = 0 \cdot 666 (0 \cdot 0230/2)^{0 \cdot 67} = 0 \cdot 033 \text{ m}$$
or
$$\underline{\underline{33 \cdot 0 \text{ mm}}}$$

Triangular Notch

For a triangular notch of angle 2θ, the flow dQ through the thin element of cross-section (Fig. 5.20) is given by:

$$dQ = C_D (D - h) 2 \tan \theta \, dh \sqrt{2gh} \qquad (5.44)$$

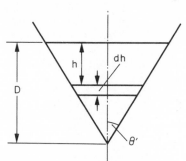

FIG. 5.20. Triangular notch.

The total rate of flow:

$$Q = 2 C_D \tan \theta \sqrt{2g} \int_0^D (Dh^{1/2} - h^{3/2}) dh$$

$$= 2 C_D \tan \theta \sqrt{2g} (\tfrac{2}{3} D^{5/2} - \tfrac{2}{5} D^{5/2})$$

$$= (8/15) C_D \tan \theta \sqrt{2g} \, D^{5/2} \qquad (5.45)$$

For a 90° notch for which $C_D = 0 \cdot 6$, and using SI units:

$$Q = 1 \cdot 42 D^{2 \cdot 5} \qquad (5.46)$$

Thus for a rectangular notch the rate of discharge is proportional to the liquid depth raised to a power of 1·5, and for a triangular notch to a power of 2·5. A triangular notch will, therefore, handle a wider range of flowrates. In general the index of D is a function of the contours of the walls of the notch, and any desired relation between flowrate and depth can be obtained by a suitable choice of contour. It is sometimes convenient to employ a notch in which the rate of discharge is directly proportional to the depth of liquid above the bottom of the notch. It can be shown that the notch must have curved walls giving a large width to the bottom of the notch and a comparatively small width towards the top.

While open channels are not used frequently for the flow of material other than cooling water between plant units, a weir is frequently installed for controlling the flow within the unit itself, for instance in a distillation column or reactor.

5.3.9. Other Methods of Measuring Flow Rates

The meters which have been described so far depend for their operation on the conversion of some of the kinetic energy of the fluid into pressure energy or vice versa. They form by far the largest class of flowmeters. Other meters are used for special purposes, and brief reference will now be made to a few of these.

Hot-wire Anemometer

If a heated wire is immersed in a fluid, the rate of loss of heat will be a function of the flowrate. In the hot-wire anemometer a fine wire whose electrical resistance has a high temperature coefficient is heated electrically. Under equilibrium conditions the rate of loss of heat is then proportional to $I^2\Omega$, where Ω is the resistance of the wire and I is the current flowing.

Either the current or the resistance (and hence the temperature) of the wire is maintained constant. The following is an example of a method in which the resistance is maintained constant. The wire is incorporated as one of the resistances of a Wheatstone network (Fig. 5.21) in which the other three resistances have low temperature coefficients. The circuit is

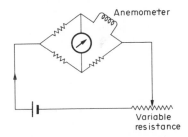

FIG. 5.21. Typical circuit for a hot-wire anemometer.

balanced when the wire is immersed in the stationary fluid but, when the fluid is set in motion, the rate of loss of heat increases and the temperature of the wire falls. Its resistance therefore changes and the bridge is thrown out of balance. The balance can be restored by increasing the current so that the temperature and resistance of the wire are brought back to their original values; the other three resistances will be unaffected by the change in current because of their low temperature coefficients. The current flowing in the wire is then measured using either an ammeter or a voltmeter. The rate of loss of heat is found to be approximately proportional to $\sqrt{u\rho + b'}$, so that

$$a'\sqrt{u\rho + b'} = I^2\Omega \quad \text{under equilibrium conditions} \tag{5.47}$$

where u is the velocity of the fluid (m/s), ρ the density (kg/m^3), and a' and b' are constant for a given meter.

Since the resistance of the wire is maintained constant:

$$u\rho = I^4\Omega^2/a'^2 - b' = a''I^4 - b' \tag{5.48}$$

where $a'' = \Omega^2/a'^2$ remains constant, i.e. the mass rate of flow per unit area is a function of the fourth power of the current, which can be accurately measured.

The hot-wire anemometer is very accurate even for very low rates of flow. It is one of the most convenient instruments for the measurement of the flow of gases at low velocities; accurate readings are obtained for velocities down to about 0·03 m/s. If the ammeter has a high natural frequency, pulsating flows can be measured. Platinum wire is commonly used.

The Magnetic Meter

The principle of the magnetic meter is shown in Fig. 5.22. In this instrument the fluid flows through a magnetic field and the voltage generated is then a direct measure of the rate of flow.

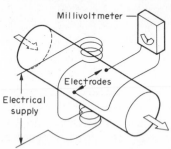

FIG. 5.22. Magnetic meter.

These meters have been very useful for handling liquid metal coolants where the conductivity is high though they have now been developed for liquids such as tap water which have a poor conductivity. Designs are now available for all sizes of tubes, and the instrument is particularly suitable for pharmaceutical and biological purposes where contamination of the fluid must be avoided.

This meter has a number of special features which should be noted. Thus it gives rise to a negligible drop in pressure for the fluid being metered; fluids containing a high percentage of solids can be metered and the flow can be in either direction. On the other hand, it requires a high frequency a.c. supply (200 Hz) and the walls must be made of a non-conducting material such as PTFE.

Quantity Meters

The meters which have been described so far give an indication of the rate of flow of fluid; the total amount passing in a given time must be obtained by integration. Orifice meters are frequently fitted with integrating devices. A number of instruments is available, however, for measuring directly the total quantity of fluid which has passed. An average rate of flow can then be obtained by dividing the quantity by the time of passage.

Gas Meters

A simple quantity meter which is used for the measurement of the flow of gas in an accessible duct is the anemometer (Fig. 5.23). A shaft carrying radial vanes or cups supported in low friction bearings is caused to rotate by the passage of the gas; the relative velocity between the gas stream and the surface of the vanes is low because the frictional resistance of the shaft is small. The number of revolutions of the spindle is counted automatically, using a gear train connected to a series of dials. The meter must be calibrated and should be checked at frequent intervals because the friction of the bearings will not necessarily remain constant. The anemometer is useful for gas velocities above about 0·15 m/s.

FIG. 5.23. Vane anemometer.

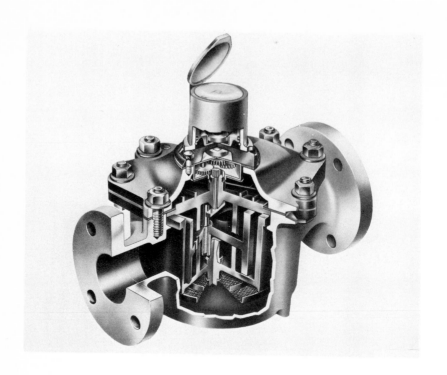

FIG. 5.25. Water meter.

Quantity meters, suitable for the measurement of the flow of gas through a pipe, include the standard wet and dry meters. In the wet meter, the gas fills a rotating segment and an equal volume of gas is expelled from another segment (Fig. 5.24). The dry gas meter employs a pair of bellows. Gas enters one of the bellows and automatically expels gas from the other; the

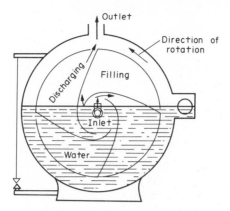

FIG. 5.24. Wet gas meter.

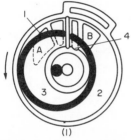

(1)

Spaces 1 and 3 are receiving fluid from the inlet port A, and spaces 2 and 4 are discharging through the outlet port B

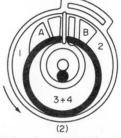

(2)

The piston has advanced and space 1, in connection with the inlet port, has enlarged, and space 2, in connection with the outlet port, has decreased, while spaces 3 and 4, which have combined, are about to move into position to discharge through the outlet port

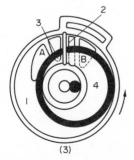

(3)

Space 1 is still admitting fluid from the inlet port and space 3 is just opening up again to the inlet port, while spaces 2 and 4 are discharging through the outlet port

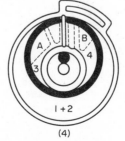

(4)

Fluid is being received into space 3 and discharged from space 4, while spaces 1 and 2 have combined and are about to begin discharging as the piston moves forward again to occupy position as shown in (1)

FIG. 5.26. The oscillating piston meter.

number of cycles is counted and recorded on a series of dials. Both of these are positive displacement meters and therefore do not need frequent calibration. Gas meters usually appear very bulky for the quantities they are measuring. This is because the linear velocity of a gas in a pipe is normally very high compared with that of a liquid, and the large volume is needed so that the speed of the moving parts can be reduced and wear minimised.

Liquid Meters

In one form of meter shown in Fig. 5.25, which is capable of handling flows between 5 cm³/s and 0·015 m³/s, the flow of the liquid results in the positive displacement of a rotating element, the cumulative flow being obtained by gearing to a counter.

The body in the form of a cylindrical chamber is fitted with a radial partition and a central hub. The circular piston has a slot gap in the circumference to fit over the partition and a peg on its upper face to control movement around the central hub. A rolling seal is thus formed between the piston and hub and between the piston and main chamber. The chamber is thus split into four spaces as shown in Fig. 5.26. The fluid enters the bottom of the chamber through one port and leaves by the other port, the piston forming a movable division between the inlet and outlet.

In a cycle of operation the liquid enters port *A* and fills the spaces 1 and 3, thus forcing the piston to oscillate counterclockwise opening spaces 2 and 4 to port *B*. Because of the partition, the piston moves downwards so that space 3 is cut off from port *A* and becomes space 4. Further movement allows the exit port to be uncovered, and the measured volume between hub and piston is then discharged. The outer space 1 increases until the piston moves upwards over the partition and space 1 becomes space 2 when a second metered volume is discharged by the filling of the inner space 3.

Further details of these and other measuring equipment are given in Ower[2] and Linford[5] in the *Process Industries and Controls Handbook*[6] and in the *Instrument Manual*.[7]

5.4. FURTHER READING

MILLER, J. T.: *The Revised Course on Industrial Instrument Technology* (United Trade Press, 1973).
MILLER, J. T. (ed.): *Instrument Manual*, 4th edn. (United Trade Press, 1971).
WIGHTMAN, E. J.: *Instrumentation in Process Control* (Butterworths, London 1972).

5.5. REFERENCES

1. COULSON, J. H. and RICHARDSON, J. F.: *Chemical Engineering*, vol. 3 (Pergamon Press, Oxford 1971).
2. OWER, E.: *Measurement of Air Flow* (Chapman & Hall, 1949).
3. SMITH, B. D.: *Design of Equilibrium Stage Processes* (McGraw-Hill, New York, 1963).
4. British Standard 1042: *Code for Flow Measurement* (1943).
5. LINFORD, A.: *Flow Measurement and Meters* (Spon, 1949).
6. CONSIDINE, D. M.: *Process Industries and Controls Handbook* (McGraw-Hill, 1957).
7. MILLER, J. T. (ed.): *Instrument Manual*, 4th edn. (United Trade Press, 1971).

5.6. NOMENCLATURE

		Units in SI System	Dimensions in M, L, T, θ, I
A	Area perpendicular to direction of flow	m²	L^2
A_f	Area of rotameter float	m²	L^2
A_0	Area of orifice	m²	L^2
B	Width of rectangular channel or notch	m	L

		Units in SI System	Dimensions in $\mathbf{M, L, T, \theta, I}$
C_c	Coefficient of contraction	—	—
C_D	Coefficient of discharge	—	—
C'	Constant for venturi meter	m^2	$\mathbf{L^2}$
D	Depth of liquid or above bottom of notch	m	$\mathbf{L}$
e	Error		
F	Energy dissipated per unit mass of fluid	J/kg	$\mathbf{L^2T^{-2}}$
G	Mass rate of flow	kg/s	$\mathbf{MT^{-1}}$
g	Acceleration due to gravity	m/s^2	$\mathbf{LT^{-2}}$
h_i	Difference between impact and static heads on pitot tube	m	$\mathbf{L}$
h_m	Reading on manometer	m	$\mathbf{L}$
h_0	Fall in head over orifice meter	m	$\mathbf{L}$
h_v	Fall in head over converging cone of venturi meter	m	$\mathbf{L}$
I	Electric current	A	$\mathbf{I}$
k	Numerical constant used as index for compression	—	—
L	Length of weir	m	$\mathbf{L}$
M	Molecular weight	kg/kmol	—
n	Number of end contractions	—	—
P	Pressure	N/m^2	$\mathbf{ML^{-1}T^{-2}}$
ΔP	Pressure difference	N/m^2	$\mathbf{ML^{-1}T^{-2}}$
Q	Volumetric rate of flow	m^3/s	$\mathbf{L^3T^{-1}}$
$\mathbf{R}$	Universal gas constant	8·314 kJ/kmol K	$\mathbf{L^2T^{-2}\theta^{-1}}$
T	Absolute temperature	K	$\boldsymbol{\theta}$
u	Mean velocity	m/s	$\mathbf{LT^{-1}}$
V_f	Volume of rotameter float	m^3	$\mathbf{L^3}$
v	Volume per unit mass of fluid	m^3/kg	$\mathbf{M^{-1}L^3}$
W_s	Shaft work per unit mass	J/kg	$\mathbf{L^2T^{-2}}$
w	Pressure ratio P_2/P_1	—	—
w_c	Critical pressure ratio	—	—
z	Distance in vertical direction	m	$\mathbf{L}$
z_m	Vertical distance between level of manometer liquid and axis of venturi meter	m	$\mathbf{L}$
α	Constant in expression for kinetic energy of fluid	—	—
γ	Ratio of specific heats at constant pressure and volume	—	—
ρ	Density of fluid	kg/m^3	$\mathbf{ML^{-3}}$
ρ_m	Density of manometer fluid	kg/m^3	$\mathbf{ML^{-3}}$
ρ_f	Density of rotameter float	kg/m^3	$\mathbf{ML^{-3}}$
θ	Half angle of triangular notch	—	—
Ω	Electrical resistance	ohm	$\mathbf{ML^2T^{-3}I^{-2}}$

CHAPTER 6

Pumping of Fluids

6.1. INTRODUCTION

For the pumping of liquids or gases from one vessel to another or through long pipes, some form of mechanical pump is usually employed. The energy required by the pump will depend on the height through which the fluid is raised, the pressure required on delivery, the length and diameter of the pipe, the rate of flow, together with the physical properties of the fluid, particularly its viscosity and density. The pumping of liquids such as sulphuric acid or petroleum products like benzene or naphtha from bulk store to process buildings, or the pumping of fluids round reaction units and through heat exchangers, are typical illustrations of the use of pumps. The pumping of crude oil or natural gas over very long distances, or of fluids from one section of a works to another, are further examples. On the one hand, it may be necessary to inject reactants or catalyst into a reactor at a low but accurately controlled rate and on the other to pump cooling water to a power station or refinery at a very high rate. The fluid may be a gas or liquid of low viscosity, or it may be a highly viscous liquid, possibly with non-Newtonian characteristics. It may be clear or it may contain suspended particles and be very corrosive. All these factors influence the choice of pump.

Because of the wide variety of requirements, many different types are in use including centrifugal, piston, gear, screw, and peristaltic pumps, though in the chemical and petroleum industries the centrifugal type is much the most important and is discussed in some detail later in the chapter.

Understanding the criteria for pump selection, determination of size and power requirements, assessing the positioning of pumps in relation to pipe systems form the main features of this chapter, though for greater detail the reader is referred to specialist publications included in section 6.8.

Pump design and construction is a specialist field, and manufacturers should always be consulted. In general, pumps used for circulating gases work at higher speeds than those used for liquids, and lighter valves are used. Moreover, the clearances between moving parts are smaller on gas pumps because of the much lower viscosity of gases, giving rise to an increased tendency for leakage to occur. When a pump is used to provide a vacuum, it is even more important to guard against leakage.

The work done by the pump is found by setting up an energy balance equation. If W_s is the shaft work done by unit mass of fluid on the surroundings, then $-W_s$ is the shaft work done on the fluid by the pump. From equations developed in Chapter 2:

from equation 2.55:
$$-W_s = (\Delta u^2/2\alpha) + g\,\Delta z + \int_1^2 v\,dP + F \tag{6.1}$$

and from equation 2.56:
$$-W_s = (\Delta u^2/2\alpha) + g\,\Delta z + \Delta H - q \tag{6.2}$$

In any practical system, the pump will not be 100 per cent efficient, and more energy must be supplied by the motor driving the pump than is given by $-W_s$. If liquids are considered to be incompressible, there is no change in specific volume from the inlet to the delivery side of the pump. The physical properties of gases are, however, considerably influenced by the pressure, and the work done in raising the pressure of a gas is influenced by the rate of heat

124

flow between the gas and the surroundings. Thus if the process is carried out adiabatically all the energy added to the system appears in the gas and its temperature rises. If an ideal gas is compressed and then cooled to its initial temperature, its enthalpy will be unchanged and the whole of the energy supplied by the compressor is dissipated to the surroundings. However, if the compressed gas is allowed to expand it will absorb heat and is therefore capable of doing work at the expense of heat energy from the surroundings.

6.2. PUMPING EQUIPMENT FOR LIQUIDS

As already indicated, the liquids used in the chemical industries differ considerably in physical and chemical properties, and it has been necessary to develop a wide variety of pumping equipment. The two main forms are the positive displacement type and the centrifugal pumps. In the former, the volume of liquid delivered is directly related to the displacement of the piston element and therefore increases directly with speed and is not appreciably influenced by the pressure. In this group are the reciprocating piston pump and the rotary gear pump, both of which are commonly used for delivery against high pressures and where nearly constant delivery rates are required. The centrifugal type depends on giving the liquid a high kinetic energy which is then converted as efficiently as possible into pressure energy. For some applications such as the handling of liquids which are particularly corrosive or contain abrasive solids in suspension, compressed air is used as the motive force instead of a mechanical pump. An illustration of the use of this form of equipment is the blowing of the contents of a reaction mixture from one vessel to another.

The following factors influence the choice of pump for a particular operation.

(1) The quantity of liquid to be handled. This primarily affects the size of the pump and determines whether it is desirable to use a number of pumps in parallel.
(2) The head against which the liquid is to be pumped. This will be determined by the difference in pressure and vertical height of the downstream and upstream reservoirs and by the frictional losses which occur in the delivery line. The suitability of a centrifugal pump and the number of stages required will largely be determined by this factor.
(3) The nature of the liquid to be pumped. For a given throughput the viscosity largely determines the friction losses and hence the power required. The corrosive nature will determine the material of construction both for the pump and the packing. With suspensions, the clearances in the pump must be large compared with the size of the particles.
(4) The nature of the power supply. If the pump is driven by an electric motor or internal combustion engine, a high-speed centrifugal or rotary pump will be preferred as it can be coupled directly to the motor. Simple reciprocating pumps can be connected to steam or gas engines.
(5) If the pump is used only intermittently corrosion troubles are more likely than with continuous working.

The cost and mechanical efficiency of the pump must always be considered, and it may be advantageous to select a cheap pump and pay higher replacement or maintenance costs rather than to install a very expensive pump of high efficiency.

6.2.1. The Reciprocating Pump

The Piston Pump

The piston pump consists of a cylinder with a reciprocating piston connected to a rod which passes through a gland at the end of the cylinder as indicated in Fig. 6.1. The liquid enters from the suction line through a suction valve and is discharged through a delivery valve. These pumps may be single-acting, with the liquid admitted only to the portion of the cylinder in front of the piston or double-acting, in which case the feed is admitted to both sides of the piston. The majority of pumps are of the single-acting type.

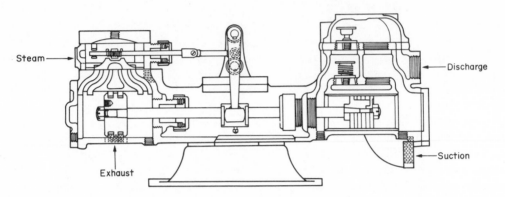

FIG. 6.1. A typical steam-driven piston pump.

The velocity of the piston varies in an approximately sinusoidal manner and the volumetric rate of discharge of the liquid shows corresponding fluctuations. In a single-cylinder pump the delivery will rise from zero as the piston begins to move forward to a maximum when the piston is fully accelerated at approximately the mid point of its stroke; the delivery will then gradually fall off to zero. If the pump is single-acting there will be an interval during the return stroke when the cylinder will fill with liquid and the delivery will remain zero. On the other hand, in a double-acting pump the delivery will be similar in the forward and return strokes. In many cases, however, the cross-sectional area of the piston rod may be significant compared with that of the piston and the volume delivered during the return stroke will therefore be less than that during the forward stroke. A more even delivery is obtained if several cylinders are suitably compounded. If two double-acting cylinders are used there will be a lag between the deliveries of the two cylinders, and the total delivery will then be the sum of the deliveries from the individual cylinders. Typical curves of delivery rate for a single-cylinder (simplex) pump are shown in Fig. 6.2a. The delivery from a two-cylinder (duplex) pump in which both the cylinders are double-acting is shown in Fig. 6.2b ; the broken lines indicate the deliveries from the individual cylinders and the unbroken line indicates the total delivery. It will be seen that the delivery is much smoother than that obtained with the simplex pump, the minimum delivery being equal to the maximum obtained from a single cylinder.

The theoretical delivery of a piston pump is equal to the total swept volume of the cylinders. The actual delivery may be less than the theoretical value because of leakage past the piston and the valves or because of inertia of the valves. In some cases, however, the actual discharge is greater than theoretical value because the momentum of the liquid in the delivery line and sluggishness in the operation of the delivery valve may result in continued delivery during a portion of the suction stroke. The volumetric efficiency, which is defined as the ratio of the actual discharge to the swept volume, is normally greater than 90 per cent.

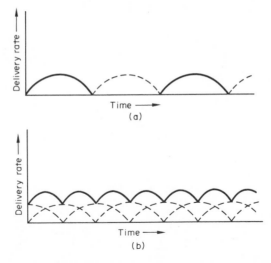

FIG. 6.2. Delivery from (a) simplex, and (b) duplex pumps.

The size of the suction and delivery valves is determined by the throughput of the pump. Where the rate of flow is high, two or more valves may be used in parallel.

The piston pump can be directly driven by steam, in which case the piston rod is common to both the pump and the steam engine. Alternatively, an electric motor or an internal combustion engine may supply the motive power through a crankshaft; because the load is very uneven, a heavy flywheel should then be fitted and a regulator in the steam supply may often provide a convenient form of speed control.

The pressure at the delivery of the pump is made up of the following four components:

(1) The static pressure at the delivery point.
(2) Pressure required to overcome the frictional losses in the delivery pipe.
(3) The pressure for the acceleration of the fluid at the commencement of the delivery stroke.

The liquid in the delivery line is accelerated and retarded in phase with the motion of the piston and therefore the whole of the liquid must be accelerated at the commencement of the delivery stroke and retarded at the end of it. Every time the fluid is accelerated, work has to be done on it and therefore in a long delivery line the expenditure of energy is very large since the excess kinetic energy of the fluid is not usefully recovered during the suction stroke. Due to the momentum of the the fluid, the pressure at the pump may fall sufficiently low for separation to occur. The pump is then said to *knock*. The flow in the delivery line can be evened out and the energy at the beginning of each stroke reduced, by the introduction of an air vessel at the pump discharge. This consists of a sealed vessel which contains air at the top and liquid at the bottom. When the delivery stroke commences, liquid is pumped into the air vessel and the air is compressed. When the discharge from the pump decreases towards the end of the stroke, the pressure in the air vessel is sufficiently high for some of the liquid to be expelled into the delivery line. If the air vessel is sufficiently large and is fitted close to the pump, the velocity of the liquid in the delivery line can be maintained approximately constant. The frictional losses are also reduced by the incorporation of an air vessel because the friction loss under turbulent conditions is approximately proportional to the linear velocity in the pipe raised to the power 1·8; i.e. the reduced friction losses during the period of minimum discharge do not compensate for the greatly increased losses when the pump is delivering at maximum rate. Further, the maximum stresses set up in the pump are reduced by the use of an air vessel.

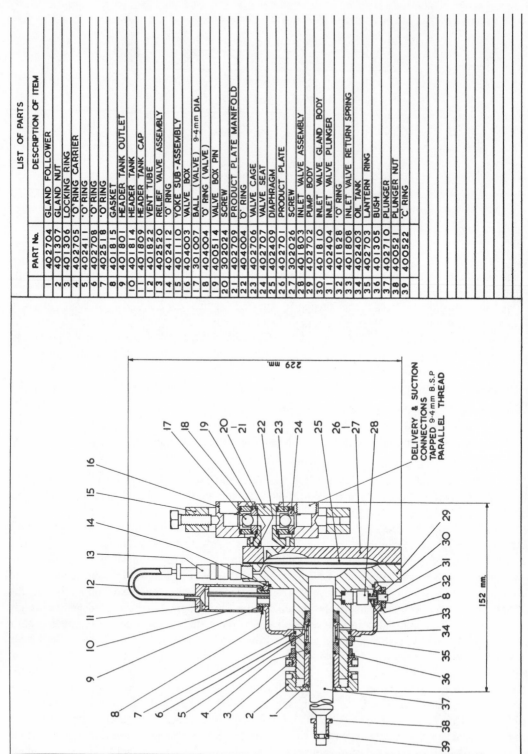

LIST OF PARTS

	PART No.	DESCRIPTION OF ITEM
1	402704	GLAND FOLLOWER
2	401307	GLAND NUT
3	401306	LOCKING RING
4	402705	'O'RING CARRIER
5	402411	'O'RING
6	402708	'O'RING
7	402518	'O'RING
8	401815	GASKET
9	401801	HEADER TANK OUTLET
10	401814	HEADER TANK
11	401809	HEADER TANK CAP
12	401822	VENT TUBE
13	402520	RELIEF VALVE ASSEMBLY
14	402412	'O'RING
15	401110	YOKE SUB-ASSEMBLY
16	404003	VALVE BOX
17	301007	BALL (VALVE) 9·4mm DIA.
18	404004	'O'RING (VALVE)
19	400514	VALVE BOX PIN
20	302024	SCREW
21	402709	PRODUCT PLATE MANIFOLD
22	404004	'O'RING
23	402706	VALVE CAGE
24	402707	VALVE SEAT
25	402409	DIAPHRAGM
26	402701	PRODUCT PLATE
27	302026	SCREW
28	401803	INLET VALVE ASSEMBLY
29	402702	PUMP BODY
30	401810	INLET VALVE GLAND BODY
31	402404	INLET VALVE PLUNGER
32	401828	'O'RING
33	401808	INLET VALVE RETURN SPRING
34	402403	OIL TANK
35	402703	LANTERN RING
36	401305	BUSH
37	402710	PLUNGER
38	400521	PLUNGER NUT
39	400522	'C'RING

DELIVERY & SUCTION CONNECTIONS TAPPED 9·4 mm B.S.P PARALLEL THREAD

229 mm

152 mm.

FIG. 6.3. Section through a low pressure diaphragm pump head.

128

Air vessels are also incorporated in the suction line for a similar reason. Here they may be of even greater importance because the pressure drop along the suction line is necessarily limited to rather less than one atmosphere if the suction tank is at atmospheric pressure. The flowrate may be limited if part of the pressure drop available must be utilised in accelerating the fluid in the suction line; the air vessel should therefore be sufficiently large for the flowrate to be maintained approximately constant.

The piston type of pump is comparatively simple in construction and operates with a high efficiency over a wide range of operating conditions. It is a positive displacement pump which will operate against a high head and does not require priming. The delivery is, however, uneven, and this presents an uneven load on the driving mechanism.

For most operations in the chemical industry, the centrifugal pump is preferred and standard pumps will give flowrates of up to 0·1 m³/s at pressures up to 1 MN/m². The very largest pumps now available and used for pumping water or oil through main pipelines will deliver up to 5 m³/s at pressures of 5 MN/m². In general, the larger the capacity of the pump, the greater is the pressure which it will develop. Positive acting, including piston, pumps will give very much higher pressures and are used for pumping small quantities of liquid to high pressures.

The Plunger or Ram Pump

This pump is the same in principle as the piston type but differs in that the gland is at one end of the cylinder making its replacement easier than with the standard piston type. The sealing of piston and ram pumps has been much improved but, because of the nature of the fluids frequently used, care in selecting and maintaining the seal is very important. The piston or ram pump may be used for injections of small quantities of inhibitors to polymerisation units or of corrosion inhibitors to high pressure systems. They are also used for boiler feed water applications.

The Diaphragm Pump

The diaphragm pump has been developed for handling corrosive liquids and those containing suspensions of abrasive solids. It is in two sections separated by a diaphragm of rubber, leather, or plastic material. In one section a plunger or piston operates in a cylinder in which a non-corrosive fluid is displaced. The movement of the fluid is transmitted by means of the flexible diaphragm to the liquid to be pumped. The only moving parts of the pump that are in contact with the liquid are the valves, and these can be specially designed to handle the material. In some cases the movement of the diaphragm is produced by direct mechanical action, as shown in Fig. 6.3.

Example 6.1. A single-acting reciprocating pump has a cylinder diameter of 110 mm and a stroke of 230 mm. The suction line is 6 m long and 50 mm in diameter and the level of the water in the suction tank is 3 m below the cylinder of the pump. What is the maximum speed at which the pump can run without an air vessel if separation is not to occur in the suction line? The piston undergoes approximately simple harmonic motion. Atmospheric pressure is equivalent to a head of 10·36 m of water and separation occurs at an absolute pressure corresponding to a head of 1·20 m of water.

Solution. The tendency for separation to occur will be greatest at:

(a) the inlet to the cylinder because here the static pressure is a minimum and the head required to accelerate the fluid in the suction line is a maximum;

(b) the commencement of the suction stroke because the acceleration of the piston is then a maximum.

Let the maximum permissible speed of the pump be N Hz.

Angular velocity of the driving mechanism $= 2\pi N$ radians/s

Acceleration of piston $= 0.5 \times 0.230(2\pi N)^2 \cos(2\pi Nt)$ m/s^2

Maximum acceleration (i.e. when $t = 0$) $= 4.54N^2$ m/s^2

Maximum acceleration of the liquid in the suction pipe

$$= (0.110/0.05)^2 \times 4.54N^2 = 21.97N^2 \text{ m/s}^2$$

Accelerating force acting on the liquid

$$= 21.97N^2(\pi/4)(0.050)^2 \times 6 \times 1000 \text{ N}$$

Pressure drop in suction line due to acceleration $= 21.97N^2 \times 6 \times 1000$ N/m^2

$$= 1.32 \times 10^5 N^2 \text{ N/m}^2$$

or $\qquad 1.32 \times 10^5 N^2/(1000 \times 9.81) = 13.44N^2$ m water

Pressure head at cylinder when separation is about to occur,

$$1.20 = 10.36 - 3.0 - 13.44N^2 \text{ m water}$$

$$\therefore \qquad \underline{\underline{N = 0.675 \text{ Hz}}}$$

6.2.2. Positive Displacement Rotary Pumps

The Gear Pump

The gear pump (Fig. 6.4) is the most widely used of the positive action rotary pumps. Two gear wheels operate inside a casing. One of the gear wheels is driven and the other rotates in mesh with it. The liquid is carried round in the space between consecutive teeth and the casing and is then ejected as the teeth come into mesh. The pump has no valves and depends for its seal on the small clearance between the gear wheels and the case. It is a positive displacement pump and will deliver against high pressures of the order of 35 MN/m^2. The delivery is almost independent of pressure and priming is unnecessary.

The main advantage of the gear pump over the reciprocating pump is that it gives an even delivery and can be directly connected to an electric motor drive. It will handle liquids of very high viscosities, and is extensively used in the oil industry for pumping viscous residual

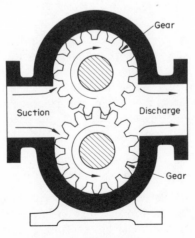

FIG. 6.4. Gear pump.

FIG. 6.5. Flow inducer.

(a)

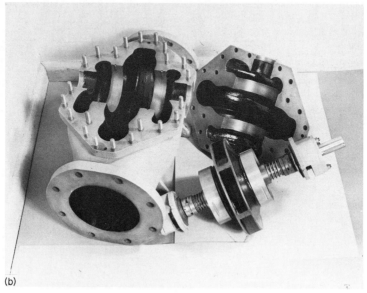

(b)

FIG. 6.6. A horizontally split Hydrostream centrifugal pump.

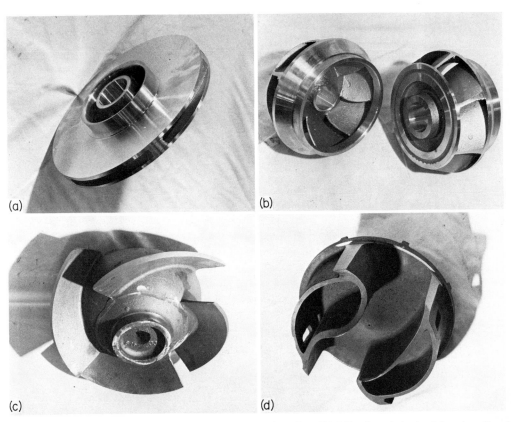

FIG. 6.7. Types of impeller; (*a*) fully shrouded radial flow impeller; (*b*) fully shrouded mixed flow impeller; (*c*) semi-open mixed flow impeller; (*d*) solid passing impeller.

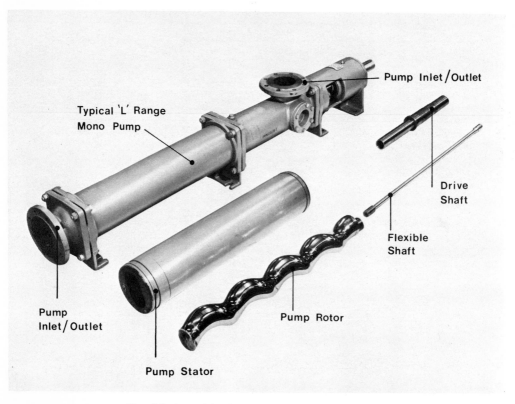

Pump Inlet/Outlet

Typical 'L' Range
Mono Pump

Drive
Shaft

Flexible
Shaft

Pump
Inlet/Outlet

Pump Rotor

Pump Stator

FIG. 6.9. Assembly of a typical 'L' range Mono pump.

oils from distillation plant. Because the spaces between the gear teeth are comparatively small, the pump cannot be used for suspensions.

The Flow Inducer or Peristaltic Pump

This is a special form of pump in which a length of silicone rubber or other elastic tubing is compressed in stages by means of a rotor as shown in Fig. 6.5. The tubing is fitted to a curved track mounted concentrically with a rotor carrying three rollers. As the rollers rotate, they flatten the tube against the track at the points of contact. These 'flats' move the fluid by positive displacement, and the flow can be precisely controlled by the speed of the motor.

These pumps have been particularly useful for biological fluids where all forms of contact must be avoided. They have been increasingly used and are suitable for emulsions, creams, and similar fluids. They are much used in laboratories and small plants where the freedom from glands, avoidance of aeration, and corrosion resistance are really valuable. With current units, capacities from 5 to 360 cm^3/s are available using tubing of from 3 to 25 mm bore. The control is such that these pumps may conveniently be used as metering pumps for dosage processes.

The Mono Pump

Another example of a positive acting rotary pump is the Mono pump shown in Figs. 6.8 and 6.9 in which a specially shaped helical metal worm rotates in a stator made of rubber or other similar material, the liquid being forced through the space between the stator and the rotor. The mono pump gives a uniform flow and is quiet in operation. It will pump against high pressures; the higher the required pressure, the longer are the stator and the rotor and the greater the number of turns. The pump can handle corrosive and gritty liquids and is extensively used for feeding slurries to filter presses. It must never be run dry.

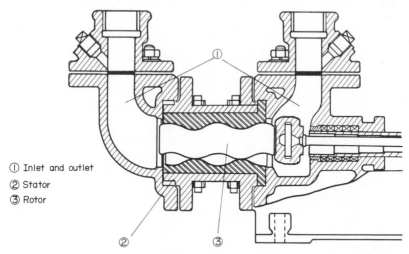

① Inlet and outlet
② Stator
③ Rotor

FIG. 6.8. Mono pump.

The Metering Pump

Metering pumps are positive displacement pumps driven by constant speed electric motors. They are used where a constant and accurately controlled rate of delivery of a liquid is required, and they will maintain this constant rate irrespective of changes in the pressure against which they operate. The pumps are usually of the plunger type for low throughput and high-pressure applications; for large volumes and lower pressures a diaphragm is used. In either case the rate of delivery is controlled by adjusting the stroke of the piston element, and this can be done whilst the pump is in operation. A single-motor driver may operate

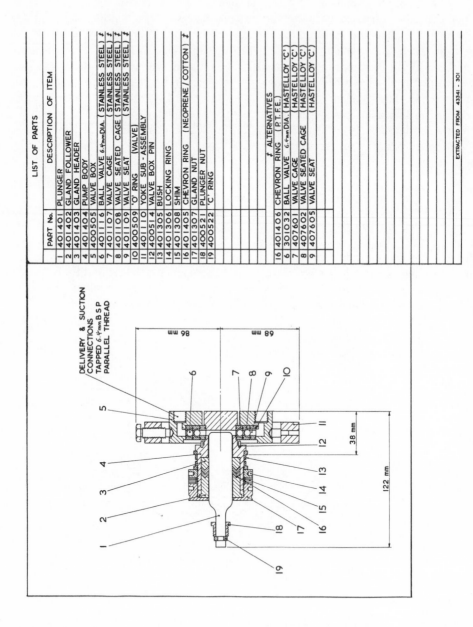

LIST OF PARTS

	PART No.	DESCRIPTION OF ITEM
1	401401	PLUNGER
2	401402	GLAND FOLLOWER
3	401403	GLAND HEADER
4	401404	PUMP BODY
5	400505	VALVE BOX
6	401116	BALL VALVE 6·4mm DIA. (STAINLESS STEEL) ‡
7	401107	VALVE CAGE (STAINLESS STEEL) ‡
8	401108	VALVE SEATED CAGE (STAINLESS STEEL) ‡
9	401109	VALVE SEAT (STAINLESS STEEL) ‡
10	400509	'O' RING (VALVE)
11	401110	YOKE SUB - ASSEMBLY
12	400514	VALVE BOX PIN
13	401305	BUSH
14	401306	LOCKING RING
15	401308	SHIM
16	401405	CHEVRON RING (NEOPRENE / COTTON) ‡
17	401307	GLAND NUT
18	400521	PLUNGER NUT
19	400522	'C' RING

‡ ALTERNATIVES

16	401406	CHEVRON RING (P.T.F.E.)
6	301032	BALL VALVE 6·4mm DIA. (HASTELLOY 'C')
7	407601	VALVE CAGE (HASTELLOY 'C')
8	407602	VALVE SEATED CAGE (HASTELLOY 'C')
9	407605	VALVE SEAT (HASTELLOY 'C')

EXTRACTED FROM 43341 - 301

DELIVERY & SUCTION
CONNECTIONS
TAPPED 6·4mm B S P
PARALLEL THREAD

98 mm

89 mm

38 mm

122 mm

FIG. 6.10. Type HM/SS14C pump head. (Metering Pumps Ltd.)

several individual pumps and in this way give control of the actual flows and of the flow ratio of several streams at the same time. The output may be controlled from zero to maximum flowrate either manually on the pump or remotely.

These pumps may be used for the dosing of works effluents and water supplies, and the feeding of reactants, catalysts, or inhibitors to reactors at controlled rates. The diaphragm and piston-type pumps are illustrated in Figs. 6.3 and 6.10, respectively.

These pumps provide a simple method for controlling flowrate, but must be constructed to precision standards of engineering. They can be made from many corrosion-resistant materials.

6.2.3.   The Centrifugal Pump

The centrifugal pump is by far the most widely used type in the chemical and petroleum industries. It will pump liquids of very wide range of properties and suspensions with a high solids content including, e.g. cement slurries, and may be constructed from a very wide range of corrosion resistant materials. The whole pump casing may be constructed from plastics such as polypropylene or it may be fitted with a corrosion-resistant lining. Because it operates at high speed, it may be directly coupled to an electric motor and it will give a high flowrate for its size.

In this type of pump (Fig. 6.6) the fluid is fed to the centre of a rotating impeller and is thrown outward by centrifugal action. As a result of the high speed of rotation the liquid acquires a high kinetic energy and the pressure difference between the suction and delivery sides arises from the conversion of kinetic energy into pressure energy.

The impeller (Fig. 6.7) consists of a series of curved vanes so shaped that the flow within the pump is as smooth as possible. The greater the number of vanes on the impeller, the greater is the control over the direction of motion of the liquid and hence the smaller are the losses due to turbulence and circulation between the vanes. In the open impeller the vanes are fixed to a central hub, whereas in the closed type the vanes are held between two supporting plates and leakage across the impeller is reduced. As will be seen later, the angle of the tips of the blades very largely determines the operating characteristics of the pump.

The liquid enters the casing of the pump, normally in an axial direction, and is picked up by the vanes of the impeller. In the simple type of centrifugal pump, the liquid discharges into a volute, a chamber of gradually increasing cross-section with a tangential outlet. A volute type of pump is shown in Fig. 6.11a. In the turbine pump (Fig. 6.11), the liquid flows from the moving vanes of the impeller through a series of fixed vanes forming a diffusion ring. This gives a more gradual change in direction to the fluid and more efficient conversion of kinetic energy into pressure energy than is obtained with the volute type. The angle of the leading edge of the fixed vanes should be such that the fluid is received without shock. The liquid flows along the surface of the impeller vane with a certain velocity whilst the tip of the vane is moving relative to the casing of the pump. The direction of motion of the liquid relative to

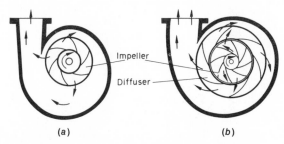

FIG. 6.11. Radial flow pumps: (a) with volute; (b) with diffuser vanes.

the pump casing—and the required angle of the fixed vanes—is found by compounding these two velocities. In Fig. 6.12, u_v is the velocity of the liquid relative to the vane and u_t is the tangential velocity of the tip of the vane; compounding these two velocities gives the resultant velocity u_2 of the liquid. It is apparent therefore that the required vane angle in the diffuser is dependent on the throughput, the speed of rotation, and the angle of the impeller blades. The pump will therefore operate at maximum efficiency only over a narrow range of conditions.

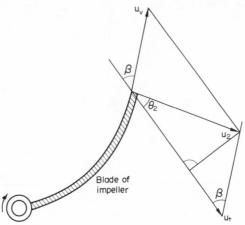

FIG. 6.12. Velocity diagram.

Virtual Head of a Centrifugal Pump

The maximum pressure is developed when the whole of the excess kinetic energy of the fluid is converted into pressure energy. As indicated below, the head is proportional to the square of the radius and to the speed, and is of the order of 30 m for a single-stage centrifugal pump; for higher pressures multistage pumps must be used. Consider the liquid which is rotating at a distance of between r and $r + dr$ from the centre of the pump (Fig. 6.13). The mass of this element of fluid dM is given by $2\pi r\, dr\, b\rho$, where ρ is the density of the fluid and b is the width of the element of fluid.

If the fluid is travelling with a velocity u and at an angle θ to the tangential direction, the angular momentum of this mass of fluid

$$= dM(ur \cos \theta)$$

The torque acting on the fluid $d\tau$ is equal to the rate of change of angular momentum with

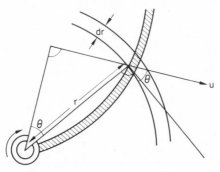

FIG. 6.13. Virtual head.

time, as it goes through the pump:

$$\therefore \quad d\tau = dM \frac{\partial}{\partial t}(ur \cos\theta)$$

$$= 2\pi rb\rho \, dr \frac{\partial}{\partial t}(ur \cos\theta) \tag{6.3}$$

The volumetric rate of flow of liquid through the pump:

$$Q = 2\pi rb(\partial r/\partial t) \tag{6.4}$$

$$\therefore \quad d\tau = Q\rho \, d(ur \cos\theta) \tag{6.5}$$

The total torque acting on the liquid in the pump is therefore obtained by integrating $d\tau$ between the limits denoted by suffix 1 and suffix 2, where suffix 1 refers to the conditions at the inlet to the pump and suffix 2 refers to the conditions at the discharge.

Thus

$$\tau = Q\rho(u_2 r_2 \cos\theta_2 - u_1 r_1 \cos\theta_1) \tag{6.6}$$

The power p developed by the pump is equal to the product of the torque and the angular velocity ω:

$$\therefore \quad p = Q\rho\omega(u_2 r_2 \cos\theta_2 - u_1 r_1 \cos\theta_1) \tag{6.7}$$

The power can also be expressed as the product Ghg, where G is the mass rate of flow of liquid through the pump, g is the acceleration due to gravity, and h is termed the virtual head developed by the pump.

Thus

$$Ghg = Q\rho\omega(u_2 r_2 \cos\theta_2 - u_1 r_1 \cos\theta_1)$$

i.e.

$$h = \omega(u_2 r_2 \cos\theta_2 - u_1 r_1 \cos\theta_1)/g \tag{6.8}$$

Since u_1 will be approximately zero, the virtual head

$$h = \omega u_2 r_2 \cos\theta_2/g \tag{6.9}$$

where g, ω, and r_2 are known in any given instance, and u_2 and θ_2 are to be expressed in terms of known quantities.

From the geometry of Fig. 6.12:

$$u_v \sin\beta = u_2 \sin\theta_2 \tag{6.10}$$

and

$$u_t = u_v \cos\beta + u_2 \cos\theta_2 \tag{6.11}$$

(where β is the angle between the tip of the blade of the impeller and the tangent to the direction of its motion. If the blade curves backwards, β lies between 0 and $\pi/2$ and if it curves forwards, β lies between $\pi/2$ and π).

The volumetric rate of flow through the pump Q is equal to the product of the area available for flow at the outlet of the impeller and the radial component of the velocity,

i.e.

$$Q = 2\pi r_2 bu_2 \sin\theta_2$$

$$= 2\pi r_2 bu_v \sin\beta \quad \text{(from equation 6.10)} \tag{6.12}$$

$$\therefore \quad u_v = Q/2\pi r_2 b \sin\beta \tag{6.13}$$

Thus

$$h = \omega r_2(u_t - u_v \cos\beta)/g \quad \text{(from equations 6.9 and 6.11)}$$

$$= \frac{\omega}{g} r_2\left(r_2\omega - \frac{Q}{2\pi r_2 b \tan\beta}\right) \quad \text{(since } u_t = r_2\omega)$$

$$= \frac{r_2^2 \omega^2}{g} - \frac{Q\omega}{2\pi bg \tan\beta} \tag{6.14}$$

The virtual head developed by the pump is therefore independent of the density of the fluid, and the pressure will thus be directly proportional to the density. For this reason a centrifugal pump needs priming. If the pump is initially full of air, the pressure developed is reduced by a factor equal to the ratio of the density of air to that of the liquid, and is insufficient to drive the liquid through the delivery pipe.

For a given speed of rotation there is a linear relation between the head developed and the rate of flow. If the tips of the blades of the impeller are inclined backwards, β is less than $\pi/2$, $\tan \beta$ is positive, and therefore the head decreases as the throughput increases. If β is greater than $\pi/2$ (i.e. the tips of the blades are inclined forwards), the head increases as the delivery increases. The angle of the blade tips therefore profoundly affects the performance and characteristics of the pump. For radial blades the head should be independent of the throughput.

Specific Speed

If θ_2 remains approximately constant, u_v, u_t, and u_2 will all be directly proportional to one another, and since $u_t = r_2\omega$, these velocities are proportional to r_2; thus u_v will vary as $r_2\omega$.

The output from a pump is a function of its linear dimensions, the shape, number, and arrangement of the impellers, the speed of rotation, and the head against which it is operating. From equation 6.12, for a radial pump with $\beta = \pi/2$ and $\sin \beta = 1$:

$$Q \propto 2\pi r_2 b u_v$$
$$Q \propto 2\pi r_2 b r_2 \omega$$
$$Q \propto r_2^2 b \omega \tag{6.15}$$

When $\tan \beta = \tan \pi/2 = \infty$, then from equation 6.14:

$$h = r_2^2 \omega^2/g \tag{6.16}$$
$$gh = r_2^2 \omega^2$$

or
$$(gh)^{3/4} = r_2^{3/2} \omega^{3/2} \tag{6.17}$$

For a series of geometrically similar pumps, b is proportional to the radius r_2, and thus, from equation 6.15:

$$Q \propto r_2^3 \omega \tag{6.18}$$

or
$$Q^{1/2} \propto r_2^{3/2} \omega^{1/2} \tag{6.19}$$

Eliminating r_2 between equations 6.17 and 6.19:

$$Q^{1/2}/(gh)^{3/4} = \omega^{1/2}/\omega^{3/2}$$

or
$$\omega Q^{1/2}/(gh)^{3/4} = \text{constant} = N_s \text{ for geometrically similar pumps} \tag{6.20}$$

Criteria for Similarity

The dimensionless quantity $\omega Q^{1/2}/(gh)^{3/4}$ is a characteristic for a particular type of centrifugal pump, and, noting that the angular velocity is proportional to the speed N, this group may be rewritten as:

$$N_s = NQ^{1/2}/(gh)^{3/4} \tag{6.21}$$

and is constant for geometrically similar pumps. N_s is defined as the specific speed and is frequently used to classify types of centrifugal pumps. Specific speed may be defined as the speed of the pump which will produce unit flow Q against unit head h under conditions of maximum efficiency.

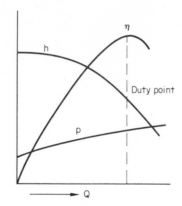

FIG. 6.14. Radial flow pump characteristics.

Operating Characteristics

The operating characteristics of a pump are conveniently shown by plotting the head h, power p, and efficiency η against the flow Q as shown in Fig. 6.14. It is important to note that the efficiency reaches a maximum and then falls, whilst the head at first falls slowly with Q but eventually falls off rapidly. The optimum conditions for operation are shown as the duty point, i.e. the point where the head curve cuts the ordinate through the point of maximum efficiency.

A set of curves for h, η, and Q for various speeds are shown in Fig. 6.15 from which it is seen that the efficiency remains reasonably constant over a range of speeds. A more general indication of the variation of efficiency with specific speed is shown in Fig. 6.16 for different

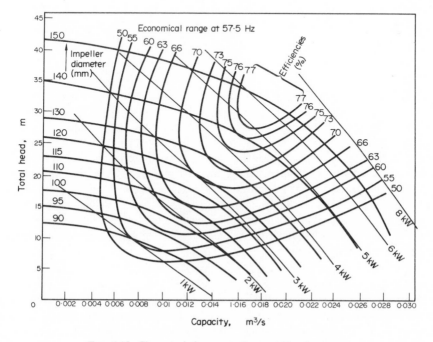

FIG. 6.15. Characteristic curves for centrifugal pump.

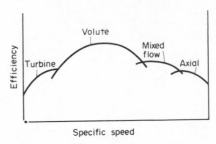

FIG. 6.16. Specific speed and efficiency.

types of centrifugal pumps. The power developed by a pump is proportional to $Qgh\rho$:

i.e.
$$p \propto r_2^2 b\omega r_2^2 \omega^2 \rho$$

or
$$p \propto r_2^4 b\omega^3 \rho \tag{6.22}$$

so that $Q \propto \omega$; $h \propto \omega^2$; $p \propto \omega^3$ from equations 6.15, 6.16 and 6.22.

Cavitation

In designing any installation in which a centrifugal pump is used, careful attention must be paid to check the minimum pressure which will arise at any point. If this pressure is less than the vapour pressure at the pumping temperature, vaporisation will occur and the pump may not be capable of developing the required suction head. Moreover, if the liquid contains gases, these may come out of solution giving rise to pockets of gas. This phenomenon is known as *cavitation* and may result in mechanical damage to the pump as the bubbles collapse. The tendency for cavitation to occur is accentuated by any sudden changes in the magnitude or direction of the velocity of the liquid in the pump. The onset of cavitation is accompanied by a marked increase in noise and vibration as the vapour bubbles collapse, and also a loss of head.

Suction Head

Pumps may be arranged so that the inlet is under a suction head or the pump may be fed from a tank. These two systems alter the duty point curves as shown in Fig. 6.17. In developing such curves the normal range of liquid velocities is 1.5 to 3 m/s, but lower values are used for pump suction lines.

For any pump, the manufacturers specify the minimum value of the *net positive suction head* (NPSH) which must exist at the suction point of the pump. The NPSH is the amount by which the pressure at the suction point of the pump, expressed as a head of the liquid to be pumped, must exceed the vapour pressure of the liquid. For any installation this must be calculated, taking into account the absolute pressure of the liquid, the level of the pump, and the velocity and friction heads in the suction line. The NPSH must allow for the fall in pressure occasioned by the further acceleration of the liquid as it flows on to the impeller and for irregularities in the flow pattern in the pump. If the required value of NPSH is not obtained, partial vaporisation is liable to occur, with the result that both suction head and delivery head may be reduced. The loss of suction head is the more important because it may cause the pump to be starved of liquid.

Consider the system shown in Fig. 6.18, in which the pump is taking liquid from a reservoir at an absolute pressure P_0 in which the liquid level is at a height h_0 above the suction point of the pump. Then if the liquid in the reservoir can be regarded as at rest, the *absolute* pressure head h_i at the suction point of the pump is obtained by applying the energy or momentum balance:

$$h_i = \frac{P_0}{\rho g} + h_0 - \frac{u_i^2}{2g} - h_f \tag{6.23}$$

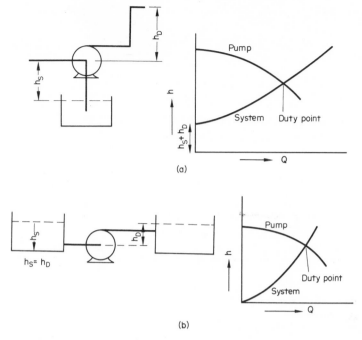

(a)

(b)

FIG. 6.17. Effect of suction head: (a) systems with suction lift and friction; (b) systems with friction losses only.

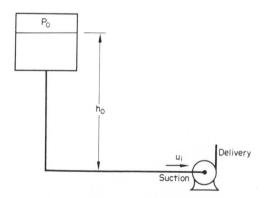

FIG. 6.18. Suction system of centrifugal pump.

where h_f is the head lost in friction, and u_i is the velocity at the inlet of the pump. If the vapour pressure of the liquid is P_v, the NPSH is given by the difference between the *total* head at the suction inlet and the head corresponding to the vapour pressure of the liquid at the pump inlet

$$\text{NPSH} = \left(h_i + \frac{u_i^2}{2g} \right) - \frac{P_v}{\rho g} \qquad (6.24)$$

$$= \frac{P_0}{\rho g} - \frac{P_v}{\rho g} + h_0 - h_f \qquad (6.25)$$

where P_v is the vapour pressure of the liquid being pumped.

If cavitation and loss of suction head does occur, it can sometimes be cured by increasing the pressure in the system, either by alteration of the layout to provide a greater hydrostatic pressure or a reduced pressure drop in the suction line. Sometimes, slightly closing the valve on the pump delivery or reducing the pump speed by a small amount may be effective.

Generally a small fast-running pump will require a larger NPSH than a larger slow-running pump.

The efficiency of a centrifugal pump and the head which it is capable of developing are dependent upon providing a good seal between the rotating shaft and the casing of the pump. When the pump is fitted with the usual type of packed gland, maintenance costs are often very high, especially when organic liquids of low viscosity are being pumped. A considerable reduction in expenditure on maintenance can be effected at the price of a small increase in initial cost by fitting the pump with a mechanical seal, in which the sealing action is achieved as a result of contact between two opposing faces, one stationary and the other rotating. In Fig. 6.19, a mechanical seal is shown in position in a centrifugal pump. The stationary seat A is held in position by means of the clamping plate D. The seal is made with the rotating face on a carbon ring B. The seal between the rotating face and the shaft is made by means of the wedge ring C, usually made of polytetrafluoroethylene (PTFE). The drive is through the retainer E secured to the shaft, usually by Allen screws. Compression between the fixed and rotating faces is provided by the spiral springs F.

It is advantageous to ensure that the seal is fed with liquid which removes any heat generated at the face. In the illustration this is provided by the connection G.

Centrifugal pumps must be fitted with good bearings since there is a tendency for an axial thrust to be produced if the suction is taken only on one side of the impeller. This thrust can be balanced by feeding back a small quantity of the high-pressure liquid to a specially designed thrust bearing. By this method the risk of air leaking into the pump at the gland and

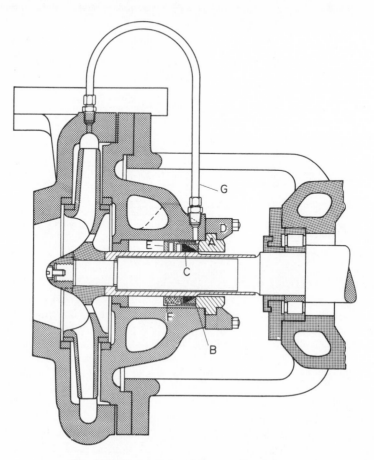

FIG. 6.19. Mechanical seal for centrifugal pump.

reducing the pressure developed is minimised. The glandless centrifugal pump, which is used extensively for corrosive liquids, works on a similar principle, and the use of pumps without glands has been increasing both in the nuclear power industry and in the chemical process industry. A typical centrifugal pump of this type is shown in Fig. 6.20. Such a unit is totally enclosed and lubrication is provided by the fluid handled.

A small volume of fluid is led from the discharge branch to the motor cover and passes through the motor and shaft clearance to the eye of the impeller. If the fluid is very hot, a water-cooled heat exchanger is fitted to the motor. The detailed parts list gives some

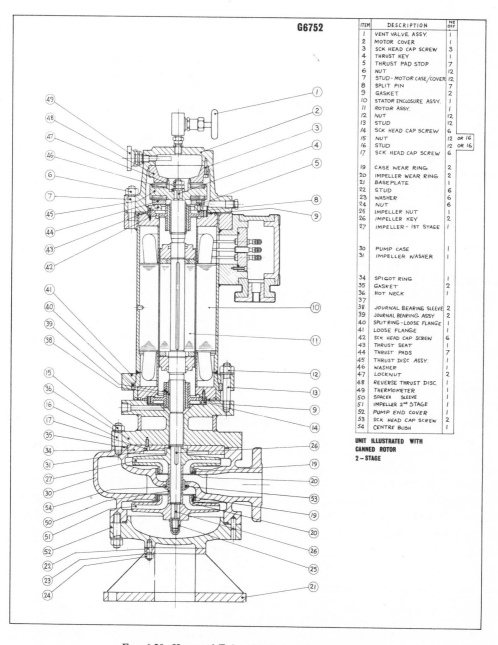

G6752

ITEM	DESCRIPTION	Nº OFF
1	VENT VALVE ASSY.	1
2	MOTOR COVER	1
3	SCK HEAD CAP SCREW	3
4	THRUST KEY	1
5	THRUST PAD STOP	7
6	NUT	12
7	STUD-MOTOR CASE/COVER	12
8	SPLIT PIN	7
9	GASKET	2
10	STATOR ENCLOSURE ASSY.	1
11	ROTOR ASSY.	1
12	NUT	12
13	STUD	12
14	SCK HEAD CAP SCREW	6
15	NUT	12 OR 16
16	STUD	12 OR 16
17	SCK HEAD CAP SCREW	6
19	CASE WEAR RING	2
20	IMPELLER WEAR RING	2
21	BASEPLATE	1
22	STUD	6
23	WASHER	6
24	NUT	6
25	IMPELLER NUT	1
26	IMPELLER KEY	2
27	IMPELLER-1ST STAGE	1
30	PUMP CASE	1
31	IMPELLER WASHER	1
34	SPIGOT RING	1
35	GASKET	2
36	HOT NECK	1
37		
38	JOURNAL BEARING SLEEVE	2
39	JOURNAL BEARING ASSY.	2
40	SPLIT RING-LOOSE FLANGE	1
41	LOOSE FLANGE	1
42	SCK HEAD CAP SCREW	6
43	THRUST SEAT	1
44	THRUST PADS	7
45	THRUST DISC ASSY.	1
46	WASHER	1
47	LOCKNUT	2
48	REVERSE THRUST DISC	1
49	THERMOMETER	1
50	SPACER SLEEVE	1
51	IMPELLER 2ND STAGE	1
52	PUMP END COVER	1
53	SCK HEAD CAP SCREW	2
54	CENTRE BUSH	1

UNIT ILLUSTRATED WITH
CANNED ROTOR
2–STAGE

FIG. 6.20. Hayward Tyler glandless motor pump unit.

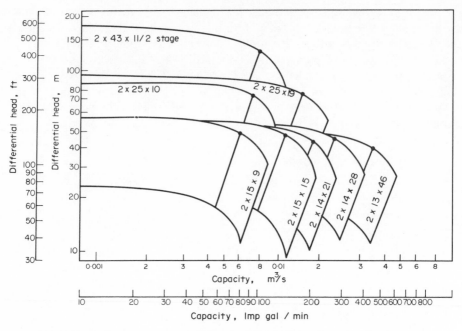

FIG. 6.21. Performance characteristics of a family of glandless centrifugal pumps.

indication of the construction and Fig. 6.21 gives the performance characteristics of one group of pumps of this type.

A ram pump without glands is sometimes used for handling corrosive liquids such as sulphuric acid. This operates with a very small clearance between the ram and cylinder, and the very small leak which arises is taken back to the suction side. No gland is used, and with very fine clearances a high efficiency is attained.

Centrifugal pumps can be made of a wide range of materials, and in many cases the impeller and the casing are covered with resistant material. Thus stainless steel, nickel, rubber, polypropylene, stoneware, and carbon are all used. When the pump is used with suspensions the ports and spaces between the vanes must be made sufficiently large to eliminate the risk of blockage. This does mean, however, that the efficiency of the pump is reduced. The *Vacseal pump*, developed by the International Combustion Company for pumping slurries, will handle suspensions containing up to 50 per cent by volume of solids. The whole impeller may be rubber covered and has three small vanes, as shown in Fig. 6.22. The back plate of the impeller has a second set of vanes of larger diameter. The pressure at the gland is thereby reduced below atmospheric pressure and below the pressure in the suction line; there is, therefore, no risk of the gritty suspension leaking into the gland and bearings. If leakage does occur, air will enter the pump. As mentioned previously, this may reduce the pressure which the pump can deliver, but this is preferable to damaging the bearings by allowing them to become contaminated with grit. This is another example of the necessity for tolerating rather low efficiencies in pumps that handle difficult materials.

The Advantages and Disadvantages of the Centrifugal Pump

The main advantages are:

(1) It is simple in construction and can, therefore, be made in a wide range of materials.
(2) There is a complete absence of valves.

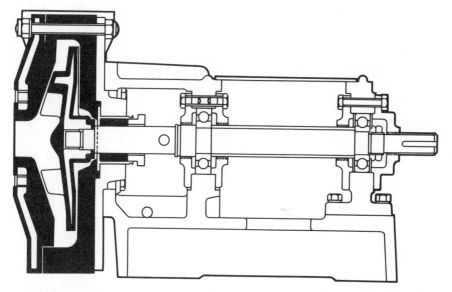

Fig. 6.22. Sectioned arrangement of a 38 mm moulded rubber V-type Vacseal pump.

(3) It operates at high speed (up to 67 Hz) and, therefore, can be coupled directly to an electric motor. In general, the higher the speed the smaller the pump and motor for a given duty.

(4) It gives a steady delivery.

(5) Maintenance costs are lower than for any other type of pump.

(6) No damage is done to the pump if the delivery line becomes blocked, provided it is not run in this condition for a prolonged period.

(7) It is much smaller than other pumps of equal capacity. It can, therefore, be made into a sealed unit with the driving motor and immersed in the suction tank.

(8) Liquids containing high proportions of suspended solids are readily handled.

The main disadvantages are:

(1) The single-stage pump will not develop a high pressure. Multistage pumps will develop greater heads but they are very much more expensive and cannot readily be made in corrosion-resistant material because of their greater complexity. It is generally better to use very high speeds in order to reduce the number of stages required.

(2) It operates at a high efficiency over only a limited range of conditions: this applies especially to turbine pumps.

(3) It is not usually self-priming.

(4) If a non-return valve is not incorporated in the delivery or suction line, the liquid will run back into the suction tank as soon as the pump stops.

(5) Very viscous liquids cannot be handled efficiently.

Example 6.2. It is required to pump cooling water from a storage pond to a condenser in a process plant situated 10 m above the level of the pond. 200 m of 75 mm i.d. pipe is available and the pump has the characteristics given below. The head lost in the condenser is equivalent to 16 velocity heads based on the flow in the 75 mm pipe.

If the friction factor $\phi = 0.003$, calculate the rate of flow and the power to be supplied to the pump assuming an efficiency of 50 per cent.

| Discharge (m^3/s) | 0·0028 | 0·0039 | 0·0050 | 0·0056 | 0·0059 |
| Head developed (m) | 23·2 | 21·3 | 18·9 | 15·2 | 11·0 |

Solution. The head to be developed, $h = 10 + 4\phi(l/d)(u^2/g) + (16/2)(u^2/g)$

$$= 10 + 4 \times 0·003(200/0·075)u^2/g + 8u^2/g$$

$$= 10 + 4·08u^2 \text{ m water}$$

Discharge, $Q = (\pi d^2/4)u = 0·0014\pi u \text{ m}^3/s$

$$\therefore \qquad u = Q/(0·0014\pi) = 232·5Q \text{ m/s}$$

$$\therefore \qquad h = 10 + 4·08(232·5Q)^2$$

$$= 10 + 2·205 \times 10^5 Q^2 \text{ m water}$$

Values of Q and h are plotted in Fig. 6.23 and the discharge at the point of intersection between the pump characteristic equation and the line of the above equation is 0·00523 m^3/s. The head developed is thus, $h = 10 + 2·205 \times 10^5 \times 0·00523^2$

$$= 15·96, \text{ say } 16 \text{ m water}$$

Power required $= 0·00523 \times 1000 \times 16 \times 9·81/0·50$

$$= 1642 \text{ W} \quad \text{or} \quad \underline{\underline{1·64 \text{ kW}}}$$

$$h_f = \phi \cdot 8 \left(\tfrac{l}{d}\right)\left(\tfrac{u^2}{2g}\right) \quad \text{Coulson}$$

$$= \epsilon' \left(\tfrac{l}{d}\right)\left(\tfrac{u^2}{2g}\right) \quad \text{Moody}$$

$$= 4\epsilon \left(\tfrac{l}{d}\right)\left(\tfrac{u^2}{2g}\right) \quad \text{Fann}$$

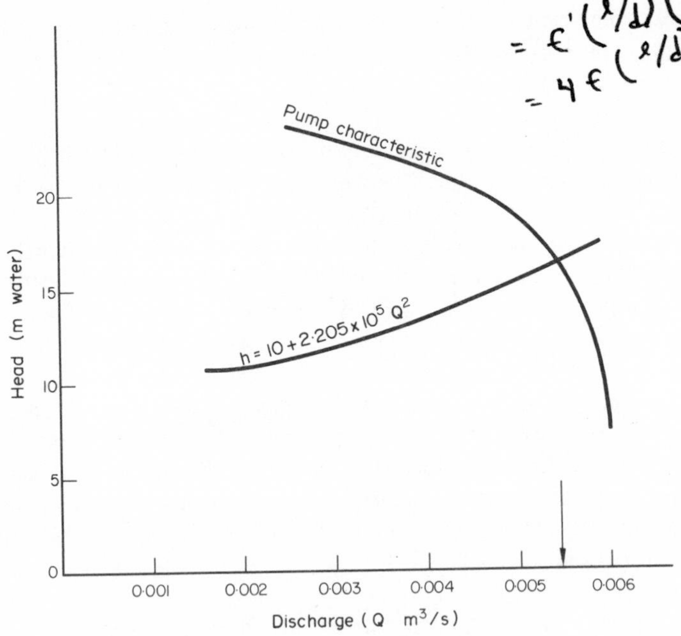

FIG. 6.23. Data for Example 6.2.

6.3. THE USE OF COMPRESSED AIR FOR PUMPING

Compressed gas is sometimes used for transferring liquid from one position to another in a chemical works, but more particularly for emptying vessels. It is frequently more convenient to apply pressure by means of compressed gas rather than to install a pump, particularly when the liquid is corrosive or contains solids in suspension. Furthermore, to an increasing extent chemicals are being delivered in tankers and are discharged by the application of gas under pressure. For instance, phthalic anhydride is now distributed in heated tankers which are discharged into storage vessels by connecting them to a supply of compressed nitrogen or carbon dioxide.

Several devices have been developed to eliminate the necessity for manual operation of valves, and the automatic acid elevator is an example of equipment incorporating such a device. However, such equipment is becoming less important now that it is possible to construct centrifugal pumps in a wide range of corrosion-resistant materials. The air-lift pump makes more efficient use of the compressed air and is used for pumping corrosive liquids. Although it is not used extensively in the chemical industry, it is used for pumping oil from wells, and the principle is incorporated into a number of items of chemical engineering equipment, including the climbing film evaporator.

6.3.1. The Air-lift Pump

In the air-lift pump (Fig. 6.24) a high efficiency is obtained by using the air expansively so that it expands to atmospheric pressure in contact with the liquid. It can be regarded as a U-tube in a state of dynamic equilibrium. One limb containing only liquid is relatively short and is connected to the feed tank, whilst air is injected near the bottom of the longer limb which therefore contains a mixture of lower density (liquid and air). If the air is introduced sufficiently rapidly, liquid will flow from the short to the long limb and be discharged into the delivery tank. The rate of flow will depend on the difference in density and will, therefore, rise as the air rate is increased, but will reach a maximum because the frictional resistance increases with the volumetric rate of flow.

The liquid-feed limb is known as the *submergence line* and the line carrying the aerated mixture is the *rising main*. The ratio of the submergence (h_s) to the total height of rising main above the air injection point ($h_r + h_s$) is known as the submergence ratio, $[1 + (h_r/h_s)]$.

If a mass M of liquid is raised through a net height h_r by a mass m of air, the net work done on the liquid is Mgh_r. If the pressure of the entering air is P, the work done by the air in expanding isothermally to atmospheric pressure P_a is given by:

$$W = P_a v_a m \ln (P/P_a) \tag{6.26}$$

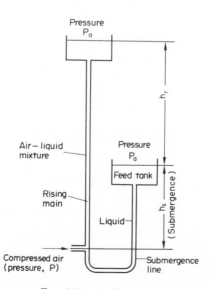

FIG. 6.24. Air-lift pump.

where v_a is the specific volume of air at atmospheric pressure. The expansion will be almost exactly isothermal because of the intimate contact between the liquid and the air.

The efficiency of the pump η is therefore given by:

$$\eta = \frac{Mgh_r}{mP_a v_a \ln (P/P_a)} \tag{6.27}$$

The mass of air required to pump unit mass of liquid is, therefore, given by:

$$\frac{m}{M} = \frac{gh_r}{\eta P_a v_a \ln (P/P_a)} \tag{6.28}$$

If all losses in the operation of the pump were neglected, the pressure at the point of introduction of the compressed air would be equal to atmospheric pressure together with the pressure due to the column of liquid of height h_s, the vertical distance between the liquid level in the suction tank, and the air inlet point. Therefore,

$$P_a = h_a \rho g \quad \text{(say)} \tag{6.29}$$

and

$$P = (h_a + h_s) \rho g \tag{6.30}$$

where ρ is the density of the liquid.

Thus from equation 6.28 the mass of air required to pump unit mass of liquid (m/M) would be equal to:

$$\frac{h_r g}{P_a v_a \ln [(h_s + h_a)/h_a]} \tag{6.31}$$

This is the minimum air requirement for the pump if all losses are neglected. It will be seen that (m/M) decreases as h_s increases; if h_s is zero (m/M) is infinite and therefore the pump will not work. A high submergence h_s is therefore desirable.

Example 6.3. An air-lift pump is used for raising $750 \text{ cm}^3/\text{s}$ of a liquid of specific gravity $1 \cdot 2$ to a height of 20 m. Air is available at 450 kN/m^2. Assuming isentropic compression of the air, what is the power requirement of the pump, if the efficiency is 30 per cent? $(\gamma = 1 \cdot 4.)$

Solution. Work done by pump $= (750 \times 10^{-6})(1 \cdot 2 \times 1000) \times 20 \times 9 \cdot 81$

$$= 176 \cdot 6 \text{ W}$$

Actual work of expansion of air $= 176 \cdot 6/0 \cdot 30 = 588 \cdot 6 \text{ W}$

Volume of air required per second at STP is given by:

$$588 \cdot 6 = 101,300 V \ln (450/101 \cdot 3)$$
$$V = 0 \cdot 0039 \text{ m}^3/\text{s}$$

∴

Work done in the isentropic compression of $0 \cdot 0039 \text{ m}^3$ air (from equation 6.42)

$$= 101,300 \times 0 \cdot 0039 (1 \cdot 4/0 \cdot 4)[(450/101 \cdot 3)^{0 \cdot 4/1 \cdot 4} - 1]$$
$$= 1383(1 \cdot 528 - 1) = 730 \text{ J}$$

∴   Power required $= 730 \text{ J/s}$ or $\underline{0 \cdot 730 \text{ kW}}$

6.4. PUMPING EQUIPMENT FOR GASES

Essentially the same basic types of mechanical equipment are used for handling gases and liquids, though the construction may be very different in the two cases. Under the normal range of operating pressures, the density of a gas is considerably less than that of a liquid so that higher speeds of operation can be employed and lighter valves fitted to the delivery and suction lines. Because of the lower viscosity of a gas there is a greater tendency for leakage to occur, and therefore gas compressors are designed with smaller clearances between the moving parts. Further differences in construction are necessitated by the decrease in volume of gas as it is compressed, and this must be allowed for in the design. Since a large proportion of the energy of compression appears as heat in the gas, there will normally be a considerable increase in temperature which may limit the operation of the compressor unless suitable cooling can be effected. For this reason gas compression is often carried out in a number of stages and the gas is cooled between each stage. Any gas which is not expelled from the cylinder at the end of compression (the clearance volume) must be expanded again to the inlet pressure before a fresh charge can be admitted. This continual compression and expansion of the residual gas results in loss of efficiency because neither the compression nor the expansion can be carried out completely reversibly. With liquids this factor has no effect on the efficiency because the residual fluid is not compressed.

The principal types of compressors for gases are given below:

6.4.1. The Reciprocating Piston Compressor

This type of compressor which may consist of from one to twelve stages is the only one capable of developing very high pressures (e.g. over 350 MN/m^2 as required for polythene manufacture). A single-stage, two-cylinder unit is shown in Fig. 6.25. It operates at relatively

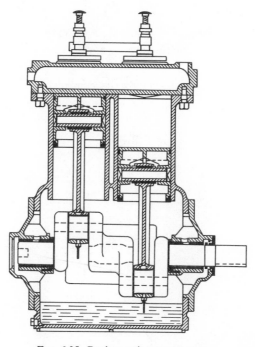

FIG. 6.25. Reciprocating compressor.

slow speed and is being replaced wherever possible by the centrifugal type which can now develop all but the very highest pressures.

6.4.2. Rotary Blowers and Compressors

These can be divided into two classes—those which develop a high compression ratio and those which have very low ratios. The former include the sliding-vane type in which the compression ratio is achieved by eccentric mounting of the rotor (Fig. 6.26) and the *Nash Hytor pump* (Fig. 6.27) in which the compression ratio is achieved by means of a specially shaped casing and a liquid seal which rotates with the impeller. In the sliding-vane unit the rotor is slotted to take the sliding blades which subdivide the crescent-shaped space between the rotor and cylinder. On rotation the blades move out trapping the gas which is compressed during rotation and discharged at the delivery port as shown.

The Nash Hytor (Fig. 6.27) type, which is also known as the *liquid-ring pump*, is a positive displacement type with a specially shaped casing and a liquid seal which rotates with the impeller. The shaft and impeller are the only moving parts and there is no sliding contact, so that no lubricants are required and the gas under compression is not contaminated. With this arrangement the liquid leaves and re-enters the impeller cells in the manner of a piston. The

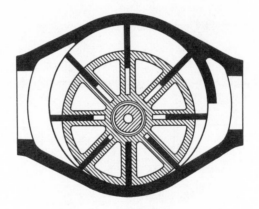

FIG. 6.26. Vane-type blower.

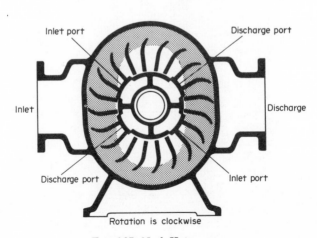

FIG. 6.27. Nash Hytor pump.

service liquid is supplied at a pressure equal to the discharge pressure of the gas and is drawn in automatically to compensate for that discharged from the ports. During compression the energy is converted to heat so that this is nearly an isothermal process. Downstream the liquid is separated from the gas and recirculated with any necessary make up.

A different kind of unit is represented by the *cycloidal* or *Rootes blower*. In this type compression is achieved by the rotation of two elements as shown in Fig. 6.28. These rotors turn in opposite directions, so that as each pass the inlet it takes in air which is compressed between impeller and the wall and is then expelled outwards.

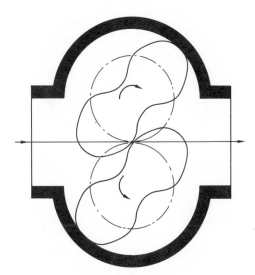

FIG. 6.28. Rootes blower.

6.4.3. Centrifugal Blowers and Compressors, including Turbocompressors (Fig. 6.29)

These depend on the conversion of kinetic energy into pressure energy. Fans are used for low pressures, and can be made to handle very large quantities of gases. For the higher pressure ratios now in demand, multistage centrifugal compressors are mainly used particularly for the requirements of high capacity chemical plants. Thus in catalytic reforming, petrochemical separation plants (ethylene manufacture), ammonia plants with a production rate of 12 kg/s, and for the very large capacity needed for natural gas fields, this type of compressor is now supreme. These units now give flowrates up to 140 m^3/s and pressures up to 5·6 MN/m^2 with the newest range going to 40 MN/m^2. It is important to accept that the very large units offer considerable savings over multiple units and that their reliability is remarkably high. The power required is also very high; thus a typical compressor operating on the process gas stream in the catalytic production of ethylene from naphtha will take 10 MW for a 6·5 kg/s plant.

Figure 6.30 shows the section of a ten-stage centrifugal compressor of the horizontally split type. This demands very high quality engineering at all stages, and the seal system is clearly vital to success. The oil film seal seen in Fig. 6.31 consists of L-shaped rings which are free to follow the radial motion of the shaft while remaining concentric to the shaft. The rings do not rotate and therefore do not wear. Oil is injected between the two rings and the shaft at a pressure greater than the gas pressure to be handled, and this oil prevents leakage past the seal rings on the housing.

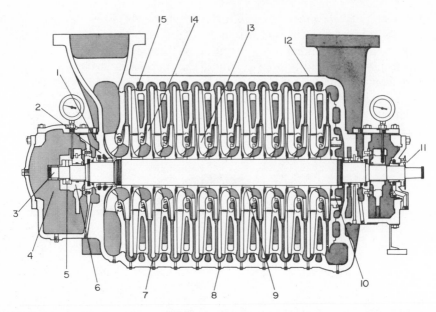

FIG. 6.30. Section of a ten-stage centrifugal compressor of the horizontal split type. 1, Isolation chamber eliminates external gas leakage to atmosphere. No fire or health hazard. 2, Shelf seals can positively eliminate or control leakage. Porting can collect gas leakage or maintain a set reference level. 3, Shelf extensions permit tandem drive of as many as five cases. 4, Bearing chambers are integral with the case to assure permanent built-in alignment. 5, Thrust bearing is double-faced type to accurately locate rotor and to handle residual thrust loads not cancelled by balancing drum. Also handles surge thrust loads. 6, Integral bearing construction allows maximum pressure rise across the case. High built-in safety factor. 7, Return bends machined, 180° turning passages from the diffuser to the diaphragm. 8, Interstage drains to wash down or remove condensate. 9, Interstage seals of labyrinth type are easily replaceable. Roll out the old—roll in the new. 10, Balancing drum. The eye of the impeller is subjected to inlet pressure—the back of the impeller to discharge pressure. Balancing drum is a rotating wall used to counteract thrust differential. 11, Horizontally split for maximum accessibility. 12, Maximum compression ratio. Up to ten stages per case. 13, Guide vanes are of either the fixed or adjustable type. Gas from the diaphragm is guided into the impeller eye at precisely the correct angle. 14, Impellers when assembled on the rotor become the primary pressure producing element in the compressor. 15, Diaphragms are the separation walls between stages. They form open diffuser passages and return passages to distribute and direct gas to the next higher stage impeller.

6.5. VACUUM-PRODUCING EQUIPMENT

A great deal of chemical plant, particularly in the area of distillation, operates under low pressures, and pumping equipment is required to create the vacuum and to maintain the low pressure. Vacuum pumps must take in gas at low pressure where the volume is very large, and discharge at atmospheric pressure, and a high compression ratio is necessary. A rotary pump is preferred to a piston type because of capacity and ease of providing a suitable drive. An alternative method very commonly used with chemical plant is the use of a *steam-jet ejector* in which the low-pressure gas is entrained with the high-pressure steam. This unit is described later.

The *sliding-vane* and the *liquid-ring* types are typical of rotary vacuum pumps. The sliding-vane pump illustrated in Fig. 6.32 consists of an eccentrically mounted rotor slotted to take the sliding vanes as shown. On rotation the vanes move outwards and the vapour is sucked in at the position of maximum area between rotor and casing and discharged through the delivery port at conditions of minimum area. Such a unit will give a pressure down to about $1·3 \text{ N/m}^2$.

Liquid-ring pumps, Fig. 6.33, are rotary displacement units in which the sealing liquid, which is usually water, is made to act as the piston. The impeller is mounted eccentrically and

FIG. 6.29. A typical Reavell turboblower.

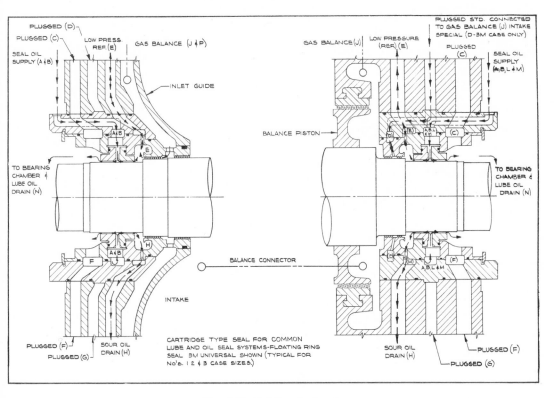

FIG. 6.31. Details of oil seal.

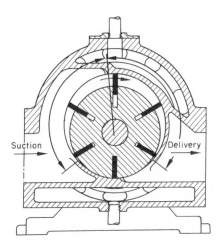

FIG. 6.32. The sliding vane rotary compressor and vacuum pump.

the liquid circulates within the case and part leaves with the exit gas stream. During
rotation, the liquid enters and leaves the individual sections acting as a piston. The sealing
liquid is fed in at a pressure equal to the discharge gas stream pressure and is separated in an
external tank and recirculated. Here the energy of the compression operation is taken up by
the gas, and the process is practically isothermal. Care must be taken to see that cavitation
does not occur, and this can arise if the vapour fraction in the cells is greater than 50 per cent.

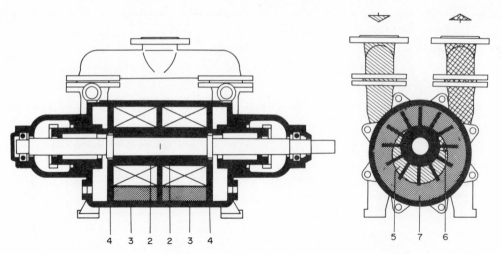

FIG. 6.33. Single-stage, single-action liquid ring gas pump: 1, shaft; 2, impeller; 3, casing; 4, guide plate; 5, suction port; 6, discharge port; 7, liquid ring.

The vapour pressure of the water restricts the lower limit of pressure in a straight unit to about $3\cdot3$ kN/m^2 and a liquid with a lower boiling point must be used if a lower pressure is required. This is usually achieved by operating the liquid ring pump in series with an additional device such as an ejector.

The steam-jet ejector, illustrated in Fig. 6.34, is very commonly used in the process industries since it has no moving parts and will handle large volumes of vapour at low pressures. The operation of the unit is shown in Fig. 6.35. The steam is fed at constant pressure P_1 and expands nearly isentropically along AC; mixing of the steam and sucked vapour occurs along CE and vapour compression continues to the throat F and to the exit at G. The steam required increases with the compression ratio so that a single-stage unit will operate down to about 17 kN/m^2 corresponding to a compression ratio of 6:1. For lower final pressures multistage units are used, and Fig. 6.36 shows the relationship between the number of stages, steam pressure, and vacuum required. If cooling water is applied between the stages an improved performance will be obtained.

As a guide, a single-stage unit gives a vacuum to $13\cdot5$ kN/m^2, a double stage from $3\cdot4$ to $13\cdot5$ kN/m^2, and a three-stage unit from $0\cdot67$ to $2\cdot7$ kN/m^2. For very low pressures, a diffusion pump is used with a rotary pump as the first stage unit. The principle of operation is that the gas diffuses into a stream of oil or mercury and is driven out of the pump by molecular bombardment. These are very specialised units, and information and advice on their application and operation should be sought direct from the manufacturers.

6.6. POWER REQUIRED FOR COMPRESSION OF GASES

Consider the compression of unit mass of gas. If the volume of the gas changes by an amount dv at a pressure P, the net work done on the gas $-\delta W$ for a reversible change is given by:

$$-\delta W = -P\,dv$$
$$= v\,dP - d(Pv) \tag{6.32}$$

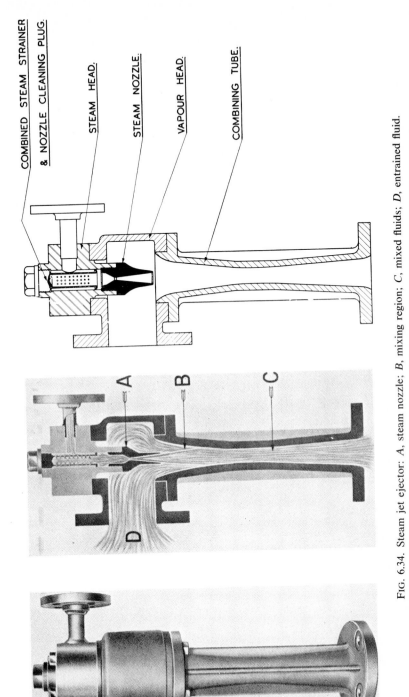

COMBINED STEAM STRAINER
& NOZZLE CLEANING PLUG.

STEAM HEAD.

STEAM NOZZLE.

VAPOUR HEAD.

COMBINING TUBE.

Fig. 6.34. Steam jet ejector: *A*, steam nozzle; *B*, mixing region; *C*, mixed fluids; *D*, entrained fluid.

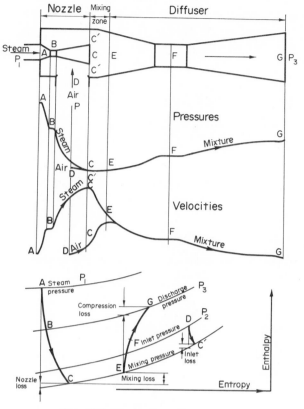

FIG. 6.35. Ejector flow phenomena.

and for an irreversible change by:

$$- \delta W = - P \, dv + \delta F$$
$$= v \, dP - d(Pv) + \delta F \tag{6.33}$$

This equation is identical with equation 2.6 given in Chapter 2.

Considering first the work done in a reversible compression, since this refers to the ideal condition for which the work of compression is a minimum, a reversible compression would have to be carried out at an infinitesimal rate and therefore is not obtainable in practice. The actual work done will be greater than that calculated, not only because of irreversibility but also because of frictional loss and leakage in the compressor. These two factors are difficult to separate and will therefore be allowed for in the overall efficiency of the machine.

The total work of compression from a pressure P_1 to a pressure P_2 is found by integrating equation 6.32. For an ideal gas undergoing an isothermal compression:

$$- \int_1^2 P \, dv = - W = P_1 v_1 \ln (P_2/P_1) \quad \text{(from equation 2.69)}$$

For the isentropic compression of an ideal gas,

$$- \int_1^2 P \, dv = - W = \frac{\gamma}{\gamma - 1} (P_2 v_2 - P_1 v_1) - (P_2 v_2 - P_1 v_1) \quad \text{(from equation 2.71)}$$

$$= \frac{1}{\gamma - 1} (P_2 v_2 - P_1 v_1) \tag{6.34}$$

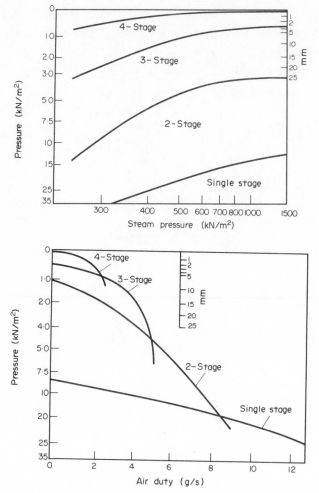

FIG. 6.36. (a) Diagram showing most suitable number of ejector stages for varying vacua and steam pressures. (b) Comparative performance curves of ejectors.

Thus
$$\int_1^2 v \, dP = -\gamma \int_1^2 P \, dv \quad \text{(comparing equations 2.71 and 6.34)}$$

for isentropic conditions and

$$\therefore \qquad -W = \frac{1}{\gamma - 1} P_1 v_1 \left[\left(\frac{P_2}{P_1} \right)^{(\gamma-1)/\gamma} - 1 \right] \qquad (6.35)$$

Under these conditions the whole of the energy of compression appears as heat in the gas.

For the isentropic compression of a mass m of gas:

$$-Wm = \frac{1}{\gamma - 1} P_1 V_1 \left[\left(\frac{P_2}{P_1} \right)^{(\gamma-1)/\gamma} - 1 \right] \qquad (6.36)$$

where V_1 is the volume of a mass m of gas at a pressure P_1.

If the conditions are intermediate between isothermal and isentropic, we must write k in place of γ, where $\gamma > k > 1$.

If the gas deviates appreciably from the ideal gas laws over the range of conditions considered, the work of compression is most conveniently calculated from the change in the thermodynamic properties of the gas.

Thus $\qquad\qquad dU = T\,dS - P\,dv$ (from equation 2.5)

$\therefore \qquad\qquad -\delta W = -P\,dv = dU - T\,dS$ $\qquad\qquad\qquad\qquad\qquad$ (6.37)

Under isothermal conditions: $\qquad -W = \Delta U - T\,\Delta S$ $\qquad\qquad\qquad\qquad$ (6.38)

Under isentropic conditions: $\qquad\qquad -W = \Delta U$ $\qquad\qquad\qquad\qquad\qquad\quad$ (6.39)

The above equations give the work done during a simple compression of gas in a cylinder and do not take account of the work done either during the admission of the gas prior to compression or during the expulsion of the compressed gas.

Suppose that, after the compression of a volume V_1 of gas at a pressure P_1 to a pressure P_2, the whole of the gas is expelled at constant pressure P_2 and a fresh charge of gas is admitted at a pressure P_1. The cycle can be followed in Fig. 6.37, where the pressure P is plotted as ordinate against the volume V as abscissa.

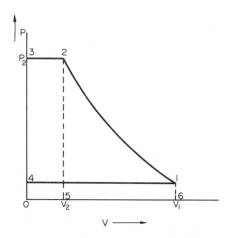

FIG. 6.37. Single-stage compression cycle—no clearance.

Point 1 represents the initial condition of the gas (P_1 and V_1).
Line 1–2 represents the compression of gas to pressure P_2, volume V_2.
Line 2–3 represents the expulsion of the gas at a constant pressure P_2.
Line 3–4 represents a sudden reduction in the pressure in the cylinder from P_2 to P_1. As the whole of the gas has been expelled, this can be regarded as taking place instantaneously.
Line 4–1 represents the suction stroke of the piston, during which a volume V_1 of gas is admitted at constant pressure, P_1.
It will be noted that the mass of gas in the cylinder varies during the cycle. The work done by the compressor during each phase of the cycle is as follows:

Compression	$-\displaystyle\int_1^2 P\,dV$	(Area 1–2–5–6)
Expulsion	$P_2 V_2$	(Area 2–3–0–5)
Suction	$-P_1 V_1$	$-$(Area 4–0–6–1)

The total work done per cycle

$$= -\int_1^2 P \, dV + P_2 V_2 - P_1 V_1 \quad \text{(Area 1–2–3–4)}$$

$$= \int_1^2 V \, dP \tag{6.40}$$

The work of compression for an ideal gas per cycle

$$= P_1 V_1 \ln (P_2/P_1) \tag{6.41}$$

under isothermal conditions.

Under isentropic conditions, work of compression

$$= P_1 V_1 \frac{\gamma}{\gamma - 1} \left[\left(\frac{P_2}{P_1} \right)^{(\gamma-1)/\gamma} - 1 \right] \tag{6.42}$$

Again, working in terms of the thermodynamic properties of the gas,

$$v \, dP = d(Pv) - P \, dv$$
$$= d(Pv) + dU + T \, dS$$
$$= dH - T \, dS \tag{6.43}$$

For an isothermal process, $-mW = m(\Delta H - T \, \Delta S)$ \hfill (6.44)

For an isentropic process, $-mW = m \, \Delta H$ \hfill (6.45)

where m is the mass of gas compressed per cycle.

6.6.1. Clearance Volume

In practice, it is not possible to expel the whole of the gas from the cylinder at the end of the compression; the volume remaining in the cylinder after the forward stroke of the piston is termed the clearance volume. The volume displaced by the piston is termed the swept volume, and therefore the total volume of the cylinder is made up of the clearance volume plus the swept volume. The clearance c is defined as the ratio of the clearance volume to the swept volume.

A typical cycle for a compressor with a finite clearance volume can be followed by reference to Fig. 6.38. A volume V_1 of gas at a pressure P_1 is admitted to the cylinder; its condition is represented by point 1.

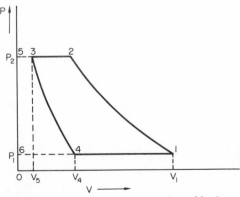

FIG. 6.38. Single-stage compression cycle—with clearance.

Line 1–2 represents the compression of the gas to a pressure P_2 and volume V_2.

Line 2–3 represents the expulsion of gas at constant pressure P_2, so that the volume remaining in the cylinder is V_3.

Line 3–4 represents an expansion of this residual gas to the lower pressure P_1 and volume V_4 during the return stroke.

Line 4–1 represents the introduction of fresh gas into the cylinder at constant pressure P_1. The work done on the gas during each stage of the cycle is as follows.

Compression $\qquad\qquad\qquad\qquad -\displaystyle\int_{V_1}^{V_2} P\,dV$

Expulsion $\qquad\qquad\qquad\qquad P_2(V_2 - V_3)$

Expansion $\qquad\qquad\qquad\qquad -\displaystyle\int_{V_3}^{V_4} P\,dV$

Suction $\qquad\qquad\qquad\qquad -P_1(V_1 - V_4)$

The total work done during the cycle is equal to the sum of these four components. It is represented by the area 1–2–3–4, which is equal to area 1–2–5–6 less area 4–3–5–6. If the compression and expansion are taken as isentropic, the work done per cycle therefore

$$= P_1 V_1 \frac{\gamma}{\gamma-1}\left[\left(\frac{P_2}{P_1}\right)^{(\gamma-1)/\gamma} - 1\right] - P_1 V_4 \frac{\gamma}{\gamma-1}\left[\left(\frac{P_2}{P_1}\right)^{(\gamma-1)/\gamma} - 1\right]$$

$$= P_1(V_1 - V_4)\frac{\gamma}{\gamma-1}\left[\left(\frac{P_2}{P_1}\right)^{(\gamma-1)/\gamma} - 1\right] \tag{6.46}$$

Thus, theoretically, the clearance volume does not affect the work done per unit mass of gas, since $V_1 - V_4$ is the volume admitted per cycle. It does, however, influence the quantity of gas admitted and therefore the work done per cycle. In practice, however, compression and expansion are not reversible, and losses arise from the compression and expansion of the clearance gases. This effect is particularly serious at high compression ratios.

Now V_4 is not known explicitly, but can be calculated in terms of V_3, the clearance volume.

For isentropic conditions:
$$V_4 = V_3(P_2/P_1)^{1/\gamma}$$
and
$$V_1 - V_4 = (V_1 - V_3) + V_3 - V_3(P_2/P_1)^{1/\gamma}$$
$$= (V_1 - V_3)\left[1 + \frac{V_3}{V_1 - V_3} - \frac{V_3}{V_1 - V_3}\left(\frac{P_2}{P_1}\right)^{1/\gamma}\right]$$

Now $V_1 - V_3$ is the swept volume, V_s, say; and $V_3/(V_1 - V_3)$ is the clearance c.

$\therefore \qquad\qquad V_1 - V_4 = V_s[1 + c - c(P_2/P_1)^{1/\gamma}] \tag{6.47}$

The total work done on the fluid per cycle is therefore

$$P_1 V_s \frac{\gamma}{\gamma-1}\left[\left(\frac{P_2}{P_1}\right)^{(\gamma-1)/\gamma} - 1\right]\left[1 + c - c\left(\frac{P_2}{P_1}\right)^{1/\gamma}\right] \tag{6.48}$$

The factor $[1 + c - c(P_2/P_1)^{1/\gamma}]$ is called the theoretical volumetric efficiency and is a measure of the effect of the clearance on an isentropic compression. The actual volumetric efficiency will be affected, in addition, by the inertia of the valves and leakage past the piston.

The gas is frequently cooled during compression so that the work done per cycle is less than that given by equation 6.48, γ is replaced by some smaller quantity k. The greater the rate of heat removal, the less is the work done. The isothermal compression is usually taken as the condition for least work of compression, but clearly the energy consumption can be reduced below this value if the gas is artificially cooled below its initial temperature as it is compressed. This is not a practicable possibility because of the large amount of energy required to refrigerate the cooling fluid. It can be seen that the theoretical volumetric efficiency decreases as the rate of heat removal is increased since γ is replaced by the smaller quantity k.

In practice the cylinders are usually water-cooled. The work of compression is thereby reduced though the effect is usually small. The reduction in temperature does, however, improve the mechanical operation of the compressor and makes lubrication easier.

6.6.2. Multistage Compressors

If the required pressure ratio P_2/P_1 is large, it is not practicable to carry out the whole of the compression in a single cylinder because of the high temperatures which would be set up and the adverse effects of clearance volume on the efficiency. Further, lubrication would be difficult due to carbonisation of the oil, and there would be a risk of causing oil mist explosions in the cylinders when gases containing oxygen were being compressed. The mechanical construction also would be difficult because the single cylinder would have to be strong enough to withstand the final pressure and yet large enough to hold the gas at the initial pressure P_1. In the multistage compressor, the gas passes through a number of cylinders of gradually decreasing volume and can be cooled between the stages. The maximum pressure ratio normally obtained in a single cylinder is 10 but values above 6 are unusual.

The operation of the multistage compressor can conveniently be followed again on a pressure–volume diagram (Fig. 6.39). The effect of clearance volume will be neglected at first. The area 1–2–3–4 represents the work done in compressing isentropically from P_1 to P_2 in a single stage. The area 1–2–5–4 represents the necessary work for an isothermal compression. Now consider a multistage isentropic compression in which the intermediate pressures are P_{i1}, P_{i2}, etc. The gas will be assumed to be cooled to its initial temperature in an interstage cooler before it enters each cylinder.

Line 1–2 represents the suction stroke of the first stage where a volume V_1 of gas is admitted at a pressure P_1.

Line 2–6 represents an isentropic compression to a pressure P_{i1}.

Line 6–7 represents the delivery of the gas from the first stage at a constant pressure P_{i1}.

Line 7–8 represents the suction stroke of the second stage. The volume of the gas has been

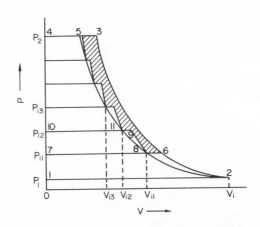

FIG. 6.39. Multistage compression cycle with interstage cooling.

reduced in the interstage cooler to V_{i1}; that which would have been obtained as a result of an isothermal compression to P_{i1}.

Line 8–9 represents an isentropic compression in the second stage from a pressure P_{i1} to a pressure P_{i2}.

Line 9–10 represents the delivery stroke of the second stage.

Line 10–11 represents the suction stroke of the third stage. Point 11 again lies on the line 2–5, representing an isothermal compression.

It is seen that the overall work done on the gas is intermediate between that for a single stage isothermal compression and that for an isentropic compression. The net saving in energy is shown as the shaded area in Fig. 6.39.

The total work done per cycle W'

$$= P_1 V_1 \frac{\gamma}{\gamma - 1} \left[\left(\frac{P_{i1}}{P_1} \right)^{(\gamma-1)/\gamma} - 1 \right] + P_{i1} V_{i1} \frac{\gamma}{\gamma - 1} \left[\left(\frac{P_{i2}}{P_{i1}} \right)^{(\gamma-1)/\gamma} - 1 \right] + \cdots$$

for an isentropic compression.

For perfect interstage cooling:

$$P_1 V_1 = P_{i1} V_{i1} = P_{i2} V_{i2} = \cdots$$

$$\therefore \qquad W' = P_1 V_1 \frac{\gamma}{\gamma - 1} \left[\left(\frac{P_{i1}}{P_1} \right)^{(\gamma-1)/\gamma} + \left(\frac{P_{i2}}{P_{i1}} \right)^{(\gamma-1)/\gamma} + \cdots - n \right]$$

where n is the number of stages.

It is now required to find how the total work per cycle W' is affected by the choice of the intermediate pressures P_{i1}, P_{i2}, etc. The work will be a minimum when

$$\frac{\partial W'}{\partial P_{i1}} = \frac{\partial W'}{\partial P_{i2}} = \frac{\partial W'}{\partial P_{i3}} = \cdots = 0.$$

When

$$\frac{\partial W'}{\partial P_{i1}} = 0$$

$$P_1 V_1 \frac{\gamma}{\gamma - 1} \left[\frac{\gamma - 1}{\gamma} \left(\frac{P_{i1}}{P_1} \right)^{(\gamma-1)/\gamma} P_{i1}^{-1} + \frac{1 - \gamma}{\gamma} \left(\frac{P_{i2}}{P_{i1}} \right)^{(\gamma-1)/\gamma} P_{i1}^{-1} \right] = 0$$

i.e. $$\frac{P_{i1}}{P_1} = \frac{P_{i2}}{P_{i1}} \qquad (6.49)$$

The same procedure is then adopted for obtaining the optimum value of P_{i2} and hence:

$$\frac{P_{i2}}{P_{i1}} = \frac{P_{i3}}{P_{i2}} \qquad (6.50)$$

Thus the intermediate pressures should be arranged so that the compression ratio is the same in each cylinder; and equal work is then done in each cylinder.

The minimum work of compression in a compressor of n stages is therefore

$$P_1 V_1 \frac{\gamma}{\gamma - 1} \left[n \left(\frac{P_2}{P_1} \right)^{(\gamma-1)/n\gamma} - n \right] = n P_1 V_1 \frac{\gamma}{\gamma - 1} \left[\left(\frac{P_2}{P_1} \right)^{(\gamma-1)/n\gamma} - 1 \right] \qquad (6.51)$$

The effect of clearance volume can now be taken into account. If the clearances in the successive cylinders are $c_1, c_2, c_3, \ldots$, the theoretical volumetric efficiency of the first cylinder

$$= 1 + c_1 - c_1 (P_{i1}/P_1)^{1/\gamma}$$

Assuming that the same compression ratio is used in each cylinder, then the theoretical volumetric efficiency of the first stage is:

$$1 + c_1 - c_1 (P_2/P_1)^{1/n\gamma}$$

If the swept volumes of the cylinders are $V_{s1}, V_{s2}, \ldots$, the volume of gas admitted to the first cylinder

$$= V_{s1}[1 + c_1 - c_1(P_2/P_1)^{1/n\gamma}] \tag{6.52}$$

The same mass of gas passes through each of the cylinders and, therefore, if the interstage coolers are assumed perfectly efficient, the ratio of the volumes of gas admitted to successive cylinders is $(P_1/P_2)^{1/n}$. The volume of gas admitted to the second cylinder then

$$= V_{s2}\left[1 + c_2 - c_2\left(\frac{P_2}{P_1}\right)^{1/n\gamma}\right] = V_{s1}\left[1 + c_1 - c_1\left(\frac{P_2}{P_1}\right)^{1/n\gamma}\right]\left(\frac{P_1}{P_2}\right)^{1/n}$$

$$\therefore \qquad \frac{V_{s1}}{V_{s2}} = \frac{1 + c_2 - c_2(P_2/P_1)^{1/n\gamma}}{1 + c_1 - c_1(P_2/P_1)^{1/n\gamma}}\left(\frac{P_2}{P_1}\right)^{1/n} \tag{6.53}$$

In this manner the swept volume of each cylinder can be calculated in terms of V_{s1} and $c_1, c_2, \ldots$, and the cylinder dimensions determined.

When the gas does not behave as an ideal gas, the change in its condition can be followed on a temperature–entropy or an enthalpy–entropy diagram. The intermediate pressures $P_{i1}, P_{i2}, \ldots$, are then selected so that the enthalpy change (ΔH) is the same in each cylinder.

Several opposing factors will influence the number of stages selected for a given compression. The larger the number of cylinders the greater is the mechanical complexity. Against this must be balanced the higher theoretical efficiency, the smaller mechanical strains set up in the cylinders, and the moving parts and the greater ease of lubrication at the lower temperatures that are experienced. Compressors with as many as nine stages are used for very high pressures.

6.6.3. Compressor Efficiencies

The efficiency quoted for a compressor is usually either an isothermal efficiency or an isentropic efficiency. The isothermal efficiency is the ratio of the work required for an ideal isothermal compression to the energy actually expended in the compressor. The isentropic efficiency is defined in a corresponding manner on the assumption that the whole compression is carried out in a single cylinder. Since the energy expended in an isentropic compression is greater than that for an isothermal compression, the isentropic efficiency is always the greater of the two. Clearly the efficiencies will depend on the heat transfer between the gas undergoing compression and the surroundings and on how closely the process approaches a reversible compression.

The efficiency of the compression will also be affected by a number of other factors which are all connected with the mechanical construction of the compressor. Thus the efficiency will be reduced as a result of leakage past the piston and the valves and because of throttling of the gas at the valves. Further, the mechanical friction of the machine will lower the efficiency and the overall efficiency will be affected by the efficiency of the driving motor and transmission.

Example 6.4. A single-acting air compressor supplies $0 \cdot 1 \, \text{m}^3/\text{s}$ of air (at STP) compressed to $380 \, \text{kN/m}^2$ from $101 \cdot 3 \, \text{kN/m}^2$. If the suction temperature is $289 \, \text{K}$, the stroke is $0 \cdot 25 \, \text{m}$, and the speed is $4 \cdot 0 \, \text{Hz}$, what is the cylinder diameter? Assume the cylinder clearance is 4 per cent and compression and re-expansion are isentropic $(\gamma = 1 \cdot 4)$. What are the theoretical power requirements for the compression?

Solution. Volume per stroke $= (0 \cdot 1/4 \cdot 0)(289/273) = 0 \cdot 0264 \, \text{m}^3$

Compression ratio $= (380/101 \cdot 3) = 3 \cdot 75$

The swept volume is given by equation 6.47:

$$0.0264 = V_s[1 + 0.04 - 0.04(3.75)^{1/1.4}]$$

$$\therefore \qquad V_s = 0.0264/(1.04 - 0.04 \times 2.7) = 0.0283 \text{ m}^3$$

Thus cross-sectional area of cylinder $= 0.0283/0.25 = 0.113 \text{ m}^2$ and diameter

$$= [0.113/(\pi/4)]^{0.5} = \underline{0.38 \text{ m}}$$

From equation 6.46, work of compression per cycle

$$= 101,300 \times 0.0264[1.4/(1.4 - 1.0)][(3.75)^{0.4/1.4} - 1]$$

$$= 9360(1.457 - 1) = 4278 \text{ J}$$

Theoretical power requirements

$$= (4278 \times 4) = 17,110 \text{ W} \quad \text{or} \quad \underline{17.1 \text{ kW}}$$

Example 6.5. Air at 290 K is compressed from 101·3 kN/m² to 2065 kN/m² in a two-stage compressor operating with a mechanical efficiency of 85 per cent. The relation between pressure and volume during the compression stroke and expansion of the clearance gas is $PV^{1.25} = \text{constant}$. The compression ratio in each of the two cylinders is the same, and the interstage cooler may be assumed 100 per cent efficient. If the clearances in the two cylinders are 4 per cent and 5 per cent respectively, calculate:

(a) the work of compression per kg of air compressed;
(b) the isothermal efficiency;
(c) the isentropic efficiency ($\gamma = 1.4$), and
(d) the ratio of the swept volumes in the two cylinders.

Solution. Overall compression ratio $= 2065/101.3 = 20.4$

Specific volume of air at 290 K $= (22.4/28.8)(290/273) = 0.826 \text{ m}^3/\text{kg}$

From equation 6.51, work of compression

$$= 101.3 \times 0.826 \times 2[1.25/(1.25 - 1)][(20.4)^{0.25/2.5} - 1]$$

$$= 836.7(1.351 - 1)$$

$$= 293.7 \text{ kJ/kg}$$

Energy supplied to the compressor, i.e. work of compression

$$= 293.7/0.85 = \underline{345.5 \text{ kJ/kg}}$$

From equation 6.41, work done in isothermal compression of 1 kg of gas

$$= 101.3 \times 0.826 \ln 20.4$$

$$= 83.7 \times 3.015 = 252.3 \text{ kJ/kg}$$

Isothermal efficiency $\qquad = 100 \times 252.3/345.5 = \underline{73\%}$

From equation 6.42, work done in isentropic compression of 1 kg of gas

$$= 101.3 \times 0.826(1.4/0.4)[(20.4)^{0.4/1.4} - 1]$$

$$= 292.9(2.36 - 1) = 398.3 \text{ kJ/kg}$$

Isentropic efficiency $\qquad = 100 \times 398.3/345.5 = \underline{115\%}$

From equation 6.52, volume swept out in first cylinder in compression of 1 kg of gas is given by:

$$0.826 = V_{s_1}[1 + 0.04 - 0.04(20.4)^{1/(2 \times 1.25)}]$$
$$= V_{s_1} \times 0.906$$

$\therefore \qquad \qquad V_{s_1} = 0.912 \text{ m}^3/\text{kg}$

Similarly, the swept volume of the second cylinder is given by

$$0.826(1/20.4)^{0.5} = V_{s_2}[1 + 0.05 - 0.05(20.4)^{1/(2 \times 1.25)}]$$
$$0.183 = 0.883\, V_{s_2}$$

$\therefore \qquad \qquad V_{s_2} = 0.207 \text{ m}^3/\text{kg}$

$\therefore \qquad \qquad V_{s_1}/V_{s_2} = \underline{\underline{4.41}}$

6.7. POWER REQUIREMENTS FOR PUMPING THROUGH PIPELINES

A fluid will flow of its own accord so long as the energy per unit mass of fluid decreases in the direction of flow. It can be made to flow in the opposite direction only by the action of some external agent, such as a pump which supplies energy and increases the pressure at the upstream end of the system.

The energy balance from equation 2.55 is:

$$(\Delta u^2/2\alpha) + g\, \Delta z + \int_1^2 v\, dP + W_s + F = 0$$

The work done on unit mass of fluid is $-W_s$ and the total rate at which energy must be transferred to the fluid is $-GW_s$, when the mass rate of flow is G:

$$-GW_s = G\left(\Delta u^2/2\alpha + g\, \Delta z + \int_1^2 v\, dP + F\right) \tag{6.54}$$

6.7.1. Liquids

F can be calculated directly for the flow of liquid through a uniform pipe. If a liquid is pumped through a height Δz from one open tank to another and none of the kinetic energy is recoverable as pressure energy, the fluid pressure is the same at both ends of the system and $\int_1^2 v\, dP$ is zero. The power requirement, is, therefore:

$$G(u^2/2\alpha + g\, \Delta z + F) \tag{6.55}$$

6.7.2. Gases

If a gas is pumped under turbulent flow conditions from a reservoir at a pressure P_1 to a second reservoir at a higher pressure P_2 through a uniform pipe of cross-sectional area A by means of a pump situated at the upstream end, the power required is:

$$G\left(u^2/2 + \int_1^2 v\, dP + F\right) \tag{6.56}$$

neglecting pressure changes arising from change in vertical height.

In order to make the gas flow, the pump must raise the pressure at the upstream end of the pipe to some value P_3, which is greater than P_2. The required value of P_3 will depend somewhat on the conditions of flow in the pipe. Thus for isothermal conditions, P_3 may be calculated from equation 4.8, since the downstream pressure P_2 and the mass rate of flow G are known. For non-isothermal conditions, the appropriate equation, such as 4.18 or 4.25, must be used.

The power requirement is then that for compression of the gas from pressure P_1 to P_3 and for imparting the necessary kinetic energy to it. Under normal conditions, however, the kinetic energy term is negligible. Thus for an isothermal efficiency of compression of η, the power required is:

$$(1/\eta)GP_1v_1 \ln (P_3/P_1) \qquad (6.57)$$

If a fluid is to be pumped between two points, the diameter of the pipeline should be chosen so that the overall cost of operation is a minimum. The smaller the diameter, the lower is the initial cost of the line but the greater is the cost of pumping; an economic balance must, therefore, be achieved.

The initial cost of a pipeline and the depreciation and maintenance costs will be approximately proportional to the diameter raised to a power of between 1·0 and 1·5. The power for pumping an incompressible fluid at a given rate G is made up of two parts:

(1) that necessitated by the difference in static pressure and vertical height at the two ends of the system (this is independent of the diameter of the pipe);
(2) that attributable to the kinetic energy of the fluid and the work done against friction. If the kinetic energy is small, this is equal to:

$$G4(R/\rho u^2)(l/d)u^2 \qquad (6.58)$$

which is proportional to $d^{-4\cdot5 \text{ to} -5}$ for turbulent flow, since $u \propto d^{-2}$ and $R/\rho u^2 \propto u^{-0\cdot25 \text{ to} 0}$, according to the roughness of the pipe.

The power requirement can, therefore, be calculated as a function of d and the cost obtained. The total cost per annum is then plotted against the diameter of pipe and the optimum conditions are given by the minimum on the curve.

Example 6.6. 600 cm³/s water at 320 K is pumped through a 40 mm i.d. pipe, through a length of 150 m in a horizontal direction, and up through a vertical height of 10 m. In the pipe there are a control valve, equivalent to 200 pipe diameters, and other pipe fittings equivalent to 60 pipe diameters. Also in the line is a heat exchanger across which the head lost is 2 m water. Assuming the main pipe has a roughness of 0·0002 m, what power must be supplied to the pump if it is 60% efficient?

Solution. Area for flow $= (0\cdot040)^2 \pi/4 = 0\cdot0012 \text{ m}^2$

Flow of water $= 600 \times 10^{-6} \text{ m}^3/\text{s}$

∴   velocity $= 600 \times 10^{-6}/0\cdot0012 = 0\cdot50 \text{ m/s}$

At 320 K, $\mu = 0\cdot65 \text{ mN s/m}^2 = 0\cdot65 \times 10^{-3} \text{ N s/m}^2$

$\rho = 1000 \text{ kg/m}^3$

∴ $Re = 0\cdot040 \times 0\cdot50 \times 1000/0\cdot65 \times 10^{-3} = 30{,}780$

∴ $(R/\rho u^2) = 0\cdot004$ for a relative roughness of $e/d = 0\cdot0002/0\cdot040 = 0\cdot005$

Equivalent length of pipe $= 150 + 10 + (260 \times 0.040) = 170.4$ m

$\therefore$
$$h_f = 4(R/\rho u^2)(l/d)(u^2/g)$$
$$= 4 \times 0.004(170.4/0.040)(0.50^2/9.81)$$
$$= 1.74 \text{ m}$$

$\therefore$ Total head to be developed $= 1.74 \times 10 + 2 = 13.74$ m

Mass flow of water $\qquad = 600 \times 10^{-6} \times 1000 = 0.60$ kg/s

$\therefore$ Power required $\qquad = 0.60 \times 13.74 \times 9.81 = 80.9$ W

$\therefore$ Power to be supplied $\qquad = 80.9 \times 100/60 = \underline{\underline{135 \text{ W}}}$

The kinetic energy head, $u^2/2g$ has been neglected as this represents only some $[0.5^2/(2 \times 9.81)] = 0.013$ m.

Example 6.7. Hydrogen is pumped from a reservoir at 2 MN/m² through a clean horizontal mild steel pipe 50 mm in diameter and 500 m long. The downstream pressure is also 2 MN/m² and the pressure of the gas is raised to 2·25 MN/m² by a pump at the upstream end of the pipe. The conditions of flow are isothermal and the temperature of the gas is 295 K. What is the flow rate and what is the effective rate of working of the pump if it operates with an efficiency of 60 per cent?

Viscosity of hydrogen $= 0.009$ mNs/m² at 295 K.

Solution. Viscosity of hydrogen $= 0.009$ mN s/m² or 9×10^{-6} N s/m²

Density of hydrogen at the mean pressure of 2·27 MN/m²
$$= (2/22.4)(2250/101.3)(273/295) = 1.83 \text{ kg/m}^3$$

Firstly, an approximate value of G is obtained by neglecting the kinetic energy of the fluid. Taking P_1 and P_2 as the pressures at the upstream and downstream ends of the pipe, 2.5×10^6 and 2.0×10^6 N/m², then in equation 4.11:

$$P_1 - P_2 = 4(R/\rho u^2)(l/d)\rho_m u_m^2$$

or
$$0.5 \times 10^6 = 4(R/\rho u^2)(500/0.050) \times 1.83 u_m^2 \text{ N/m}^2$$

$\therefore$
$$(R/\rho u^2)u_m^2 = 6.83$$

$\therefore$
$$(R/\rho u^2)Re^2 = 6.83 \times 0.050^2 \times 1.83^2/(9 \times 10^{-6})^2$$
$$= 7.02 \times 10^8$$

Taking the roughness of the pipe surface, e as 0·00005 m,

$$e/d = 0.001 \text{ and } Re = 5.7 \times 10^5 \text{ from Fig. 3.8}$$

$\therefore$
$$4G/\pi\mu d = 5.7 \times 10^5$$

$\therefore$
$$G = 5.7 \times 10^5(\pi/4) \times 9 \times 10^{-6} \times 0.050$$
$$= 0.201 \text{ kg/s}$$

From Fig. 3.7, $(R/\rho u^2) = 0.0024$

Taking the kinetic energy of the fluid into account, equation 4.10 may be used:

$$(G/A)^2 \ln(P_1/P_2) + (P_2 - P_1)\rho_m + 4(R/\rho u^2)(l/d)(G/A)^2 = 0$$

Using the value of $(R/\rho u^2)$ obtained by neglecting the kinetic energy,

$$(G/A)^2 \ln (2 \cdot 5/2 \cdot 0) - (0 \cdot 5 \times 10^6 \times 1 \cdot 83) + [4 \times 0 \cdot 0024(500/0 \cdot 05)](G/A)^2 = 0$$

$$0 \cdot 223(G/A)^2 - 915,000 + 96 \cdot 0(G/A)^2 = 0$$

$$\therefore \qquad (G/A) = 97 \cdot 5$$

$$\therefore \qquad G = 97 \cdot 5(\pi/4)0 \cdot 050^2$$

$$= 0 \cdot 200 \text{ kg/s}$$

Thus, as is commonly the case, when the pressure drop is a relatively small proportion of the total pressure, the change in kinetic energy is negligible compared with the frictional losses. This would not be true had the pressure drop been much greater.

The power requirements for the pump can now be calculated from equation 6.57.

$$\text{Power} = GP_m v_m \ln (P_1/P_2)/\eta$$

$$= 0 \cdot 200 \times 2 \cdot 25 \times 10^6 \times (1/1 \cdot 83) \times 0 \cdot 223/0 \cdot 60$$

$$= 9 \cdot 14 \times 10^4 \text{ W} \quad \text{or} \quad \underline{\underline{91 \cdot 4 \text{ kW}}}$$

6.8.   FURTHER READING

HOLLAND, F. A. and CHAPMAN, F. S.: *Pumping of Liquids* (Reinhold Publishing Corp., New York, 1966).

ENGINEERING EQUIPMENT USERS' ASSOCIATION: *Vacuum Producing Equipment*, EEUA Handbook No. 11 (Constable, London, 1961).

ENGINEERING EQUIPMENT USERS' ASSOCIATION: *Electrically Driven Glandless Pumps*, EEUA Handbook No. 26 (Constable, London, 1968).

ENGINEERING EQUIPMENT USERS' ASSOCIATION: *Guide to the Selection of Rotodynamic Pumps*, EEUA Handbook No. 30 (Constable, London, 1972).

LAZARKIENICZ, S. and TROSKOLANSKI, A. T.: *Impeller Pumps* (Pergamon Press, Oxford, 1965).

M. W. KELLOGG CO.: *Design of Piping Systems*, 2nd edn. (Wiley, Chichester, 1964).

6.9.   NOMENCLATURE

		Units in SI system	Dimensions in M, L, T, θ
A	Cross-sectional area	m^2	L^2
b	Width of pump impeller	m	L
c	Clearance in cylinder, i.e. ratio clearance volume/swept volume	—	—
d	Diameter	m	L
F	Energy degraded due to irreversibility per unit mass of fluid	J/kg	L^{2}T^{-2}
G	Mass flow rate	kg/s	MT^{-1}
g	Acceleration due to gravity	m/s^2	LT^{-2}
H	Enthalpy per unit mass	J/kg	L^{2}T^{-2}
h	Head	m	L
h_a	Heat equivalent to atmospheric pressure	m	L
h_f	Friction head	m	L
h_i	Head at pump inlet	m	L
h_o	Height of liquid above pump inlet	m	L
h_r	Height through which liquid is raised	m	L
h_s	Submergence of air lift pump	m	L
l	Length	m	L
m	Mass of gas	kg	M
M	Mass of liquid	kg	M
N	Revolutions per unit time	Hz	T^{-1}
N_s	Specific speed of a pump	—	—
n	Number of stages of compression	—	—
P	Pressure	N/m^2	ML^{-1}T^{-2}

		Units in SI system	Dimensions in M, L, T, θ.
P_a	Atmospheric pressure	N/m²	$ML^{-1}T^{-2}$
P_i	Intermediate pressure	N/m²	$ML^{-1}T^{-2}$
P_v	Vapour pressure of liquid	N/m²	$ML^{-1}T^{-2}$
P_0	Pressure at pump suction tank	N/m²	$ML^{-1}T^{-2}$
p	Power	W	$ML^{2}T^{-3}$
Q	Volumetric rate of flow	m³/s	$L^{3}T^{-1}$
q	Heat added from surrounding per unit mass of fluid	J/kg	$L^{2}T^{-2}$
R	Shear stress at pipe wall	N/m²	$ML^{-1}T^{-2}$
r	Radius	m	L
S	Entropy per unit mass	J/kg K	$L^{2}T^{-2}\theta_{-}^{-1}$
T	Temperature	K	θ
t	Time	s	T
U	Internal energy per unit mass	J/kg	$L^{2}T^{-2}$
u	Velocity of fluid	m/s	LT^{-1}
u_i	Velocity at pump inlet	m/s	LT^{-1}
u_t	Tangential velocity of tip of vane of compressor	m/s	LT^{-1}
u_v	Velocity of fluid relative to vane of compressor	m/s	LT^{-1}
u_1	Velocity at inlet to centrifugal pump	m/s	LT^{-1}
u_2	Velocity at outlet to centrifugal pump	m/s	LT^{-1}
V	Volume	m³	L^{3}
V_s	Swept volume	m³	L^{3}
v	Specific volume	m³/kg	$M^{-1}L^{3}$
v_a	Specific volume at atmospheric pressure	m³/kg	$M^{-1}L^{3}$
W	Net work done per unit mass	J/kg	$L^{2}T^{-2}$
W_s	Shaft work per unit mass	J/kg	$L^{2}T^{-2}$
W'	Work of compressor per cycle	J	$ML^{2}T^{-2}$
z	Height	m	L
α	Correction factor for kinetic energy of fluid	—	—
β	Angle between tangential direction and blade of impeller at its tip	—	—
γ	Ratio of specific heat at constant pressure to specific heat at constant volume	—	—
ρ	Density	kg/m³	ML^{-3}
θ	Angle between tangential direction and direction of motion of fluid	—	—
τ	Torque	Nm	$ML^{2}T^{-2}$
ω	Angular velocity	rad/s	T^{-1}
Δ	Finite difference in quantity	—	—
η	Efficiency	—	—
Π	Dimensionless group	—	—
ϕ	Friction factor	—	—

CHAPTER 7

Heat Transfer

7.1. INTRODUCTION

In the majority of chemical processes heat is either given out or absorbed, and in a very wide range of chemical plant, fluids must often be either heated or cooled. Thus in furnaces, evaporators, distillation units, driers, and reaction vessels one of the major problems is that of transferring heat at the desired rate. Alternatively, it may be necessary to prevent the loss of heat from a hot vessel or steam pipe. The control of the flow of heat in the desired manner forms one of the most important sections of chemical engineering. Provided that a temperature difference exists between two parts of a system, heat transfer will take place in one or more of three different ways.

Conduction. In a solid, the flow of heat by conduction is the result of the transfer of vibrational energy from one molecule to another, and in fluids it occurs in addition as a result of the transfer of kinetic energy. Heat transfer by conduction may also arise from the movement of free electrons. This process is particularly important with metals and accounts for their high thermal conductivities.

Convection. Heat transfer by convection is attributable to macroscopic motion of the fluid and therefore is confined to liquids and gases. In natural convection it is caused by differences in density arising from temperature gradients in the system. In forced convection, it is due to eddy currents in a fluid in turbulent motion.

Radiation. All materials radiate thermal energy in the form of electromagnetic waves. When this radiation falls on a second body it may be partially reflected, transmitted, or absorbed. It is only the fraction that is absorbed that appears as heat in the body.

7.2. BASIC CONSIDERATIONS

In many of the applications of heat transfer in chemical engineering, each of the mechanisms of conduction, convection, and radiation is involved. In the majority of heat-exchanger units, the process is complicated in that the heat has to pass through a number of intervening layers before it reaches the material whose temperature is to be raised and the form of the equation may be complex.

As an example take the problem of transferring heat to oil in a crude oil still from a flame obtained by burning waste refinery gas. The heat from the flame is transferred by a combination of radiation and convection to the outer surface of the pipes, then passes through the walls by conduction, and, finally, is transferred to the boiling oil by convection. After prolonged usage, solid deposits may form on both the inner and outer walls of the pipes, and these will then contribute additional resistance to the transfer of heat. The simplest form of equation which represents this heat transfer operation can be written as:

$$Q = UA\Delta T \tag{7.1}$$

where Q is the heat transferred per unit time, A the area available for the flow of heat, ΔT the difference in temperature between the flame and the boiling oil, and U is known as the overall heat transfer coefficient (W/m^{2}K in SI units).

At first sight, equation 7.1 implies that the relationship between Q and ΔT is linear.

Whereas this is approximately so, over limited ranges of temperature difference for which U is nearly constant, in practice U may well be influenced both by the temperature difference and by the absolute value of the temperatures.

If it is required to know the area needed for the transfer of heat at a specified rate, the temperature difference ΔT, and the value of the overall heat-transfer coefficient must be known. Thus the calculation of the value of U is a key requirement in any design problem in which heating or cooling is involved. A large part of the study of heat transfer is therefore devoted to the evaluation of this coefficient.

The value of the coefficient will depend on the mechanism by which heat is transferred, on the fluid dynamics of both the heated and the cooled fluid, on the properties of the materials through which the heat must pass, and on the geometry of the fluid paths. In solids, heat is normally transferred by conduction; some materials such as metals have a high thermal conductivity, whilst others such as ceramics have a low conductivity. Transparent solids like glass also transmit radiant energy particularly in the visible part of the spectrum.

Liquids also transmit heat readily by conduction, though circulating currents are frequently set up and the resulting convective transfer may be considerably greater than the transfer by conduction. Many liquids also transmit radiant energy. Gases are poor conductors of heat and circulating currents are difficult to suppress; convection is therefore much more important than conduction in a gas. Radiant energy is transmitted with only limited absorption in gases and, of course, without any absorption *in vacuo*. Radiation is the only mode of heat transfer which does not require the presence of an intervening medium.

If the heat is being transmitted through a number of media in series, the overall heat transfer coefficient can be broken down into individual coefficients h each relating to a single medium. This is as shown in Fig. 7.1. It is assumed that there is good contact between each pair of elements so that the temperature is the same on the two sides of each junction.

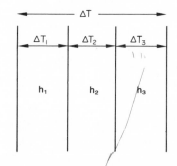

FIG. 7.1. Heat transfer through a composite wall.

Suppose heat is being transferred through three media, each of area A, that the individual coefficients for each of the media are h_1, h_2, and h_3, and the corresponding temperature changes are ΔT_1, ΔT_2, and ΔT_3. If there is no accumulation of heat in the media, the heat transfer rate Q will be the same through each. Three equations, analogous to equation 7.1 can therefore be written:

$$Q = h_1 A \, \Delta T_1$$
$$Q = h_2 A \, \Delta T_2 \tag{7.2}$$
$$Q = h_3 A \, \Delta T_3$$

Rearranging:

$$\Delta T_1 = \frac{Q}{A} \frac{1}{h_1}$$

$$\Delta T_2 = \frac{Q}{A} \frac{1}{h_2}$$

$$\Delta T_3 = \frac{Q}{A} \frac{1}{h_3}$$

Adding:

$$\Delta T_1 + \Delta T_2 + \Delta T_3 = \frac{Q}{A} \left(\frac{1}{h_1} + \frac{1}{h_2} + \frac{1}{h_3}\right) \tag{7.3}$$

Noting that $\Delta T_1 + \Delta T_2 + \Delta T_3 = $ total temperature difference ΔT:

then

$$\Delta T = \frac{Q}{A} \left(\frac{1}{h_1} + \frac{1}{h_2} + \frac{1}{h_3}\right) \tag{7.4}$$

But from equation 7.1:

$$\Delta T = \frac{Q}{A} \frac{1}{U} \tag{7.5}$$

Comparing equations 7.4 and 7.5:

$$\frac{1}{U} = \frac{1}{h_1} + \frac{1}{h_2} + \frac{1}{h_3} \tag{7.6}$$

The reciprocals of the heat transfer coefficients are resistances, and equation 7.6 therefore illustrates that the resistances are additive.

In some cases, particularly for the radial flow of heat through a thick pipe wall or cylinder, the area for heat transfer is a function of position. Thus the area for transfer applicable to each of the three media could differ and may be A_1, A_2 and A_3. Equation 7.3 then becomes:

$$\Delta T_1 + \Delta T_2 + \Delta T_3 = Q\left(\frac{1}{h_1 A_1} + \frac{1}{h_2 A_2} + \frac{1}{h_3 A_3}\right) \tag{7.7}$$

Equation 7.7 must then be written in terms of one of the area terms A_1, A_2, and A_3 or sometimes in terms of a mean area. Since Q and ΔT must be independent of the particular area considered, the value of U will vary according to which area is used as the basis. Thus equation 7.7 can be written, for example:

$$Q = U_1 A_1 \Delta T \quad \text{or} \quad \Delta T = Q/U_1 A_1$$

This will then give U_1 as:

$$\frac{1}{U_1} = \frac{1}{h_1} + \frac{A_1}{A_2}\left(\frac{1}{h_2}\right) + \frac{A_1}{A_3}\left(\frac{1}{h_3}\right) \tag{7.8}$$

In the above analysis it is assumed that the heat flowing per unit time through each of the media is the same.

Now that the overall coefficient U has been broken down into its component parts, each of the individual coefficients h_i, h_2, and h_3 must be evaluated. This can be done from a knowledge of the nature of the heat-transfer process in each of the media. A study will therefore be made of how these individual coefficients can be calculated for conduction, convection, and radiation.

Mean Temperature Difference

If heat is being transferred from one fluid to a second fluid through the wall of a vessel and the temperature is the same throughout the bulk of each of the fluids, there is no difficulty in specifying the overall temperature difference ΔT. Frequently, however, each fluid is flowing through a heat exchanger such as a pipe or a series of pipes in parallel, and its temperature

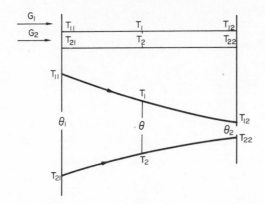

FIG. 7.2. Mean temperature difference for co-current flow.

changes as it flows, and consequently the temperature difference is continuously changing. If the two fluids are flowing in the same direction (co-current flow), the temperature of the two streams progressively approach one another (Fig. 7.2). In these circumstances the outlet temperature of the heating fluid must always be higher than that of the cooling fluid. If the fluids are flowing in opposite directions (countercurrent flow), the temperature difference will show less variation throughout the heat exchanger (Fig. 7.3). In this case it is possible for the cooling liquid to leave at a higher temperature than the heating liquid, and one of the great advantages of countercurrent flow is that it is possible to extract a higher proportion of the heat content of the heating fluid. It will now be shown how to calculate the appropriate value of the temperature difference for co-current and for countercurrent flow. It will be assumed that the overall heat transfer coefficient U remains constant everywhere in the heat exchanger.

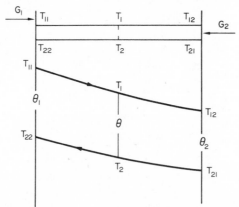

FIG. 7.3. Mean temperature difference for countercurrent flow.

It is necessary to find the average value of the temperature difference θ_m to be used in the general equation:

$$Q = UA\theta_m \tag{7.9}$$

Figure 7.3 shows the temperature conditions for the fluids flowing in opposite directions, a condition known as countercurrent flow.

Outside stream (specific heat C_{p1}) at rate G_1 falls in temperature from T_{11} to T_{12}.

Inside stream (specific heat C_{p2}) at rate G_2 rises in temperature from T_{21} to T_{22}.

Over a small element of area dA where the temperatures of the streams are T_1 and T_2. The temperature difference:

$$\theta = T_1 - T_2$$
$$\therefore \quad d\theta = dT_1 - dT_2$$

Heat given out by hot stream $\quad = dQ = - G_1 C_{p1}\, dT_1$

Heat taken up by cold stream $\quad = dQ = G_2 C_{p2}\, dT_2$

$$\therefore \quad d\theta = -\frac{dQ}{G_1 C_{p1}} - \frac{dQ}{G_2 C_{p2}} = -dQ\left(\frac{G_1 C_{p1} + G_2 C_{p2}}{G_1 C_{p1} \times G_2 C_{p2}}\right) = -\psi\, dQ \quad \text{(say)}$$

$$\therefore \quad \theta_1 - \theta_2 = \psi Q$$

Over this element $\quad U\, dA\, \theta = dQ$

$$\therefore \quad U\, dA\, \theta = -d\theta/\psi$$

If U can be taken as constant:

$$\therefore \quad -\psi U \int_0^A dA = \int_{\theta_1}^{\theta_2} d\theta/\theta$$

$$\therefore \quad -\psi U A = -\ln(\theta_1/\theta_2)$$

But $Q = UA\theta_m$ by definition of θ_m.

$$\therefore \quad \theta_1 - \theta_2 = \psi Q = \psi UA\, \theta_m = \ln\frac{\theta_1}{\theta_2}(\theta_m)$$

$$\therefore \quad \theta_m = \frac{\theta_1 - \theta_2}{\ln(\theta_1/\theta_2)} \tag{7.10}$$

θ_m is known as the *logarithmic mean temperature difference*.

If the two fluids flow in opposite directions on each side of the tube, countercurrent flow is taking place and the general shape of the temperature profile along the tube is as shown in Fig. 7.4. A similar analysis will show that this gives the same expression for θ_m, the logarithmic mean temperature difference. For the same terminal temperatures it is important to note that the value of θ_m for countercurrent flow is appreciably greater than the value for co-current flow. This is seen from the temperature profiles, for with co-current flow the cold fluid cannot be heated to a higher temperature than the exit temperature of the hot fluid and is illustrated in Example 7.1.

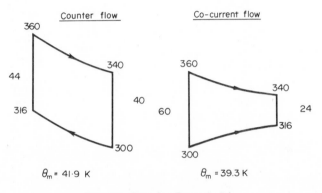

FIG. 7.4. Data for Example 7.1.

Example 7.1. A heat exchanger is required to cool 20 kg/s of water from 360 K to 340 K by means of 25 kg/s water entering at 300 K. If the overall coefficient of heat transfer is constant at $2 \text{ kW/m}^2\text{K}$, calculate the surface area required in (a) a countercurrent concentric tube exchanger, and (b) a co-current flow concentric tube exchanger.

Solution. Heat load: $Q = 20 \times 4 \cdot 18(360 - 340) = 1672 \text{ kW}$

The cooling water outlet temperature is given by:

$$1672 = 25 \times 4 \cdot 18(\theta_2 - 300) \quad \text{or} \quad \theta_2 = 316 \text{ K}$$

(a) *Counterflow*

$$\theta_m = (44 - 40)/\ln(44/40) = 41 \cdot 9 \text{ K}$$

Heat transfer area:
$$A = Q/U\theta_m$$
$$= 1672/(2 \times 41 \cdot 9)$$
$$= \underline{\underline{19.95 \text{ m}^2}}$$

(b) *Co-current flow*

$$\theta_m = (60 - 24)/\ln(60/24) = 39 \cdot 3 \text{ K}$$

Heat transfer area.
$$A = 1672/(2 \times 39 \cdot 3)$$
$$= \underline{\underline{21.27 \text{ m}^2}}$$

7.3. HEAT TRANSFER BY CONDUCTION

This important mechanism of heat transfer will now be considered in more detail for the flow of heat through a plane wall of thickness x as shown in Fig. 7.5.

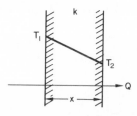

FIG. 7.5. Conduction of heat through a plane wall.

The rate of heat flow Q over the area A is given by:

$$Q = kA(T_1 - T_2)/x \tag{7.11}$$

which over a small distance dx may be written as:

$$Q = -kA\left(\frac{dT}{dx}\right) \tag{7.12}$$

the negative sign indicating that the temperature gradient is in the opposite direction to the flow of heat. k is the thermal conductivity of the material of the wall.

It will be seen in Table 7.1 that metals have very high thermal conductivities, non-metallic solids lower values, non-metallic liquids low values, and gases very low values. It is important to note that amongst metals, stainless steel has a low value, that water has a very high value for liquids (due to partial ionisation), and that hydrogen has a high value for gases (due to the high mobility of the molecules). With gases, k decreases with increase in molecular weight and increases with the temperature. In addition, for gases the dimensionless *Prandtl group* $C_p\mu/k$ is approximately constant (C_p is the specific heat at constant pressure and μ is the viscosity) and can be used to evaluate k at high temperatures where it is difficult to determine experimentally because of the formation of convection currents. k does not vary significantly with pressure until this is reduced to so low a value that the mean free path of the molecules becomes comparable with the dimensions of the vessel; further reduction of pressure then causes k to decrease.

Typical values for Prandtl numbers are as follows:

Air	0·71	n-Butanol	50
Oxygen	0·63	Light oil	600
Ammonia (gas)	1·38	Glycerol	1000
Water	5 to 10	Polymer melts	10^4
		Mercury	0·02

The low conductivity of heat insulating materials, such as cork, glass wool, etc., is largely accounted for by their high proportion of air space. The flow of heat through the materials is governed mainly by the resistance of the air spaces, which should be sufficiently small for convection currents to be suppressed.

It is convenient to rearrange equation 7.11 in the form:

$$Q = \frac{(T_1 - T_2)A}{(x/k)} \tag{7.13}$$

where x/k is known as the *thermal resistance* and k/x is the *transfer coefficient*.

Example 7.2. Find the heat loss per square metre of surface through a brick wall 0·5 m thick when the inner surface is at 400 K and the outside at 300 K. The thermal conductivity of the brick may be taken as 0·7 W/mK.

Solution. From equation 7.11:

$$Q = 0·7 \times 1(400 - 300)/0·5$$
$$\underline{\underline{= 140 \text{ W/m}^2}}$$

7.3.1. Thermal Resistances in Series

It has been seen earlier that thermal resistances may be added together in series for the case of heat transfer through a complete section formed from different media.

Figure 7.6 shows a composite wall made up of three materials with thermal conductivities k_1, k_2, and k_3, with thicknesses as shown and with the temperatures T_1, T_2, T_3, and T_4 at the faces. Applying equation 7.11 to each section in turn and noting that the same quantity of heat Q must pass through each area A:

$$T_1 - T_2 = \frac{x_1}{k_1 A} Q, \quad T_2 - T_3 = \frac{x_2}{k_2 A} Q \quad \text{and} \quad T_3 - T_4 = \frac{x_3}{k_3 A} Q$$

By addition:

$$(T_1 - T_4) = \left(\frac{x_1}{k_1 A} + \frac{x_2}{k_2 A} + \frac{x_3}{k_3 A} \right) Q \tag{7.14}$$

TABLE 7.1. *Thermal conductivities*

	Temp (K)	k (Btu/h ft² °F/ft)	k (W/mK)		Temp (K)	k (Btu/h ft² °F/ft)	k (W/mK)
Solids—Metals				*Liquids*			
Aluminium	573	133	230	Acetic acid 50%	293	0.20	0·35
Cadmium	291	54	94	Acetone	303	0·10	0·17
Copper	373	218	377	Aniline	273–293	0·1	0·17
Iron (wrought)	291	35	61	Benzene	303	0·09	0·16
Iron (cast)	326	27·6	48	Calcium chloride brine 30%	303	0·32	0·55
Lead	373	19	33	Ethyl alcohol 80%	293	0·137	0·24
Nickel	373	33	57	Glycerol 60%	293	0·22	0·38
Silver	373	238	412	Glycerol 40%	293	0·26	0·45
Steel 1% C	291	26	45	n-Heptane	303	0·08	0·14
Tantalum	291	32	55	Mercury	301	4·83	8·36
				Sulphuric acid 90%	303	0·21	0·36
Admiralty metal	303	65	113	Sulphuric acid 60%	303	0·25	0·43
Bronze	—	109	189	Water	303	0·356	0·62
Stainless Steel	293	9·2	16	Water	333	0·381	0·66
Solids—Non-metals				*Gases*			
Asbestos sheet	323	0·096	0·17	Hydrogen	273	0·10	0·17
Asbestos	273	0·09	0·16	Carbon dioxide	273	0·0085	0·015
Asbestos	373	0·11	0·19	Air	273	0·014	0·024
Asbestos	473	0·12	0·21	Air	373	0·018	0·031
Bricks (alumina)	703	1·8	3·1	Methane	273	0·017	0·029
Bricks (building)	293	0·4	0·69	Water vapour	373	0·0145	0·025
Magnesite	473	2·2	3·8	Nitrogen	273	0·0138	0·024
Cotton wool	303	0·029	0·050	Ethylene	273	0·0097	0·017
Glass	303	0·63	1·09	Oxygen	273	0·0141	0·024
Mica	323	0·25	0·43	Ethane	273	0·0106	0·018
Rubber (hard)	273	0·087	0·15				
Sawdust	293	0·03	0·052				
Cork	303	0·025	0·043				
Glass wool	—	0·024	0·041				
85% Magnesia	—	0·04	0·070				
Graphite	273	87	151				

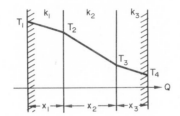

FIG. 7.6. Conduction of heat through a composite wall.

or

$$Q = \frac{T_1 - T_4}{\Sigma(x_1/k_1 A)}$$

$$= \frac{\text{Total driving force}}{\text{Total (thermal resistance/area)}} \quad (7.15)$$

Example 7.3. A furnace is constructed with 0·20 m of firebrick, 0·10 m of insulating brick, and 0·20 m of building brick. The inside temperature is 1200 K and the outside temperature 330 K. If the thermal conductivities are as shown in Fig. 7.7, find the heat loss per unit area and the temperature at the junction of the firebrick and the insulating brick.

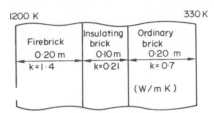

FIG. 7.7. Data for Example 7.3.

Solution. From equation 7.15

$$Q = (1200 - 330) \Big/ \left[\left(\frac{0·20}{1·4 \times 1}\right) + \left(\frac{0·10}{0·21 \times 1}\right) + \left(\frac{0·20}{0·7 \times 1}\right) \right]$$

$$= 870/(0·143 + 0·476 + 0·286) = 870/0·905$$

$$= \underline{\underline{961 \text{ W/m}^2}}$$

(Temperature drop over firebrick)/(Total temperature drop) $= 0·143/0·905$

∴   Temperature drop over firebrick $= 870 \times 0·143/0·905 = 137$ K

Hence temperature at this plane $= \underline{\underline{1063 \text{ K}}}$

7.3.2.   Conduction Through a Thick-Walled Tube

The conditions for heat flow through a thick-walled tube when the temperature on the inside and outside are held constant are shown in Fig. 7.8. Here the area for heat flow is proportional to the radius so that the temperature gradient is inversely proportional to the radius.

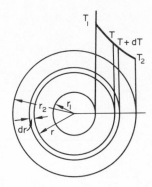

FIG. 7.8. Conduction through thick-walled tube or spherical shell.

The heat flow at any radius r is given by:

$$Q = -k2\pi rl \,(dT/dr) \qquad (7.16)$$

where l is the length of tube.

Integrating between limits r_1 and r_2:

$$Q \int_{r_1}^{r_2} \frac{dr}{r} = -2\pi lk \int_{T_1}^{T_2} dT$$

or

$$Q = \frac{2\pi lk(T_1 - T_2)}{\ln(r_2/r_1)} \qquad (7.17)$$

This equation can be put into the form of equation 7.11:

i.e.

$$Q = \frac{k(2\pi r_m l)(T_1 - T_2)}{r_2 - r_1} \qquad (7.18)$$

where $r_m = (r_2 - r_1)/\ln(r_2/r_1)$, and is known as the logarithmic mean radius. For thin-walled tubes it will be sufficient to use the arithmetic mean radius r_a, giving:

$$Q = 2\pi klr_a(T_1 - T_2)/(r_2 - r_1) \qquad (7.19)$$

7.3.3.   Conduction to a Sphere

For heat conduction through the shell of a hollow sphere, the heat flow at radius r is given by:

$$Q = -k4\pi r^2(dT/dr) \qquad (7.20)$$

Integrating between limits r_1 and r_2:

$$Q \int_{r_1}^{r_2} \frac{dr}{r^2} = -4\pi k \int_{T_1}^{T_2} dT$$

$$\therefore \qquad Q = \frac{4\pi k(T_1 - T_2)}{(1/r_1)(1/r_2)} \qquad (7.21)$$

An important application of this process is heat transfer by conduction through a stationary fluid surrounding a spherical particle or droplet of radius r. If the temperature difference $T_1 - T_2$ is spread over a very large distance so that $r_2 = \infty$ and T_1 is the temperature of the surface of the drop:

$$\frac{Qr}{(4\pi r^2)(T_1 - T_2)k} = 1$$

or $\qquad\qquad\qquad\qquad\qquad hd/k = Nu = 2 \qquad\qquad\qquad\qquad\qquad$ (7.22)

where $Q/4\pi r^2(T_1 - T_2) = h$ is known as the heat transfer coefficient and hd/k is a dimensionless group known as the *Nusselt number* (Nu) and d is the diameter of the particle or droplet.

Heat transfer by conduction to particles is often important in equipment such as fluidised beds, rotary kilns, spray driers and plasma-jet spraying-torches. Equations may be developed to predict the rate of change of diameter d of evaporating droplets. If the latent heat of vaporisation is provided by heat conducted through a hotter stagnant gas to the droplet surface, and heat transfer is the rate controlling step, it is shown in Volume 2[1] that d^2 decreases linearly with time. A closely related and important practical problem is the prediction of the residence time required in a combustion chamber to ensure virtually complete burning of oil droplets. Complete combustion is desirable to obtain maximum utilisation of energy and to minimise pollution of the atmosphere by partially burned oil droplets. Here a droplet is surrounded by a flame and heat conducted back from the flame to the droplet surface provides the heat to vaporise the oil and sustain the surrounding flame. Again d^2 decreases approximately linearly with time but derivation of the equation is more complex due to mass transfer effects, steep temperature gradients[2] and natural convection.[3]

If there is relative motion between the particle and surrounding fluid then usually, as discussed later in this chapter, heat transfer between particle and fluid is enhanced by convection and the Nusselt number is greater than 2. A practical problem in which there is initially a very high relative velocity between the particle and the fluid is in the processing of powders in high temperature plasmas.[4] The plasmas referred to here are partially ionised gases with temperatures around 10,000 K formed by electric discharges such as arcs. There is an increasing industrial use of the technique of plasma spraying in which powders are injected into a high velocity plasma jet so that they are both melted and projected at velocities of several hundred metres per second on to a surface. The molten particles with diameters typically of the order $10-100\ \mu m$ impinge to form an integral layer on the surface. Applications include the building up of worn shafts, e.g. of pumps, and the deposition of erosion resistant ceramic layers on centrifugal pump impellers and other equipment prone to erosion damage. When a powder particle first enters the plasma jet, the relative velocity may be hundreds of metres per second and heat transfer to the particle is enhanced by convection. Often however, and more particularly for smaller particles, the particle is quickly accelerated to essentially the same velocity as the plasma jet[1] and conduction becomes the main mechanism of heat transfer from plasma to particle. Certainly from a design point of view, neglecting the convective contribution will ease calculations and give a more conservative and safer estimate of the size of the largest particle which can be melted before it strikes the surface. In the absence of complications due to non-continuum conditions discussed later, the value of $Nu = hd/k$ is therefore taken as 2·0 (see equation 7.22). One complication which arises in the application of this equation to powder heating in high temperature plasmas lies in the dependence of k, the thermal conductivity of the gas or plasma surrounding the particle, on temperature. For example, the temperature of the particle surface may be 1000 K say, whilst that of the plasma away from the particle may be about 10,000 K or even higher. The thermal conductivity of argon increases by a factor of about 20 over this range of temperature and that of nitrogen gas passes through a pronounced peak at about 7100 K due to dissociation–recombination effects. Thus the temperature at which the thermal conductivity k is evaluated will have a pronounced effect on the value of h obtained from $Nu = 2 \cdot 0$. A mean value of k would seem appropriate where:

$$(k)_{\text{mean}} = \frac{\int_{T_2}^{T_1} k \, dT}{T_1 - T_2}$$

Some workers have correlated experimental results in terms of k at the arithmetic mean temperature, and some at the temperature of the bulk plasma. Experimental validation of the true effective thermal conductivity is difficult because of the high temperatures, small particle sizes and variations in velocity and temperature in plasma jets.

In view of the high temperatures involved in plasma devices and the T^4 dependence of radiation heat transfer discussed later in this chapter, it is surprising at first sight that conduction is more significant than radiation in heating particles in plasma spraying. The explanation lies in the small values of d and relatively high values of k for the gas, both of which contribute to high values of h for a given value of Nu. Also the emissivities of most gases are, as seen later in the chapter, rather low.

In situations where the surrounding fluid behaves as a non-continuum fluid, e.g. at very high temperatures and/or at low pressures, it is possible to have values of Nu less than 2. A gas begins to exhibit non-continuum behaviour when the mean free path between collisions of gas molecules or atoms with each other is greater than about 1/100 of the characteristic size of the surface considered. The molecules or atoms are then sufficiently far apart on average for the gas to begin to lose the character of a homogeneous or continuum fluid which is normally assumed in the majority of chemical engineering heat transfer or fluid dynamics problems. For example, consider a particle of diameter 25 μm; particles of this size can be encountered in, for example, oil-burner sprays, pulverised coal flames, and in plasma spraying. With particle of such a size in air at room temperature and atmospheric pressure, the mean free path of gas molecules is about $0 \cdot 06$ μm and the air then behaves as a continuum fluid. If, however, the temperature were, say, 1800 K, as in a flame, then the mean free path would be about $0 \cdot 33$ μm, i.e. greater than 1/100 of the particle diameter. Non-continuum effects, leading to values of Nu lower than 2 would then be likely according to theory.[5, 6] The exact value of Nu would depend on the surface accommodation coefficient. This is a difficult parameter to measure for the examples considered here, and hence experimental confirmation of the theory is difficult. At the still higher temperatures that exist in thermal plasma devices, non-continuum effects should be more pronounced and there is limited evidence that values of Nu below 1 are obtained.[4] In general, non-continuum effects, leading in particular to values of Nu less than 2, would be more likely at high temperatures, low pressures, and small particle sizes. Thus there is an interest in these effects in the aerospace industry, for example when considering the behaviour of small particles present in rocket engine exhausts.

In this section heat transfer to a spherical particle by conduction through the surrounding fluid has been the prime consideration. In many practical situations the flow of heat from the surface to the internal parts of the particle is of importance. For example, if the particle is a poor conductor then the rate at which the particulate material reaches some desired average temperature may be limited by conduction inside the particle rather than by conduction to the outside surface of the particle. This problem involves unsteady state transfer of heat which is the subject of the next section.

7.3.4. Unsteady State Transfer of Heat

In the problems which have been considered so far, it has been assumed that the conditions at any point in the system remain constant with respect to time. The case of heat transfer by conduction in a medium in which the temperature is changing with time, will now be treated. This problem is of importance in the calculation of the temperature distribution in a body which is being heated or cooled. In an element of dimensions dx by dy by dz, let the temperature at the point (x, y, z) be θ and at the point $(x + dx, y + dy, z + dz)$ be $\theta + d\theta$.

Assuming that the thermal conductivity k is constant and that no heat is generated in the medium, the rate of conduction of heat through the element

$$= -k \, dy \, dz (\partial\theta/\partial x)_{yz} \text{ in the } x\text{-direction}$$

$$= -k \, dz \, dx (\partial\theta/\partial y)_{zx} \text{ in the } y\text{-direction}$$

and

$$= -k \, dx \, dy (\partial\theta/\partial z)_{xy} \text{ in the } z\text{-direction}$$

Now the rate of change of heat content of the element will be equal to minus the rate of increase of heat flow from (x, y, z) to $(x + dx, y + dy, z + dz)$.

Thus rate of change of heat content of element

$$= k \, dy \, dz \left(\frac{\partial^2\theta}{\partial x^2}\right)_{yz} dx + k \, dz \, dx \left(\frac{\partial^2\theta}{\partial y^2}\right)_{zx} dy + k \, dx \, dy \left(\frac{\partial^2\theta}{\partial z^2}\right)_{xy} dz$$

$$= k \, dx \, dy \, dz \left[\left(\frac{\partial^2\theta}{\partial x^2}\right)_{yz} + \left(\frac{\partial^2\theta}{\partial y^2}\right)_{zx} + \left(\frac{\partial^2\theta}{\partial z^2}\right)_{zy}\right] \tag{7.23}$$

But the rate of increase of heat content is also equal to the product of the heat capacity of the element and the rate of rise of temperature.

Thus

$$k \, dx \, dy \, dz \left[\left(\frac{\partial^2\theta}{\partial x^2}\right)_{yz} + \left(\frac{\partial^2\theta}{\partial y^2}\right)_{zx} + \left(\frac{\partial^2\theta}{\partial z^2}\right)_{xy}\right] = C_p\rho \, dx \, dy \, dz \frac{\partial\theta}{\partial t}$$

i.e.

$$\frac{\partial\theta}{\partial t} = \frac{k}{C_p\rho} \left[\left(\frac{\partial^2\theta}{\partial x^2}\right)_{yz} + \left(\frac{\partial^2\theta}{\partial y^2}\right)_{zx} + \left(\frac{\partial^2\theta}{\partial z^2}\right)_{xy}\right]$$

$$= D_H \left[\left(\frac{\partial^2\theta}{\partial x^2}\right)_{yz} + \left(\frac{\partial^2\theta}{\partial y^2}\right)_{zx} + \left(\frac{\partial^2\theta}{\partial z^2}\right)_{xy}\right] \tag{7.24}$$

where $D_H = k/C_p\rho$, is known as the *thermal diffusivity*.

The above partial differential equation is most conveniently solved by the use of the Laplace transform of temperature with respect to time. As an illustration of the method of solution, the problem of the unidirectional flow of heat in a continuous medium will be considered. The basic differential equation for the process is:

$$\frac{\partial\theta}{\partial t} = D_H \frac{\partial^2\theta}{\partial x^2} \tag{7.25}$$

This equation cannot be integrated directly, since the temperature θ is expressed as a function of two independent variables, distance x and time t. The method of solution involves transforming the equation so that the Laplace transform of θ with respect to time is used in place of θ. The equation then involves only the Laplace transform $\bar{\theta}$ and the distance x. The Laplace transform of θ is defined by the relation:

$$\bar{\theta} = \int_0^\infty \theta \, e^{-pt} \, dt \tag{7.26}$$

where p is a parameter.

Thus $\bar{\theta}$ is obtained by operating on θ with respect to t. Since no operation with respect to x is involved:

$$\frac{\partial^2\bar{\theta}}{\partial x^2} = \overline{\frac{\partial^2\theta}{\partial x^2}} \tag{7.27}$$

Now

$$\frac{\overline{\partial \theta}}{\partial t} = \int_0^\infty \frac{\partial \theta}{\partial t} e^{-pt} \, dt$$

$$= [\theta \, e^{-pt}]_0^\infty + p \int_0^\infty e^{-pt} \theta \, dt$$

$$= -\theta_{t=0} + p\bar{\theta} \qquad (7.28)$$

Then, taking the Laplace transforms of each side of equation 7.25:

$$\frac{\partial \theta}{\partial t} = D_H \frac{\partial^2 \theta}{\partial x^2}$$

i.e. $$p\bar{\theta} - \theta_{t=0} = D_H \frac{\partial^2 \bar{\theta}}{\partial x^2} \quad \text{(from equations 7.27 and 7.28)}$$

i.e. $$\frac{\partial^2 \bar{\theta}}{\partial x^2} - \frac{p}{D_H} \bar{\theta} = -\frac{\theta_{t=0}}{D_H}$$

If the temperature everywhere is constant initially, $\theta_{t=0}$ is a constant and the equation may be integrated as a normal second-order differential equation since p is not a function of x.

Thus $$\bar{\theta} = B_1 \, e^{\sqrt{(p/D_H)}x} + B_2 \, e^{-\sqrt{(p/D_H)}x} + \theta_{t=0} p^{-1} \qquad (7.29)$$

and $\therefore$ $$\frac{\partial \bar{\theta}}{\partial x} = B_1 \sqrt{\frac{p}{D_H}} e^{\sqrt{(p/D_H)}x} - B_2 \sqrt{\frac{p}{D_H}} e^{-\sqrt{(p/D_H)}x} \qquad (7.30)$$

The temperature θ, corresponding to the above transform $\bar{\theta}$, can now be found by reference to tables of the Laplace transform. It is first necessary, however, to evaluate the constants B_1 and B_2 using the boundary conditions for the particular problem because these constants will in general involve the parameter p which was introduced in the transformation.

Consider the particular problem of the unidirectional flow of heat through a body with plane parallel faces a distance l apart. The heat flow is normal to these faces and the temperature of the body is initially constant throughout. The temperature scale will be so chosen that this uniform initial temperature is zero. At time, $t = 0$, one face (at $x = 0$) will be brought into contact with a source at a constant temperature θ' and the other face (at $x = l$) will be assumed to be perfectly insulated thermally.

The boundary conditions are therefore:

$$t = 0, \ \theta = 0$$

$$t > 0, \theta = \theta' \quad \text{when } x = 0$$

$$t > 0 \qquad \frac{\partial \theta}{\partial x} = 0 \quad \text{when } x = l$$

Thus $$\bar{\theta}_{x=0} = \int_0^\infty \theta' \, e^{-pt} \, dt = \frac{\theta'}{p}$$

and $$\left(\frac{\partial \bar{\theta}}{\partial x}\right)_{x=l} = 0$$

Substitution of these boundary conditions in equations 7.29 and 7.30 gives:

$$B_1 + B_2 = \frac{\theta'}{p}$$

and $$B_1 \, e^{\sqrt{(p/D_H)}l} - B_2 \, e^{-\sqrt{(p/D_H)}l} = 0 \qquad (7.31)$$

Hence

$$B_1 = \frac{\dfrac{\theta'}{p} e^{-\sqrt{(p/D_H)}l}}{e^{\sqrt{(p/D_H)}l} + e^{-\sqrt{(p/D_H)}l}}$$

and

$$B_2 = \frac{\dfrac{\theta'}{p} e^{\sqrt{(p/D_H)}l}}{e^{\sqrt{(p/D_H)}l} + e^{-\sqrt{(p/D_H)}l}}$$

Then

$$\theta = \frac{e^{(l-x)\sqrt{(p/D_H)}} + e^{-(l-x)\sqrt{(p/D_H)}}}{e^{\sqrt{(p/D_H)}l} + e^{-\sqrt{(p/D_H)}l}} \frac{\theta'}{p}$$

$$= \frac{\theta'}{p} (e^{(l-x)\sqrt{(p/D_H)}} + e^{-(l-x)\sqrt{(p/D_H)}})(1 + e^{-2\sqrt{(p/D_H)}l})^{-1}(e^{-\sqrt{(p/D_H)}l})$$

$$= \frac{\theta'}{p} (e^{-x\sqrt{(p/D_H)}} + e^{-(2l-x)\sqrt{(p/D_H)}})(1 - e^{-2l\sqrt{(p/D_H)}} + \cdots + (-1)^N e^{-2Nl\sqrt{(p/D_H)}} + \cdots)$$

$$= \sum_{N=0}^{N=\infty} \frac{\theta'}{p} (-1)^N (e^{-(2lN+x)\sqrt{(p/D_H)}} + e^{-\{2(N+1)l-x\}\sqrt{(p/D_H)}}) \qquad (7.32)$$

The temperature θ is then obtained from tables of inverse Laplace transforms (Vol 3[7]) and is given by:

$$\theta = \sum_{N=0}^{N=\infty} (-1)^N \theta' \left(\text{erfc} \frac{2lN + x}{2\sqrt{D_H t}} + \text{erfc} \frac{2(N+1)l - x}{2\sqrt{D_H t}} \right) \qquad (7.33)$$

where

$$\text{erfc } x = \frac{2}{\sqrt{\pi}} \int_x^\infty e^{-\xi^2} \, d\xi$$

Values of erfc x are to be found in mathematical tables.[8]
Equation 7.33 can be written in the form:

$$\frac{\theta}{\theta'} = \sum_{N=0}^{N=\infty} (-1)^N \left\{ \text{erfc} \left[Fo_l^{-1/2} \left(N + \frac{1}{2}\frac{x}{l} \right) \right] + \text{erfc} \left[Fo_l^{-1/2} \left((N+1) - \frac{1}{2}\frac{x}{l} \right) \right] \right\} \qquad (7.34)$$

where $Fo_l = (D_H t / l^2)$ and is known as the *Fourier number*.

Thus

$$\frac{\theta}{\theta'} = f\left(Fo_l, \frac{x}{l} \right) \qquad (7.35)$$

The numerical solution to the above problem is then obtained by inserting the appropriate values for the physical properties of the system and using as many terms in the series as are necessary for the degree of accuracy required. In most cases, the above series converge quite rapidly.

This method of solution of problems of unsteady flow is particularly useful because it is applicable when there are discontinuities in the physical properties of the material.[9] The boundary conditions, however, become a little more complicated, but the problem is intrinsically no more difficult.

Example 7.4. Calculate the time taken for the distant face of a brick wall, of thermal diffusivity $D_H = 0.0043$ cm²/s and thickness $l = 0.45$ m, to rise from 295 to 375 K. The whole wall is initially at a constant temperature of 295 K and the near face is suddenly raised to 900 K and maintained at this temperature. Assume that all the flow of heat is perpendicular to the faces of the wall and that the distant face is perfectly insulated.

Solution. The temperature at any distance x from the near face at time t is given by:

$$\theta = \sum_{N=0}^{N=\infty} (-1)^N \theta' \{erfc[(2lN + x)/(2\sqrt{D_H t})] + erfc[(2(N+1)l - x)/(2\sqrt{D_H t})]\}$$

(equation 7.33)

The temperature at the distant face is therefore given by:

$$\theta = \sum_{N=0}^{N=\infty} (-1)^N \theta' \, 2 \, erfc[(2N+1)l)/(2\sqrt{D_H t})]$$

Choosing the temperature scale so that the initial temperature is everywhere zero:

$$\therefore \qquad \theta/2\theta' = (375 - 295)/(2(900 - 295)) = 0.066$$

$$D_H = 0.0042 \text{ cm}^2/\text{s} \qquad \therefore \quad \sqrt{D_H} = 0.065$$

$$l = 0.45 \text{ m} \quad \text{or} \quad 45 \text{ cm}$$

Thus

$$0.066 = \sum_{N=0}^{N=\infty} (-1)^N \, erfc[45(N+1)/(2 \times 0.065 t^{0.5})]$$

$$= \sum_{N=0}^{N=\infty} (-1)^N \, erfc[385(N+1)/t^{0.5}]$$

$$= erfc(385 t^{-0.5}) - erfc(1155 t^{-0.5}) + erfc(1925 t^{-0.5}) - \cdots$$

An approximate answer is obtained by taking the first term only.

Then

$$385 t^{-0.5} = 1.30$$

$$\therefore \qquad t = 87\,700 \text{ s}$$

$$= 87.7 \text{ ks} = 24.4 \text{ h}$$

Use of Dimensionless Groups

The exact mathematical solution of problems involving unsteady thermal conduction may be very difficult, and sometimes impossible, especially where bodies of irregular shapes are concerned, and other methods are therefore required.

When a body of characteristic linear dimension L, initially at a uniform temperature θ_0, is exposed suddenly to surroundings at a temperature θ', the temperature distribution at any time t is found by applying the method of dimensional analysis to be:

$$\frac{\theta' - \theta}{\theta' - \theta_0} = f\left(\frac{hL}{k}, \quad D_H \frac{t}{L^2}, \quad \text{and} \quad \frac{x}{L}\right)$$

(7.36)

where x is the distance from the surface and h is the heat transfer coefficient at the surface. Where the resistance to heat transfer at the surface can be neglected, e.g. where the temperature of the surface of the body is maintained at the temperature θ', the temperature will no longer be a function of h and only three dimensionless groups exist:

i.e.

$$\frac{\theta' - \theta}{\theta' - \theta_0} = f\left(D_H \frac{t}{L^2}, \frac{x}{L}\right) = f\left(Fo_L, \frac{x}{L}\right)$$

(7.37)

or

$$\frac{\theta' - \theta}{\theta' - \theta_0} = f\left(D_H \frac{t}{x^2}, \frac{x}{L}\right) = f\left(Fo_x, \frac{x}{L}\right)$$

(7.38)

Curves connecting these groups have been plotted by a number of workers for bodies of various shapes. The method is limited, however, to those shapes which have been studied experimentally.

In Fig. 7.9, the value of $(\theta' - \theta)/(\theta' - \theta_0)$ is plotted to give the temperature θ at the centre

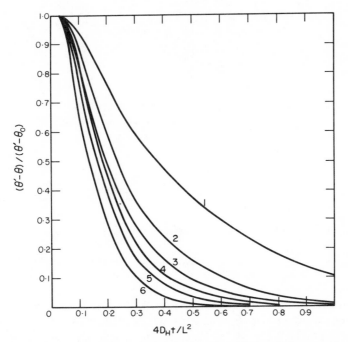

FIG. 7.9. Cooling curve for bodies of various shapes: 1, slab (L = thickness); 2, square bar (L = side); 3, long cylinder (L = diameter); 4, cube (L = length = diameter); 5, cylinder (L = length = diameter); 6, sphere (L = diameter).

of bodies of various shapes, initially at a uniform temperature θ_0, at a time t after the surfaces have been suddenly altered to and maintained at a constant temperature θ'. In this case (x/L) is constant at 0·5 and the results are shown as a function of the particular value $(4D_H t/L^2)$ of the Fourier number Fo_x. The results are taken from reference 8.

A general method of estimating the temperature distribution in a body of any shape consists of replacing the heat flow problem by the analogous electrical problem and measuring the electrical potentials at various points. The heat capacity per unit volume $C_p\rho$ is represented by an electrical capacitance, and the thermal conductivity k by an electrical conductivity. The method can be used to take account of variations in the thermal properties over the body.

Schmidt's Method

Numerical methods have been developed by replacing the differential equation by a finite difference equation. Thus in a problem of unidirectional flow of heat,

$$\frac{\partial \theta}{\partial t} \approx \frac{\theta_{x(t+\Delta t)} - \theta_{x(t-\Delta t)}}{2\,\Delta t} \approx \frac{\theta_{x(t+\Delta t)} - \theta_{xt}}{\Delta t}$$

$$\frac{\partial^2 \theta}{\partial x^2} \approx \left(\frac{\theta_{(x+\Delta x)t} - \theta_{xt}}{\Delta x} - \frac{\theta_{xt} - \theta_{(x-\Delta x)t}}{\Delta x} \right) \Big/ \Delta x$$

$$= \frac{\theta_{(x+\Delta x)t} + \theta_{(x-\Delta x)t} - 2\theta_{xt}}{(\Delta x)^2}$$

where θ_{xt} is the value of θ at time t and distance x from the surface, and the other values of θ are at intervals Δx and Δt as shown in Fig. 7.10.

Substituting these values in equation 7.25:

$$\theta_{x(t+\Delta t)} - \theta_{x(t-\Delta t)} = D_H \frac{2\,\Delta t}{(\Delta x)^2} \left(\theta_{(x+\Delta x)t} + \theta_{(x-\Delta x)t} - 2\theta_{xt} \right) \qquad (7.39)$$

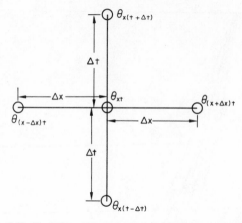

FIG. 7.10. Variation of temperature with time and distance.

and
$$\theta_{x(t+\Delta t)} - \theta_{xt} = D_H \frac{\Delta t}{(\Delta x)^2} (\theta_{(x+\Delta x)t} + \theta_{(x-\Delta x)t} - 2\theta_{xt}) \qquad (7.40)$$

Thus, if the temperature distribution at time t, is known, the corresponding distribution at time $t + \Delta t$ can be calculated by the application of equation 7.40 over the whole extent of the body in question. The intervals Δx and Δt are so chosen that the required degree of accuracy is obtained.

A graphical method of procedure is due to Schmidt.[10] Suppose that the temperature distribution at time t is represented by the curve shown in Fig. 7.11. If the points representing the temperatures at $x - \Delta x$ and $x + \Delta x$ are joined by a straight line, the distance θ_a is given by:

$$\theta_a = \frac{\theta_{(x+\Delta x)t} + \theta_{(x-\Delta x)t}}{2} - \theta_{xt}$$

$$= \frac{(\Delta x)^2}{2D_H \Delta t} (\theta_{x(t+\Delta t)} - \theta_{xt}) \quad \text{(from equation 7.40)} \qquad (7.41)$$

Thus θ_a represents the change in θ_{xt} after a time interval Δt, such that:

$$\Delta t = (\Delta x)^2/2D_H. \qquad (7.42)$$

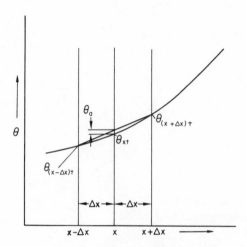

FIG. 7.11. Schmidt's method.

If this simple construction is carried out over the whole of the body, the temperature distribution after time Δt is obtained. The temperature distribution after an interval $2 \Delta t$ is then obtained by a repetition of this procedure.

Southwell's relaxation method[11] can also be applied to determine the temperature distribution at any time by using the finite difference equation in the form of equation 7.39.

Example 7.5. The previous example will be solved by Schmidt's method.

Solution. The development of the temperature profile is shown in Fig. 7.12. At time $t = 0$ the temperature is constant at 295 K throughout and the temperature of the hot face is raised to 900 K. The problem will be solved by taking relatively large intervals for Δx.

Choosing $\Delta x = 50$ mm, the construction shown in Fig. 7.11 is carried out starting at the hot face.

Points corresponding to temperature after a time interval Δt are marked 1, after a time interval $2 \Delta t$ by 2, and so on. Because the second face is perfectly insulated, the temperature gradient must be zero at this point. Thus, in obtaining temperatures at $x = 450$ mm it is assumed that the temperature at $x = 500$ mm will be the same as at $x = 400$ mm, i.e. horizontal lines are drawn on the diagram. It is seen that the temperature is less than 375 K after time $24 \Delta t$ and greater than 375 K after time $26 \Delta t$.

Thus
$$t \approx 25\Delta t$$

From equation 7.42: $\Delta t = 5 \cdot 0^2/(2 \times 0 \cdot 0042) = 2976$ s

Thus required time $= 25 \times 2976 = 74400$ s

or
$$74 \cdot 4 \text{ ks} = 20 \cdot 7 \text{ h}$$

This value is quite close to the value obtained by calculation, especially so in view of the coarse increments in Δx.

Biot Modulus or Biot Number

When solids or liquid drops are being heated by a surrounding fluid, for example in a rotary kiln, fluidised bed or in a plasma jet, then it is often necessary to predict the time needed for the interior of the particle to reach some specified temperature. If the solids are in the form of small particles and/or the solid is a good thermal conductor, then the rate at which the centre of the particle reaches a specified temperature will be controlled by the rate at which heat arrives at the particle surface. Conversely, with larger particles and solids of low thermal conductivity, internal conduction will be the controlling factor. In some situations internal and external heat transfer may be of similar importance. A useful means of analysing such situations quantitatively is to evaluate the so-called *Biot modulus* (or *Biot number*).

For a sphere, Biot modulus $Bi = hr/k_p$

where h is the external heat transfer coefficient which could include contributions from conduction, convection and radiation, r the radius of sphere and k_p the thermal conductivity of the particle (not that of the surrounding fluid). Note that if the external heat transfer is purely by conduction, then as seen previously:

$$Nu = hd/k = 2 \qquad\qquad \text{(equation 7.22)}$$

Hence
$$h = 2k/d = k/r$$

and
$$Bi = \frac{k}{k_p} = \frac{\text{thermal conductivity of fluid}}{\text{thermal conductivity of particle}}$$

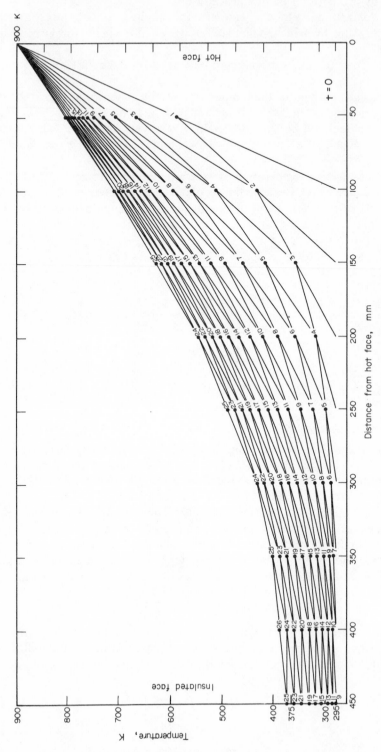

Fig. 7.12. Development of temperature profile.

Thus the higher the Biot number the more likely it is that internal conduction will be the factor controlling the time taken for the centre of a particle to reach some specified temperature. To take the melting of powders in plasma spraying as an example, the Biot numbers can range from around 0·005 to over 5, depending on the compositions of the particles and plasma.

If both the Fourier number and the Biot modulus are evaluated then temperature response or Gurney–Lurie charts[12] can be used to ease the calculation of the time dependent temperature profiles for the surface and inside of the particle. Note that in reference[12] the charts are plotted in terms of the reciprocal of the Biot modulus. Charts are available for other shapes besides spheres, for example, cubes, slabs, and cylinders.

7.3.5. Temperature Distribution with Internal Heat Source

If an electric current flows through a wire then the heat generated internally will result in a temperature distribution between the central axis and the surface of the wire. This type of problem will also arise in chemical or nuclear reactors where heat is generated internally. It is necessary to determine the temperature distribution in such a system and the maximum temperature which will occur.

Suppose the temperature at the surface of the wire to be T_0 and the rate of heat generation per unit volume to be Q_G. Consider unit length of a cylindrical element of radius r. Then the heat generated must be transmitted in an outward direction by conduction so that:

$$-k2\pi r(dT/dr) = \pi r^2 Q_G$$

Hence

$$dT/dr = -Q_G r/2k$$

Integrating:

$$T = -(Q_G r^2/4k) + C$$

But $T = T_0$ when $r = r_0$ the radius of wire.

$\therefore$

$$T = T_0 + Q_G(r_0^2 - r^2)/4k$$

$$T - T_0 = \frac{Q_G r_0^2}{4k}\left(1 - \frac{r^2}{r_0^2}\right) \tag{7.43}$$

This gives a parabolic distribution of temperature and the maximum temperature will occur at the axis of the wire where $(T - T_0) = Q_G r_0^2/4k$. The average temperature difference $(T - T_0)_{av} = Q_G r_0^2/8k$.

Since $Q_G \pi r_0^2$ is the rate of heat release per unit length of the wire, then, putting T_1 as the temperature at the centre:

$$T_1 - T_0 = \text{rate of heat release per unit length}/4\pi k \tag{7.44}$$

Example 7.6. A fuel channel in a natural uranium reactor is 5 m long and has a heat release of 0·25 MW. If the thermal conductivity of the uranium is 33 W/m K, what is the temperature difference between the surface and the centre of the uranium element, assuming that the heat release is uniform along the rod?

Solution. Heat release = $0·25 \times 10^6$ W

$$= 0·25 \times 10^6/5 = 5 \times 10^4 \text{ W/m}$$

Thus, from equation 7.44: $T_1 - T_0 = 5 \times 10^4/(4\pi \times 33)$

$$= \underline{\underline{121 \text{ K}}}$$

Non-Uniform Heat Release

It should be noted that the temperature difference is independent of the diameter of the fuel rod for a cylindrical geometry and uniform heat release per unit volume.

In practice the assumption of the uniform heat release per unit length of the rod is never valid since the neutron flux and hence the heat flux varies sinusoidally along the length of the channel. In the simplest case of an unreflected reactor where the neutron flux may be taken as zero at the ends of the fuel element, the conditions become as shown in Fig. 7.13.

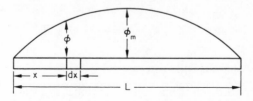

FIG. 7.13. Variation of neutron flux along length of fuel rod.

The neutron flux may be represented by $\phi = \phi_m \sin(\pi x/L)$.

Since the heat flux is proportional to the neutron flux, the heat dQ developed per unit time in a differential element of the fuel rod dx may be written as:

$$dQ = C \sin\left(\frac{\pi x}{L}\right) dx$$

The total heat generated by the rod Q is then given by

$$Q = C \int_0^L \sin\left(\frac{\pi x}{L}\right) dx = \frac{2CL}{\pi}$$

Thus $C = \pi Q/2L$. Then the heat release per unit length at any point is given by:

$$\frac{dQ}{dx} = \frac{\pi Q}{2L} \sin\left(\frac{\pi x}{L}\right)$$

Substituting into equation 7.44 gives:

$$T_1 - T_0 = \frac{(\pi Q/2L)\sin(\pi x/L)}{4\pi k}$$

It should be noted that when $x = 0$ or $x = L$, then $T_1 - T_0$ is zero as would be expected since the neutron flux is also zero under these conditions.

7.4. HEAT TRANSFER BY CONVECTION

7.4.1. Determination of Film Coefficients for Convection

Heat transfer by convection occurs as a result of the movement of fluid on a macroscopic scale in the form of eddies or circulating currents. If the currents arise from the heat transfer process itself, *natural convection* occurs. An example of a natural convection process is the heating of a vessel containing liquid by means of a heat source such as a gas flame situated underneath. The liquid at the bottom of the vessel becomes heated and expands and rises because its density has become less than that of the remaining liquid. Cold liquid of higher density takes its place and a circulating current is set up.

In *forced convection* the circulating currents are produced by an external agency such as an agitator in a reaction vessel or as a result of turbulent flow in a pipe. In general, the magnitude of the circulation in forced convection is greater, and higher rates of heat transfer are obtained than in natural convection.

In most cases where convective heat transfer is taking place from a surface to a fluid, the circulating currents die out in the immediate vicinity of the surface and a film of fluid, free of turbulence, covers the surface. In this film, heat transfer is by thermal conduction and, as the thermal conductivity of most fluids is low, the main resistance to transfer lies there. Thus an increase in the velocity of the fluid over the surface gives rise to improved heat transfer mainly because the thickness of the film is reduced. As a guide, the film coefficient increases as (fluid velocity)^n, where $0 \cdot 6 < n < 0 \cdot 8$, depending upon the geometry.

If the resistance to transfer is regarded as lying within the film covering the surface, the rate of heat transfer Q is given by equation 7.11 as:

$$Q = kA(T_1 - T_2)/x$$

The effective thickness x is not generally known but is approximately proportional to $Re^{-0 \cdot 2}$ (see Chapter 3) and therefore the equation is usually rewritten in the form:

$$Q = hA(T_1 - T_2) \tag{7.45}$$

where h is the heat transfer coefficient for the film and $(1/h)$ is the thermal resistance.

7.4.2. Application of Dimensional Analysis to Heat-Transfer by Convection

So many factors influence the value of h that it is almost impossible to determine their individual effects by direct experimental methods. By arranging the variables in a series of dimensionless groups, the problem is made more manageable in that the number of groups is significantly less than the number of parameters. It is found that the heat transfer rate per unit area q is dependent on those physical properties which affect flow pattern—viscosity μ and density ρ—, the thermal properties of the fluid—the specific heat capacity C_p and the thermal conductivity k—a linear dimension of the surface l, the velocity of flow u of the fluid over the surface, the temperature difference ΔT and the factor determining the natural circulation effort caused by the expansion of the fluid on heating—the product of the coefficient of cubical expansion β and the acceleration due to gravity g. Writing this as a functional relationship:

$$q = \varphi[u, l, \rho, \mu, C_p, \Delta T, \beta g, k] \tag{7.46}$$

Dimensions of the variables in length **L**, mass **M**, time **T**, temperature θ, heat **H**:

q	Heat transferred/unit area and unit time	$HL^{-2}T^{-1}$
u	Velocity	LT^{-1}
l	Linear dimension	L
μ	Viscosity	$ML^{-1}T^{-1}$
ρ	Density	ML^{-3}
k	Thermal conductivity	$HT^{-1}L^{-1}\theta^{-1}$
C_p	Specific heat capacity at constant pressure	$HM^{-1}\theta^{-1}$
ΔT	Temperature difference	θ
(βg)	Coefficient of thermal expansion times the acceleration due to gravity	$LT^{-2}\theta^{-1}$

It will be noted that both temperature and heat are taken as fundamental units and heat is not here expressed in terms of **M, L, T**.

With nine parameters and four dimensions equation 7.46 will be rearranged in four dimensionless groups.

Using the Π-theorem for solution of the equation:

Taking as the recurring set: $\qquad\qquad$ $l, \rho, \mu, \Delta T, k$

The non-recurring variables are: $\qquad$ $q, u, (\beta g), C_p$

Then:

$$l \equiv \mathbf{L} \qquad\qquad \mathbf{L} = l$$

$$\rho \equiv \mathbf{ML}^{-3} \qquad\qquad \mathbf{M} = \rho \mathbf{L}^3 = \rho l^3$$

$$\mu \equiv \mathbf{ML}^{-1}\mathbf{T}^{-1} \qquad \mathbf{T} = \mathbf{ML}^{-1}\mu^{-1} = \rho l^3 l^{-1}\mu^{-1} = \rho l^2 \mu^{-1}$$

$$\Delta T \equiv \boldsymbol{\theta} \qquad\qquad \boldsymbol{\theta} = \Delta T$$

$$k \equiv \mathbf{HL}^{-1}\mathbf{T}^{-1}\boldsymbol{\theta}^{-1} \qquad \mathbf{H} = k\mathbf{LT}\boldsymbol{\theta} = kl\rho l^2 \mu^{-1}\Delta T = kl^3 \rho\mu^{-1}\Delta T$$

The Π groups will then be:

$$\Pi_1 = q\mathbf{H}^{-1}\mathbf{L}^2\mathbf{T} = qk^{-1}l^{-3}\rho^{-1}\mu\,\Delta T^{-1}l^2\rho l^2 \mu^{-1} = qk^{-1}l\,\Delta T^{-1}$$

$$\Pi_2 = u\mathbf{L}^{-1}\mathbf{T} = ul^{-1}\rho l^2 \mu^{-1} = u\rho l\mu^{-1}$$

$$\Pi_3 = C_p\mathbf{H}^{-1}\mathbf{M}\boldsymbol{\theta} = C_p k^{-1}l^{-3}\rho^{-1}\mu\,\Delta T^{-1}\rho l^3\,\Delta T = C_p k^{-1}\mu$$

$$\Pi_4 = \beta g\mathbf{L}^{-1}\mathbf{T}^2\boldsymbol{\theta} = \beta gl^{-1}\rho^2 l^4 \mu^{-2}\Delta T = \beta g\,\Delta T\rho^2\mu^{-2}l^3$$

So that the relation 7.46 becomes:

$$\frac{ql}{k\,\Delta T} = \frac{hl}{k} = \phi\left(\frac{lu\rho}{\mu}\right)\left(\frac{C_p\mu}{k}\right)\left(\frac{\beta g\,\Delta Tl^3\rho^2}{\mu^2}\right) \tag{7.47}$$

or $\qquad\qquad\qquad\qquad$ $Nu = \phi(Re)(Pr)(Gr)$

This general equation involves the use of four dimensionless groups but may frequently be simplified for design purposes. In equation 7.47,

$\qquad\qquad\qquad$ hl/k $\quad$ is known as the *Nusselt* group Nu,

$\qquad\qquad\qquad$ $lu\rho/\mu$ $\quad$ the *Reynolds* group Re,

$\qquad\qquad\qquad$ $C_p\mu/k$ $\quad$ the *Prandtl* group Pr, and

$\qquad\qquad$ $\beta g\,\Delta Tl^3\rho^2/\mu^2$ $\quad$ the *Grashof* group Gr

For conditions in which only natural convection occurs the velocity is dependent solely on the buoyancy effects, represented by the Grashof number and the Reynolds group can be omitted. Again, when forced convection occurs the effects of natural convection are usually negligible and the Grashof number may be omitted. Thus:

for natural convection: $\qquad\qquad$ $Nu = f(Pr, Gr)$ $\qquad\qquad\qquad\qquad$ (7.48)

and for forced convection: $\qquad\qquad$ $Nu = f(Re, Pr)$ $\qquad\qquad\qquad\qquad$ (7.49)

Convection in Gases

For most gases over a wide range of temperature and pressure, the product $C_p\mu/k$ is constant and the Prandtl group may often be omitted, making the design equations for calculation of film coefficients with gases simpler.

7.4.3. Forced Convection in Tubes

The results of a number of workers who have used a variety of gases such as air, carbon dioxide, and steam and of others who have used liquids such as water, acetone, kerosene, and benzene can be expressed in the general form of equation 7.49 as:

$$Nu = 0.023\,Re^{0.8}\,Pr^n \tag{7.50}$$

where n has a value of 0.4 for heating and 0.3 for cooling. In this equation all of the physical properties are taken at the mean bulk temperature of the fluid $(T_i + T_0)/2$, where T_i and T_0 are the inlet and outlet temperatures. Equation 7.50 is valid for Reynolds numbers greater than 10,000 and has been tested for values of the Prandtl group lying between 0.7 and 160. The difference in the value of the index for heating and cooling occurs because in the former case the film temperature will be greater than the bulk temperature and in the latter case less. Conditions in the film, particularly the viscosity of the fluid, exert an important effect on the heat transfer process.

An alternative equation which is in many ways more convenient has been proposed by Colburn[13] and includes the Stanton number $(St = h/C_p\rho u)$ instead of the Nusselt number. This equation is:

$$St\,Pr^{0.67} = 0.023 Re^{-0.2} \tag{7.51}$$

It will be noted that:

$$h/C_p\rho u = (hd/k)(\mu/ud\rho)(k/C_p\mu)$$

i.e.

$$St = Nu\,Re^{-1}\,Pr^{-1} \tag{7.52}$$

Thus, multiplying equation 7.51 by $Re\,Pr^{0.33}$

$$Nu = 0.023\,Re^{0.8}\,Pr^{0.33} \tag{7.53}$$

which is a form of equation 7.50.

Again, the physical properties are taken at the bulk temperature except the viscosity in the Reynolds group which is evaluated at the mean film temperature taken as $(T_{surface} + T_{bulk\ fluid})/2$. The group on the left-hand side of equation 7.51 is known as the j_h factor for heat transfer and the equation is often written as:

$$j_h = 0.023\,Re^{-0.2} \tag{7.54}$$

Writing a heat balance for the flow through a tube of diameter d and length l with a rise in temperature for the fluid from T_i to T_0:

$$h\pi d l\,\Delta T = \frac{\pi d^2}{4}\,C_p u\rho(T_0 - T_i)$$

or

$$St = \frac{h}{C_p\rho u} = \frac{d(T_0 - T_i)}{4l\,\Delta T} \tag{7.55}$$

where ΔT is the mean temperature difference between the bulk fluid and the walls.

Viscous Liquids

With very viscous liquids there will be a marked difference at any cross-section between the viscosity of the fluid adjacent to the surface and the value at the axis or at the bulk temperature of the fluid. Sieder and Tate[14] have presented a modified equation for these conditions including a viscosity correction term:

$$Nu = 0.027\,Re^{0.8}\,Pr^{0.33}(\mu/\mu_s)^{0.14} \tag{7.56}$$

where μ is the viscosity at the bulk temperature and μ_s the viscosity at the wall or surface. This equation may also be written in the form of the Colburn equation 7.51.

Turbulent Flow of Gases

When the above equations are applied to heating or cooling of gases for which the Prandtl group usually has a value of about 0.74, substitution of $Pr = 0.74$ in equation 7.50 gives the equation:

$$Nu = 0.02\,Re^{0.8} \tag{7.57}$$

Special Equation for Water

Water is very frequently used as the cooling liquid, and if the variation in value of physical properties with temperature is included, then equation 7.50 becomes:

$$h = 150(1 + 0.011\,T)u^{0.8}/d^{0.2} \tag{7.58}$$

where d is in inches, u in ft/s, and T in degrees F, and h is then in Btu/h ft^2 °F. If T is in degrees K, d in m, u in m/s, then:

$$h = 1063(1 + 0.00293\,T)u^{0.8}/d^{0.2} \text{ W/m}^2 \text{ K}. \tag{7.59}$$

There is a very big difference in the values of h for water and air for the same linear velocity. This is shown in Figs. 7.14–7.16 and Table 7.2.

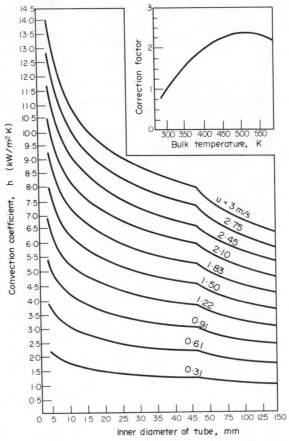

FIG. 7.14. Film coefficients of convection for flow of water through a tube.

Effect of Tube Length and Diameter

The effect of length to diameter ratio (l/d) on the value of the heat transfer coefficient can be seen in Fig. 7.17. It is important at low Reynolds numbers but ceases to be significant at a Reynolds number of about 10^4.

Streamline Flow

Although heat transfer to a fluid in streamline flow takes place solely by conduction, it is convenient to consider it here so that the results can be compared with those for turbulent

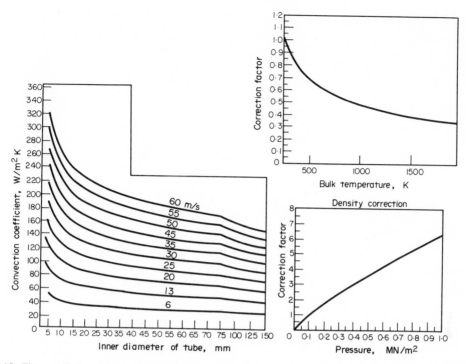

FIG. 7.15. Film coefficients of convection for flow of air through a tube, velocity curves (288 K, 101·3 kN/m²).

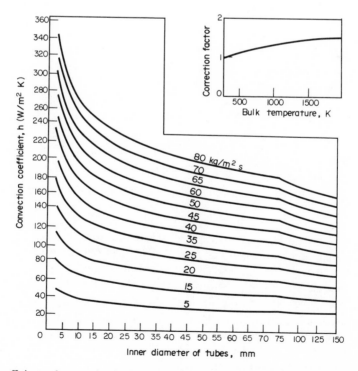

FIG. 7.16. Film coefficients of convection for flow of air through a tube, mass flow curves (288 K, 101·3 kN/m²).

TABLE 7.2. *Film coefficients for air and water (289 K and 101·3 kN/m²)*[15]

Inside diameter of tube		Velocity		Mass velocity		h	
(in)	(mm)	(ft/s)	(m/s)	(lb/ft² h)	(kg/m² s)	(Btu)h ft² °F	(W/m² K)
Air							
1	25	20	6·1	5500	7·15	6	34·1
		80	24·4	22,000	28·6	17.8	101·1
		140	42·7	40,000	52·0	28	159·0
		200	61·0	55,000	71·5	37	210·1
2	50	20	6·1	5500	7·15	5·2	29·5
		80	24·4	22,000	28·6	15·8	89·7
		140	42·7	40,000	52·0	24·2	137·4
		200	61·0	55,000	71·5	32·4	184·0
3	75	20	6·1	5500	7·15	4·6	26·1
		80	24·4	22,000	28·6	14·2	80·6
		140	42·7	40,000	52·0	22·2	126·1
		200	61·0	55,000	71·5	29·8	169·2
Water							
1	25	2	0·61	450,000	585	440	2498
		4	1·22	900,000	1170	770	4372
		8	2·44	1,800,000	2340	1340	7609
2	50	2	0·61	450,000	585	380	2158
		4	1·22	900,000	1170	670	3804
		8	2·44	1,800,000	2340	1160	6586
3	75	2	0·61	450,000	585	360	2044
		4	1·22	900,000	1170	620	3520
		8	2·44	1,800,000	2340	1080	6132

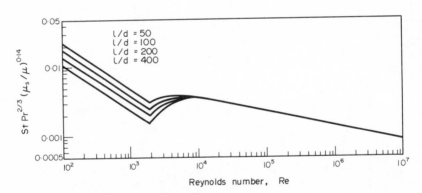

FIG. 7.17. Effect of length:diameter ratio on heat transfer coefficient.

flow. It has been seen in Chapter 3 that for streamline flow through a tube the velocity distribution across a diameter is parabolic (Fig. 7.18). If a liquid enters a section heated on the outside, the fluid near the wall will be at a higher temperature than that in the centre and its viscosity will be lower. The velocity of the fluid near the wall will therefore be greater in the heated section, and correspondingly less at the centre. The velocity distribution will therefore be altered, as shown. If the fluid enters a section where it is cooled, the same reasoning will show that the distribution in velocity will be altered to that shown. With a gas the conditions are reversed, because of the increase of viscosity with temperature. The heat transfer problem is therefore complex.

In Chapter 9 equations will be derived for h based on the assumptions of a parabolic distribution of velocity and of heat transfer by radial conduction. h will be shown to be

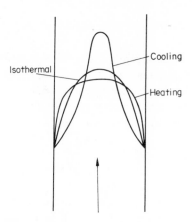

FIG. 7.18. Effect of heat transfer on velocity distribution.

infinite at the entrance and to fall slowly along the pipe. For values of $(Re\,Pr\,d/l)$ greater than 12, the following empirical equation is applicable:

$$Nu = 1\cdot62\left(Re\,Pr\frac{d}{l}\right)^{1/2} = 1\cdot75\left(\frac{GC_p}{kl}\right)^{1/2} \qquad (7.60)$$

where $G = (\pi d^2/4)\rho u$, i.e. the mass rate of flow.

The product $Re\,Pr$ is termed the Peclet number Pe. Thus:

$$Pe = \frac{ud\rho}{\mu}\frac{C_p\mu}{k} = \frac{C_p\rho ud}{k}$$

Thus equation 7.60 can be written:

$$Nu = 1\cdot62\left(Pe\frac{d}{l}\right)^{1/3} \qquad (7.61)$$

In this equation the temperature difference is taken as the arithmetic mean of the terminal values, i.e.:

$$[(T_w - T_1) + (T_w - T_2)]/2$$

where T_w is the temperature of the tube wall which is taken as constant.

If the liquid is heated almost to the wall temperature T_w (i.e. when GC_p/kl is very small) then, on equating the heat gained by the liquid to that transferred from the pipe:

$$GC_p(T_2 - T_1) = \pi dlh(T_2 - T_1)/2$$

or

$$h = 2GC_p/\pi dl \qquad (7.62)$$

For values of $(Re\,Pr\,d/l)$ less than about 17, the Nusselt group becomes approximately constant at $4\cdot1$ (see Chapter 9).

Viscous Liquids

Experimental values of h for viscous oils are greater than those given by equation 7.60 for heating and less for cooling. This is due to the large variation of viscosity with temperature and the correction introduced for turbulent flow may also be used here, giving the equation as:

$$Nu\left(\frac{\mu_s}{\mu}\right)^{0\cdot14} = 1\cdot86\left(Re\,Pr\frac{d}{l}\right)^{1/3} = 2\cdot01\left(\frac{GC_p}{kl}\right)^{1/3} \qquad (7.63)$$

or
$$Nu\left(\frac{\mu_s}{\mu}\right)^{0\cdot14} = 1\cdot86\left(Pe\frac{d}{l}\right)^{1/3}$$
(7.64)

When $(GC_p/kl) < 10$, the outlet temperature closely approaches that of the wall and equation 7·48 applies. These equations have been obtained with tubes from about 10 mm to 40 mm diameter and the length of unheated tube preceding the heated section is important. They are not entirely consistent since for very small values of ΔT the constants in equations 7.60 and 7.63 should be expected to be the same. It is important to note, when using these equations for design purposes, that the error may be as much as ±25 per cent for turbulent flow and greater for streamline conditions.

With laminar flow there is a marked influence of tube length and the curves shown in Fig. 7.17 show the parameter l/d from 50 to 400.

Whenever possible, streamline conditions of flow are avoided in heat-exchange equipment because of the very low heat transfer coefficients which are obtained. With very viscous liquids, however, turbulent conditions can be produced only if a very high pressure drop across the plant is permissible. In the chemical industry, streamline flow in heat exchangers is most commonly experienced with heavy oils and brines at low temperatures. Since the viscosity of these materials is critically dependent on temperature, the equations would not be expected to apply with a high degree of accuracy.

7.4.4. Forced Convection Outside Tubes

If a fluid passes at right angles across a single tube, the distribution of velocity around the tube will not be uniform. In the same way the rate of heat flow around a hot pipe across which air is passed is not uniform but is a maximum at the front and back and a minimum at the sides, where the rate is only some 40 per cent of the maximum. The general picture is shown in Fig. 7.19 but for design purposes reference is made to the average value.

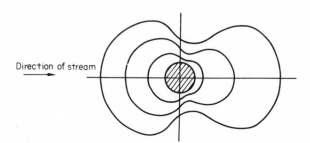

FIG. 7.19. Distribution of film heat transfer coefficient round a cylinder with flow normal to axis for three different values of Re.

Flow Across Single Cylinders

A number of workers (Reiher,[16] Hilpert,[17] Griffiths and Awbery[18]) have studied the flow of a hot gas past a single cylinder varying from a thin wire to a tube of 150 mm diameter. They have used temperatures up to 1073 K and air velocities up to 30 m/s with Reynolds numbers $(d_o u \rho/\mu)$ from 1000 to 100,000 (d_o is the cylinder diameter, or the outside tube diameter). Their results may be expressed by the relation:

$$Nu = 0\cdot26 \, Re^{0\cdot6} \, Pr^{0\cdot3}$$
(7.65)

Taking Pr as 0·74 for gases, this reduces to

$$Nu = 0\cdot24 \, Re^{0\cdot6}$$
(7.66)

Davis[19] has also worked with water, paraffin, and light oils and obtained similar results. For very low values of Re (from 0·2 to 200) with liquids the data are better represented by the equation:

$$Nu = 0·86 \, Re^{0·43} \, Pr^{0·3} \qquad (7.67)$$

In each case the physical properties of the fluid are measured at the mean film temperature T_f, taken as the average of the surface temperature T_w and the mean fluid temperature, T_m where $T_m = (T_1 + T_2)/2$.

Flow at Right Angles to Tube Bundles

One of the great difficulties with this system is that the area for flow is continually changing. Moreover the degree of turbulence is considerably less for banks of tubes in line, as at (a) than for staggered tubes, as at (b) in Fig. 7.20. With the small bundles which are common in the chemical industry the selection of the true mean area for flow is further complicated by the change in number of tubes in the rows.

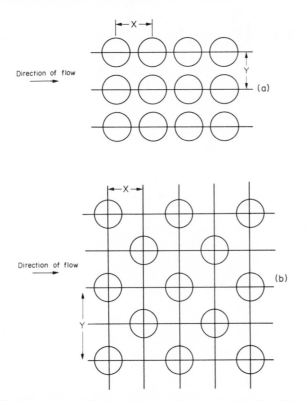

FIG. 7.20. Arrangements of tubes in heat exchangers: (a) in-line; (b) staggered.

The results of a number of workers for the flow of gases across tube banks can be put in the form:

$$Nu = 0·33 \, C_h \, Re_{max}^{0·6} \, Pr^{0·3} \qquad (7.68)$$

where C_h depends on the geometrical arrangement of the tubes (Table 7.3). Grimison[20] proposed this form of expression to correlate the data of Huge[21] and Pierson[22] who worked with small electrically heated tubes in rows of ten deep. Other workers have used similar equations. Some correction factors have been given by Pierson for bundles with less than ten rows but there are insufficient reported data from commercial exchangers to fix these values with accuracy. Thus for five rows a factor of 0·92 and for eight rows 0·97 is suggested.

TABLE 7.3.[15]   Values of C_h and C_f

| | X = 1·25d_o | | | | X = 1·5d_o | | | |
| | In-line | | Staggered | | In-line | | Staggered | |
Re_{max}	C_h	C_f	C_h	C_f	C_h	C_f	C_h	C_f
				Y = 1·25d_o				
2000	1·06	1·68	1·21	2·52	1·06	1·74	1·16	2·58
20,000	1·00	1·44	1·06	1·56	1·00	1·56	1·05	1·74
40,000	1·00	1·20	1·03	1·26	1·00	1·32	1·02	1·50
				Y = 1·5d_o				
2000	0·95	0·79	1·17	1·80	0·95	0·97	1·15	1·80
20,000	0·96	0·84	1·04	1·10	0·96	0·96	1·02	1·16
40,000	0·96	0·74	0·99	0·88	0·96	0·85	0·98	0·96

These equations are based on the maximum velocity through the bundle. Thus for an in-line arrangement as in Fig. 7.20, $G'_{max} = G' Y/(Y - d_o)$, where Y is the pitch of the pipes at right angles to direction of flow; it is more convenient here to use the mass flow rate per unit area G' in place of velocity. For staggered arrangements the maximum velocity may be based on the distance between the tubes in a horizontal line or on the diagonal of the tube bundle, whichever is the less.

It has been suggested that for in-line arrangements the constant in equation 7.68 should be reduced to 0·26, but there is insufficient evidence from commercial exchangers to settle this question.

With liquids the same equation may be used, but for Re less than 2000, there is insufficient published work to justify an equation. McAdams,[23] however, has given a curve for h for a bundle with staggered tubes ten rows deep.

An alternative approach has been suggested by Kern[24] who works in terms of the hydraulic mean diameter for flow *parallel* to the tubes:

i.e. $d_e = 4 \times$ Free area for flow/Wetted perimeter

$$= 4 \left(\frac{Y^2 - \pi d_o^2/4}{\pi d_o} \right)$$

for a square pitch as shown in Fig. 7.21. Then the maximum cross-flow area A_s is given by:

$$A_s = d_s l_B C'/Y$$

where C' is clearance, l_B the baffle spacing, and d_v the internal diameter of the shell.

The mass rate of flow per unit area G'_s is then given as rate of flow divided by A_s, and the film coefficient is obtained from the Nusselt type expression:

$$\frac{h_o d_e}{k} = 0·36 \left(\frac{d_e G'_s}{\mu} \right)^{0·55} \left(\frac{C_p \mu}{k} \right)^{1/3} \left(\frac{\mu}{\mu_s} \right)^{0·14} \tag{7.69}$$

There are insufficient published data to assess the relative merits of equations 7.68 and 7.69.

For 19 mm tubes on 25 mm square pitch

$$d_e = 4\{[25^2 - (\pi/4)19^2]/\pi 19\}$$

$$= 22·8 \text{ mm}$$

$$= 0·023 \text{ m}$$

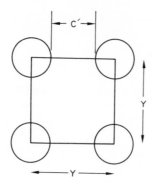

FIG. 7.21. Tube arrangement.

Example 7.7. 4·0 kg/s of nitrobenzene is to be cooled from 400 to 315 K by heating a stream of benzene from 305 to 345 K.

Two tubular heat exchangers are available each with a 0·44 m i.d. shell fitted with 166 tubes, 19·0 mm o.d. and 15·0 mm i.d., each 5·0 m long. The tubes are arranged in two passes on 25 mm square pitch with a baffle spacing of 150 mm. There are two passes on the shell side and operation is to be countercurrent. With benzene passing through the tubes, the anticipated film coefficient on the tube side is 1000 W/m² K.

What value of scale resistance could be allowed if these units were used?

For nitrobenzene:

$$C_p = 2380 \text{ J/kg K}$$

$$k = 0·15 \text{ W/m K}$$

$$\mu = 0·70 \text{ mN s/m}^2$$

Solution. (i) Tube side coefficient.

$$h_i = 1000 \text{ W/m}^2 \text{ K based on inside area}$$

or $\qquad 1000 \times 15·0/19·0 = 790$ W/m² K based on outside area

(ii) Shell side coefficient.

Area for flow = shell diameter × baffle spacing × clearance/pitch
$$= 0·44 \times 0·150 \times 0·006/0·025 = 0·0158 \text{ m}^2$$

Hence $\qquad\qquad\qquad G'_s = 4·0/0·0158 = 253·2 \text{ kg/m}^2\text{s}$

Taking $\mu/\mu_s = 1$,

$$h = (k/d_e)0·36(d_e G'_s/\mu)^{0·55}(C_p\mu/k)^{0·33}$$

$$d_e = 4\left[\left(25^2 - \frac{\pi \times 19·0^2}{4}\right)\middle/ \pi \times 19·0\right] = 22·8 \text{ mm} \quad \text{or} \quad 0·023 \text{ m}$$

∴ $\qquad h = (0·15/0·023)0·36(0·023 \times 253·2/0·70 \times 10^{-3})^{0·55}(2380 \times 0·70 \times 10^{-3}/0·15)^{0·33}$
$$= 2·35 \times 143 \times 2·23 = 750 \text{ W/m}^2 \text{ K}$$

(iii) Overall coefficient.

Log mean temperature difference,

$$\Delta T_m = [(400 - 345) - (315 - 305)]/[\ln (400 - 345)/(315 - 305)]$$

$$= 26·4 \text{ K} \qquad \text{(see Example 7.1)}$$

True mean $\quad = \Delta T_m \times F = 26 \cdot 4 \times 0 \cdot 8 = 21 \cdot 1$ K

$\qquad\qquad$ (see section on multipass exchangers)

Heat load: $Q = 4 \cdot 0 \times 2380(400 - 315) = 8 \cdot 09 \times 10^5$ W

Area of each tube $= 0 \cdot 0598$ m²/m

Thus $\qquad U_0 = Q/A_0 \Delta T_m F = 8 \cdot 09 \times 10^5/(166 \times 5 \cdot 0 \times 0 \cdot 0598 \times 21 \cdot 1)$

$\qquad\qquad\quad = 386 \cdot 2$ W/m² K

(iv) Scale resistance.

If scale resistance is R_d,

$$R_d = \frac{1}{386 \cdot 2} - \frac{1}{750} - \frac{1}{1000} = \underline{0 \cdot 00026 \text{ m}^2 \text{ K/W}}$$

This is a rather low value, though the heat exchangers would probably be used for this duty.

Effect of Baffles on Transfer Coefficients

To increase the velocity over the tubes it is common practice to fit baffles across the bundle. The commonest form is shown in Fig. 7.22, where it is seen that the cut-away section is about 25 per cent of the total area. With such an arrangement, the flow pattern becomes more complex and the extent of leakage between the tubes and the baffle, and between the baffle and the inside of the shell of the exchanger, complicates the problem. For more detail the reader is referred to papers by Short,[25] Donohue,[26] and Tinker.[27] The various methods are all concerned with developing a method of calculating the true area of flow and of assessing the probable influence of leaks. When using baffles, the value of h, as found from equation 7.68, is commonly multiplied by 0·6 to allow for leakage.

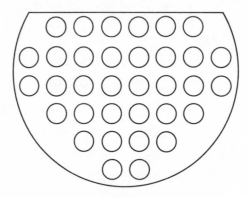

FIG. 7.22. Baffle for heat exchanger.

Heat Transfer and Pressure Drop over Tube Bundles

The drop in pressure for the flow of a fluid across a tube bundle (Table 7.4) may be important because of the small pressure head available and because by good design it is possible to get a better heat transfer for the same drop in pressure. $-\Delta P_f$ depends on the velocity u_t through the minimum area of flow and in Chapter 3 an equation proposed by Grimison was[20] given as:

$$-\Delta P_f = C_f j \rho u_t^2/6 \qquad\qquad\text{(equation 3.73)}$$

where C_f depends on the geometry of the tube layout and j is the number of rows of tubes. It is found that the ratio of C_h, the heat transfer factor in equation 7.68, to C_f is greater for the in-line arrangement but that the actual heat transfer is greater for the staggered arrangement.

TABLE 7.4.[15] *Ratio of heat transfer to friction for tube bundles*

	$X = 1\cdot25d_o$			$X = 1\cdot5d_o$		
	C_h	C_f	C_h/C_f	C_h	C_f	C_h/C_f
In-line						
$Y = 1\cdot25d_o$	1	1·44	0·69	1	1·56	0·64
$Y = 1\cdot5d_o$	0·96	0·84	1·14	0·96	0·96	1·0
Staggered						
$Y = 1\cdot25d_o$	1·06	1·56	0·68	1·05	1·74	0·60
$Y = 1\cdot5d_o$	1·04	1·10	0·95	1·02	1·16	0·88

To calculate the drop in pressure $-\Delta P_f$ over the tube bundles of a heat exchanger the following equation may also be used:[24]

$$-\Delta P_f = \frac{f'G_s'^2(N+1)d_v}{2\rho d_e} \tag{7.70}$$

where f' is the friction factor given in Fig. 7.23 (–), G_s' the mass velocity through bundle [see equation 7.69] (kg/m²s), N the number of baffles in unit (–), d_v the inside shell diameter (m), ρ density of fluid (kg/m³), d_e the equivalent diameter, (m), and ΔP_f the drop in pressure (N/m²).

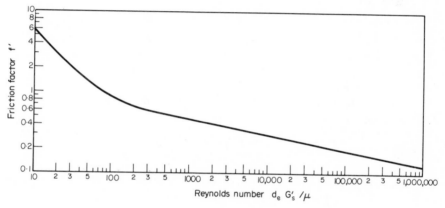

FIG. 7.23. Friction factor for flow over tube bundles.

Example 7.8. 15 kg/s of benzene is cooled by passing the stream through the shell side of a tubular heat exchanger, 1 m i.d., fitted with 5 m tubes, 19 mm o.d. arranged on a 25 mm square pitch with 6 mm clearance. If the baffle spacing is 0·25 m (19 baffles), what will be the pressure drop over the tube bundle? ($\mu = 0\cdot5$ mN s/m².)

Solution. Cross-flow area: $A_s = 1\cdot0\times0\cdot25\times0\cdot006/0\cdot025 = 0\cdot06$ m²

Mass flow: $\qquad\qquad\qquad G_s' = 15/0\cdot06 = 250$ kg/m²s

Equivalent diameter:

$$d_e = 4[0\cdot025^2 - (\pi/4)0\cdot019^2]/(\pi\times0\cdot019) = 0\cdot0229 \text{ m}$$

Reynolds number through the tube bundle $= 250\times0\cdot0229/0\cdot5\times10^{-3} = 11450$

From Fig. 7.23: $\qquad\qquad\qquad f' = 0\cdot280$

Density of benzene $\qquad\qquad = 881$ kg/m³

From equation 7.70:

$$-\Delta P_f = 0\cdot280 \times 250^2 \times 20 \times 1\cdot0/(2 \times 881 \times 0\cdot0229)$$
$$= 8674 \text{ N/m}^2$$

or
$$8674/(881 \times 9\cdot81) = \underline{\underline{1\cdot00 \text{ m of benzene}}}$$

7.4.5. Flow in Non-circular Sections

For the heat transfer for fluids flowing in non-circular ducts, such as rectangular ventilating ducts, the equations developed for turbulent flow inside a circular pipe may be used if some equivalent diameter is used in the place of d.

The use of the hydraulic mean diameter d_e has already been discussed in Chapter 3.

The data for heating and cooling water in turbulent flow in rectangular ducts are reasonably well expressed by the use of equation 7.50 in the form:

$$\frac{hd_e}{k} = 0\cdot023 \left(\frac{d_eG'}{\mu}\right)^{0\cdot8} \left(\frac{C_p\mu}{k}\right)^{0\cdot4} \tag{7.71}$$

Whilst the experimental data of Cope and Bailey[28] are somewhat low, data by Washington and Marks[29] for heating air in ducts are well represented by this equation.

Annular Sections between Concentric Tubes

Concentric tube heat exchangers are widely used because of their simplicity of construction and the ease with which additions may be made to increase the area. They are also used to give turbulent conditions where the quantity flowing is small.

In presenting equations for the film coefficient in the annulus, one of the difficulties has been to select the best equivalent diameter to use. When considering the film on the outside of the inner tube, Davis[30] has proposed the equation:

$$\frac{hd_1}{k} = 0\cdot031 \left(\frac{d_1G'}{\mu}\right)^{0\cdot8} \left(\frac{C_p\mu}{k}\right)^{0\cdot33} \left(\frac{\mu}{\mu_s}\right)^{0\cdot14} \left(\frac{d_2}{d_1}\right)^{0\cdot15} \tag{7.72}$$

Carpenter *et al.*[31] suggested using the hydraulic mean diameter $d_e = d_2 - d_1$ in the Sieder and Tate equation 7.56 and recommend the expression:

$$\frac{hd_e}{k}\left(\frac{\mu_s}{\mu}\right)^{0\cdot14} = 0\cdot027 \left(\frac{d_eG'}{\mu}\right)^{0\cdot8} \left(\frac{C_p\mu}{k}\right)^{0\cdot33} \tag{7.73}$$

Their results, which were obtained using a small annulus, were somewhat below those given by equation 7.73 for values of d_eG/μ less than 10,000, but this may have been because the flow was not fully turbulent: with an index on the Reynolds group of 0·9, the equation fitted the points much better. There is little to choose between these two equations, but they both give rather high values for h.

For the viscous region Carpenter's results are reasonably well covered by the equation:

$$\frac{hd_e}{k}\left(\frac{\mu_s}{\mu}\right)^{0\cdot14} = 2\cdot01 \left(\frac{GC_p}{kl}\right)^{0\cdot33}$$
$$= 1\cdot86 \left[\left(\frac{d_eG'}{\mu}\right)\left(\frac{C_p\mu}{k}\right)\left(\frac{d_1 + d_2}{l}\right)\right]^{1/3}$$

This equation is the same as equation 7.63 with d_e in the place of d.

These results have all been obtained with small units and mainly with water as the fluid in the annulus, and no entirely satisfactory solution has yet been obtained.

Flow over Flat Plates

For the turbulent flow of a fluid over a flat plate the Colburn type of equation may be used with a different constant:

$$j_h = 0\cdot037 Re^{-0\cdot2} \qquad (7.74)$$

where the physical properties are taken as for equation 7.54 and the characteristic dimension in the Reynolds group is the actual distance along the plate. This equation therefore gives a point value for j_h.

7.4.6. Convection Heat Transfer to Spherical Particles

Earlier in this chapter, consideration was given to the problem of heat transfer by conduction through a surrounding fluid to spherical particles or droplets. Relative motion between the fluid and particle or droplet causes an increase in heat transfer due to convection. Many experimenters have correlated their data in the form:

$$Nu = 2 + \beta'' Re^n Pr^m \qquad (7.75)$$

where values of β'', a numerical constant, and exponents n and m are found by experiment. Here $Nu = hd/k$ and Re, the Reynolds number for the particle, $= du\rho/\mu$, u is the relative velocity between particle and fluid, and d is the particle diameter. As the relative velocity goes to zero, Re goes to zero and the equation reduces to $Nu = 2$ for pure conduction.

Rowe *et al.*,[32] having analysed a large number of previous studies on this subject and provided further experimental data, conclude that for particle Reynolds numbers in the range 20–2000, equation 7.75 be written as:

$$Nu = 2\cdot0 + \beta'' Re^{0\cdot5} Pr^{0\cdot33} \qquad (7.76)$$

β'' lies between $0\cdot4$ and $0\cdot8$ and has a value of $0\cdot69$ for air and $0\cdot79$ for water. In some practical situations the relative velocity between particle and fluid may change due to particle acceleration or deceleration and the value of Nu can then be time-dependent.

For mass transfer, which is considered in more detail in Chapters 8 and 9, an analogous equation applies with the Sherwood number replacing the Nusselt number and the Schmidt number replacing the Prandtl number.

7.4.7. Natural Convection

If a beaker containing water rests on a hot plate, the water at the bottom of the beaker becomes hotter than that at the top. Since the density of the hot water is lower than that of the cold, the water in the bottom rises and heat is transferred by natural convection. In the same way air in contact with a hot plate will be heated by natural convection currents, the air near the surface being hotter and of lower density than that some distance away. In these cases there is no external agency providing forced convection currents, and the transfer of heat occurs at a correspondingly lower rate since the natural convection currents move rather slowly.

For these processes which depend on buoyancy effects, it has been seen from equation 7.48 that the rate of heat transfer would be expected to follow a relation:

$$Nu = \psi(Pr, Gr) \qquad (\text{equation } 7.48)$$

Measurements by Schmidt[10] of the upward air velocity near a 300 mm vertical plate show that the velocity rises rapidly to a maximum at a distance of about 2 mm from the plate and then falls rapidly. However, the temperature evens out at about 10 mm from the plate. Temperature measurements around horizontal cylinders have been made by Ray.[33]

Horizontal Surfaces

Natural convection from horizontal surfaces to air, nitrogen, hydrogen, and carbon dioxide, and to liquids (water, aniline, carbon tetrachloride, glycerol) has been studied by a number of workers (Davis,[34] Ackerman,[35] Saunders,[36] and others). Most of the results are for thin wires and tubes up to about 50 mm diameter; the temperature differences used are up to about 1100 K with gases and about 85 K with liquids. The general form of the results is shown in Fig. 7.24, where log Nu is plotted against log $(Pr\,Gr)$ for streamline conditions. The curve can be represented by the relation:

$$Nu = C'(Pr\,Gr)^n \qquad (7.77)$$

where n is 0.25 for values of $Gr\,Pr$ between 10^3 and 10^8. For turbulent conditions n is found to be about 0.33, and hence the heat transfer coefficient becomes independent of the extent of the surface.

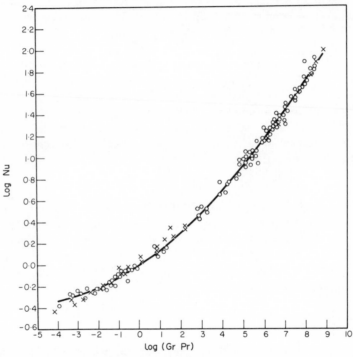

Fig. 7.24. Natural convection from horizontal tubes.

Vertical Surfaces

Work by Saunders[36,37] and others can also be expressed by equation 7.77 with values of C' as given in Table 7.5. The results are for air with surfaces up to about 1 m in height and for liquids with surfaces up to only about 0.3 m high. Under these conditions turbulence occurs at a value of $Gr\,Pr$ of about 2×10^9. Again n is 0.25 for streamline and 0.33 for turbulent conditions.

TABLE 7.5. *Values of C' for natural convection for use in equation 7.77*

Nature of surface	Characteristic dimension	Streamline	Turbulent
Horizontal or vertical cylinders	Diameter	0.47	0.10
Vertical planes or vertical cylinders of large diameter	Height	0.56	0.12
Horizontal planes facing upwards	Mean length of side	0.54	0.14
Horizontal planes facing downwards	Mean length of side	0.25	Not reached

In using these equations the values of the physical properties ρ, C_p, k, μ are taken at the mean of the surface and bulk temperatures. The coefficient of cubical expansion β is taken as $1/T$, where T is the absolute temperature.

Simplified Equations for Natural Convection to Air

For the special case of convection from a hot body to air these equations may be simplified. Thus for streamline flow equation 7.77 may be written as:

$$\frac{hl}{k} = C' \left(\frac{\beta g \, \Delta T l^3 \rho^2}{\mu^2} \frac{C_p \mu}{k} \right)^{1/4} \quad \text{or} \quad h = C' \left(\frac{\Delta T}{l} \right)^{1/4} k \left(\frac{\beta g \rho^2 C_p}{\mu k} \right)^{1/4} \tag{7.78}$$

Over a wide range in temperature the group $k(\beta g \rho^2 C_p/\mu k)^{1/4}$ remains reasonably constant. The heat transfer coefficient h may thus be obtained from the relation:

$$h = C'(2{\cdot}45)\left(\frac{\Delta T}{l} \right)^{1/4}$$

where ΔT is in K, l is in metres, and h is in W/m^2 K.

For turbulent conditions:

$$\frac{hl}{k} = C' \left(\frac{\beta g \, \Delta T l^3 \rho^2}{\mu^2} \frac{C_p \mu}{k} \right)^{1/2} \quad \text{or} \quad h = C'(\Delta T)^{1/4} \left(\frac{k^2 \, \Delta T^{1/4} \beta g \rho^2 C_p}{\mu} \right)^{1/2} \tag{7.79}$$

h is then independent of l, and since $k^2 \, \Delta T^{1/4} \beta g \rho^2 C_p/\mu$ is approximately constant, h is proportional to $\Delta T^{0 \cdot 25}$.

For the following geometrical conditions, equations can be obtained as shown below. ΔT is in K and h is in W/m^2 K.

	Streamline	Turbulent
Horizontal or vertical pipes	$h = 1{\cdot}18(\Delta T/d_0)^{1/4}$	$1{\cdot}65\Delta T^{1/4}$
Vertical planes	$1{\cdot}35(\Delta T/l)^{1/4}$	$2{\cdot}00\Delta T^{1/4}$
Horizontal planes facing upwards	$1{\cdot}31(\Delta T/l)^{1/4}$	$2{\cdot}33\Delta T^{1/4}$
Horizontal planes facing downwards	$0{\cdot}59(\Delta T/l)^{1/4}$	

Fluid Layers between Two Surfaces

For the transfer of heat from a hot surface across a thin layer of fluid to a parallel cold surface:

$$\frac{Q}{Q_k} = \frac{h \, \Delta T}{(k/x) \, \Delta T} = \frac{hx}{k} = Nu \tag{7.80}$$

where Q_k is the rate at which heat would be transferred by pure thermal conduction between the layers, a distance x apart, and Q is the actual rate.

For $(Gr \, Pr)$ equal to 10^3, the heat transferred is approximately equal to that due to conduction alone, but for $(Gr \, Pr)$ from 10^4 to 10^6, the heat transferred is given (Fig. 7.25) by

$$Q/Q_k = 0{\cdot}15(Gr \, Pr)^{0 \cdot 25} \tag{7.81}$$

In this equation the characteristic dimension to be used for the Grashof group is x, the distance between the planes, and the heat transfer is independent of surface area, provided that the linear dimensions of the surfaces are large compared with x. For higher values of $(Gr \, Pr)$, Q/Q_k is proportional to $(Gr \, Pr)^{1/3}$, showing that the heat transferred is not entirely by convection and is not influenced by the distance x between the surfaces.

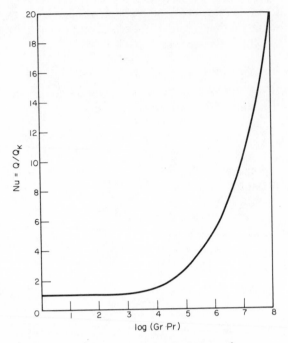

FIG. 7.25. Natural convection between surfaces.

A similar form of analysis has been given by Kraussold[38] for air between two concentric cylinders. It is important to note from this general analysis that a single layer of air will not be a good insulator because convection currents set in before it becomes 25 mm thick. The good insulating properties of porous materials are attributable to the fact that they offer a series of very thin layers of air in which convection currents do not arise.

7.5. HEAT TRANSFER BY RADIATION

A heated body emits energy in the form of electromagnetic waves. This energy is radiated in all directions and on falling on a second body is partially absorbed, partially reflected, and partially transmitted as indicated in Fig. 7.26. The fraction of the incident radiation absorbed is known as the *absorptivity a*, and the fraction reflected the *reflectivity* of the body. The amount transmitted will therefore depend on these two properties. If the amount transmitted is negligible the material is termed *opaque*.

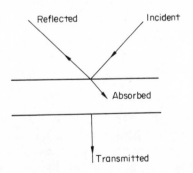

FIG. 7.26. Reflection, absorption, and transmission of radiation.

7.5.1. Kirchhoff's Law

Suppose two bodies **A** and **B** of areas A_1 and A_2 to be in a large enclosure from which no energy is lost to the outside. Then the energy absorbed by **A** from the enclosure will be $A_1 a_1 I'$, where I' is the rate at which radiation is falling on **A** per unit area, and a_1 is the absorptivity. The energy given out by **A** will be $E_1 A_1$, where E_1, the *emissive power*, is the energy emitted per unit area per unit time. At equilibrium these quantities will be equal. In the same way the energy emitted by **B** will equal the energy received. Thus

$$I'A_1 a_1 = A_1 E_1 \quad \text{and} \quad I'A_2 a_2 = A_2 E_2$$

or
$$E_1/a_1 = E_2/a_2 = E/a \quad \text{for any other body}$$

Thus for all bodies the ratio of the emissive power to the absorptivity will be the same. The maximum possible value for E will occur when a has its maximum value of unity. This condition applies to a *black body* which is defined as one which absorbs all the radiation falling on it, i.e. its absorptivity, $a_b = 1$.

Now the *emissivity* e of a body is defined as the ratio of its emitting power to that of a black body; i.e. $e = E/E_b$. Since E/a is constant for all bodies, $E/a = E_b/a_b$.

Thus
$$e = E/E_b = a/a_b$$

Then, since $a_b = 1$, the emissivity e of any body is equal to its absorptivity a.

7.5.2. Energy Emitted by a Black Body

The energy emitted per unit time by a black body depends only on its temperature. This energy is given out over a range of wavelengths and the general distribution of energy from a black body at various temperatures is shown in Fig. 7.27. At any temperature there is a wavelength Z_m at which the maximum energy is emitted. The energy at wavelengths less than this falls off rapidly, but for wavelengths greater than Z_m the drop in energy is much less. It will be seen that the total emission increases rapidly with temperature and that the higher the

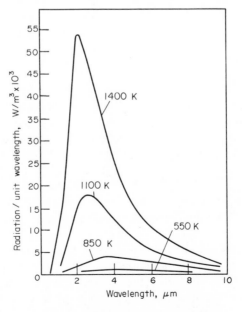

FIG. 7.27. Emission from a black body.

temperature the smaller is the wavelength at which the maximum occurs and the more pronounced is the peak.

The wavelength of the energy radiated at temperature below 820 K is too great for it to be visible, but at higher temperatures the wavelengths correspond to those of the visible spectrum and the colour of a radiating body under these conditions is a good indication of its temperature.

The total energy emitted per unit area per unit time is given by the *Stefan–Boltzmann law*:

$$E = \sigma T^4 \qquad (7.82)$$

where T is the absolute temperature and σ is the Stefan–Boltzmann constant which has the value 1.73×10^{-9} Btu/h ft^2 °R^4 or 5.67×10^{-8} W/m^2 K^4. The wavelength Z_m at which the maximum energy is radiated was found by Wien to vary inversely as the absolute temperature. Taking the temperature in degrees, $TZ_m = 0.00288$, where the wavelength is in metres.

If $E'_Z \, dZ$ is the radiation between wavelengths Z and $Z + dZ$, the total radiation is given by:

$$E = \int_0^\infty E'_Z \, dZ \qquad (7.83)$$

It can be shown that this integral is equal to σT^4, so that:

$$E = \sigma T^4 \qquad \text{(equation 7.82)}$$

Emissivity e

The ratio of the energy emitted by a body to that emitted by a black body at the same temperature has been defined above as the emissivity. Strictly this should be taken for each wavelength, since the ratio e will not remain constant over a wide range of wavelengths. A *grey body* is defined as one which has a constant value of e, so that for any temperature range it radiates the same proportion of the energy radiated by a black body. Similarly it will have a constant absorptivity.

Values of e have been measured for many materials, and it is found that for most industrial non-metallic surfaces and for non-polished metals e may be taken as about 0.9; for highly polished surfaces such as copper or aluminium values of e as low as 0.03 are obtained. A small cavity in a body has an effective emissivity of unity and therefore behaves as a black body.

7.5.3. Heat Transferred by Radiation

A body of emissivity e at an absolute temperature T_1 emits energy $e\sigma T_1^4$ per unit area. If the surroundings are black, they reflect back none of this radiation, but if they are at an absolute temperature T_2 they will emit radiation σT_2^4. If the body is grey it will absorb a fraction e, so that the net radiation per unit area from the grey body will be:

$$q = e\sigma(T_1^4 - T_2^4) \qquad (7.84)$$

This relation will still be true where the grey body is so small that a negligible proportion of its radiation is reflected back to it from the surroundings, e.g. a body radiating to the atmosphere.

For a material that does not behave as a grey body but as a selective emitter, the absorptivity of the surface at T_1 for radiation from surroundings at T_2 will be a_{T_2}. This will not be equal to its emissivity (e_{T_1}) at T_1 but to its emissivity at T_2, i.e. e_{T_2}. Under these conditions the general equation for the net exchange of heat becomes:

$$q = \sigma(e_{T_1} T_1^4 - e_{T_2} T_2^4) \qquad (7.85)$$

Transfer Coefficient for Radiation

The net heat transfer from unit surface of a grey body at a temperature T_1 to a black enclosure at T_2 can be written as:

$$q = h_r(T_1 - T_2)$$

Thus

$$h_r = \frac{q}{T_1 - T_2} = \frac{\sigma e}{T_1 - T_2}(T_1^4 - T_2^4) \qquad (7.86)$$

and h_r may be looked upon as a radiation transfer coefficient. Equation 7.86 is also applicable if the surroundings are not black, provided that the body is small and none of its radiation is reflected back to it.

Example 7.9. Calculate the total heat loss by radiation and convection from an unlagged horizontal steam pipe, 50 mm o.d. at 377 K to air at 283 K.

Solution. From equation 7.84 assuming an emissivity of 0·9:

loss by radiation: $q_r = 5·67 \times 10^{-8} \times 0·9(377^4 - 283^4)$

$$= 704 \text{ W/m}^2$$

loss by natural convection, assuming streamline flow:

$$= 1·18 \,(\Delta T/d_0)^{0·25} \times \Delta T$$
$$= 1·18 \,(94/0·050)^{0·25} \times 94$$
$$= 732 \text{ W/m}^2$$

The total loss is therefore

$$(704 + 732) = \underline{\underline{1436 \text{ W/m}^2}}$$

In practice the heat loss from an unlagged pipe will be rather higher, especially if the pipe is vertical when the convection currents become stronger.

Intensity of Radiation in a Given Direction

If experiments are made on the intensity of radiation emitted by a small plane surface in a given direction it is found that the strongest radiation is in a line normal to the surface. This intensity I normal to the plane is defined as the radiation given out per unit time per unit area of a small surface dA over unit solid angle. Then the intensity of radiation at an angle α to the normal is seen from Fig. 7.28 to be $I \cos \alpha$. This form of relation shows the similarity of

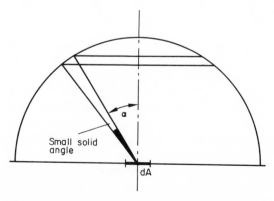

FIG. 7.28. Intensity of radiation in a given direction.

the transfer process to the radiation of light. It is followed accurately by black bodies but metallic surfaces may show wide divergences.

Radiation intensity normal to surface, I = energy per unit area per unit time per unit solid angle = $dq/d\omega$.

Radiation from small surface dA, through small solid angle $d\omega$, at angle α to normal is $I\,dA\cos\alpha\,d\omega$.

$$\therefore \text{ Total energy emitted in forward direction} = \int_{\alpha=0}^{\alpha=\pi/2} I\,dA\cos\alpha\,d\omega$$

But $d\omega = 2\pi r\sin\alpha\,rd\alpha/r^2$, where r is radius of described circle.

$$\therefore \text{ Emissive power } (\sigma T^4 \text{ for black body}) = \int_0^{\pi/2} I\cos\alpha\,2\pi\sin\alpha\,d\alpha = \pi I \qquad (7.87)$$

7.5.4. Geometric or Angle Factors

In many cases when two *non-black* bodies are situated in fairly close proximity, part of the energy emitted by one body will be reflected back to it by the second body and will then be partly reabsorbed and partly reflected again. Thus the heat undergoes a series of internal reflections and absorptions.

The simplest case to consider is the heat exchange between two large parallel plates. In this case, all of the heat radiated by the one surface will fall on the other. In many cases, however, the second surface will intercept only part of the radiation. The fraction of the total radiation from surface 1 which is intercepted by surface 2 is referred to as the *geometric* or *angle factor* F_{12} and its value depends on the geometrical arrangement of the two surfaces. For simple configurations the geometric factors can be calculated, but for more complex arrangements they are most readily determined by experimenting with models.

The geometric factor is best understood by considering the projected area which an element of the radiating surface *sees*. Thus, with large parallel plates the field of view of any element of surface 1 remote from the edges consists entirely of surface 2 and therefore the geometric factor F_{12} is unity; and, of course, F_{21} is also unity. Furthermore, since the element does not radiate directly on to any part of its own surface, the geometric factor with respect to itself, F_{11} (and F_{22}) is zero. This latter condition also applies to any convex surface and therefore for all convex surfaces F_{11} is zero.

From the above definition of geometric factor, it follows that the sum of the geometric factors with respect to a surface must be unity:

i.e.
$$F_{11} + F_{12} + \cdots = 1 \qquad (7.88)$$

This is known as the *summation rule*.

Two particular cases of radiation between surfaces are of particular practical interest and are capable of analytical solution. The first is the radiation between two long concentric cylinders and the other is the radiation between concentric spheres. In both cases, the inner surface (1) is convex and therefore F_{11} is zero. Since all the radiation from this surface falls on the concave outer surface, F_{12} is unity. However, part of the radiation from surface 2 falls on the inner surface and part falls on itself. The fraction of the field of view of an element of the concave surface which is occupied by the inner convex surface is A_1/A_2 and therefore $F_{21} = A_1/A_2$ and $F_{22} = 1 - (A_1/A_2)$ in both cases.

The net transfer of heat can be considered by summing the individual components. This is done below for two parallel plates, but a second method, the net radiation method, is more general in its application.

For transfer between two large parallel plates, each of which has a grey surface, it will be shown that the heat transfer rate per unit area is given by:

$$q = \frac{e_1 e_2 \sigma}{e_1 + e_2 - e_1 e_2}(T_1^4 - T_2^4) \tag{7.89}$$

For long concentric cylinders or concentric spheres, the heat transfer rate per unit area of surface 1 is given by:

$$q = \frac{e_1 e_2 \sigma}{e_2 + e_1(1 - e_2)A_1/A_2}(T_1^4 - T_2^4) \tag{7.90}$$

Radiation between Parallel Plates—Multiple Reflection Method

Consider the radiation of heat between two parallel grey surfaces which are large compared with their distance apart. Then only a negligible fraction of the total heat transmitted will be lost to the surroundings at the ends, and all the heat radiated by one surface will be intercepted by the other, i.e. the geometric factors will both be unity.

Then, per unit area and unit time:

heat radiated from surface 1	$= \sigma e_1 T_1^4$	(A)
of this, amount absorbed by surface 2	$= \sigma e_1 T_1^4 e_2$	(B)
and amount reflected by surface 2	$= \sigma e_1 T_1^4 (1 - e_2)$	(C)
of this, amount re-absorbed by surface 1	$= \sigma e_1 T_1^4 (1 - e_2) e_1$	(D)
and amount re-reflected by surface 1	$= \sigma e_1 T_1^4 (1 - e_2)(1 - e_1)$	(E)
and, of this, amount absorbed by surface 2	$= \sigma e_1 T_1^4 (1 - e_2)(1 - e_1) e_2$	(F)

Thus, as a result of each complete cycle of internal reflections, it is seen by comparing (B) and (F) that the absorption is reduced by a factor $(1 - e_1)(1 - e_2)$.

As the energy suffers an infinite number of reflections:

total transfer of energy from surface 1 to surface 2 per unit area and unit time =

$$= \sigma e_1 e_2 T_1^4[1 + (1 - e_1)(1 - e_2) + (1 - e_1)^2(1 - e_2)^2 + \cdots \text{ to } \infty]$$

$$= \sigma e_1 e_2 T_1^4 \frac{1}{1 - (1 - e_1)(1 - e_2)}$$

$$= \frac{\sigma e_1 e_2}{e_1 + e_2 - e_1 e_2} T_1^4$$

Since this expression is symmetrical in e_1 and e_2, the total transfer of energy from surface 2 to surface 1 per unit area and unit time

$$= \frac{e_1 e_2 \sigma}{e_1 + e_2 - e_1 e_2} T_2^4$$

Thus net energy transferred per unit area and unit time:

$$q = \frac{e_1 e_2 \sigma}{e_1 + e_2 - e_1 e_2}(T_1^4 - T_2^4) \tag{equation 7.89}$$

Radiation between Parallel Plates—Net Radiation Method

Let Q_{o1} be the total radiation per unit time leaving surface 1, and Q_{i1} be the total incident radiation. Then in this case:

$$Q_{o1} = Q_{i2} \quad \text{and} \quad Q_{o2} = Q_{i1}$$

Now Q_{o1} consists of emitted radiation Q_{e1} and reflected radiation Q_{r1}.

Thus $\qquad Q_{o1} = Q_{e1} + Q_{r1}$

But $\qquad Q_{r1} = (1 - e_1)Q_{i1}$

Thus $\qquad Q_{o1} = Q_{e1} + (1 - e_1)Q_{i1}$

$$\left. \begin{aligned} &= Q_{e1} + (1 - e_1)Q_{o2} \end{aligned} \right\} \qquad (7.91)$$

Similarly $\qquad Q_{o2} = Q_{e2} + (1 - e_2)Q_{o1}$

Hence $\qquad Q_{o1} = \dfrac{Q_{e1} + (1 - e_1)Q_{e2}}{1 - (1 - e_1)(1 - e_2)}$ and $\qquad Q_{o2} = \dfrac{Q_{e2} + (1 - e_2)Q_{e1}}{1 - (1 - e_1)(1 - e_2)}$

The net transfer of radiation per unit time

$$Q = Q_{o1} - Q_{o2} = \frac{e_2 Q_{e1} - e_1 Q_{e2}}{1 - (1 - e_1)(1 - e_2)}$$

But $\qquad Q_{e1} = A\sigma e_1 T_1^4$ and $\qquad Q_{e2} = A\sigma e_2 T_2^4$

Thus $\qquad q = \dfrac{Q}{A} = \dfrac{e_1 e_2 \sigma}{e_1 + e_2 - e_1 e_2}(T_1^4 - T_2^4)$ $\qquad$ (equation 7.89)

Net Radiation Method for other Surfaces

In the general case, the radiation incident on a surface will have come partly from itself and partly from the other surface, the proportion being determined by the geometric factors. Thus:

$$Q_{i2} = Q_{o1}F_{12} + Q_{o2}F_{22}$$

$$Q_{i1} = Q_{o2}F_{21} + Q_{o1}F_{11}$$

Again $\qquad Q_{o1} = Q_{e1} + (1 - e_1)Q_{i1}$ $\qquad$ (equation 7.91)

i.e. $\qquad Q_{o1} = Q_{e1} + (1 - e_1)(Q_{o2}F_{21} + Q_{o1}F_{11})$

Similarly $\qquad Q_{o2} = Q_{e2} + (1 - e_2)(Q_{o1}F_{12} + Q_{o2}F_{22})$

Solving for Q_{o1} and Q_{o2}:

$$Q_{o1} = \frac{Q_{e1}[1 - (1 - e_2)F_{22}] + Q_{e2}(1 - e_1)F_{21}}{[1 - (1 - e_2)F_{22}][1 - (1 - e_1)F_{11}] - (1 - e_1)(1 - e_2)F_{12}F_{21}}$$

and $\qquad Q_{o2} = \dfrac{Q_{e2}[1 - (1 - e_1)F_{11}] + Q_{e1}(1 - e_2)F_{12}}{[1 - (1 - e_2)F_{22}][1 - (1 - e_1)F_{11}] - (1 - e_1)(1 - e_2)F_{12}F_{21}}$

Now the net rate of transfer of heat is given by:

$$Q = Q_{o1}F_{12} - Q_{o2}F_{21}$$

$$= \frac{Q_{e1}F_{12}[1 - (1 - e_2)(F_{22} + F_{21})] - Q_{e2}F_{21}[1 - (1 - e_1)(F_{11} + F_{12})]}{[1 - (1 - e_2)F_{22}][1 - (1 - e_1)F_{11}] - (1 - e_1)(1 - e_2)F_{12}F_{21}}$$

Substituting $Q_{e1} = A_1\sigma e_1 T_1^4$ and $Q_{e2} = A_2\sigma e_2 T_2^4$ and noting from the *summation rule*, equation 7.88 that $F_{11} + F_{12}$ and $F_{21} + F_{22}$ are both unity:

$$Q = \frac{e_1 e_2 \sigma (A_1 F_{12} T_1^4 - A_2 F_{21} T_2^4)}{[1 - (1 - e_2)F_{22}][1 - (1 - e_1)F_{11}] - (1 - e_1)(1 - e_2)F_{12}F_{21}} \qquad (7.92)$$

Since there can be no net heat transfer between the surfaces when T_1 and T_2 are equal:

$$A_1 F_{12} = A_2 F_{21} \qquad (7.93)$$

This is known as the *reciprocal rule*.

Long Concentric Cylinders and Concentric Spheres

For long concentric cylinders and concentric spheres:

$$F_{11} = 0 \quad F_{12} = 1 \quad F_{21} = A_1/A_2 \quad F_{22} = 1 - (A_1/A_2)$$

Thus

$$Q = \frac{A_1 e_1 e_2 \sigma}{e_2 + e_1(1 - e_2)A_1/A_2}(T_1^4 - T_2^4)$$

and the heat transfer rate per unit area of surface 1 (Q/A_1) is given by:

$$q = \frac{e_1 e_2 \sigma}{e_2 + e_1(1 - e_2)A_1/A_2}(T_1^4 - T_2^4) \qquad \text{(equation 7.90)}$$

7.5.5. Gas Radiation

Most of the simple monatomic and diatomic gases such as helium, hydrogen, oxygen, and nitrogen are transparent to thermal radiation, but some polyatomic gases—notably carbon dioxide, water vapour, carbon monoxide, ammonia, and hydrocarbons—absorb a considerable amount of radiation of certain frequencies. These gases, which are industrially very important, radiate appreciably in the same wave bands. In contrast with the behaviour of solids a considerable thickness of gas is required to absorb a large fraction of the radiation falling on it. Thus if I' is the intensity of the incident radiation, the intensity I after the radiation has passed through a layer of thickness x is given by:

$$I = I' e^{-m'x} \qquad (7.94)$$

where m' will, in general, vary with the wavelength and is approximately proportional to the partial pressure of the gas for any one wavelength, i.e. to the number of molecules per unit volume. The absorption will then be a function of the product $P_g l$, where P_g is the partial pressure of the gas and l is the equivalent thickness of the gas stream. The absorptivity of a gas is conveniently plotted against temperature for a constant value of $P_g l$. The thickness l is assumed the same in all directions but this will only be true for the case of a hemisphere where the gas is radiating to the mid-point of its base. It has been found that for a wide range of conditions l may be taken as 3·4 times the gas volume divided by the area of the retaining walls. A few values on this basis are given in Table 7.6.

TABLE 7.6. *Equivalent dimensions for hemispherical radiation*

Shape	Characteristic dimension L	Equivalent dimension for hemispherical radiation
Sphere	Diameter	0·57L
Infinite cylinder radiating to walls	Diameter	0·86L
Space between infinite parallel planes	Distance apart	1·70L
Space outside bank of tubes with centres on equilateral triangle; clearance = diameter of tube	Clearance	2·89L
Ditto, with tube centres on square pitch; tube diameter = clearance	Clearance	3·49L

Curves showing the emissivity of water vapour at a total pressure of one atmosphere, (Fig. 7.29) are taken from the work of Hottel.[39] The net radiant heat exchange between a gas at T_1 and unit area of enclosure at T_2 acting as a black body will be:

$$q = \sigma(e_g T_1^4 - a_g T_2^4) \qquad (7.95)$$

where e_g and a_g are the emissivity and absorptivity of the gas.

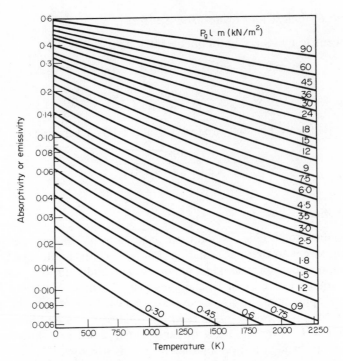

FIG. 7.29. Absorptivity and emissivity of water vapour.

If the enclosure acts as a non-black surface some of the radiation falling on it is reflected back; of this, part is absorbed by the gas and part by the surface. For these general conditions it is usually sufficiently accurate to take the emissivity of the enclosure as e', where $e' = (e_s + 1)/2$. This is permissible because e_s, the emissivity of the surface, will probably lie between 0·7 and 1. The radiation exchange will then be:

$$q = e'\sigma(e_g T_1^4 - a_g T_2^4) \tag{7.96}$$

Example 7.10. A cylindrical peep-hole in a 0·3 m thick furnace wall is 0·15 m in diameter. The cylindrical surface of the peep-hole behaves as a diffuse reflector and heat losses through it to the material of the furnace wall are negligible. If the inside surfaces of the furnace are at a uniform temperature of 1350 K and the outside surroundings are at 285 K, determine the heat loss by radiation through the peep-hole.

The angle factor F_{12} for radiation between two parallel discs of equal radii in opposite location (i.e. a line joining the centres is normal to the discs) is given by:

$$F_{12} = [1 + 2x^2 - \sqrt{(1 + 4x^2)}]/2x^2$$

where x is the ratio of the disc radius to the distance of separation of the discs.

The peep-hole connects two uniform temperature enclosures, one at 1350 K and one at 285 K; hence the ends of the peep-hole may be treated as plane black body surfaces at these two temperatures. (Stefan–Boltzmann constant = $5·67 \times 10^{-8}$ W/m² K⁴.)

Solution. (a) Determination of angle factors (Fig. 7.30).
The formula given for the angle-factor between surfaces (1) and (2) may be used:

$$x = r/z = 0·075/0·3 = 0·25$$

∴

$$F_{12} = F_{21} = [1 + 0·125 - \sqrt{(1 + 0·25)}]/0·125 = 0·056$$

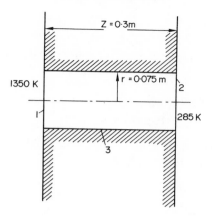

FIG. 7.30. Angle factors for peep-hole.

Since (1) and (2) are plane surfaces,

$$F_{11} = F_{22} = 0$$

$\therefore$ By the *summation rule* (equation 7.88):

$$F_{13} = 1 - F_{12} - F_{11} = 1 - 0.056 - 0 = 0.944$$

By the *reciprocal rule* (equation 7.93):

$$A_3 F_{31} = A_1 F_{13}$$

$\therefore$ $$F_{13}\pi r^2/2\pi rz = F_{31} = 0.944 \times 0.075/2 \times 0.3 = 0.118$$

By symmetry: $$F_{23} = F_{13} \quad \text{and} \quad F_{32} = F_{31}$$

By the summation rule:

$$F_{33} = 1 - F_{31} - F_{32} = 1 - 2 \times 0.118$$
$$= 0.764$$

(b) Heat balances. Since (1) and (2) are black surfaces, all incident radiation is absorbed and the only outgoing radiation is the emitted radiation.
Using Stefan's law:

$$Q_{o1} = Q_{e1} = \pi r^2 \times 5.67 \times 10^{-8} T_1^4$$
$$= 0.075^2 \pi \times 5.67 \times 10^{-8}(1350)^4 = 3328 \text{ W}$$

Similarly: $$Q_{o2} = Q_{e2} = 0.075^2 \pi \times 5.67 \times 10^{-8}(285)^4 = 6.6 \text{ W}$$

Since surface (3) is perfectly insulated, the total outgoing radiation (reflected + emitted) is equal to the incident radiation, or

$$Q_{i3} = Q_{o3} \tag{1}$$

However, from the definition of angle-factors:

total incident radiation, $$Q_{i3} = Q_{o1}F_{13} + Q_{o2}F_{23} + Q_{o3}F_{33} \tag{2}$$

From equations (1) and (2):

$$Q_{o3} = (Q_{o1}F_{13} + Q_{o2}F_{23})/(1 - F_{33})$$
$$= 0.944(3328 + 6.6)/(1 - 0.764) = 13338 \text{ W}$$

Similarly:
$$Q_{i2} = Q_{o1}F_{12} + Q_{o2}F_{22} + Q_{o3}F_{32}$$
$$= 3328 \times 0 \cdot 056 + 6 \cdot 6 \times 0 + 13{,}338 \times 0 \cdot 118$$
$$= 1760 \cdot 3 \text{ W}$$

Net radiation from the peep-hole $= Q_{i2} - Q_{o2}$
$$= 1760 \cdot 3 - 6 \cdot 6 = \underline{1753 \cdot 7 \text{ W}}$$

Alternatively, the radiation entering by surface (1) is 3328 W and, by definition of F_{12}, the proportion of this leaving by surface (2) is:

$$Q_{e1}F_{12} = 3328 \times 0 \cdot 056 = 186 \text{ W}$$

The rest, 3142 W hits surface (3).

Similarly: radiation entering by surface (2) is 6·6 W

proportion leaving by surface (1) is 0·4 W

proportion hitting surface (3) is 6·2 W

∴   Total radiation incident on surface (3) $= 3142 + 6 \cdot 2 = 3148 \cdot 2$ W

Since all radiation from surface (3) is emitted or reflected uniformly, by symmetry, half of this incident radiation leaves from each end.

∴   Net radiation to atmosphere $= (0 \cdot 5 \times 3148 \cdot 2) + 186 - 6 \cdot 6$
$$= \underline{\underline{1753 \cdot 5 \text{ W}}}$$

7.6. HEAT TRANSFER IN THE CONDENSATION OF VAPOURS

When a saturated vapour is brought into contact with a cool surface, heat is transferred from the vapour to the surface and a film of condensate is produced.

In considering the heat that is transferred in such a case, the method which was first put forward by Nusselt[40] and later modified by Jakob and others will be followed. If the vapour is condensing on a vertical surface, the condensate film flows downwards under the influence of gravity, but is retarded by the viscosity of the liquid. This flow will normally be streamline and the heat will flow through the film by conduction. Nusselt supposed that the temperature of the film at the cool surface was equal to that of the surface, and at the other side was at the temperature of the vapour. Actually there must be some small difference in temperature between the vapour and the film, but this may generally be neglected except where non-condensable gas is present in the vapour.

7.6.1. Calculation of Film Coefficient

It is first necessary to calculate the velocity distribution in a liquid which is flowing down a surface inclined at an angle to the horizontal.

Consider a plane surface of unit width, inclined at an angle ϕ to the horizontal. At a distance x from the top of the surface, the thickness of the liquid layer is s, say (Fig. 7.31).

In an element of condensate of thickness dx situated at a distance x from the top of the surface, the accelerating force in the X-direction, acting on the liquid at a distance greater than y from the surface, $= (s - y) \, dx \rho g \sin \phi$. The retarding force is made up of the drag of the gas or vapour on the free surface and the drag at the inner boundary of the element. The shear stress R' at the free surface will be small, except at high vapour velocities and will be

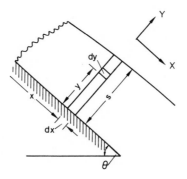

FIG. 7.31. Flow of liquid over a surface.

neglected at present. Thus the retarding force in the X-direction $= \mu(\partial u_y/\partial y)\,dx$, where u_y is the velocity at a distance y from the surface. Under equilibrium conditions therefore:

$$(s - y)\,dx\rho g\,\sin\phi = \mu\frac{\partial u_y}{\partial y}\,dx \tag{7.97}$$

Thus

$$du_y = \frac{\rho g\,\sin\phi}{\mu}(s - y)\,dy$$

and

$$u_y = \frac{\rho g\,\sin\phi}{\mu}(sy - \tfrac{1}{2}y^2) + \text{constant} \tag{7.98}$$

Since the liquid in contact with the surface must be at rest, $u_y = 0$ when $y = 0$, and the constant is, therefore, zero.

The mass rate of flow G of liquid over the surface can now be calculated.

$$G = \int_0^s \left(\frac{\rho g\,\sin\phi}{\mu}(sy - \tfrac{1}{2}y^2)\right)\rho\,dy$$

$$= \frac{\rho^2 g\,\sin\phi}{\mu}\left(\frac{s^3}{2} - \frac{s^3}{6}\right)$$

$$= (\rho^2 g\,\sin\phi\,s^3)/3\mu$$

so that the mean velocity of the fluid:

$$u_m = (\rho g\,\sin\phi\,s^2)/3\mu \tag{7.99}$$

For a vertical surface: $\sin\phi = 1$ and $u_m = (\rho g s^2)/3\mu$

The maximum velocity, which occurs at the free surface:

$$= (\rho g\,\sin\phi\,s^2)/2\mu$$

and is 1·5 times the mean velocity of the liquid.

Since the liquid is produced by condensation, the thickness of the film will be zero at the top and will gradually increase towards the bottom. Under stable conditions the difference in the mass rates of flow at distances x and $x + dx$ from the top of the surface will result from condensation over the small element of the surface of length dx (Fig. 7.32).

If the thickness of the liquid film increases from s to $s + ds$ in that distance, the increase in the mass rate of flow of liquid dG

$$= \frac{d}{ds}\left(\frac{\rho^2 g\,\sin\phi\,s^3}{3\mu}\right)ds$$

$$= (\rho^2 g\,\sin\phi\,s^2\,ds)/\mu$$

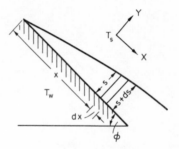

FIG. 7.32. Condensation of an inclined surface.

If the vapour temperature is T_s and the wall temperature is T_w, the heat transferred by thermal conduction to an element of surface of length dx

$$= k(T_s - T_w)\,dx/s$$

where k is the thermal conductivity of the condensate.
Thus the mass rate of condensation on this small area of surface

$$= k(T_s - T_w)\,dx/s\lambda$$

where λ is the latent heat of vaporisation of the liquid.

Thus $$k(T_s - T_w)\,dx/s\lambda = (\rho^2 g \sin \phi\, s^2\, ds)/\mu$$

On integration, this equation gives:

$$\mu k(T_s - T_w)x = \rho^2 g \sin \phi s^4 \lambda/4$$

since $s = 0$ when $x = 0$.

Thus $$s = \sqrt[4]{\frac{4\mu kx(T_s - T_w)}{g \sin \phi\, \lambda \rho^2}} \tag{7.100}$$

Now the heat transfer coefficient h at $x = x$, $= k/s$.

Thus $$h = \sqrt[4]{\frac{\rho^2 g \sin \phi \lambda k^3}{4\mu x(T_s - T_w)}} \tag{7.101}$$

and $$Nu = \frac{hx}{k} = \sqrt[4]{\frac{\rho^2 g \sin \phi\, \lambda x^3}{4\mu k(T_s - T_w)}} \tag{7.102}$$

These expressions give point values of h and Nu_x at $x = x$. It is seen that the coefficient decreases from a theoretical value of infinity at the top as the condensate film thickens. The mean value of the heat transfer coefficient over the whole surface, between $x = 0$ and $x = x$ is given by:

$$h_m = \frac{1}{x}\int_0^x h\,dx = \frac{1}{x}\int_0^x Kx^{-1/4}\,dx \quad \text{(where } K \text{ is independent of } x)$$

$$= \frac{1}{x}K\frac{x^{3/4}}{\tfrac{3}{4}} = \frac{4}{3}Kx^{-1/4} = 4h/3$$

$$= 0.943\sqrt[4]{\frac{\rho^2 g \sin \phi\, \lambda k^3}{\mu x\, \Delta T_f}} \tag{7.103}$$

where ΔT_f is the temperature difference across the condensate film.

For a vertical surface, $\sin \phi = 1$ and:

$$h_m = 0.943 \sqrt[4]{\frac{\rho^2 g \lambda k^3}{\mu x \, \Delta T_f}} \qquad (7.104)$$

7.6.2. Dimensionless Form of Nusselt Equation

Suppose the vapour to condense on the outside of a vertical tube of diameter d_o. Then the hydraulic mean diameter for the film

$$= 4 \times \text{flow area/wetted perimeter} = 4S/b \quad \text{(say)}$$

If G is the mass rate of flow of condensate, the mass rate of flow per unit area G' is G/S. The Reynolds number for the condensate film is then given by

$$Re = (4S/b)(G/S)/\mu = 4G/\mu b = 4M/\mu$$

where M is the mass rate of flow of condensate per unit length of perimeter, i.e.

$$M = G/\pi d_o$$

Then for streamline conditions in the film, $4M/\mu \not> 2100$. Thus

$$h_m = \frac{Q}{A \, \Delta T_f} = \frac{G\lambda}{bl \, \Delta T_f} = \frac{\lambda M}{l \, \Delta T_f}.$$

From equation 7.104:

$$h_m = 0.943 \left(\frac{k^3 \rho^2 g}{\mu} \frac{\lambda}{l \, \Delta T_f} \right)^{1/4} = 0.943 \left(\frac{k^3 \rho^2 g}{\mu} \frac{h_m}{M} \right)^{1/4}$$

whence

$$h_m \left(\frac{\mu^2}{k^3 \rho^2 g} \right)^{1/2} = 1.47 \left(\frac{4M}{\mu} \right)^{-1/2} \qquad (7.105)$$

For horizontal tubes, Nusselt gives:

$$h_m = 0.72 \left(\frac{k^3 \rho^2 g \lambda}{d_o \mu \, \Delta T_f} \right)^{1/4} \qquad (7.106)$$

This can be rearranged to give:

$$h_m \left(\frac{\mu^2}{k^3 \rho^2 g} \right)^{1/3} = 1.19 \left(\frac{4M}{\mu} \right)^{-1/3} \qquad (7.107)$$

where M is the mass rate of flow per unit length of tube.

This is approximately the same as equation 7.105 for vertical tubes and is a universal equation for condensation if it is remembered that for vertical tubes $M = G/\pi d_o$ and for horizontal tubes $M = G/l$, where l is the length of the tube. Comparison of the two equations shows that, provided the length is more than three times the diameter, the horizontal tube will give a higher transfer coefficient for the same temperature conditions.

For j vertical rows of horizontal tubes, equation 7.106 is modified to give:

$$h_m = 0.72 \left(\frac{k^3 \rho^2 g \lambda}{j d_o \mu \, \Delta T_f} \right)^{1/4} \qquad (7.108)$$

7.6.3. Experimental Values

In testing Nusselt's equation it is important to see that the conditions comply with the requirements of the theory. In particular, it is necessary for the condensate to form a uniform film on the tubes, for the drainage of this film to be by gravity, and the flow streamline.

Although some of these requirements have probably not been entirely fulfilled, results for pure vapours such as steam, benzene, toluene, diphenyl, ethanol, etc., are sufficiently near to give support to the theory. Some results by Haselden and Prosad[41] for condensing oxygen and nitrogen vapours on a vertical surface, where precautions were taken to see that the conditions were met, are in very good agreement with Nusselt's theory. The results for most of the workers are within 15 per cent for horizontal tubes but tend to be substantially higher than the theoretical for vertical tubes. Some typical values are given in Table 7.7 taken from McAdams[23] and elsewhere.

TABLE 7.7. *Average values of film coefficients h_m for condensation of pure saturated vapours on horizontal tubes*

Vapour	Value of h_m (Btu/h ft² °F)	Value of h_m (W/m² K)	Range of ΔT_f (K)
Steam	1700–5000	9650–28,400	1–11
Steam	3200–6500	18,170–36,910	4–37
Benzene	242–381	1375–2163	23–37
Diphenyl	225–400	1278–2270	4–15
Toluene	193–241	1100–1368	31–40
Methanol	500–600	2840–3407	8–16
Ethanol	320–450	1817–2555	6–22
Propanol	250–300	1920–1700	13–20
Oxygen	573–1400	3253–7950	0·08–2·5
Nitrogen	400–1009	2270–5730	0·15–3·5
Ammonia	1000	5680	—
Freon-12	200–400	1136–2271	—

When considering commercial equipment, there are several factors which prevent the true conditions of Nusselt's theory being met. Thus the temperature of the tube wall will not be constant, and for a vertical condenser with a ratio of ΔT at the bottom to ΔT at the top of five, the film coefficient should be increased by about 15 per cent.

7.6.4. Influence of Vapour Velocity

A high vapour velocity upwards tends to increase the thickness of the film and thus reduce h though the film may sometimes be disrupted mechanically as a result of the formation of small waves. For the downward flow of vapour Ten Bosch[42] has shown that h increases considerably at high vapour velocities and may increase to two or three times the value given by the Nusselt equation. It must be remembered that when a large fraction of the vapour is condensed there may be a considerable change in velocity over the surface.

Under conditions of high vapour velocity Carpenter and Colburn[43] have shown that turbulence may occur with low values of the Reynolds number, e.g. 200–400. When the vapour velocity is high there will be an appreciable drag on the condensate film and equation 7.97 becomes:

$$(s - y) \, dx \rho g \sin \phi = \mu \frac{\partial u_y}{\partial y} \, dx + R' \, dx \qquad (7.109)$$

where R' is the shear stress produced by the vapour at the surface of the condensate. The expression obtained for the heat transfer coefficient by using equation 7.109 is difficult to manage.

Carpenter and Colburn have put forward a simple experimental correlation of their results for condensation at varying vapour velocities on the inner surface of a vertical tube:

$$h_m = 0.065 G'_m \sqrt{\frac{C_p \rho k (R'/\rho_v u^2)}{\mu \rho_v}} \qquad (7.110)$$

where

$$G'_m = \sqrt{(G'^2_1 + G'_1 G'_2 + G'^2_2)/3}$$

and u is the velocity calculated from G'_m.

In this equation C_p, k, ρ, and μ are properties of the condensate and ρ_v refers to the vapour. G'_1 is the mass rate of flow per unit area at the top of the tube and G'_2 the corresponding value at the bottom.

As pointed out by Colburn[45], the group $C_p \rho k / \mu \rho_v$ does not vary much for a number of organic vapours so that a plot of h_m and G'_m will provide a simple approximate correlation with separate lines for steam and for organic vapours as shown in Fig. 7.33. Whilst this must be regarded as an empirical approximation it is very useful for obtaining a good idea of the effect of vapour velocity.

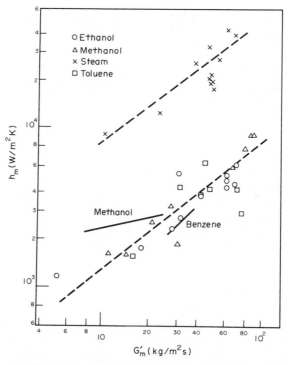

FIG. 7.33. Average heat transfer data of Carpenter (shown as points) compared with those of Tepe and Mueller[44] (solid lines). Dashed lines represent equation 7.110.

7.6.5. Turbulence in the Film

If Re is greater than 2100 during condensation on a vertical tube the mean coefficient h_m will increase. Data of Kirkbride[46] and Badger[47, 48] for the condensation of diphenyl vapour and Dowtherm on nickel tubes were expressed in the form:

$$h_m (\mu^2/k^3 \rho^2 g)^{1/3} = 0.0077(4M/\mu)^{0.4} \qquad (7.111)$$

Comparing equation 7.105 for streamline flow of condensate and equation 7.111 for turbulent flow, it is seen that, with increasing Reynolds number, h decreases with streamline

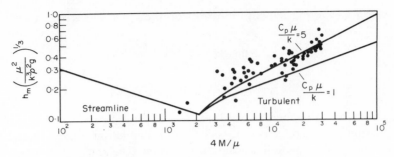

FIG. 7.34. Effects of turbulence in condensate film.

flow but increases with turbulent flow. These results are indicated in Fig. 7.34 but they are as yet incomplete.

7.6.6. Dropwise Condensation

In the discussion so far it has been assumed that the condensing vapour, on coming into contact with the cold surface, wets the tube so that a continuous film of condensate is formed. If the droplets initially formed do not wet the surface, after growing slightly they will fall from the tube exposing fresh condensing surface. This gives what is known as dropwise condensation and, since the heat does not have to flow through a film by conduction, much higher transfer coefficients are obtained. Steam is the only pure vapour for which definite dropwise condensation has been obtained, and values of h from 40 to 114 kW/m² K have been obtained, with much higher values on occasions. This question has been discussed by Drew, Nagle, and Smith[40] who have shown that there are many materials which make the surface non-wettable; of these, only those which are firmly held to the surface are of any practical use. Mercaptans and oleic acid have been used to promote dropwise condensation, but at present there is little practical application of this technique. Exceptionally high values of h will not give a corresponding increase in the overall coefficient, since for a condenser with steam, a value of about 11 kW/m² K can be obtained with film condensation. On the other hand, it may be helpful in experimental work to reduce the thermal resistance on one side of a surface to negligible value.

7.6.7. Condensation of Mixed Vapours

In the previous discussion it has been assumed that the vapour was a pure material, such as steam or organic vapour. If it contains a proportion of non-condensable gas and is cooled below its dew point, a layer of condensate is formed on the surface with a mixture of non-condensable gas and vapour above it. The heat flow from the vapour to the surface then takes place by two methods. First sensible heat is passed to the surface because of the temperature difference. Secondly, since the concentration of vapour in the main stream is greater than that in the gas film at the condensate surface, vapour molecules diffuse to the surface and there condense, giving up their latent heat. The actual rate of condensation is then determined by the combination of these two effects, and its calculation requires a knowledge of mass transfer by diffusion discussed in Chapter 8.

Vapour and Permanent Gas

In this treatment the design of a cooler-condenser for a vapour and permanent gas by the method of Colburn and Hougen[50] is considered. The method requires the point to point calculation of the condensate–vapour interface conditions T_c and P_s. This calculation is performed by a trial and error solution of the equation:

$$q_v \quad + \quad q_\lambda \quad = \quad q_c \quad = U \, \Delta T \tag{7.112}$$

$$h_g(T_s - T_c) + k_G\lambda(P_g - P_s) = h_o(T_c - T_{cm}) = U \, \Delta T \tag{7.113}$$

where the first term q_v represents the sensible heat transferred to the condensing surface, the second term q_λ the latent heat transferred by the diffusing vapour molecules, and the third term q_c the heat transferred from the condensing surface through the pipe wall, dirt and scales, and water film to the cooling medium, h_g is the heat transfer coefficient over the gas film, h_o the conductance of the combined condensate film, tube wall, dirt and scale films, and the cooling medium film, U the overall heat transfer coefficient, T_s the vapour temperature, T_c the surface temperature of condensate, T_w the temperature of condensation surface, T_{cm} the cooling medium temperature, ΔT the overall temperature difference $= (T_s - T_m)$, P_g the partial pressure of diffusing vapour, P_s the vapour pressure at T_c, λ the latent heat of vaporisation per unit mass, and k_G the mass transfer coefficient in mass per unit time, unit area, unit partial pressure difference.

To evaluate the required condenser area, point values of the group $U \, \Delta T$ as a function of q must be determined by a trial and error solution of equation 7.113. A plot of q against $1/U \, \Delta T$ upon graphical integration will yield the required condenser area. The method takes into account point variations in temperature difference, overall coefficient and mass velocities and consequently produces a reasonably good value for surface area required.

Individual terms in equation 7.113 will now be examined to enable a trial solution to proceed. Values for h_g and k_G are most conveniently obtained from the Chilton and Colburn[51] analogy discussed in Chapter 8. Thus:

$$h_g = \frac{j_h G' C_p}{(C_p\mu/k)^{0.67}}$$

$$k_G = \frac{j_d G'}{P_{Bm}(\mu/\rho D)^{0.67}}$$

Values of j_h and j_d are obtained from a knowledge of the Reynolds number at a given point in the condenser. The combined conductance h_o is evaluated by determining the condensate film coefficient h_c from the Nusselt equation and combining this with the dirt and tube wall conductances and a cooling medium film conductance predicted from the Sieder–Tate relationships. Generally, h_o may be considered constant throughout the exchanger.

From a knowledge of h_g, k_G, and h_o and for a given T_s and T_m values of the condensate surface temperature T_c can now be estimated until equation 7.113 is satisfied. The calculations are repeated, and in this manner several point values of the group $U \, \Delta T$ throughout the condenser may be obtained.

Mixture of Two Vapours

The design of a cooler condenser for the case of condensation of two vapours is more complicated than the preceding single vapour–permanent gas case.[52] For an example the reader is referred to the work of Jeffreys.[53]

7.7. HEAT TRANSFER TO BOILING LIQUIDS

In chemical plants, liquids are boiled either on submerged surfaces or on the inside of vertical tubes. Mechanical agitation may be applied in the first case and in the second the liquid may be driven through the tubes by means of an external pump. The boiling of liquids under either of these conditions normally leads to the formation of vapour first in the form of bubbles and later as a distinct vapour phase above a liquid interface. The conditions for boiling on the submerged surface will be discussed in this chapter and the problems arising with boiling inside tubes will be considered in Volume 2,[54] on evaporators. Much of the fundamental work on the ideas of boiling has been presented by Jakob,[55] and more recently by Rohsenow[56, 57] and by Forster.[58] The boiling of solutions in which a solid phase is separated after evaporation has proceeded to a sufficient extent will be considered in Volume 2.[54]

For a bubble to be formed in a liquid, for instance steam in water, it is necessary for a surface of separation to be produced. Kelvin has shown that, as a result of the surface tension between the liquid and vapour, the vapour pressure on the inside of a concave surface will be less than that at a plane surface. As a result, the vapour pressure P_r inside the bubble is less than the saturation vapour pressure P_s at a plane surface. The relation between P_r and P_s is:

$$P_r = P_s - (2\sigma/r) \tag{7.114}$$

where r is the radius of curvature of the bubble, and σ is the surface tension.

Hence the liquid must be superheated near the surface of the bubble, the extent of the superheat increasing with decrease in the radius of the bubble. On this basis it follows that very small bubbles are difficult to form without excessive superheat. The formation of bubbles is made much easier by the fact that they will form on curved surfaces or on irregularities on the heating surface, so that only a small degree of superheat is normally required.

Nucleation at much lower values of superheat is believed to arise from the presence of existing nuclei such as non-condensing gas bubbles or from the effect of the shape of the cavities in the surface. Of these, the current discussion on the influence of cavities is the most promising. In many cavities the angle θ will be greater than 90° and the effective contact angle, which includes the contact angle of the cavity β, will be considerably greater $[= \theta + (180 - \beta)/2]$, so that a much-reduced superheat is required to give nucleation. Thus the size of the mouth of the cavity and the shape of the cavity plays a significant part in nucleation.[59]

It follows that for boiling to occur a small difference in temperature must exist between the liquid and the vapour. Jakob and Fritz[60] have measured the temperature distribution for water boiling above an electrically heated hot plate. The temperature dropped very steeply from about 383 K on the actual surface of the plate to 374 K about 0·1 mm from it. Beyond this the temperature was reasonably constant till the water surface was reached. The mean superheat of the water above the temperature in the vapour space was about 0·5 K and this changed very little with the rate of evaporation. At higher pressures this superheating became smaller becoming 0·2 K at 5 MN/m² and 0·05 K at 101 MN/m². The temperature drop from the heating surface, however, depends very much on the rate of heat transfer and on the nature of the surface. Thus in order to maintain a heat flux of about 25·2 kW/m² a temperature difference of only 6 K was required with a rough surface as against 10·6 K with a smooth surface. The heat transfer coefficient on the boiling side is therefore dependent on the nature of the surface and on the difference in temperature available. For water boiling on copper plates Jakob and Fritz give the following coefficients for a constant temperature difference of 5·6 K, with different surfaces:

(1) Surface after 8 h (28·8 ks) use and 48 h (172·8 ks)
        immersion in water $h = 7950 \text{ W/m}^2 \text{ K}$
(2) Freshly sandblasted $h = 3860 \text{ W/m}^2 \text{ K}$
(3) Sandblasted surface after long use $h = 2560 \text{ W/m}^2 \text{ K}$
(4) Chromium plated $h = 1990 \text{ W/m}^2 \text{ K}$

The initial surface, with freshly cut grooves, gave much higher figures than case (1) above.

The nature of the surface will make a marked difference to the physical form of the bubble and the area actually in contact with the surface, as shown in Fig. 7.35.

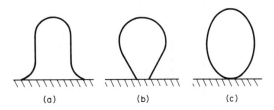

(a)   (b)   (c)

FIG. 7.35. Shapes of bubbles: (a) screen surface—thin oil layer; (b) chromium plated and polished surface; (c) screen surface—clean.

(a) *Non-wettable surface.* Here the vapour bubbles spread out thus reducing the area available for heat transfer from the hot surface to the liquid.

(b) *Partially wettable surface*—which is the commonest form. Here the bubbles rise from a larger number of sites and the rate of transfer is increased.

(c) *Entirely wetted surface*, such as that formed by a screen. This gives the minimum area of contact between vapour and surface and the bubbles leave the surface when still very small. It will therefore follow that if the liquid has detergent properties this may give rise to much higher rates of heat transfer.

7.7.1. Types of Boiling

In boiling liquids on a submerged surface it is found that the heat transfer coefficient depends very much on the temperature difference between the hot surface and the boiling liquid. The general relation between the temperature difference and heat transfer coefficient was first presented by Nukiyama[61] who boiled water on an electrically heated wire. His results have been confirmed and extended by others, and Fig. 7.36 shows the data of Farber and Scorah.[62] This relationship is complex and is best considered in stages.

Interface Evaporation

Here the bubbles of vapour formed on the heated surface move to the vapour–liquid interface by natural convection and exert very little agitation on the liquid. The results are given by:

$$Nu = 0 \cdot 61 (Pr \, Gr)^{1/4} \qquad (7.115)$$

which may be compared with the expression for natural convection:

$$Nu = C'(Pr \, Gr)^n \qquad \text{(equation 7.77)}$$

where $n = 0 \cdot 25$ for streamline conditions and $n = 0 \cdot 33$ for turbulent conditions.

Nucleate Boiling

At higher values of ΔT the bubbles form more rapidly and from more centres of nucleation. Under these conditions the bubbles exert an appreciable agitation on the liquid and the heat transfer coefficient rises rapidly. This is the most important region for boiling in industrial equipment.

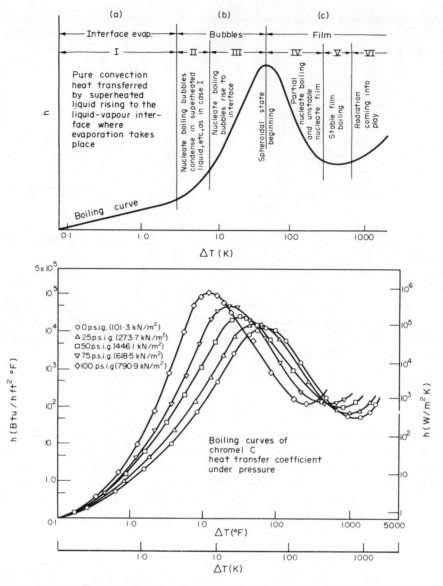

FIG. 7.36. Heat transfer results of Farber and Scorah[62].

Maximum Coefficient and Film Boiling

With a sufficiently high value of ΔT the bubbles are formed so rapidly that they cannot get away from the hot surface, and they therefore form a blanket over the surface. This means that the liquid is prevented from flowing on to the surface by the bubbles of vapour and the coefficient falls. The maximum coefficient occurs during nucleate boiling but in an unstable region for operation. In passing from the nucleate boiling region to the film boiling region, two critical changes occur in the process. The first manifests itself in a decrease in the heat flux, the second is the prelude to stable film boiling. The intermediate region is generally known as the transition region. It should be noted that the first change in the process is an important hydrodynamic phenomenon which is common to other two-phase systems, e.g. *flooding* in countercurrent gas–liquid or vapour–liquid systems.

With very high values of ΔT the heat transfer coefficient rises again because of heat transfer by radiation. These very high values are rarely achieved in practice and the aim will be to operate the plant at a temperature difference a little below the value giving the maximum heat transfer coefficient.

7.7.2. Heat Transfer Coefficients and Heat Flux

The values of the heat transfer coefficients for low values of temperature difference have been given by equation 7.115. Figure 7.37 shows the values of h and for q for water boiling on a submerged surface. Whilst the actual values vary somewhat between observers, they all give a maximum for a temperature difference of about 22 K. The maximum value of h is about 51 kW/m² K and the flux about 1100 kW/m².

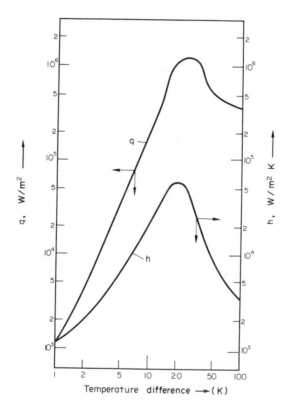

FIG. 7.37. Effect of temperature difference on heat flux and heat transfer coefficient to water boiling at 373 K on a submerged surface.

Similar results have been obtained by Bonilla and Perry,[63] Insinger and Bliss,[64] and others for a number of organic liquids such as benzene, alcohols, acetone, and carbon tetrachloride. The data in Table 7.8 for liquids boiling at atmospheric pressure show that the maximum heat flux is much smaller with organic liquids than with water and the temperature difference at this condition is rather higher. In practice the critical value of ΔT may be exceeded. Sauer et al.[66] found that the overall transfer coefficient U for boiling ethyl acetate with steam at 377 kN/m² was only 14 per cent of that when the steam pressure was reduced to 115 kN/m².

In considering the problem of nucleate boiling, the nature of the surface, the pressure, and the temperature difference must be taken into consideration as well as the actual physical properties of the liquid.

TABLE 7.8. *Maximum heat flux for various liquids boiling at atmospheric pressure*

Liquid	Surface	Critical ΔT (K)	Maximum flux (kW/m²)
Water	Chromium	25	910
50 mol% ethanol-water	Chromium	29	595
Ethanol	Chromium	33	455
n-Butanol	Chromium	44	455
iso-Butanol	Nickel	44	370
Acetone	Chromium	25	455
iso-Propanol	Chromium	33	340
Carbon tetrachloride	Copper	—	180
Benzene	Copper	—	170–230

Effect of Surface

Apart from the question of scale, the nature of the clean surface has a pronounced influence on the rate of boiling. Thus Bonilla and Perry boiled ethanol at atmospheric pressure and a temperature difference of 23 K, and found the heat flux at atmospheric pressure to be 850 kW/m² for polished copper, 450 for gold plate, and 370 for fresh chromium plate, and only 140 for old chromium plate. This wide fluctuation means that care must be taken in judging the anticipated heat flux, since the high values that may be obtained initially will not be obtained in practice because of tarnishing of the surface.

Effect of Temperature Difference

Cryder and Finalborgo,[65] boiled a number of liquids on a horizontal brass surface, both at atmospheric and at reduced pressure. Some of their results are shown in Fig. 7.38, where the coefficient for the boiling liquid h is plotted against the temperature difference between the

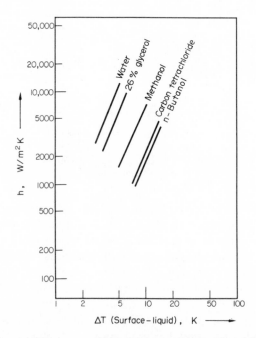

FIG. 7.38. Effect of temperature difference on heat transfer coefficient for boiling liquids (Cryder and Finalborgo[65]).

hot surface and the liquid. The points for the various liquids in Fig. 7.38 lie on nearly parallel straight lines, which can be represented by:

$$h = \text{constant} \times \Delta T^{2 \cdot 5} \qquad (7.116)$$

This value for the index of ΔT has been found by other workers, but Jakob and Linke[67] found values as high as 4 for some of their work. It is important to note that this value of $2 \cdot 5$ is true only for temperature differences up to 19 K.

In some ways it is more convenient to show the results in the form of heat flux versus temperature difference. This is shown in Fig. 7.39, where some results from a number of workers are given.

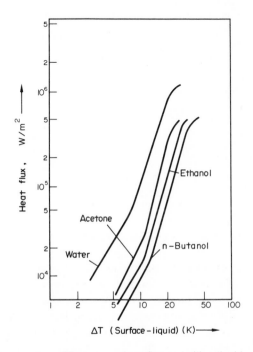

FIG. 7.39. Effect of temperature difference on heat flux to boiling liquids (Bonilla and Perry[63]).

Effect of Pressure

Cryder and Finalborgo[65] found that h fell off regularly as the pressure and hence the boiling point was reduced, according to the relation $h = \text{constant} \times B^{T''}$, where T'' is numerically equal to the temperature in K and B is a constant. Combining this with equation 7.116, they gave their results for h in the empirical form:

$$h = \text{constant} \times \Delta T^{2 \cdot 5} B^{T''}$$

or, using SI units: $\log (h/5 \cdot 67) = a' + 2 \cdot 5 \log \Delta T + b'(T'' - 273)$ $\qquad (7.117)$

If a' and b' are given the following values, h is expressed in W/m² K:

	a'	b'		a'	b'
Water	$-0 \cdot 96$	$0 \cdot 025$	Kerosene	$-4 \cdot 13$	$0 \cdot 022$
Methanol	$-1 \cdot 11$	$0 \cdot 027$	10% Na$_2$SO$_4$	$-1 \cdot 47$	$0 \cdot 029$
CCl$_4$	$-1 \cdot 55$	$0 \cdot 022$	24% NaCl	$-2 \cdot 43$	$0 \cdot 031$

The values of a' will apply only to a given apparatus but a value of b' of 0·025 is of more general application. If h_n is the coefficient at some standard boiling point T_n, and h at some other temperature T, equation 7.117 can be rearranged to give:

$$\log h/h_n = 0·025(T'' - T''_n) \tag{7.118}$$

for a given material and temperature difference.

As the pressure is raised above atmospheric pressure the film coefficient increases for a constant temperature difference. Cichelli and Bonilla[68] have examined this problem for pressures up to the critical value for the vapour, and have shown that ΔT for maximum rate of boiling decreases with the pressure. They obtained a single curve (Fig. 7.40) by plotting

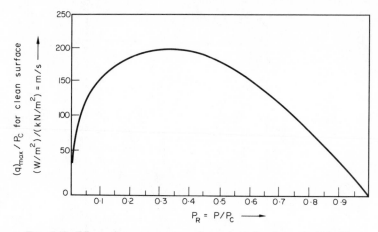

FIG. 7.40. Effect of pressure on maximum heat flux in nucleate boiling.

q_{max}/P_c against P_R, where P_c is the critical pressure and P_R the reduced pressure $= P/P_c$. This curve represented the data for water, ethanol, benzene, propane, n-pentane, n-heptane, and several mixtures with water. For water their results cover only a small range of P/P_c because of the high value of P_c. For the organic liquids investigated, they were able to show that the maximum value of heat flux q occurs at a pressure P of about one-third of the critical pressure P_c. The range of physical properties of the organic liquids is not wide (Table 7.9), and further data are required to substantiate the above relation.

TABLE 7.9. Heat transfer coefficients for boiling liquids

Liquid	Boiling point (K)	ΔT (K)	h (Btu/h ft² °F)	h (W/m² K)
Water	372	4·7	1614	9164
	372	2·9	480	2725
	326	8·8	830	4713
	326	6·1	235	1335
Methanol	337	8·9	845	4798
	337	5·6	260	1475
	306	14·4	515	2929
	306	9·3	158	900
Carbon tetrachloride	349	12·6	613	3480
	349	7·2	191	1085
	315	20·1	370	2100
	315	11·8	125	710

7.7.3. Analysis of the Heat Transfer Process Based on Bubble Characteristics

Why are such high values of heat flux obtained with the boiling process? It was once thought that the bubbles themselves were carriers of latent heat which was added to the liquid by their movement. It has now been shown, by determining the numbers of bubbles, that this mechanism would result in the transfer of only a moderate part of the heat that is actually transferred. The current views of the mechanism are that the high flux arises from the agitation produced by the bubbles, and two rather different explanations have been put forward. Rohsenow[56, 59] bases his argument on the condition of the bubble on leaving the hot surface. Thus by calculating the velocity and size of the bubble he is able to derive an expression for the heat transfer coefficient in the form of a Nusselt type equation, relating the Nusselt group to the Reynolds and Prandtl groups. Forster and Zuber[69, 70] however, argue that the important velocity is that of the growing bubble, and this is the term used to express the velocity. In either case the bubble movement is vital for obtaining a high flux. The liquid adjacent to the surface is agitated and exerts a mixing action by pushing hot liquid from the surface to the bulk of the stream.

Analysis of Heat Flux by Rohsenow [49, 50]

The size of a bubble at the instant of breakaway from the surface has been determined by Fritz[71] who showed that d_b was given by the expression:

$$d_b = C_1\phi\left(\frac{2\sigma}{g[\rho_l - \rho_v]}\right)^{1/2} \tag{7.119}$$

where σ is the surface tension, ρ_l and ρ_v the density of the liquid and vapour, ϕ is the contact angle, and C_1 is a constant depending on conditions.

The flowrate of vapour per unit area as bubbles u_b is given by:

$$u_b = fn\pi d_b{}^3/6 \tag{7.120}$$

where f is the frequency of bubble formation at each bubble site and n is the number of sites of nucleation per unit area.

The heat transferred by the bubbles q_b is to a good approximation given by:

$$q_b = \tfrac{1}{6}\pi d_b{}^3 fn\rho_v\lambda \tag{7.121}$$

where λ is the latent heat of vaporisation.

Jakob has shown that for heat flux rates up to $3\cdot2\,\text{kW/m}^2$ the product fd_b is constant and that the total heat flow per unit area q is proportional to n. From equation 7.121 it is seen that q_b is proportional to n at a given pressure, so that $q \propto q_b$.

Hence
$$q = C_2(\pi/6)d_b{}^3 fn\rho_v\lambda \tag{7.122}$$

Substituting from equations 7.120 and 7.122 the mass flow per unit area

$$\rho_v u_b = fn(\pi/6)d_b{}^3\rho_v = q/C_2\lambda$$

Then a Reynolds number for the bubble flow which represents the term for agitation may be defined by:

$$Re_b = d_b\rho_v u_b/\mu_l$$

$$= C_1\phi\left(\frac{2\sigma}{g(\rho_l - \rho_v)}\right)^{1/2}\left(\frac{q}{C_2\lambda}\right)\frac{1}{\mu_l}$$

$$= C_3\phi\frac{q}{\lambda\mu_l}\left(\frac{\sigma}{g(\rho_l - \rho_v)}\right)^{1/2} \tag{7.123}$$

The Nusselt group for bubble flow $Nu_b = h_b C_1 \dfrac{\phi}{k_l} \left(\dfrac{2\sigma}{g(\rho_l - \rho_v)} \right)^{1/2}$

$$= C_4 h_b \dfrac{\phi}{k_l} \left(\dfrac{\sigma}{g(\rho_l - \rho_v)} \right)^{1/2} \qquad (7.124)$$

Then a final correlation is obtained in the form:

$$Nu_b = \text{constant } Re_b{}^n Pr^m \qquad (7.125)$$

or $\qquad\qquad Nu_b = \text{constant} \left[\dfrac{C_3 \phi q}{\mu_l \lambda} \left(\dfrac{\sigma}{g(\rho_l - \rho_v)} \right)^{1/2} \right]^n \left(\dfrac{C_l \mu_l}{k_l} \right)^m \qquad (7.126)$

where n and m have been found from experimental data to be 2/3 and -0.7 respectively.

Analysis of Heat Flux by Forster and Zuber[69, 70]

These workers employ a similar basic approach but use the radial rate of growth dr/dt for the bubble velocity in the Reynolds group. They are able to show that:

$$\dfrac{dr}{dt} = \dfrac{\Delta T C_l \rho_l}{2\lambda \rho_v} \left(\dfrac{\pi D_{Hl}}{t} \right)^{1/2} \qquad (7.127)$$

where D_{Hl} is the thermal diffusivity $(k_l/C_l\rho_l)$ of the liquid. Using this method they present a final correlation in the form of equation 7.125.

Although these two forms of analysis give rise to somewhat similar expressions, the basic terms are evaluated in quite different ways and the final expressions show many differences. Some data (Addoms[72]) fit the Rohsenow equation reasonably well and other data fit Forster's equation.

These expressions all indicate the importance of the bubbles on the rate of transfer, but as yet they have not been used for design purposes. Insinger and Bliss[64] made the first approach by dimensional analysis and McNelly[73] has subsequently obtained a more satisfactory result. He assumed that the flux q could be expressed in the form:

$$h \propto q^a k^b C^c \lambda^d \mu^e \sigma^f P^g d^h \left(\dfrac{\rho_l}{\rho_v} - 1 \right)^i \qquad (7.128)$$

The influence of ΔT is taken into account by using the flux q, and the last term allows for the change in volume when the liquid vaporises. He was able to obtain the following expression, the numerical values of the indices being deduced from existing data:

$$\dfrac{hd}{k_l} = 0.225 \left(\dfrac{C_l \mu_l}{k_l} \right)^{0.69} \left(\dfrac{qd}{\lambda \mu} \right)^{0.69} \left(\dfrac{Pd}{\sigma} \right)^{0.31} \left(\dfrac{\rho_l}{\rho_v} - 1 \right)^{0.33} \qquad (7.129)$$

7.7.4. Sub-cooled Boiling

If bubbles are formed in a liquid which is much below its boiling point, then the bubbles will collapse in the bulk of the liquid. Thus if a liquid flows over a very hot surface then the bubbles formed are carried away from the surface by the liquid and sub-cooled boiling exists. Under these conditions a very large number of small bubbles are formed and a very high heat flux is obtained. Some results for these conditions are given in Fig. 7.41.

If a liquid flows through a tube heated on the outside then the heat flux q will increase with ΔT as shown in Fig. 7.41. Beyond a certain value of ΔT the increase in q is very rapid. If the velocity through the tube is increased, then a similar plot is obtained with a higher value of q at low values of ΔT and then the points follow the first line. Over the first section forced convection boiling exists where an increase in Reynolds number does not bring about a very great increase in q because the bubbles are themselves producing agitation in the boundary

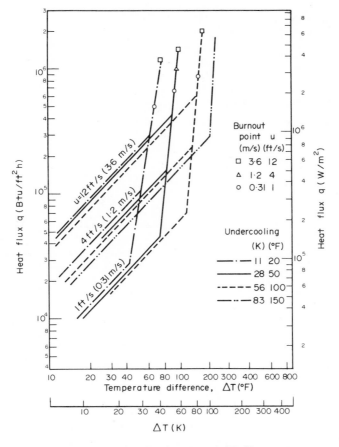

FIG. 7.41. Heat flux in sub-cooled boiling.

layer near the wall. Over the steep section sub-cooled boiling exists where the velocity is not important provided it is sufficient to remove the bubbles rapidly from the surface. In the same way mechanical agitation of a liquid boiling on a submerged surface will not markedly increase the heat flux.

7.7.5. Kettles and Reboilers

For vaporising the liquid at the bottom of a distillation column a reboiler is used, as shown in Fig. 7.42. The liquid from the still enters the boiler at the base, and, after flowing over the tubes, passes out over a weir. The vapour formed, together with entrained liquid, passes from the top of the unit to the column. The liquid flow may be either by gravity or by forced circulation. In such equipment, provision is made for expansion of the tubes either by having a floating head as shown, or by arranging the tubes in the form of a hairpin bend (Fig. 7.43). A vertical reboiler may also be used with steam condensing on the outside of the tube bundle. With all such installations the calculation of the pressure head available to provide flow with a gravity circulation is complex since a two phase medium exists in many of the pipes. Some suggestions are given in Kern[24] and in Chapter 4. With all systems it is undesirable to vaporise more than a small percentage of the feed since a good liquid flow over the tubes is necessary to avoid scale building up. A maximum safe flux for organic materials is $20 \cdot 5 \, \text{kW/m}^2$ with natural circulation and 35 with forced circulation.

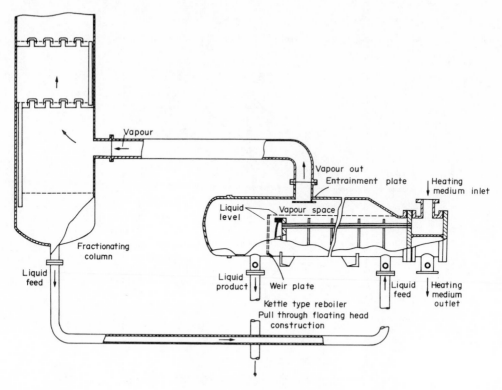

FIG. 7.42. Reboiler on distillation column.

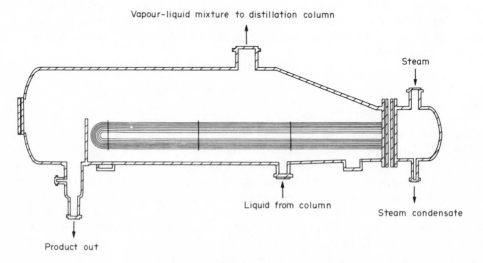

FIG. 7.43. Reboiler with hairpin tubes.

7.8. HEAT TRANSFER IN REACTION VESSELS

A simple jacketed pan or kettle is very commonly used in the chemical industries as a reaction vessel. In many cases, such as in nitration or sulphonation reactions, heat has to be removed or added to the mixture in order either to control the rate of reaction or to bring it to completion. The addition or removal of heat is conveniently arranged by passing steam or water through a jacket fitted to the outside of the vessel or through a helical coil fitted to the inside. In either case some form of agitator is used to obtain even distribution in the vessel. This may be of the anchor type for very thick mixes or a propeller or turbine if the contents are not too viscous.

7.8.1. Reaction Vessel with Helical Cooling Coil

In this case the thermal resistances to heat transfer arise from the water film on the inside of the coil, the wall of the tube, the film on the outside of the coil, and any scale that may be present on either surface. The overall transfer coefficient may be expressed by the relation:

$$\frac{1}{UA} = \frac{1}{h_i A_i} + \frac{x_w}{k_w A_w} + \frac{1}{h_o A_o} + \frac{R_o}{A_o} + \frac{R_i}{A_i} \tag{7.130}$$

where R_o and R_i are the scale resistances and the other terms have the usual meaning.

Inside Film Coefficient for a Coil

The value of h_i can be found from a form of equation 7.50:

$$\frac{h_i d}{k} = 0 \cdot 023 \left(\frac{du\rho}{\mu}\right)^{0 \cdot 8} \left(\frac{C_p \mu}{k}\right)^{0 \cdot 4} \tag{7.131}$$

if water is used in the coil, and the Sieder and Tate equation 7.56 if a viscous brine is used for cooling.

These equations have been obtained for straight tubes; with a coil somewhat greater transfer is obtained for the same physical conditions. Jeschke[74] cooled air in a 31 mm steel tube wound in the form of a helix and expressed his results in the form:

$$h_i(\text{coil}) = h_i(\text{straight pipe}) \left(1 + 3 \cdot 5 \frac{d}{d_c}\right) \tag{7.132}$$

where d is the inside diameter of the tube and d_c the diameter of the helix. Pratt[75] has examined this problem in greater detail for liquids and has given almost the same result. Combining equations 7.131 and 7.132 the inside film coefficient h_i for the coil may be calculated.

Outside Film Coefficient

The value of h_o is determined by the physical properties of the liquor and by the degree of agitation achieved. This latter quantity is difficult to express in a quantitative manner and the group $L^2 N\rho/\mu$ has been used both for this problem and for the allied one of power used in agitation. In this group L is the length of the paddle and N the revolutions per unit time. Chilton, Drew and Jebens,[76] working with a small tank only 0·3 m in diameter d_v, expressed their results in the form:

$$\frac{h_o d_v}{k} \left(\frac{\mu_s}{\mu}\right)^{0 \cdot 14} = 0 \cdot 87 \left(\frac{C_p \mu}{k}\right)^{1/3} \left(\frac{L^2 N\rho}{\mu}\right)^{0 \cdot 62} \tag{7.133}$$

where the factor $(\mu_s/\mu)^{0 \cdot 14}$ was introduced to cover the difference between the viscosity adjacent to the coil (μ_s) and that in the bulk of the liquor. They obtained a range in physical properties by using water, two oils, and glycerol.

Pratt[77] used both circular and square tanks up to 0·6 m across and a series of different arrangements of a simple paddle (Fig. 7.44). He also examined the influence of alteration in the arrangement of the coil and varied the tube diameter d_o, the gap between the turns d_g, the diameter of the helix d_c, the height of the coil d_p, and the width of the stirrer W. His final equations for cylindrical tanks were:

$$\frac{h_o d_v}{k} = 34 \left(\frac{L^2 N \rho}{\mu}\right)^{0\cdot5} \left(\frac{C_p \mu}{k}\right)^{0\cdot3} \left(\frac{d_g}{d_p}\right)^{0\cdot8} \left(\frac{W}{d_c}\right)^{0\cdot25} \left(\frac{L^2 d_v}{d_o^3}\right)^{0\cdot1} \tag{7.134}$$

and for square tanks:

$$\frac{h_o l_v}{k} = 39 \left(\frac{L^2 N \rho}{\mu}\right)^{0\cdot5} \left(\frac{C_p \mu}{k}\right)^{0\cdot3} \left(\frac{d_g}{d_p}\right)^{0\cdot8} \left(\frac{W}{d_c}\right)^{0\cdot25} \left(\frac{L^2 l_v}{d_o^3}\right)^{0\cdot1} \tag{7.135}$$

where l_v is the length of the side of the vessel.

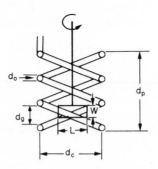

FIG. 7.44. Arrangement of coil in Pratt's work.

These give almost the same results as the earlier equations over a wide range in conditions. Cummings and West[78] have tested these results with a much larger tank of 0·45 m³ capacity and have given an expression similar to equation 7.133, but with a constant of 1·01 instead of 0·87. They used a retreating blade turbine impeller and in many cases mounted a second impeller above the first; this agitation was probably more intense than that given by the other workers. A constant of 0·9 seems a reasonable average from existing work.

Example 7.11. Toluene is continuously nitrated to mononitro-toluene in a cast-iron vessel, 1 m diameter, fitted with a propeller agitator 0·3 m diameter rotating at 2·5 Hz. The temperature is maintained at 310 K by circulating 0·5 kg/s cooling water through a stainless steel coil 25 mm o.d. and 22 mm i.d. wound in the form of a helix, 0·80 m in diameter. The conditions are such that the reacting material may be considered to have the same physical properties as 75 per cent sulphuric acid. If the mean water temperature is 290 K, what is the overall coefficient of heat transfer?

Solution. The overall coefficient U_o based on the outside area of the coil is obtained from equation 7.130:

$$\frac{1}{U_o} = \frac{1}{h_o} + \frac{x_w d_o}{k_w d_w} + \frac{d_o}{h_i d} + R_o + \frac{R_i d_o}{d}$$

where d_w is the mean diameter of the pipe.

From equations 7.131 and 7.132 the inside film coefficient for the water is given by:

$$h_i = (k/d)(1 + 3\cdot5 d/d_c)0\cdot023(du\rho/\mu)^{0\cdot8}(C_p \mu/k)^{0\cdot4}$$

In this equation:

$$u\rho = 0.5/[(\pi/4) \times 0.022^2] = 1315 \text{ kg/m}^2 \text{ s}$$
$$d = 0.022 \text{ m}$$
$$d_c = 0.80 \text{ m}$$
$$k = 0.59 \text{ W/m K}$$
$$\mu = 1.08 \text{ mN s/m}^2 \quad \text{or} \quad 1.08 \times 10^{-3} \text{ N s/m}^2$$
$$C_p = 4.18 \times 10^3 \text{ J/kg K}$$

$$\therefore \quad h_i = (0.59/0.022)(1 + 3.5 \times 0.022/0.80)$$
$$\times 0.023(0.022 \times 1315/1.08 \times 10^{-3})^{0.8}(4.18 \times 10^3 \times 1.08 \times 10^{-3}/0.59)^{0.4}$$
$$= 0.680(26{,}780)^{0.8}(7.65)^{0.4} = 5490 \text{ W/m}^2\text{K}$$

The external film coefficient is given by equation 7.133:

$$(h_o d_v/k)(\mu_s/\mu)^{0.14} = 0.87(C_p\mu/k)^{0.33}(L^2 N\rho/\mu)^{0.62}$$

For 75 per cent sulphuric acid:

$$k = 0.40 \text{ W/m K}$$
$$\mu_s = 8.6 \times 10^{-3} \text{ N s/m}^2 \text{ at } 300 \text{ K}$$
$$\mu = 6.5 \times 10^{-3} \text{ N s/m}^2 \text{ at } 310 \text{ K}$$
$$C_p = 1.88 \times 10^3 \text{ J/kg K}$$
$$\rho = 1666 \text{ kg/m}^3$$

$$\therefore \quad (h_o \times 1.0/0.40)(8.6/6.5)^{0.14} = 0.87(1.88 \times 10^3 \times 6.5 \times 10^{-3}/0.40)^{0.33}$$
$$\times (0.3^2 \times 2.5 \times 1666/6.5 \times 10^{-3})^{0.62}$$
$$2.5h_o \times 1.04 = 0.87 \times 2.29 \times 900$$
$$\therefore \quad h_o = 690 \text{ W/m}^2 \text{ K}$$

Taking $k_w = 15.9$ W/m K and R_o and R_i as 0.0004 and 0.0002 m² K/W:

$$\frac{1}{U_o} = \frac{1}{690} + \frac{0.0015 \times 0.025}{15.9 \times 0.0235} + \frac{0.025}{5490 \times 0.022} + 0.0004 + \frac{0.0002 \times 0.025}{0.022}$$
$$= 0.00145 + 0.00010 + 0.00021 + 0.00040 + 0.00023 = 0.00239$$
$$\therefore \quad \underline{\underline{U_o = 418 \text{ W/m}^2 \text{ K}}}$$

In this calculation a mean area of surface might have been used with sufficient accuracy. It is important to note the great importance of the scale terms which together form a major part of the thermal resistance.

7.8.2. Reaction Vessel with Jacket

For many purposes heating or cooling of a reaction mixture is most satisfactorily achieved by condensing steam in a jacket or passing water through it; this is commonly done in organic reactions where the mixture is too viscous for the use of coils and a high-speed agitator. Chilton et al.[76] and Cummings and West[78] have measured the transfer coefficients for this case by using an arrangement as shown in Fig. 7.45, where heat is applied to the jacket and simultaneously removed by passing water through the coil. Chilton measured the temperatures of the inside of the vessel wall, of the bulk liquid, and of the surface of the coil by means of thermocouples and thus obtained the film heat transfer coefficients directly.

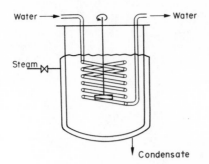

FIG. 7.45. Reaction vessel with jacket and coil.

Cummings and West used an indirect method to give the film coefficient from measurements of the overall coefficients.

Chilton expressed his results in the form:

$$\frac{h_b d_v}{k}\left(\frac{\mu_s}{\mu}\right)^{0.14} = 0.36\left(\frac{L^2 N\rho}{\mu}\right)^{2/3}\left(\frac{C_p\mu}{k}\right)^{1/3} \qquad (7.136)$$

where h_o is the film coefficient for the liquor adjacent to the wall of the vessel. Cummings and West used the same equation but gave the value of the constant as 0.40. Considering that Chilton's vessel was only 0.3 m in diameter and was fitted with a single paddle of 150 mm length, and that Cummings used a 0.45 m³ vessel with two turbine impellers, agreement between their results is remarkably good. The group $(\mu_s/\mu)^{0.14}$ is again used to cover the difference in the viscosities at the surface and in the bulk of the fluid.

Brown et al.[79] have given results of measurements on 1.5 m diameter sulphonators and nitrators of 3.4 m³ capacity as used in the dyestuffs industry. The sulphonators were of cast iron and had a wall thickness of 25.4 mm; the annular space in the jacket was also 25.4 mm. The agitator of the sulphonator was of the anchor type with a 127 mm clearance at the walls and was driven at 0.67 Hz. The nitrators were fitted with four-bladed propellers of 0.61 m diameter driven at 2 Hz. For cooling, they expressed the film coefficient h_b for the inside of the vessel by the relation:

$$\frac{h_b d_v}{k}\left(\frac{\mu_s}{\mu}\right)^{0.14} = 0.55\left(\frac{L^2 N\rho}{\mu}\right)^{2/3}\left(\frac{C_p\mu}{k}\right)^{1/4} \qquad (7.137)$$

which is very similar to that given by equation 7.136.

They also measured the film coefficient for the water jacket and gave values of 635–1170 W/m² K for water rates of 1440–9230 cm³/s, respectively, in the sulphonator. It should be noted that 7580 cm³/s corresponds to a vertical velocity of only 0.061 m/s and to a Reynolds number in the annulus of 5350. The thermal resistance of the wall of the pan was important, since with the sulphonator it accounted for 13 per cent of the total resistance at 323 K and 31 per cent at 403 K. The change in viscosity with temperature is important when considering these processes, since, for example, the viscosity of the sulphonation liquors ranged from 340 mN s/m² at 323 K to 22 mN s/m² at 403 K.

7.8.3. Heating Liquids in Tanks

It is frequently necessary to heat or cool the contents of a large batch reactor or storage tank. In this case the physical constants of the liquor may alter and the overall transfer coefficient change during the process. In practice, however, it is often possible to assume an average value of the transfer coefficient so as to simplify the calculation of the time of heating or cooling. The heating of the contents of a storage tank is commonly effected by condensing steam, either in a coil or in some form of hairpin tube heater.

For the case of a storage tank with liquor of mass m and specific heat C_p, heated by steam condensing in a helical coil, suppose that the overall transfer coefficient U is constant. Let T_s be the temperature of the condensing steam, T_1 and T_2 the initial and final temperatures of the liquor, and A the area of heat transfer surface. If T is the temperature of the liquor at any time t, the rate of transfer of heat is given by:

$$Q = mC_p \frac{dT}{dt} = UA(T_s - T)$$

$$\therefore \qquad \frac{dT}{dt} = \frac{UA}{mC_p}(T_s - T)$$

$$\therefore \qquad \int_{T_1}^{T_2} \frac{dT}{T_s - T} = \frac{UA}{mC_p} \int_0^t dt$$

$$\therefore \qquad \ln \frac{T_s - T_1}{T_s - T_2} = \frac{UA}{mC_p} t \qquad (7.138)$$

From this equation, the time t of heating from T_1 to T_2, can be found. The same analysis can be used if the steam condenses in a jacket of a reaction vessel.

7.9. SHELL AND TUBE EXCHANGERS

7.9.1. Description

These units form the most important class of heat transfer equipment because they can be constructed with very large surfaces in a relatively small volume, can be fabricated from alloy steels to resist corrosion, and can be used for heating and condensing all kinds of fluids. Figures 7.46–7.48 show various forms of construction of varying complexity, and Fig. 7.49 shows an outside view. The simplest type, shown in Fig. 7.46, has fixed tube plates at each end and the tubes are expanded into the plates. The tubes are so connected that the water or internal fluid makes several passes up and down the exchanger, thus enabling a high velocity of flow to be obtained for a given heat transfer area and given throughput of liquid. The fluid flowing in the shell is made to flow first in one sense and then in the opposite sense across the tube bundle by fitting a series of baffles along the length. These baffles are frequently of the segmental form, with about 25 per cent cut for the free space, and increase the velocity of flow across the tubes, thus giving better heat transfer. The difficulty with this type of construction is that the tube bundle cannot be removed for cleaning and no provision is made for

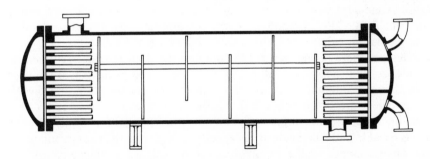

FIG. 7.46. Heat exchanger with fixed tube plates (four tube pass, one shell pass).

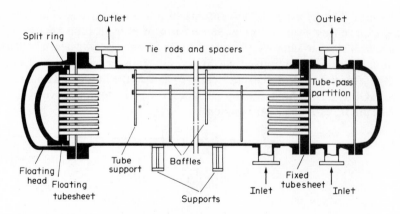

FIG. 7.47. Heat exchanger with floating head (two tube pass, one shell pass).

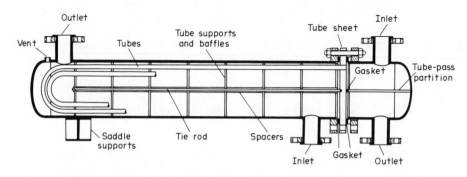

FIG. 7.48. Heat exchanger with hairpin tubes.

differential expansion between the tubes and the shell, although an expansion joint is often fitted in the shell.

In order to make removal of the tube bundle possible and to allow for considerable expansion of the tubes, a floating head exchanger is used, as shown in Fig. 7.47. In this arrangement one tube plate is fixed as before but the second is bolted to a floating head cover so that the tube bundle can move relative to the shell. This floating tube sheet is clamped between the floating head and a split backing flange in such a way that it is relatively easy to break the flanges at both ends and pull out the tube bundle. It will be seen that the shell cover at the floating head end is larger than that at the other end; this is to enable the tubes to be placed as near as possible to the edge of the fixed tube plate and leave very little unused space between the outer ring of tubes and the shell.

Another arrangement which provides for expansion involves the use of hairpin tubes, as shown in Fig. 7.48. This method is very commonly used for the reboilers on large fractionating columns where steam is condensed inside the tubes.

In these designs there is one pass for the fluid on the shell side and a number of passes on the tube side. As will be shown later it is often desirable to have two or more shell side passes, but they considerably increase the difficulty of construction and therefore several smaller exchangers are often connected together to obtain the same effect.

The essential requirements in the design of a heat exchanger are, firstly, the provision of a unit which is reliable and has the desired capacity, and, secondly, the economic consideration of making an exchanger at the minimum overall cost. This will generally involve using standard parts and fittings and making the design as simple as possible. In most cases it will

FIG. 7.49. Expanding the ends of the tubes into the tube plate of a heat exchanger bundle.

be necessary to balance the capital cost, as measured by depreciation, against the operating cost. Thus, for instance, in a condenser a high heat transfer coefficient is obtained and hence a smaller exchanger is required if a high water velocity is used through the tubes; on the other hand, the cost of pumping increases rapidly with increase in velocity and an economic balance must be struck. A typical graph showing the operating cost, depreciation, and total overall cost plotted against water velocity through the tubes for a condenser is given in Fig. 7.50.

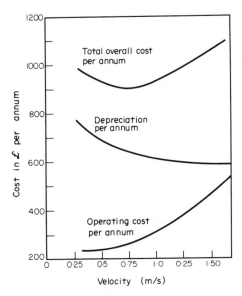

FIG. 7.50. Effect of water velocity on annual operating cost of condenser.

Tubes

These are usually 19 mm or 25·4 mm outside diameter d_o, and 14 or 16 BWG for admiralty metal or other alloy and 10 or 12 BWG for carbon steel. The pitch of the tubes is dictated largely by the necessity for cleaning the outside of the bundle if the fluid is likely to give rise to considerable scale. A square pitch is preferred as this permits cleaning round the tubes; the minimum spacing is $1·25d_o$ between centres. If space is at a premium, or if the fluids are very clean, a triangular pitch is used with a 6·35 mm space between the tubes, as this enables considerably more tubes to be put into a given shell.

Shells

These are commonly made of carbon steel, and standard pipes are used for the smaller sizes and rolled welded plate for larger sizes (say 0·4–1 m). The thickness of the shell can be calculated from the formula for thin-walled cylinders, but a minimum of 9·5 mm is used for shells over 0·33 m outside diameter and 11·1 mm for shells over 0·9 m outside diameter. Unless the exchanger works at very high pressure, the calculated thickness will usually be less than these figures, but a corrosion allowance of 3·2 mm is commonly added to all carbon steel parts; the thickness may be determined more by questions of rigidity than of simple internal pressure.

Baffles and Support Plates for Tubes

As mentioned earlier, these are fitted to increase the rate of flow over the tube bundle. The commonest type is the segmental baffle with about 25 per cent cut, as shown in Fig. 7.22. The diameter is governed by the fit of the bundle in the shell, but for good work the clearance between the baffle and the inside of the shell should be not more than 5–10 mm for vapours, and 2·5–5 mm for liquids. Baffles are not normally spaced closer than one-fifth of the shell diameter; very close spacing is avoided because the increase in heat transfer is then small compared with the increase in pressure drop. Support plates to hold the tubes in position are similar to the baffles, but the holes for the tubes are only 0·4 mm greater than the tube diameter, whereas for baffles the clearance may be twice this. These support plates are fitted at least every 1 m for 19 mm tubes and every 60 tube diameters for larger tubes.

Tie Rods

In order to keep the tube bundle straight, tie rods are fitted to the fixed tube sheet and to the baffle nearest the floating tube sheet. Usually between four and six rods of thickness 9·5–12·7 mm are necessary. These can be used to hold the baffles and support plates in position if sleeves are fitted over the rods between the baffles.

Tube Sheets

The thickness of the fixed tube sheet is frequently calculated from the relation

$$d_t = d_G \sqrt{(0 \cdot 25 P/f)} \tag{7.139}$$

where d_G is the diameter of the gasket on the tube sheet, P the design pressure, f the allowable working stress, and d_t the thickness of the sheet measured at the bottom of the partition baffle grooves. The floating tube sheet may be made $\sqrt{2}$ times as thick.

Flanges and Covers

Where dished ends are used for the cover plates, these are made to the designs supplied by the maker, but details are given in the *Code of the Tubular Exchangers Manufacturers' Association of America* (TEMA)[80] and in BS 1500.[81] Flanges are also designed from one of these codes. Extensive practical details are given in the TEMA Code which applies particularly to the petroleum industry.

7.9.2. Mean Temperature Difference in Multipass Exchangers

In an exchanger with one shell pass and several tube passes, the fluids in the tubes and shell will flow co-currently in some of the passes and countercurrently in the others. The mean temperature difference for countercurrent flow is greater than that for co-current or parallel flow, so that there is no easy way of finding the true temperature difference for the unit. The problem has been examined by Underwood[82] and by Bowman *et al.*[83] who have presented a graphical method for calculating the true mean temperature difference in terms of the nominal value of θ_m for countercurrent flow and a correction factor F. Providing the following conditions are maintained or assumed then F can be found from the curves presented in Figs. 7.51–7.54:

(a) The shell fluid temperature is uniform over the cross-section considered as a pass.
(b) There is equal heat transfer surface in each pass.
(c) U is constant over the length.
(d) The specific heats of the two fluids are constant over the temperature range considered.
(e) There is no change in phase of either fluid.
(f) Heat loss from the unit is negligible.

Then
$$Q = UAF\theta_m \tag{7.140}$$

$$Y = \frac{T_1 - T_2}{\theta_2 - \theta_1}$$

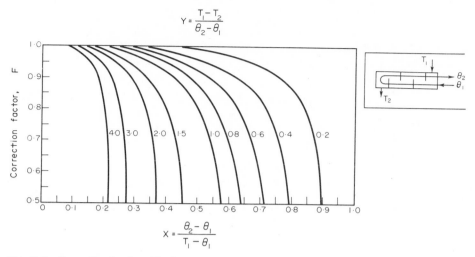

FIG. 7.51. Correction for logarithmic mean temperature difference for single shell pass exchanger.

$$Y = \frac{T_1 - T_2}{\theta_2 - \theta_1}$$

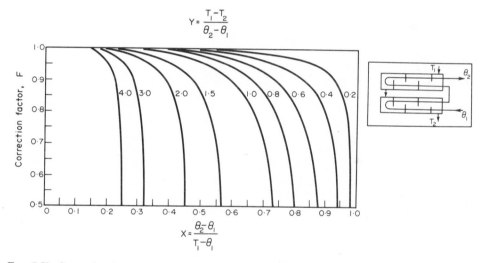

FIG. 7.52. Correction for logarithmic mean temperature difference for double shell pass exchanger.

F is expressed in relation to the two terms:

$$X = \frac{\theta_2 - \theta_1}{T_1 - \theta_1} \quad \text{and} \quad Y = \frac{T_1 - T_2}{\theta_2 - \theta_1}$$

Then if one shell side system is used Fig. 7.51 applies, for two shell side passes Fig. 7.52, for three shell side passes Fig. 7.53, and for four shell passes Fig. 7.54. For the case of a single shell side pass and two tube passes illustrated in Fig. 7.55a and b the temperature profile is as shown. Because one of the passes constitutes a parallel flow arrangement the exit temperature of the cold fluid θ_2 cannot be brought up very near to the hot fluid temperature T_1. Nor can the temperature θ_1 approach T_2 very closely except in a very large unit.

Suppose an exchanger is required to operate over the following temperature ranges:

$$T_1 = 455 \text{ K}, \quad T_2 = 372 \text{ K}$$
$$\theta_1 = 283 \text{ K}, \quad \theta_2 = 388 \text{ K}$$

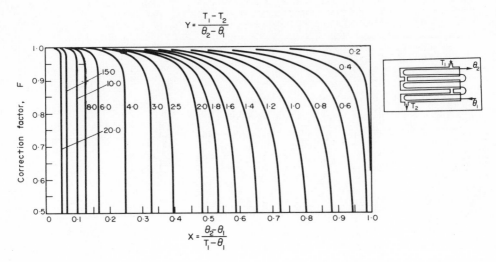

$$Y = \frac{T_1 - T_2}{\theta_2 - \theta_1}$$

FIG. 7.53. Correction for logarithmic mean temperature difference for three shell pass exchanger.

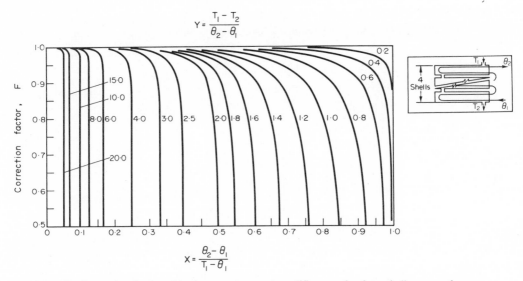

$$Y = \frac{T_1 - T_2}{\theta_2 - \theta_1}$$

FIG. 7.54. Correction for logarithmic mean temperature difference for four shell pass exchanger.

Then

$$X = \frac{\theta_2 - \theta_1}{T_1 - \theta_1} = \frac{388 - 283}{455 - 283} = 0.6$$

$$Y = \frac{T_1 - T_2}{\theta_2 - \theta_1} = \frac{455 - 372}{388 - 283} = 0.8$$

If a single shell side pass is used, then F is found to be 0·65 and with a two shell side unit F has a value of 0·95. For such a requirement the two shell side unit should be used.

To obtain maximum heat recovery from the hot fluid, θ_2 must be as high as possible. $T_2 - \theta_2$ is known as the *approach temperature*, and when $\theta_2 > T_2$ a temperature cross is said to occur and the value of F falls rapidly for a single pass on the shell side. This can be seen by taking as an illustration the values opposite where equal ranges of temperature are considered.

	Case 1				Case 2				Case 3		
	T_1 613		θ_2 463		T_1 573		θ_2 473		T_1 543 θ_2 463		
	T_2 573		θ_1 363		T_2 473		θ_1 373		T_2 443 θ_1 363		
Approach temp. $T_2 - \theta_2$	50				0				Cross of 20		
$X = \dfrac{\theta_2 - \theta_1}{T_1 - \theta_1}$	$\dfrac{100}{250} = 0\cdot4$				$\dfrac{100}{200} = 0\cdot5$				$\dfrac{100}{180} = 0\cdot55$		
$Y = \dfrac{T_1 - T_2}{\theta_2 - \theta_1}$	1				1				1		
F	0·92				0·80				0·66		

Thus as soon as a temperature cross occurs, a unit with two shell passes should be used. This can be seen because with a cross there will be some point where the temperature of the cold liquid will be greater than θ_2, so that beyond this point the stream will be cooled rather than heated. This is avoided by using the increased number of shell passes. The general form of the temperature profile for a two shell side unit is as shown in Fig. 7.55c. Longitudinal shell side baffles are rather difficult to fit and there is a serious chance of leakage. For this reason

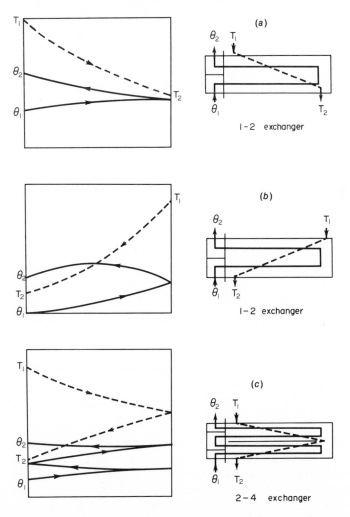

FIG. 7.55. Temperature profiles in single and double shell pass exchangers.

two exchangers arranged one below the other in series are to be preferred. The use of separate exchangers is even more desirable when a three shell side unit is required. On the very largest installations it may be necessary to link up in parallel a number of exchangers arranged as, say, sets of three in series, as indicated in Fig. 7.56. This type of arrangement is desirable for any very large unit which is likely to be unwieldy as a single system. When the total surface is much greater than 230 m², consideration should be given to using two smaller units even though the initial cost will be rather higher.

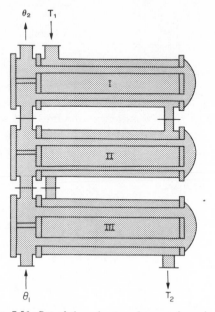

FIG. 7.56. Set of three heat exchangers in series.

Example 7.12. Using the data of Example 7.1, calculate the surface area required to effect the given duty using a multipass heat exchanger in which the cold water makes two passes through the tubes and the hot water makes a single pass outside the tubes.

Solution. As in Example 7.1,   the heat load $= 1672 \, \text{kW}$

With reference to Fig. 7.51: $T_1 = 360 \, K$, $T_2 = 340 \, K$

$$X = (\theta_2 - \theta_1)/(T_1 - \theta_1) = (316 - 300)/(360 - 300) = 0.267$$

$$Y = (T_1 - T_2)/(\theta_2 - \theta_1) = (360 - 340)/(316 - 300) = 1.25$$

$$\therefore \qquad F = 0.97$$

$$\therefore \qquad \Delta t_m = (41.9 \times 9.97) = 49.6 \, K$$

and the heat transfer area, $A = 1672/(2 \times 40.6) = \underline{20.59 \, \text{m}^2}$

7.10. DETERMINATION OF FILM COEFFICIENTS

If in any heat transfer apparatus the overall temperature difference, the area of surface and the heat transferred are measured directly, the overall transfer coefficient U is obtained from the simple relation, $Q = UA \, \Delta T$. The determination of the individual film coefficients has proved difficult even for the simplest cases, and it is quite common for equipment to be designed on the basis of practical values of U rather than from a series of film coefficients. For the important case of the transfer of heat from one fluid to another across a metal surface, two methods have been developed for measuring the film coefficients. The first requires a knowledge of the temperature difference across each film and therefore involves measuring the temperatures of both fluids and the surface of separation. With a concentric tube system it is very difficult to insert a thermocouple into the thin tube and to prevent the thermocouple wires from interfering with the flow of the fluid. Nevertheless, this method is commonly adopted, particularly when electric heating is used. It must be remembered that when the heat flux is very high, as with boiling liquids, there will be an appreciable temperature drop across the tube wall and the position of the thermocouple is then important. For this reason working with stainless steel, which has a relatively low value of thermal conductivity, is difficult.

The second method uses a technique put forward by Wilson.[84] Suppose steam to be condensing on the outside of a horizontal tube through which water is passed at various velocities. Then the overall and film transfer coefficients can be related by the equation:

$$\frac{1}{U} = \frac{1}{h_o} + \frac{x_w}{k_w} + R_i + \frac{1}{h_i} \quad \text{(from equation 7.130)}$$

if the transfer area on each side of the tube is approximately the same.

For conditions of turbulent flow the transfer coefficient for the water side $h_i = \varepsilon u^{0.8}$, R_i the scale resistance is constant, and h_o the coefficient for the condensate film is almost independent of the water velocity. Thus the above equation reduces to:

$$\frac{1}{U} = (\text{constant}) + 1/\varepsilon u^{0.8} \tag{7.141}$$

and if $1/U$ is plotted against $1/u^{0.8}$ a straight line is obtained with slope $1/\varepsilon$ and intercept equal to the value of the constant in equation 7.141. ε represents the value of the film coefficient h_i for unit water velocity. For a clean tube R_i should be nil, and hence h_o can be found from the value of the intercept.

This technique has been applied by Rhodes and Younger[85] to obtain the values of h_o for condensation of a number of organic vapours, by Pratt[77] to obtain the inside coefficient for a coiled tube, and by Coulson and Mehta[86] to obtain the coefficient for an annulus. If the results are repeated over a period of time, R_i can also be obtained by this method.

Typical values of thermal resistances and individual and overall heat transfer coefficients are given in Tables 7.10–7.13.

TABLE 7.10 *Thermal resistance of heat exchanger tubes*

Gauge (BWG)	Thickness (in.)	Values of x_w/k_w (ft² h °F/Btu)				
		Copper	Steel	Stainless Steel	Admiralty Metal	Aluminium
18	0·049	0·000018	0·00011	0·00047	0·000065	0·000031
16	0·065	0·000024	0·00014	0·00062	0·000086	0·000042
14	0·083	0·000031	0·00018	0·0008	0·00011	0·000053
12	0·109	0·000041	0·00024	0·001	0·00026	0·000070
	(mm)	Values of x_w/k_w (m² K/kW)				
18	1·24	0·0031	0·019	0·083	0·011	0·0054
16	1·65	0·0042	0·025	0·109	0·015	0·0074
14	2·10	0·0055	0·032	0·141	0·019	0·0093
12	2·77	0·0072	0·042	0·176	0·046	0·0123

TABLE 7.11 *Thermal resistances of scale deposits from water, etc.*

	(ft² h °F/Btu)	(m²K/kW)		(ft² h °F/Btu)	(m²K/kW)
Water (3 ft/s velocity, temperatures less than 50°C)*			*Steam*		
			Good quality—oil free	0·0003	0·052
			Poor quality—oil free	0·0005	0·09
Distilled	0·0005	0·09	Exhaust from reciprocating		
Sea	0·0005	0·09	engines	0·001	0·176
Clear river	0·0012	0·21			
Untreated cooling tower	0·0033	0·58	*Liquids*		
Treated cooling tower	0·0015	0·26	Treated brine	0·0015	0·264
Treated boiler feed	0·0015	0·26	Organics	0·001	0·176
Hard well	0·0033	0·58	Fuel oils	0·006	1·056
			Tars	0·01	1·76
			Gases		
(*1 m/s velocity, θ < 320 K)			Air	0·0015–0·003	0·26–0·53
			Solvent vapours	0·0008	0·14

TABLE 7.12 *Approximate overall heat transfer coefficients U for shell and tube equipment*

		Overall U	
Hot side	Cold side	Btu/h ft² °F	W/m²K
Condensers			
Steam (pressure)	Water	350–750	2000–4250
Steam (vacuum)	Water	300–600	1700–3400
Saturated organic solvents (atmospheric)	Water	100–200	570–1140
Saturated organic solvents (vacuum some non-condensable)	Water-brine	50–120	300–680
Organic solvents (atmospheric and high non-condensable)	Water-brine	20–80	110–455
Organic solvents (vacuum and high non-condensable)	Water-brine	10–50	60–300
Low boiling hydrocarbons (atmospheric)	Water	80–200	455–1140
High boiling hydrocarbons (vacuum)	Water	10–30	60–170
Heaters			
Steam	Water	250–750	1420–4250
Steam	Light oils	50–150	300–850
Steam	Heavy oils	10–80	60–455
Steam	Organic solvents	100–200	570–1140
Steam	Gases	5–50	30–300
Dowtherm	Gases	4–40	20–200
Dowtherm	Heavy oils	8–60	45–340
Evaporators			
Steam	Water	350–750	2000–4250
Steam	Organic solvents	100–200	570–1140
Steam	Light oils	80–180	455–1020
Steam	Heavy oils (vacuum)	25–75	140–425
Water	Refrigerants	75–150	425–850
Organic solvents	Refrigerants	30–100	170–570
Heat exchangers (no change of state)			
Water	Water	150–300	850–1700
Organic solvents	Water	50–150	280–850
Gases	Water	3–50	17–280
Light oils	Water	60–160	340–910
Heavy oils	Water	10–50	60–280
Organic solvents	Light oil	20–70	115–400
Water	Brine	100–200	570–1140
Organic solvents	Brine	30–90	170–510
Gases	Brine	3–50	20–280
Organic solvents	Organic solvents	20–60	115–340
Heavy oils	Heavy oils	8–50	45–280

TABLE 7.13 *Approximate film coefficients* $(h_i \text{ or } h_o)$

	Approximate film coefficients	
	Btu/h ft² °F	W/m²K
No change of state		
Water	300–2000	1700–11350
Gases	3–50	17–280
Organic solvents	60–500	340–2840
Oils	10–120	60–680
Condensing		
Steam	1000–3000	5680–17030
Organic solvents	150–500	850–2840
Light oils	200–400	1140–2270
Heavy oils (vacuum)	20–50	115–285
Ammonia	500–1000	2840–5680
Evaporation		
Water	800–200	4540–11350
Organic solvents	100–300	570–1700
Ammonia	200–400	1140–2270
Light oils	150–300	850–1700
Heavy oils	10–50	60–285

Example 7.13. 7·5 kg/s of pure isobutane is to be condensed at 332 K in a horizontal tubular exchanger using water entering at 300 K (Fig. 7.57). It is proposed to use 19·0 mm o.d. tubes of wall thickness 1·65 mm arranged on a 25 mm triangular pitch. The resistance due to scale may be taken as 0·0005 m² K/W. What is the number of tubes and the arrangement of passes to be specified?

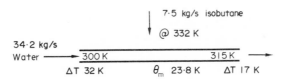

FIG. 7.57. Data for Example 7.13.

Solution. The latent heat of vaporisation of isobutane is 286 kJ/kg and hence the heat load Q is:

$$Q = 7 \cdot 5 \times 286 = 2145 \text{ kW}$$

The cooling water outlet temperature should not exceed 325 K and a value of 315 K will be used. The water requirement is then $2145/[4 \cdot 18(315 - 300)] = 34 \cdot 2 \text{ kg/s}$.

An approximate size of the unit is now obtained, assuming an overall coefficient, based on the outside area of the tubes, of 510 W/m² K.

The various temperatures are:

isobutane (shell):	in 332 K	out 332 K
water (tube side):	in 300 K	out 315 K (mean 307·5 K)
ΔT:	32 K	17 K

$$\therefore \qquad \theta_m = 32 - 17/\ln(32/17) = 23 \cdot 8 \text{ K}$$

The approximate heat transfer area is therefore:

$$A = 2145 \times 10^3/(510 \times 23 \cdot 8) = 176 \cdot 7 \text{ m}^2$$

19·0 mm tubes have an outside area of 0·059 m^2/m and hence the total length of tube is 176·7/0·059 = 2995 m.

Using the standard length of 4·88 m, number of tubes = 613. With this large quantity of water it will be possible to use a four-pass exchanger and still have a reasonable water velocity. In a shell of 0·78 m i.d., 678 tubes can be arranged in four passes using 25 mm triangular pitch and this size will be selected. This result will now be checked by calculating the individual film coefficients and hence the assumed value of the overall film coefficient.

Calculation of inside and outside film coefficients

(a) Inside coefficient, h_i

Water flow through each tube = 34·2/(678/4) = 0·202 kg/s

For a tube i.d. of (19·0 − 2 × 1·67) = 15·7 mm, cross-sectional area is 193·6 mm^2 or 0·000194 m^2

The water velocity is therefore 0·202/(1000 × 0·000194) = 1·04 m/s

From equation 7.59 therefore:

$$h_i = 1063(1 + 0·00293 × 307·5)1·04^{0·8}/0·0157^{0·2}$$
$$= 1063 × 1·89 × 1·032/0·435 = 4766 \text{ W/m}^2 \text{ K}$$

or, based on the outside diameter;

$$h_{io} = 4766 × 15·7/19·0 = 3939 \text{ W/m}^2 \text{ K}$$

or
$$3·94 \text{ kW/m}^2 \text{ K}$$

(b) Outside coefficient, h_o

The temperature drop across the condensate film ΔT_f is given by:

$$\frac{\text{Thermal resistance, water film} + \text{scale}}{\text{Total thermal resistance}} = \frac{(1/3·94) + 0·5}{(1/0·510)} = \frac{\theta_m - \Delta T_f}{\theta_m}$$

∴
$$\Delta T_f = 23·8 − 23·8(1/3·94 + 1/2·0)/(1/0·510)$$
$$= 23·8 − 23·8(5·94 × 0·510)/(3·94 × 2·0)$$
$$= 14·7 \text{ K}$$

The condensate film is thus at 332 − 14·7 = 317·3 K

From equation 7.108:

$$h_o = 0·72[(k^3\rho^2 g\lambda)/(jd_o\mu \ \Delta T_f)]^{0·25}$$

Taking $k = 0·13$ W/m K, $\rho = 508$ kg/m^3, $j = \sqrt{678} = 26$, and $\mu = 0·136 × 10^{-3}$ N s/m^2,

$$h_o = 0·72(0·13^3 × 508^2 × 9·81 × 286 × 10^3)/(26 × 19·0 × 10^{-3} × 0·136 × 10^{-3} × 14·7)^{0·25}$$
$$= 802 \text{ W/m}^2 \text{ K} \quad \text{or} \quad 0·802 \text{ kW/m}^2 \text{ K}$$

Hence the overall coefficient, based on the outside area, is:

$$(1/U) = (1/3·94) + (1/0·802) + 0·50 = 2·001$$
∴
$$U = 0·50 \text{ kW/m}^2 \text{ K} \quad \text{or} \quad 500 \text{ W/m}^2 \text{ K}$$

This is sufficiently near the assumed value. The total heat load of the condenser under these conditions will be:

$$(UA\theta_m) = 0·50 × 4·88 × 678 × 0·059 × 23·8 = 2323 \text{ kW}$$

It should be noted that this is not a complete solution in that the exit water temperature and the optimum velocity should be determined on economic grounds, although it serves as an illustration of the general method of approach.

Example 7.14. 37·5 kg/s of a crude oil is to be heated from 295 to 330 K by heat exchange with the bottom product from a distillation unit, which is to be cooled from 420 to 380 K, flowing at 29·6 kg/s (Fig. 7.58). A tubular heat exchanger is available with a 0·60 m i.d. shell and one shell side pass with two passes on the tube side. The tube bundle consists of 324 19·0 mm o.d. tubes, wall thickness 2·1 mm, each 3·65 m long arranged on 25 mm square pitch and supported by baffles with a 25 per cent cut, spaced at 0·23 m intervals.
Is this exchanger suitable?

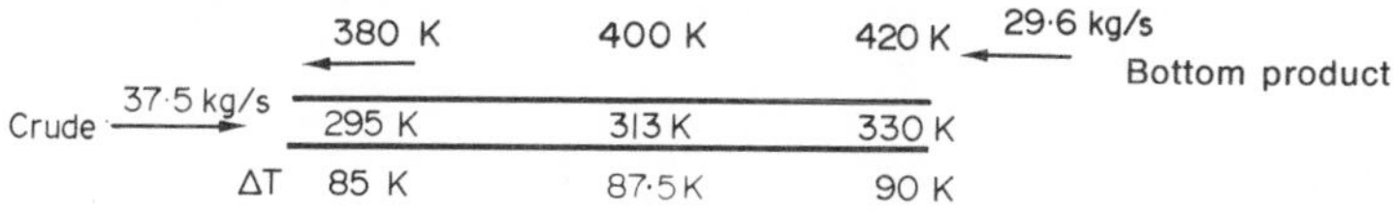

```
                        380 K          400 K        420 K  29·6 kg/s
                      ◄────                        ◄────
                                                          Bottom product
              37·5 kg/s
Crude ───────►          295 K          313 K        330 K
              ΔT        85 K           87·5K         90 K
```

FIG. 7.58. Data for Example 7.14.

Solution.

Temperatures: bottom product, in 420 K out 380 K mean 400 K

crude oil, out 330 K in 295 K mean 313 K

ΔT 90 K 85 K mean 87·5 K

For the crude oil at 313 K: $C_p = 1\cdot986$ kJ/kg K or 1986 J/kg K

$\mu = 2\cdot9$ mN s/m^2 or $2\cdot9 \times 10^{-3}$ N s/m^2

$k = 0\cdot136$ W/m K

$\rho = 824$ kg/m^3

For the product at 400 K: $C_p = 2\cdot20$ kJ/kg K or 2200 J/kg K

Heat load on the tube side: $Q = 37\cdot5 \times 1\cdot986(330 - 295) = 2607$ kW

Heat load on the shell side: $Q = 29\cdot6 \times 2\cdot20(420 - 380) = 2605$ kW

Shell side coefficient, h_o
Temperature of pipe wall $\approx (400 + 313)/2 = 256\cdot5$ K
Film temperature: $T_f = (400 + 356\cdot5)/2 = 378$ K
At T_f: $\rho = 867$ kg/m^3

$\mu = 5\cdot2$ mN s/m^2 or $5\cdot2 \times 10^{-3}$ N s/m^2

$k = 0\cdot119$ W/m K

Cross area for flow = (shell i.d. × clearance × baffle spacing)/pitch

$= 0\cdot60 \times 6\cdot4 \times 10^{-3} \times 0\cdot23/0\cdot025 = 0\cdot0353$ m^2

Hence $G'_{max} = 29\cdot6/0\cdot0353 = 838\cdot5$ kg/m^2 s

and $Re_{max} = 0\cdot019 \times 838\cdot5/(5\cdot2 \times 10^{-3}) = 3064$

∴ $(Re_{max})^{0\cdot6} = 125$

$Pr^{0\cdot3} = (2200 \times 5\cdot2 \times 10^{-3}/0\cdot119)^{0\cdot3} = (96\cdot1)^{0\cdot3} = 3\cdot94$

Taking $C_h = 1$, in equation 7.68:

$h_o \times 0\cdot019/0\cdot119 = 0\cdot33 \times 125 \times 3\cdot94$

∴ $h_o = 1018$ W/m^2 K or $1\cdot02$ kW/m^2 K

Tube side coefficient, h_{io}

Number of tubes per pass $= 162$

Cross-sectional area of tube $= (\pi/4)(19\cdot0 - 2 \times 2\cdot1)^2 = 172 \text{ mm}^2$ or $1\cdot72 \times 10^{-4} \text{ m}^2$

Mass flow: $G' = 37\cdot5/(1\cdot72 \times 10^{-4} \times 162) = 1346 \text{ kg/m}^2 \text{ s}$

$Re = 0\cdot0148 \times 1346/2\cdot9 \times 10^{-3} = 6869$ (taking $d = 14\cdot8$ mm)

$\therefore$ $Re^{0\cdot8} = 1175$

$Pr^{0\cdot4} = (1986 \times 2\cdot9 \times 10^{-3}/0\cdot136)^{0\cdot4} = 4\cdot47$

From equation 7.50:

$$h_i \times 0\cdot0148/0\cdot136 = 0\cdot023 \times 1175 \times 4\cdot47$$

$\therefore$ $h_i = 1110 \text{ W/m}^2 \text{ K}$

Based on outside area: $h_{io} = 1110 \times 14\cdot8/19\cdot0 = 865 \text{ W/m}^2 \text{ K}$ or $0\cdot865 \text{ kW/m}^2 \text{ K}$

Neglecting the thermal resistances of the wall and the scale, the clean overall coefficient based on the outside area is:

$$(1/U) = (1/1\cdot02) + (1/0\cdot865)$$
$$= 2\cdot136 \text{ K m}^2/\text{kW}$$

Outside area available: $A = 324 \times 3\cdot65 \times 0\cdot0598 = 70\cdot7 \text{ m}^2$

(where the outside area $= 0\cdot0598 \text{ m}^2/\text{m}$).

During operation, scale would form and the actual overall coefficient would not have to fall below U_d, given by:

$$1/U_d = A \, \Delta T/Q = 70\cdot7 \times 87\cdot5/2607 = 2\cdot37 \text{ K m}^2/\text{kW}$$

The maximum allowable scale resistance is then:

$$R = (1/U_d) - (1/U) = 2\cdot37 - 2\cdot136 = 0\cdot234 \text{ K m}^2/\text{kW}$$

This value is very low, as seen from Table 7.11, and, if installed, the exchanger would not give the required temperatures without frequent cleaning. It should be noted that heat losses have been neglected in making this calculation.

7.11. SPECIAL FORMS OF EQUIPMENT

A number of special types of equipment have been developed to give improved heat transfer under conditions where the standard forms of tubular units are unsatisfactory. Of these the finned tube units and the plate type exchangers will be taken as illustrations.

7.11.1. Finned Tube Units

When viscous liquids are heated in a concentric tube or standard tubular exchanger by condensing steam or hot liquid of low viscosity, the film coefficient for the viscous liquid will be much smaller than that on the hot side and it will therefore control the rate of heat transfer. This condition also arises with air or gas heaters where the coefficient on the gas side will be very low compared with that for the liquid or condensing vapour on the other side. It is often possible to obtain a much better performance by increasing the area of surface on the side with the limiting coefficient. This may be done conveniently by using a

FIG. 7.59. Tube with radial fins.

finned tube as in Fig. 7.59 which shows one typical form of such equipment which may have either longitudinal or transverse fins.

The calculation of the film coefficients on the fin side is complex because each unit of surface on the fin is less effective than a unit of surface on the tube wall. This arises because there will be a temperature gradient along the fin so that the temperature difference between the fin surface and the fluid will vary along the fin. To calculate the film coefficients it is convenient in the first place to consider the extended surface as shown in Fig. 7.60. Here a cylindrical rod of length L and cross-sectional area A and perimeter b is heated at one end by a surface at temperature T_1 and cooled throughout its length by a medium at temperature T_G so that the cold end is at a temperature T_2.

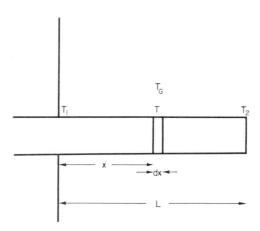

FIG. 7.60. Heat flow in rod with heat loss to surroundings.

A heat balance over a length dx at distance x from the hot end gives:

heat in = heat out along rod plus heat lost to surroundings

$$-kA\frac{dT}{dx} = \left[-kA\frac{dT}{dx} + \frac{d}{dx}\left(-kA\frac{dT}{dx}\right)dx\right] + hb\ dx(T - T_G)$$

where h is the film coefficient from fin to surroundings.

Writing the temperature difference $T - T_G$ equal to θ:

$$kA\frac{d^2T}{dx^2}dx = hb\ dx\ \theta$$

Since T_G is constant, $d^2T/dx^2 = d^2\theta/dx^2$.

$$\therefore \qquad \frac{d^2\theta}{dx^2} = \frac{hb}{kA}\theta = m^2\theta\left(\text{where } m^2 = \frac{hb}{kA}\right) \tag{7.142}$$

$$\therefore \qquad \theta = C_1e^{mx} + C_2e^{-mx} \tag{7.143}$$

In solving this equation there are three important cases:

(a) Long rod with temperature falling to that of surroundings, i.e. $\theta = 0$ when $x = \infty$.

Then
$$\theta = \theta_1 e^{-mx} \tag{7.144}$$

(b) Short rod from which heat loss from end is neglected.

$$\text{Hot end:} \quad x = 0, \quad \theta = \theta_1 = C_1 + C_2$$

$$\text{Cold end:} \quad x = L, \quad d\theta/dx = 0$$

Then
$$0 = C_1 m e^{mL} - C_2 m e^{-mL}$$

and
$$\theta_1 = C_1 + C_1 e^{2mL}$$

∴
$$C_1 = \frac{\theta_1}{1 + e^{2mL}}, \quad C_2 = \frac{\theta_1}{1 + e^{-2mL}}$$

Hence
$$\theta = \frac{\theta_1 e^{mx}}{1 + e^{2mL}} + \frac{\theta_1 e^{2mL} e^{-mx}}{1 + e^{2mL}}$$

$$= \frac{\theta_1}{1 + e^{2mL}} [e^{mx} + e^{2mL} e^{-mx}] \tag{7.145}$$

i.e.
$$\frac{\theta}{\theta_1} = \frac{e^{-mL} e^{mx} + e^{mL} e^{-mx}}{e^{-mL} + e^{mL}}$$

This may be written
$$\frac{\theta}{\theta_1} = \frac{\cosh m(L - x)}{\cosh mL} \tag{7.146}$$

(c) More accurately, allowing for heat loss from the end.

Hot end: $x = 0, \quad \theta = \theta_1 = C_1 + C_2$

Cold end: $x = L, \quad Q = hA\theta_{x=L} = -kA \left(\frac{d\theta}{dx}\right)_{x=L}$

Determination of C_1 and C_2 in equation 7.143 then gives:

$$\theta = \frac{\theta_1}{1 + J e^{-2mL}} (J e^{-2mL} e^{mx} + e^{-mx}) \tag{7.147}$$

where
$$J = \frac{km - h}{km + h}$$

or, again, noting that $\cosh x = \frac{1}{2}(e^x + e^{-x})$ and $\sinh x = \frac{1}{2}(e^x - e^{-x})$:

$$\frac{\theta}{\theta_1} = \frac{\cosh m(L - x) + (h/mk) \sinh m(L - x)}{\cosh mL + (h/mk) \sinh mL} \tag{7.148}$$

Heat Loss from Finned Tube

This is obtained in the first place by determining the heat flow into the base of the fin from the tube surface. Thus the heat flow to the root of the fin is:

$$Q_f = -kA \left(\frac{dT}{dx}\right)_{x=0} = -kA \left(\frac{d\theta}{dx}\right)_{x=0}$$

For case (a):
$$Q_f = -kA(-m\theta_1) = kA \sqrt{\frac{hb}{kA}} \theta_1 = \sqrt{hbkA} \theta_1 \tag{7.149}$$

For case (b):
$$Q_f = -kAm\theta_1 \frac{1 - e^{2mL}}{1 + e^{2mL}} = \sqrt{hbkA} \theta_1 \tanh mL \tag{7.150}$$

For case (c):
$$Q_f = \sqrt{hbkA} \theta_1 \left(\frac{1 - J e^{-2mL}}{1 + J e^{-2mL}}\right) \tag{7.151}$$

The above expressions are valid provided that the cross-section for heat flow remains constant. When it is not constant, e.g. for a radial or tapered fin, the temperature distribution is in the form of a Bessel function.

Fin Effectiveness

If the fin were such that there was no drop in temperature along its length, then the maximum rate of heat loss from the fin would be:

$$Q_{f\,\text{max}} = bLh\theta_1$$

The fin effectiveness is then given by $Q_f/Q_{f\,\text{max}}$ and for case (b) above this becomes:

$$\frac{kAm\theta_1 \tanh mL}{bLh\theta_1} = \frac{\tanh ml}{mL} \tag{7.152}$$

Example 7.15. In order to measure the temperature of a gas flowing through a copper pipe, a thermometer pocket is fitted perpendicularly through the pipe wall, the open end making very good contact with the pipe wall. The pocket is made of copper tube, 10 mm o.d. and 0·9 mm wall, and it projects 75 mm into the pipe. A thermocouple is welded to the bottom of the tube and this gives a reading of 475 K when the wall temperature is at 365 K. If the coefficient of heat transfer between the gas and the copper tube is 140 W/m² K, calculate the gas temperature. The thermal conductivity of copper may be taken as 350 W/m K (Fig. 7.61).

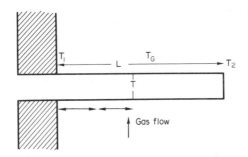

FIG. 7.61. Heat transfer to thermometer pocket.

Solution. If θ is the temperature difference $(T - T_G)$:

$$\theta = \theta_1 \cosh m(L - x)/\cosh mL$$

At $x = L$:
$$\theta = \theta_1/\cosh mL$$
$$m^2 = hb/kA$$

where perimeter: $\quad b = \pi 0\cdot010$ m, $\quad$ tube i.d. $= 8\cdot2$ mm $\quad$ or $\quad 0\cdot0082$ m

cross-area of metal: $\quad A = (\pi/4)(10\cdot0^2 - 8\cdot2^2) = 8\cdot19\pi$ mm² $\quad$ or $\quad 8\cdot19\pi \times 10^{-6}$ m²

$$\therefore \qquad m^2 = 140 \times 0\cdot010\pi/(350 \times 8\cdot19\pi \times 10^{-6})$$
$$= 488 \text{ m}^{-2}$$

$$\therefore \qquad m = 22\cdot1 \text{ m}^{-1}$$

$$\theta_1 = T_G - 365, \quad \theta_2 = T_G - 475$$

$$\theta_1/\theta_2 = \cosh mL$$

$$\therefore \qquad (T_G - 365)/(T_G - 475) = \cosh (22\cdot1 \times 0\cdot075) = 2\cdot83$$

$$\underline{\underline{T_G = 535 \text{ K}}}$$

Example 7.16. A steel tube fitted with transverse circular steel fins of constant cross-section has the following specification:

$$\text{tube o.d.:} \quad d_2 = 54 \cdot 0 \text{ mm} \quad \text{fin diameter } d_1 = 70 \cdot 0 \text{ mm}$$

$$\text{fin thickness:} \quad w = 2 \cdot 0 \text{ mm} \quad \text{number of fins/metre run} = 230$$

Determine the heat loss per metre run of the tube when the surface temperature is 370 K and the temperature of the surroundings 280 K. The heat transfer coefficient between gas and fin is 30 W/m² K and the thermal conductivity of steel is 43 W/m K.

Solution. Assume that the height of the fin is small compared with its circumference and that it may be treated as a straight fin of length $(\pi/2)(d_1 + d_2)$.

Perimeter: $b = 2\pi(d_1 + d_2)/2 = \pi(d_1 + d_2)$

Area: $A = \pi(d_1 + d_2)w/2$,   i.e. the average area at right
                                                            angles to the heat flow

Then $m = (hb/kA)^{0 \cdot 5} = \{[h\pi(d_1 + d_2)]/[k\pi(d_1 + d_2)w/2]\}^{0 \cdot 5}$
                                $= (2h/kw)^{0 \cdot 5}$
                        $= (2 \times 30/43 \times 0 \cdot 002)^{0 \cdot 5}$
                        $= 26 \cdot 42 \text{ m}^{-1}$

From equation 7.150 the heat flow is given for case (b) as:

$$Q_f = mkA\theta_1(e^{2mL} - 1)/(1 + e^{2mL})$$

In this equation: $A = \pi(70 \cdot 0 + 54 \cdot 0) \times 2 \cdot 0/2 = 390 \text{ mm}^2$ or $0 \cdot 00039 \text{ m}^2$

$$L = (d_1 - d_2)/2 = 8 \cdot 0 \text{ mm} \quad \text{or} \quad 0 \cdot 008 \text{ m}$$

$$mL = 26 \cdot 42 \times 0 \cdot 008 = 0 \cdot 211$$

$$\theta_1 = 370 - 280 = 90 \text{ K}$$

∴ $Q_f = 26 \cdot 42 \times 43 \times 3 \cdot 9 \times 10^{-4} \times 90(e^{0 \cdot 422} - 1)/(1 + e^{0 \cdot 422})$
                        $= 39 \cdot 9 \times 0 \cdot 525/2 \cdot 525 = 8 \cdot 29 \text{ W per fin}$

∴   Heat loss per metre run of tube $= 8 \cdot 29 \times 230 = 1907 \text{ W/m}$

or $\underline{\underline{1 \cdot 91 \text{ kW/m}}}$

In this case the low value of mL indicates a fin efficiency of almost $1 \cdot 0$, though where mL tends to $1 \cdot 0$ the efficiency falls to about $0 \cdot 8$.

Practical Data on Finned Tubes

A neat form of construction has been designed by the Brown Fintube Company of America. On both prongs of a hairpin tube are fitted horizontal fins which fit inside concentric tubes, joined at the base of the hairpin as shown. Units of this form can conveniently be arranged in banks to give large heat transfer surfaces. It is usual for the extended surface to be at least five times greater than the inside surface, so that the low coefficient on the fin side is balanced by the increase in surface. An indication of the surface obtained is given in Table 7.14.

A typical hairpin unit with an effective surface on the fin side of $9 \cdot 4 \text{ m}^2$ has an overall length of $6 \cdot 6 \text{ m}$, height of $0 \cdot 34 \text{ m}$, and width of $0 \cdot 2 \text{ m}$. The free area for flow on the fin side is $2 \cdot 645 \text{ mm}^2$ against $1 \cdot 320 \text{ mm}^2$ on the inside; the ratio of the transfer surface on the fin side to that inside the tubes is $5 \cdot 93 : 1$.

Fig. 7.62. APV Paraflow plate heat exchanger.

TABLE 7.14. *Data on surface of finned tube units* [a]

Pipe size outside diameter (in)	Outside surface of pipe (ft²/ft length)	Number of fins	Surface of finned pipe (ft²/ft)		(m²/m)	
			Height of fin		Height of fin	
			0·5 in	1 in	12·7 mm	25·4 mm
1	0·262	12	1·262	2·262	0·385	0·689
(25·4 mm)	(0·08 m²/m)	16	1·595	2·929	0·486	0·893
		20	1·927	3·595	0·587	1·096
1·9	0·497	20	2·164	3·830	0·660	1·167
(48·3 mm)	(0·15 m²/m)	24	2·497	4·497	0·761	1·371
		28	2·830	5·164	0·863	1·574
		36	3·498	6·497	1·066	1·980

[a] Brown Fintube Company.

The fin side film coefficient h_f has been expressed by plotting

$$\frac{h_f}{C_p G'} \left(\frac{C_p \mu}{k}\right)^{2/3} \left(\frac{\mu}{\mu_s}\right)^{-0·14} \quad \text{against} \quad \frac{d_e G'}{\mu}$$

where h_f is based on the total finside surface area (fin and tube), G' is the mass rate of flow per unit area, and d_e is the equivalent diameter, i.e.:

$$d_e = \frac{4 \times \text{cross-sectional area for flow on fin side}}{\text{total wetted perimeter for flow (fin + outside of tube + inner surface of shell tube)}}$$

Experimental work has been carried out with exchangers in which the inside tube was 48·3 mm outside diameter and was fitted with 24, 28, or 36 fins (12·5 mm by 0·89 mm) in a 6·1 m length; the finned tubes were inserted inside tubes 89 mm outside diameter. With steam on the tube side, and lube oils and kerosene on the fin side, the experimental data were well correlated by plotting:

$$\frac{h_f}{C_p G'} \left(\frac{C_p \mu}{k}\right)^{2/3} \left(\frac{\mu}{\mu_s}\right)^{-0·14} \quad \text{against} \quad \frac{d_e G'}{\mu}$$

typical values were:

$$\frac{h_f}{C_p G'} \left(\frac{C_p \mu}{k}\right)^{2/3} \left(\frac{\mu}{\mu_s}\right)^{-0·14} = 0·25 \quad 0·055 \quad 0·012 \quad 0·004$$

$$\frac{d_e G'}{\mu} = 1 \quad 10 \quad 100 \quad 1000$$

Transverse Fins

Some indication of the performance obtained with transverse finned tubes is given in Table 7.15. The figures show the heat transferred per unit length of pipe when heating air on the fin side with steam or hot water on the tube side, using a temperature difference of 100 K. The results are given for three different spacings of the fins.

7.11.2. Plate Type Exchangers

A series of plate type heat exchangers which present some special features has been developed by the APV Company of London. The general construction is shown in Fig. 7.62 from which it is seen that the exchangers consist of a series of parallel plates held firmly together between substantial head frames. The plates are one-piece pressings, frequently of stainless steel, and are spaced by rubber sealing gaskets cemented into a channel around the edge of each plate. Each plate has a number of troughs pressed out at right angles to the

TABLE 7.15. *Data on finned tubes*

Inside diam. of tube	$\frac{3}{4}$ in	1 in	$1\frac{1}{2}$ in	2 in	3 in
Outside diam. of fin	$2\frac{1}{2}$ in	$2\frac{3}{4}$ in	$3\frac{7}{8}$ in	$4\frac{5}{16}$ in	$5\frac{1}{2}$ in
No. of fins/ft run	Heat transferred (Btu/h ft)				
20	485	650	1010	1115	1440
24	505	665	1060	1170	1495
30	565	720	1190	1295	1655
Inside diam. of tube	19·1 mm	25·4 mm	38·1 mm	50·8 mm	76·2 mm
Outside diam. of fin	63·5 mm	69·9 mm	98·4 mm	109·5 mm	139·7 mm
No. of fins/m run	Heat transferred (kW/m)				
65	0·47	0·63	0·37	1·07	1·38
79	0·49	0·64	1·02	1·12	1·44
98	0·54	0·69	1·14	1·24	1·59

[a]Data taken from catalogue of G. A. Harvey and Co. Ltd. of London.

direction of flow and arranged so that they interlink with each other to form a channel of constantly changing direction and section. With normal construction the gap between the plates is 1·3–1·5 mm. Each liquid flows in alternate spaces and a large surface can be obtained in a small volume.

Because of the shape of the plates, the developed area of surface is appreciably greater than the projected area; this is shown below for the four common sizes of plate.

TABLE 7.16. *Plate areas*

Plate type	Projected area		Developed area	
	ft²	m²	ft²	m²
HT	1·00	0·09	1·35	0·13
HX	1·45	0·13	1·81	0·17
HM	2·88	0·27	3·73	0·35
HF	3·85	0·36	4·60	0·43

A high degree of turbulence is obtained even at low flowrates and the high heat transfer coefficients obtained are illustrated by the data in Table 7.17: these refer to the heating of cold water by the equal flow of hot water in an HF type exchanger (aluminium or copper), at an average temperature of 310 K.

TABLE 7.17. *Performance of plate type exchanger Type HF*[a]

Heat transferred per plate		Water flow		U based on developed area	
Btu/h °F	W/K	gal/h	cm³/s	Btu/h ft² °F	kW/m² K
3000	1580	550	700	650	3·70
4000	2110	850	1075	870	4·94
5000	2640	1250	1580	1080	6·13

[a]Courtesy of Mr. Goodman of the APV Company.

Using a stainless steel plate with a flow of 1140 cm³/s, the heat transferred is 1760 kJ/K for each plate.

The high transfer coefficient enables these exchangers to be operated with very small temperature difference, so that a high heat recovery is obtained. These units have been particularly successful in the dairying and brewing industries, where the low liquid capacity and the close control of temperature have been valuable features. A further advantage is that they are easily dismantled for inspection of the whole plate. The necessity for the long gasket is an inherent weakness, but the exchangers have been worked successfully up to 423 K and at pressures of 930 kN/m². They are now being used in the chemical and gas industries with solvents, sugar, acetic acid, ammoniacal liquor, etc.

A spiral plate exchanger is illustrated in Fig. 7.63. Here two fluids flow countercurrently through the channels formed between the spiral plates. With this form of construction the velocity may be as high as 2·1 m/s and overall transfer coefficients of 2·8 kW/m² K are frequently obtained. The size can therefore be kept relatively small and the cost becomes comparable or even less than that of tubular units, particularly when they are fabricated from alloy steels.

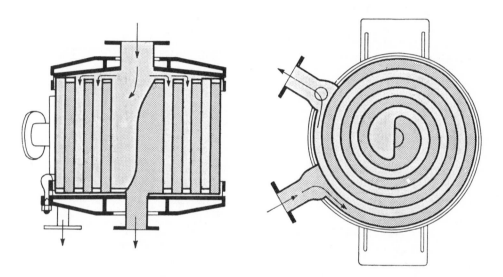

FIG. 7.63. Spiral plate heat exchanger.

7.12. LIQUID METALS AS HEAT TRANSFER MEDIA

In developing nuclear reactors there has been a strong desire to reduce the volume of the fluid in the heat transfer circuits and thus fluids with a high heat capacity per unit volume have been studied. Liquid metals such as sodium fulfil these requirements with a high value of $C_p\rho$ of about 1275 J/m³ K. Although water has a much greater value it is unsuitable because of its high vapour pressure at the desired temperatures and the corresponding need to use high-pressure piping. These liquid metals have particularly low values of the Prandtl number (of about 0·01) and they behave rather differently from the normal fluids under conditions of forced convection. Some values for typical liquid metals are given in Table 7.18.

The results of work on sodium, lithium, and mercury for forced convection in a pipe have been correlated by the expression:

$$Nu = 0·625 \, (Re \, Pr)^{0·4} \tag{7.153}$$

but the accuracy of the correlation is not very good. With values of Reynolds number of about 18,000 it is quite possible to obtain a value of h of about 11·4 kW/m² K for flow in a pipe system.

TABLE 7.18. *Properties of liquid metals*

Material	Temp (K)	Pr
Potassium	975	0·003
Sodium	975	0·004
Na/K alloy (56:44)	975	0·06
Mercury	575	0·008
Lithium	475	0·065

7.13. THERMAL INSULATION

A hot reaction or storage vessel or a steam pipe will lose heat to the atmosphere by radiation, conduction, and convection. The loss by radiation is a function of the fourth power of the absolute temperatures of the body and surroundings, and will be small for low-temperature differences but will increase rapidly as the temperature difference increases. Air is a very poor conductor, and heat loss by conduction will therefore be small except possibly through the supporting structure. On the other hand, since convection currents form very easily, the heat loss from an unlagged surface is considerable. The conservation of heat, and hence usually of steam and coal, is an economic necessity, and some form of lagging should normally be applied to hot surfaces. In furnaces, as has already been seen, the surface temperature is reduced substantially by using a series of insulating bricks which are poor conductors.

The two main requirements of a good lagging material are that it should have a low thermal conductivity and that it should suppress convection currents. The materials that are frequently used are cork, asbestos, 85 per cent magnesia, glass wool, and vermiculite. Cork is a very good insulator but it becomes charred at moderate temperatures; it is mainly used in refrigerating plants. Eighty-five per cent magnesia with asbestos, and asbestos itself, are very widely used for lagging steam pipes. Eighty-five per cent magnesia is probably the most widely used and may be applied either as a hot plastic material or in preformed sections. The preformed sections are quickly fitted and can frequently be dismantled and re-used whereas the plastic material must be applied to a hot surface and cannot be re-used. Thin sheeting is often used to protect the lagging.

The rate of heat loss per unit area is given by:

$$\frac{\text{total temperature difference}}{\text{total thermal resistance}}$$

For the case of heat loss to the atmosphere from a lagged steam pipe, the thermal resistance is due to that of the condensate film and dirt on the inside of the pipe, that of the pipe wall, that of the lagging, and that of the air film outside the lagging. Thus for unit length of a lagged pipe:

$$\frac{Q}{l} = \Sigma \, \Delta T \bigg/ \left[\frac{1}{h_i \pi d} + \frac{x_w}{k_w \pi d_w} + \frac{x_l}{k_r \pi d_m} + \frac{1}{(h_r + h_c)\pi d_s} \right] \tag{7.154}$$

where d is the inside diameter of pipe, d_w the mean diameter of pipe wall, d_m the logarithmic mean diameter of lagging, d_s the outside diameter of lagging, x_w, x_l are the pipe wall and lagging thickness respectively, k_w, k_r the thermal conductivity of the pipe wall and lagging, and h_i, h_r, h_c the inside film, radiation, and convection coefficients.

Example 7.17. A 150 mm i.d. steam pipe is carrying steam at 444 K and is lagged with 50 mm of 85 per cent magnesia. What is the heat loss to air at 294 K?

Solution. In this case:

$$d = 150 \text{ mm} \quad \text{or} \quad 0 \cdot 150 \text{ m}$$
$$d_o = 168 \text{ mm} \quad \text{or} \quad 0 \cdot 168 \text{ m}$$
$$d_w = 159 \text{ mm} \quad \text{or} \quad 0 \cdot 159 \text{ m}$$
$$d_s = 269 \text{ mm} \quad \text{or} \quad 0 \cdot 268 \text{ m}$$
$$d_m, \text{ the log mean of } d_o \text{ and } d_s = 215 \text{ mm} \quad \text{or} \quad 0 \cdot 215 \text{ m}.$$

The coefficient for condensing steam together with that for any scale will be taken as 8500 W/m² K, k_w as 45 W/m K, and k_l as 0·073 W/m K.

The temperature on the outside of the lagging is estimated at 314 K and $(h_r + h_c)$ will be taken as 10 W/m² K.

The thermal resistance are therefore:

$$1/h_i\pi d = 1/(8500 \times \pi \times 0 \cdot 150) = 0 \cdot 00025$$
$$x_w/k_w\pi d_w = 0 \cdot 009/(45 \times \pi \times 0 \cdot 159) = 0 \cdot 00040$$
$$x_l/k_l\pi d_m = 0 \cdot 050/(0 \cdot 073 \times \pi \times 0 \cdot 215) = 1 \cdot 013$$
$$1/(h_r + h_c)\pi d_s = 1/(10 \times \pi \times 0 \cdot 268) = 0 \cdot 119$$

The first two terms may be neglected and hence the total thermal resistance is 1·132 mK/W.

The heat loss per metre length = $(444 - 294)/1 \cdot 132 = 132 \cdot 5$ W/m (from equation 7.154)

The temperature on the outside of the lagging can now be checked by

$$\Delta T(\text{lagging})/\Sigma\Delta T = 1 \cdot 013/1 \cdot 132 = 0 \cdot 895$$
$$\therefore \qquad \Delta T(\text{lagging}) = 0 \cdot 895(444 - 294) = 134 \text{ K}$$

Thus the temperature on the outside of the lagging is $444 - 134 = 310$ K, which approximates to the assumed value.

Taking an emissivity of 0·9 and assuming streamline natural convection:

$$h_r = 5 \cdot 67 \times 10^{-8} \times 0 \cdot 9(310^4 - 294^4)/(310 - 294) = 7 \cdot 40 \text{ W/m}^2 \text{ K}$$

and $\qquad h_c = 1 \cdot 37(\Delta T/d_s)^{0 \cdot 25} = 1 \cdot 37[(310 - 294)/0 \cdot 268]^{0 \cdot 25} = 3 \cdot 81 \text{ W/m}^2 \text{ K}$

Thus $(h_r + h_c) = 11 \cdot 2$ W/m² K, which agrees with the assumed value. In practice it is rare for forced convection currents to be absent, and the heat loss is probably higher than this value.

If the pipe was unlagged, $(h_r + h_c)$ for $\Delta T = 150$ K would be about 20 W/m² K and the heat loss would then be:

$$Q/l = (h_r + h_c)\pi d_o \Delta T$$
$$= 20 \times \pi \times 0 \cdot 168 \times 150 = 1584 \text{ W/m}$$

or $\qquad \underline{1 \cdot 58 \text{ kW/m}}$

Under these conditions it is seen that the heat loss has been reduced to about 10 per cent of the unlagged by the addition of 50 mm of lagging.

7.13.1. Economic Thickness of Lagging

Increasing the thickness of the lagging will reduce the loss of heat and thus give a saving in the operating costs. The cost of the lagging will increase with thickness and there will be an optimum thickness when further increase does not save sufficient heat to justify the cost. In general the smaller the pipe the smaller the thickness used, but it cannot be too strongly stressed that some lagging everywhere is better than excellent lagging in some places and none in others. For temperatures of 373–423 K, and for pipes up to 150 mm diameter, Lyle[87] recommends a 25 mm thickness of 85 per cent magnesia lagging and 50 mm for pipes over 230 mm diameter. With temperatures of 470–520 K he suggests 38 mm for pipes less than 75 mm diameter and 50 mm for pipes up to 230 mm diameter.

Example 7.18. A steam main, 100 mm o.d., at 420 K is to be insulated with lagging at a cost of £10 per m³ with a thermal conductivity of 0·1 W/m K. Assuming the capital cost is depreciated over five years and interest at 10 per cent, what is the economic thickness of lagging? The value of heat may be taken as 5×10^{-4} £/MJ and ambient temperature as 285 K.

Solution. In the general conduction equation;

$$q = kA \, \Delta T / x$$

where
$$k = 0 \cdot 1 \text{ W/m K}$$
$$A = \pi D_0 L = \pi \times 0 \cdot 1 \times 1 = 0 \cdot 314 \text{ m}^2/\text{m}$$

and
$$\Delta T = (420 - 285) = 135 \text{ K}$$

If D is the outer diameter of lagging, the thickness is $0 \cdot 5(D - 0 \cdot 1)$

and
$$q = 0 \cdot 1 \times 0 \cdot 314 \times 135 / 0 \cdot 5(D - 0 \cdot 1)$$
$$= 8 \cdot 48 / (D - 0 \cdot 1) \text{ W/m}$$

Value of heat lost
$$= 8 \cdot 48 / (D - 0 \cdot 1) \times 31 \cdot 5 \times 10^6 \times 5 \times 10^{-4} / 10^6$$
$$= 0 \cdot 134 / (D - 0 \cdot 1) \text{ £/m year}$$

Volume of lagging
$$= (\pi / 4)(D^2 - 0 \cdot 1^2) = 0 \cdot 785 D^2 - 0 \cdot 0079 \text{ m}^3/\text{m}$$

∴   cost
$$= 7 \cdot 84 D^2 - 0 \cdot 079 \text{ £/m}$$

Depreciation
$$= (7 \cdot 85 D^2 - 0 \cdot 079)/5 = 1 \cdot 57 D^2 - 0 \cdot 0156 \text{ £/m year}$$

Interest
$$= 0 \cdot 785 D^2 - 0 \cdot 0079 \text{ £/m year}$$

Neglecting maintenance, annual cost $= 1 \cdot 57 D^2 - 0 \cdot 0156 + 0 \cdot 785 D^2 - 0 \cdot 0079$
$$= 2 \cdot 355 D^2 - 0 \cdot 0235 \text{ £/m}$$

Total cost
$$= \text{value of heat lost} + \text{annual cost}$$

or
$$c = 0 \cdot 134 / (D - 0 \cdot 1) + 2 \cdot 355 D^2 - 0 \cdot 0235$$
$$dc/dD = 0 \cdot 134 / (D - 0 \cdot 1)^2 + 4 \cdot 71 D$$

The cost is a minimum when $dc/dD = 0$

i.e. when
$$4 \cdot 71 D = 0 \cdot 134 / (D - 0 \cdot 1)^2$$

hence
$$D = 0 \cdot 372 \text{ m}$$

Thickness of lagging
$$= 0 \cdot 5(D - 0 \cdot 1) = \underline{\underline{0 \cdot 136 \text{ m}}}$$

7.14. FURTHER READING

BACKHURST, J. R. and HARKER, J. H.: *Process Plant Design* (Heinemann, London, 1973).
BIRD, R. B., STEWART, W. E., and LIGHTFOOT, E. N.: *Transport Phenomena* (Wiley, New York, 1960).
CHAPMAN, A. J.: *Heat Transfer*, 2nd edn. (Macmillan, New York, 1967).
COLLIER, J. G.: *Convective Boiling and Condensation* (McGraw-Hill, New York, 1972).
ECKERT, E. R. G. and DRAKE, R. M., Jr.: *Analysis of Heat and Mass Transfer* (McGraw-Hill, New York, 1972).
FRAAS, H. P. and OZISIK, M. N.: *Heat Exchanger Design* (Wiley, New York, 1965).
GEBHART, B.: *Heat Transfer*, 2nd edn. (McGraw-Hill, New York, 1971).
GROBER, H., ERK, E., and GRIGULL, U.: *Fundamentals of Heat Transfer* (McGraw-Hill, New York, 1961).
JAKOB, M.: *Heat Transfer*, Vol. 1 (Wiley, New York, 1949).
JAKOB, M.: *Heat Transfer*, Vol. II (Wiley, New York, 1957).
KAYS, W. M.: *Convective Heat and Mass Transfer* (McGraw-Hill, New York, 1966).
KAYS, W. M. and LONDON, A. L.: *Compact Heat Exchangers* (McGraw-Hill, New York, 1958).
KERN, D. Q.: *Process Heat Transfer* (McGraw-Hill, New York, 1950).
KREITH, F.: *Principles of Heat Transfer* (Feffer & Simons, Scranton, Pennsylvania, 1965).
MCADAMS, W. H.: *Heat Transmission*, 3rd edn. (McGraw-Hill, New York, 1954).
OZISIK, M. N.: *Boundary Value Problems of Heat Conduction* (Int. Textbook Co., Stanton, Penn. 1965).
SCHACK, A.: *Industrial Heat Transfer* (Chapman & Hall, London, 1965).
SCHNEIDER, P. J.: *Conduction Heat Transfer* (Addison-Wesley, Cambridge, Mass., 1955).

7.15. REFERENCES

1. COULSON, J. M. and RICHARDSON, J. F.: *Chemical Engineering*, volume 2 (Pergamon Press, Oxford, 1976).
2. LONG, V. D.: *J. Inst. Fuel.* **37** (1964) 522. A simple model of droplet combustion.
3. MONAGHAN, M. T., SIDDAL, R. G., and THRING, M. W.: *Comb and Flame* **17** (1968) 45. The influence of initial diameter on the combustion of single drops of liquid fuel.
4. WALDIE, B.: *The Chem. Engr.* No. 261, (1972) 188. Review of recent work on the processing of powders in high temperature plasmas: Pt. II, Particle dynamics, heat transfer and mass transfer.
5. TAYLOR, T. D.: *Physics of Fluids* **6** (1963) 987. Heat transfer from single spheres in a low Reynolds number slip flow.
6. FIELD, M. A., GILL, D. W., MORGAN, B. B., and HAWKSLEY, P. G. W.: Combustion of Pulverised Coal (BCURA, Leatherhead, 1967).
7. COULSON, J. M. and RICHARDSON, J. F.: *Chemical Engineering*, volume 3 (Pergamon Press, Oxford, 1975).
8. CARSLAW, H. S. and JAGER, J. C.: *Conduction of Heat in Solids* (Oxford, 1947).
9. RICHARDSON, J. F.: *Fuel* **28** (1949) 265. Spread of fire by thermal conduction.
10. SCHMIDT, E.: *Föppl Festschrift* (Springer, 1924), and *Z. ges. Kälte-Ind.* **35** (1928) 213. Versuche über den Wärmeübergang in ruhender Luft.
11. SOUTHWELL, R.: *Relaxation Methods in Theoretical Physics* (Oxford, 1946).
12. SCHNEIDER, P. J. *Temperature Response Charts* (J. Wiley, 1963).
13. COLBURN, A. P.: *Trans. Am. Inst. Chem. Eng.* **29** (1933) 174. A method of correlating forced convection heat transfer data and a comparison with fluid friction.
14. SIEDER, E. N. and TATE, G. E.: *Ind. Eng. Chem.* **28** (1936) 1429. Heat transfer and pressure drop of liquids in tubes.
15. FISHENDEN, M. and SAUNDERS, O. A.: *An Introduction to Heat Transfer* (Oxford, 1950).
16. REIHER, M.: *Mitt. Forsch.* **269** (1925) 1. Wärmeübergang von strömender Luft an Rohre und Rohrenbündel im Kreuzstrom.
17. HILPERT, R.: *Forsch. Gebiete Ingenieurw* **4** (1933) 215. Wärmeabgabe von geheizten Drahten und Rohren.
18. GRIFFITHS, E. and AWBERY, J. H.: *Proc. Inst. Mech. Eng.* **125** (1933) 319. Heat transfer between metal pipes and a stream of air.
19. DAVIS, A. H.: *Phil. Mag.* **47** (1924) 1057. Convective cooling of wires in streams of viscous liquids.
20. GRIMISON, E. D.: *Trans. Am. Soc. Mech. Eng.* **59** (1937) 583; and *ibid.* **60** (1938) 381. Correlation and utilization of new data on flow resistance and heat transfer for cross flow of gases over tube banks.
21. HUGE, E. C.: *Trans. Am. Soc. Mech. Eng.* **59** (1937) 573. Experimental investigation of effects of equipment size on convection heat transfer and flow resistance in cross flow of gases over tube banks.
22. PIERSON, O. L.: *Trans. Am. Soc. Mech. Eng.* **59** (1937) 563. Experimental investigation of influence of tube arrangement on convection heat transfer and flow resistance in cross flow of gases over tube banks.
23. MCADAMS, W. H.: *Heat Transmission* (McGraw-Hill, 1954).
24. KERN, D. Q.: *Process Heat Transfer* (McGraw-Hill, 1950).
25. SHORT, B. E.: *Univ. of Texas Pub.* No. 4324 (1943). Heat transfer and pressure drop in heat exchangers.
26. DONOHUE, D. A.: *Ind. Eng. Chem.* **41** (1949) 2499. Heat transfer and pressure drop in heat exchangers.
27. TINKER, T.: *Proceedings of the General Discussion on Heat Transfer, September, 1951*, p. 89. Analysis of the fluid flow pattern in shell and tube exchangers and the effect of flow distribution on the heat exchanger performance, (Inst. of Mech. Eng. and Am. Soc. Mech. Eng.).
28. COPE, W. F. and BAILEY, A.: *Aeronautical Research Comm. (Gt. Britain). Tech. Rept.* **43** (1933) 199. Heat transmission through circular, square, and rectangular tubes.

29. WASHINGTON, L. and MARKS, W. M.: *Ind. Eng. Chem.* **29** (1937) 337. Heat transfer and pressure drop in rectangular air passages.
30. DAVIS, E. S.: *Trans. Am. Soc. Mech. Eng.* **65** (1943) 755. Heat transfer and pressure drop in annuli.
31. CARPENTER, F. G., COLBURN, A. P., and SCHOENBORN, E. M.: *Trans. Am. Inst. Chem. Eng.* **42** (1946) 165. Heat transfer and friction of water in an annular space.
32. ROWE, P. N., CLAXTON, K. T., and LEWIS, J. B.: *Trans. Inst. Chem. Eng.* **41** (1965) T14. Heat and mass transfer from a single sphere in an extensive flowing fluid.
33. RAY, B. B.: *Proc. Indian Assoc. Cultivation of Science* **6** (1920) 95. Convection from heated cylinders in air.
34. DAVIS, A.H.: *Phil. Mag.* **44** (1922) 920. Natural convective cooling in fluids.
35. ACKERMAN, G.: *Forsch. Gebiete Ingenieurw.* **3** (1932) 42. Die Wärmeabgabe eines horizontalen geheizten Rohres an kaltes Wasser bei natürlicher Konvektion.
36. SAUNDERS, O. A.: *Proc. Roy. Soc.* **A, 157** (1936) 278. Effect of pressure on natural convection to air.
37. SAUNDERS, O. A.: *Proc. Roy. Soc.* **A, 172** (1939) 55. Natural convection in liquids.
38. KRAUSSOLD, H.: *Forsch. Gebiete Ingenieurw.* **5** (1934) 186. Wärmeabgabe von cylindrischen Flüssigkeiten bei natürlicher Konvektion.
39. HOTTEL, H. C. and MANGELSDORF, H. G.: *Trans. Am. Inst. Chem. Eng.* **31** (1935) 517. Heat transmission by radiation from non-luminous gases. Experimental study of carbon dioxide and water vapour.
40. NUSSELT, W.: *Z. Ver. deut. Ing.* **60** (1916) 541 and 569 Die Oberflächenkondensation des Wasserdampfes. See also: MONRAD, C. C. and BADGER, W. L.: *Trans. Am. Inst. Chem. Eng.* **24** (1930) 84. The condensation of vapors.
41. HASELDEN, G. G. and PROSAD, S.: *Trans. Inst. Chem. Eng.* **27** (1949) 195. Heat transfer from condensing oxygen and nitrogen vapours.
42. TEN BOSCH, M.: *Die Wärmeübertragung* (Springer, 1936).
43. CARPENTER, E. F. and COLBURN, A. P.: *Proceedings of the General Discussion on Heat Transfer, September,* (1951), p. 20. The effect of vapour velocity on condensation inside tubes. (*Inst. of Mech. Eng. and Am. Soc. Mech. Eng.*).
44. TEPE, J. B. & MUELLER, A. C.: *Chem. Eng. Prog.* **43** (1947) 267. Condensation and subcooling inside an inclined tube.
45. COLBURN, A. P.: *Proceedings of the General Discussion on Heat Transfer, September, 1951,* p. 1. Problems in design and research on condensers of vapours and vapour mixtures. (*Inst. of Mech. Eng. and Am. Soc. Mech. Eng.*).
46. KIRKBRIDE, G. C.: *Ind. Eng. Chem.* **26** (1934) 425. Heat transfer by condensing vapours on vertical tubes.
47. BADGER, W. L.: *Ind. Eng. Chem.* **22** (1930) 700. The evaporation of caustic soda to high concentrations by means of diphenyl vapour.
48. BADGER, W. L.: *Trans. Am. Inst. Chem. Eng.* **33** (1937) 441. Heat transfer coefficient for condensing Dowtherm films.
49. DREW, T. B., NAGLE, W. M., and SMITH, W. Q.: *Trans. Am. Inst. Chem. Eng.* **31** (1935) 605. The conditions for drop-wise condensation of steam.
50. COLBURN, A. P. and HOUGEN, O. A.: *Ind. Eng. Chem.* **26** (1934) 1178. Design of cooler condensers for mixtures of vapors with non-condensing gases.
51. CHILTON, T. H. and COLBURN, A. P.: *Ind. Eng. Chem.* **26** (1934) 1183. Mass transfer (absorption) coefficients.
52. REVILOCK, J. F., HURLBURT, H. Z., BRAKE, D. R., LANG, E. G., and KERN, D. Q.: *Chem. Eng. Prog.* Symposium Series No. 30, **56** (1960) 161. Heat and mass transfer analogy: An appraisal using plant scale data.
53. JEFFREYS, G. V.: *A Problem in Chemical Engineering Design* (Institution of Chemical Engineers, 1961). The manufacture of acetic anhydride.
54. COULSON, J. M. and RICHARDSON, J. F.: *Chemical Engineering,* volume 2 (Pergamon Press, Oxford, 1976).
55. JAKOB, M.: *Mech. Eng.* **58** (1936) 643, 729. Heat transfer in evaporation and condensation.
56. ROHSENOW, W. M. and Clark, J. A.: *Trans. Am. Soc. Mech. Eng.* **73** (1951) 609. A study of the mechanism of boiling heat transfer.
57. ROHSENOW, W. M.: *Trans. Am. Soc. Mech. Engrs.* **74** (1952) 969. A method of correlating heat transfer data for surface boiling of liquids.
58. FORSTER, H. K.: *J. Appl. Phys.* **25** (1954) 1067. On the conduction of heat into a growing vapor bubble.
59. GRIFFITH, P. and WALLIS, J. D.: *Chem. Eng. Prog. Symposium Series* No. 30, **56** (1960), The role of surface conditions in nuclear boiling, [and HSU, Y. Y.: On the size range of active nucleation cavities on a heating surface, *Trans. ASME J. Ht. Transfer* **84** (1962) 207.]
60. JAKOB, M. and FRITZ, W.: *Forsch. Gebiete Ingenieurw,* **2** (1931) 435. Versuche über den Verdampfungsvorgang.
61. NUKIYAMA, S.: *J. Soc. Mech. Eng. (Japan)* **37** (1934) 367. English abstract pp. S53–S54. The maximum and minimum values of the heat Q transmitted from metal to boiling water under atmospheric pressure.
62. FARBER, E. A. and SCORAH, R. L.: *Trans. Am. Soc. Mech. Eng.* **70** (1948) 369. Heat transfer to boiling water under pressure.
63. BONILLA, C. F. and PERRY, C. H.: *Trans. Am. Inst. Chem. Eng.* **37** (1941) 685. Heat transmission to boiling binary liquid mixtures.
64. INSINGER, T. H. and BLISS, H.: *Trans. Am. Inst. Chem. Eng.* **36** (1940) 491. Transmission of heat to boiling liquids.
65. CRYDER, D. S. and FINALBORGO, A. C.: *Trans. Am. Inst. Chem. Eng.* **33** (1937) 346. Heat transmission from metal surfaces to boiling liquids.
66. SAUER, E. T., COOPER, H. B. H., AKIN, G. A. and MCADAMS, W. H.: *Mech. Eng.* **60** (1938) 669. Heat transfer to boiling liquids.
67. JAKOB, M. and LINKE, W.: *Forsch. Gebiete Ing.* **4** (1933) 75. Der Wärmeübergang von einer waagerechten Platte an siedendes Wasser.

68. CICHELLI, M. T. and BONILLA, C. F.: *Trans. Am. Inst. Chem. Eng.* **41** (1945) 755. Heat transfer to liquids boiling under pressure.
69. FORSTER, H. K. and ZUBER, N.: *J. Appl. Phys.* **25** (1954) 474. Growth of a vapor bubble in a superheated liquid.
70. FORSTER, H. K. and ZUBER, N.: *A.I.Ch.E. Jl* **1** (1955) 531. Dynamics of vapor bubbles and boiling heat transfer.
71. FRITZ, W.: *Physik Z.* **36** (1935) 379. Berechnung des Maximalvolumens von Dampfblasen.
72. ADDOMS, J. N.: Massachusetts Institute of Technology, D.Sc. thesis in chemical engineering (1948). Heat transfer at high rates to water boiling outside cylinders.
73. MCNELLY, M. J.: *J. Imp. Coll. Chem. Eng. Soc.* **7** (1953) 18. A correlation of the rates of heat transfer to nucleate boiling liquids.
74. JESCHKE, D.: *Z. Ver. deut. Ing.* **69** (1925) 1526. Wärmeübergang und Druckverlust in Rohrschlangen.
75. KERN, D. Q.: *Chem. Eng. Prog.* Symposium Series No. 29, **55** (1959) 187. Optimum air fin cooler design.
76. CHILTON, T. H., DREW, T. B., and JEBENS, R. H.: *Ind. Eng. Chem.* **36** (1944) 510. Heat transfer coefficients in agitated vessels.
77. PRATT, N. H.: *Trans. Inst. Chem. Eng.* **25** (1947) 163. The heat transfer in a reaction tank cooled by means of a coil.
78. CUMMINGS, G. H. and WEST, A. S.: *Ind. Eng. Chem.* **42** (1950) 2303. Heat transfer data for kettles with jackets and coils.
79. BROWN, R. W., SCOTT, M. A. and TOYNE, C.: *Trans. Inst. Chem. Eng.* **25** (1947) 181, An investigation of heat transfer in agitated jacketed cast iron vessels.
80. *Tubular Exchanger Manufacturers' Association of America* (TEMA) (1959).
81. British Standards Institution. Fusion welded pressure vessels BS 1500 (1958).
82. UNDERWOOD, A. J. V.: *J. Inst. Petrol. Technol.* **20** (1934) 145. The calculation of the mean temperature difference in multipass heat exchangers.
83. BOWMAN, R. A., MUELLER, A. C. and NAGLE, W. M.: *Trans. Am. Soc. Mech. Eng.* **62** (1940) 283. Mean temperature difference in design.
84. WILSON, E. E.: *Trans. Am. Soc. Mech. Eng.* **37** (1915) 546. A basis for rational design of heat transfer apparatus.
85. RHODES, F. H. and YOUNGER, K. R.: *Ind. Eng. Chem.* **27** (1935) 957. Rate of heat transfer between condensing organic vapours and a metal tube.
86. COULSON, J. M. and MEHTA, R. R.: *Trans. Inst. Chem. Eng.* **31** (1953) 208. Heat transfer coefficients in a climbing film evaporator.
87. LYLE, O.: *Efficient use of Steam* (HMSO, London, 1947).

7.16. NOMENCLATURE

		Units in SI systems	Dimensions in M, L, T, θ and (M, L, T, θ, H)
A	Area available for heat transfer or area of radiating surface	m^2	L^2
A_s	Maximum cross-flow area over tube bundle	m^2	L^2
a	Absorptivity	—	—
b	Wetted perimeter of condensation surface or perimeter of fin	m	L
C	Constant	—	—
C_f	Constant for friction in flow past a tube bundle	—	—
C_h	Constant for heat transfer in flow past a tube bundle	—	—
C'	Clearance between tubes in heat exchanger	m	L
C_p	Specific heat at constant pressure	J/kg K	$L^2T^{-2}\theta^{-1}$ (*or* $HM^{-1}\theta^{-1}$)
D	Diffusivity of vapour	m^2/s	L^2T^{-1}
D_H	Thermal diffusivity $(k/C_p\rho)$	m^2/s	L^2T^{-1}
d	Diameter (internal or of sphere)	m	L
d_1, d_2	Inner and outer diameters of annulus	m	L
d_c	Diameter of helix	m	L
d_e	Hydraulic mean diameter	m	L
d_g	Gap between turns in coil	m	L
d_m	Logarithmic mean diameter of lagging	m	L
d_o	Outside diameter of tube	m	L
d_p	Height of coil	m	L
d_s	Outside diameter of lagging	m	L
d_t	Thickness of fixed tube sheet	m	L
d_v	Internal diameter of vessel	m	L
d_w	Mean diameter of pipe wall	m	L
E	Emissive power	W/m^2	MT^{-3} (*or* $HL^{-2}T^{-1}$)
E'_z	Energy emitted per unit area and unit time per unit wavelength	W/m^3	$ML^{-1}T^{-3}$ (*or* $HL^{-3}T^{-1}$)
e	Emissivity	—	—
e'	$\frac{1}{2}(e_s+1)$	—	—

		Units in SI systems	Dimensions in M, L, T, θ and (M, L, T, θ, H)
F	Geometric factor for radiation or correction factor for logarithmic mean temperature difference	—	—
f	Working stress	N/m²	$\mathbf{ML^{-1}T^{-2}}$
f'	Shell-side friction factor	—	—
G	Mass rate of flow	kg/s	$\mathbf{MT^{-1}}$
G'	Mass rate of flow per unit area	kg/m² s	$\mathbf{ML^{-2}T^{-1}}$
G'_s	Mass flow per unit area over tube bundle	kg/m² s	$\mathbf{ML^{-2}T^{-1}}$
g	Acceleration due to gravity	m/s²	$\mathbf{LT^{-2}}$
h	Heat transfer coefficient	W/m² K	$\mathbf{MT^{-3}\theta^{-1}}$ (or $\mathbf{HL^{-2}T^{-1}\theta^{-1}}$)
h_b	Film coefficient for liquid adjacent to vessel	W/m² K	$\mathbf{MT^{-3}\theta^{-1}}$ (or $\mathbf{HL^{-2}T^{-1}\theta^{-1}}$)
h_c	Heat transfer coefficient for convection	W/m² K	$\mathbf{MT^{-3}\theta^{-1}}$ (or $\mathbf{HL^{-2}T^{-1}\theta^{-1}}$)
h_f	Fin-side film coefficient	W/m² K	$\mathbf{MT^{-3}\theta^{-1}}$ (or $\mathbf{HL^{-2}T^{-1}\theta^{-1}}$)
h_m	Mean value of h over whole surface	W/m² K	$\mathbf{MT^{-3}\theta^{-1}}$ (or $\mathbf{HL^{-2}T^{-1}\theta^{-1}}$)
h_n	Heat transfer coefficient for liquid boiling at T_n	W/m² K	$\mathbf{MT^{-3}\theta^{-1}}$ (or $\mathbf{HL^{-2}T^{-1}\theta^{-1}}$)
h_r	Heat transfer coefficient for radiation	W/m² K	$\mathbf{MT^{-3}\theta^{-1}}$ (or $\mathbf{HL^{-2}T^{-1}\theta^{-1}}$)
I	Intensity of radiation	W/m²	$\mathbf{MT^{-3}}$ (or $\mathbf{HL^{-2}T^{-1}}$)
I'	Intensity of radiation falling on body	W/m²	$\mathbf{MT^{-3}}$ (or $\mathbf{HL^{-2}T^{-1}}$)
J	For fin: $(km - h)/(km + h)$	—	—
j	Number of vertical rows of tubes	—	—
j_d	"j-factor" for mass transfer	—	—
j_h	"j-factor" for heat transfer	—	—
k	Thermal conductivity	W/m K	$\mathbf{MLT^{-3}\theta^{-1}}$ (or $\mathbf{HL^{-1}T^{-1}T^{-1}\theta^{-1}}$)
k_G	Mass transfer coefficient (mass/unit area . unit time . unit partial pressure difference)	s/m	$\mathbf{L^{-1}T}$
L	Length of paddle, length of fin or characteristic dimension	m	$\mathbf{L}$
l	Length of tube or plate or distance apart of faces or thickness of gas stream	m m	$\mathbf{L}$ $\mathbf{L}$
l_B	Distance between baffles	m	$\mathbf{L}$
l_v	Length of side of vessel		
M	Mass rate of flow of condensate per unit length of perimeter for vertical pipe and per unit length of pipe for horizontal pipe	kg/s m	$\mathbf{ML^{-1}T^{-1}}$
m	Mass of liquid	kg	$\mathbf{M}$
m	For fin: $(hb/kA)^{0.5}$	m⁻¹	$\mathbf{L^{-1}}$
N	Number of revolutions in unit time	Hz	$\mathbf{T^{-1}}$
$\mathbf{N}$	Number of general term in series	—	—
n	An index	—	—
P	Pressure	N/m²	$\mathbf{ML^{-1}T^{-2}}$
P_{Bm}	Logarithmic mean partial pressure of inert gas B	N/m²	$\mathbf{ML^{-1}T^{-2}}$
P_c	Critical pressure	N/m²	$\mathbf{ML^{-1}T^{-2}}$
P_R	Reduced pressure (P/P_c)	—	—
P_r	Vapour pressure at surface of radius r	N/m²	$\mathbf{ML^{-1}T^{-2}}$
P_s	Saturation vapour pressure	N/m²	$\mathbf{ML^{-1}T^{-2}}$
p	Parameter in Laplace transform	s⁻¹	$\mathbf{T^{-1}}$
Q	Heat flow or generation per unit time	W	$\mathbf{ML^2T^{-3}}$ (or $\mathbf{HL^{-2}T^{-1}}$)
Q_e	Radiation emitted per unit time	W	$\mathbf{ML^2T^{-3}}$ (or $\mathbf{HT^{-1}}$)
Q_f	Heat flow to root of fin per unit time	W	$\mathbf{ML^2T^{-3}}$ (or $\mathbf{HT^{-1}}$)
Q_G	Rate of heat generation per unit volume	W/m³	$\mathbf{ML^{-1}T^{-3}}$ (or $\mathbf{HL^{-3}T^{-1}}$)
Q_i	Total incident radiation per unit time	W	$\mathbf{ML^2T^{-3}}$ (or $\mathbf{HT^{-1}}$)
Q_k	Heat flow per unit time by conduction in fluid	W	$\mathbf{ML^2T^{-3}}$ (or $\mathbf{HT^{-1}}$)
Q_o	Total radiation leaving surface per unit time	W	$\mathbf{ML^2T^{-3}}$ (or $\mathbf{HT^{-1}}$)
Q_r	Total heat reflected from surface per unit time	W	$\mathbf{ML^2T^{-3}}$ (or $\mathbf{HT^{-1}}$)
q	Heat flow per unit time and unit area	W/m²	$\mathbf{MT^{-3}}$ (or $\mathbf{HL^{-2}T^{-1}}$)
R	Thermal resistance	m²K/W	$\mathbf{M^{-1}T^3\theta}$ (or $\mathbf{H^{-1}T^2L\theta}$)
R_i, R_0	Thermal resistance of scale on inside, outside of tubes	m²K/W	$\mathbf{M^{-1}T^3\theta}$ (or $\mathbf{H^{-1}T^2L\theta}$)
R'	Shear stress at free surface of condensate film	N/m²	$\mathbf{ML^{-1}T^{-2}}$
r	Radius	m	$\mathbf{L}$
r_1, r_2	Radius (inner, outer) of annulus or tube	m	$\mathbf{L}$
r_a	Arithmetic mean radius	m	$\mathbf{L}$
r_m	Logarithmic mean radius	m	$\mathbf{L}$
S	Flow area for condensate film	m²	$\mathbf{L^2}$
s	Thickness of condensate film at a point	m	$\mathbf{L}$
T	Temperature	K	θ
T_c	Temperature of free surface of condensate	K	θ

		Units in SI systems	Dimensions in M, L, T, θ and (M, L, T, θ, H)
T_{cm}	Temperature of cooling medium	K	θ
T_f	Mean temperature of film	K	θ
T_G	Temperature of atmosphere surrounding fin	K	θ
T_m	Mean temperature of fluid	K	θ
T_n	Standard boiling point	K	θ
T_s	Temperature of condensing vapour	K	θ
T_w	Temperature of wall	K	θ
t	Time	s	T
U	Overall heat transfer coefficient	W/m^2 K	$MT^{-3}\theta^{-1}$ (or $HL^{-2}T^{-1}$)
u	Velocity	m/s	LT^{-1}
u_m	Maximum velocity in condensate film	m/s	LT^{-1}
u_y	Velocity at distance y from surface	m/s	LT^{-1}
W	Width of stirrer	m	L
w	Width of fin	m	L
$w_1, w_2, \ldots$	Indices in equation for heat transfer by convection	—	—
X	Distance between centres of tubes in direction of flow	m	L
X	Ratio of temperature differences used in calculation of mean temperature difference	—	—
x	Distance in direction of transfer or along surface	m	L
Y	Distance between centres of tubes at right angles to flow direction	m	L
Y	Ratio of temperature difference used in calculation of mean temperature difference	—	—
y	Distance perpendicular to surface	m	L
Z	Wavelength	m	L
Z_m	Wavelength at which maximum energy is emitted	m	L
z	Distance in third principal direction	m	L
α	Angle between normal and direction of radiation	—	—
β	Coefficient of cubical expansion	K^{-1}	θ^{-1}
ε	Coefficient relating h to $u^{0.8}$	J/s$^{0.2}$ m$^{2.8}$K	$ML^{-0.8}T^{-2.2}\theta^{-1}$ (or $HL^{-2.8}T^{-0.2}\theta^{-1}$)
λ	Latent heat of vaporisation per unit mass	J/kg	L^2T^{-2} (or HM^{-1})
μ	Viscosity	Ns/m^2	$ML^{-1}T^{-1}$
μ_s	Viscosity of fluid at surface	Ns/m^2	$ML^{-1}T^{-1}$
ρ	Density or density of liquid	kg/m^3	ML^{-3}
	Density of vapour	kg/m^3	ML^{-3}
σ	Stefan–Boltzmann constant, or	W/m^{2}K^4	$MT^{-3}\theta^{-4}$ (or $HL^{-2}T^{-1}\theta^{-4}$)
	Surface tension	J/m^2	MT^{-2}
ψ	$\dfrac{G_1c_1 + G_2c_2}{G_1c_1G_2c_2}$		
ϕ	Angle between surface and horizontal or angle of contact	—	—
θ	Temperature or temperature difference	K	θ
θ_a	Temperature difference in Schmidt method	K	θ
θ_m	Logarithmic mean temperature difference	K	θ
θ_{xt}	Temperature at $t = t$, $x = x$	K	θ
θ_0	Initial uniform temperature of body	K	θ
θ'	Temperature of source or surroundings	K	θ
$\bar{\theta}$	Laplace transform of temperature	sK	$T\theta$
ω	Solid angle	—	—
Fo	Fourier number D_Ht/L^2	—	—
Gr	Grashof number	—	—
Nu	Nusselt number	—	—
Pr	Prandtl number	—	—
Re	Reynolds number	—	—
Δ	Finite difference in a property	—	—

Suffix, w refers to wall material
Suffixes i, o refer to inside, outside of wall
Suffixes b, l, v refer to bubble, liquid, vapour
Suffixes g, b, and s refer to gas, black body, and surface.

CHAPTER 8

Mass Transfer

8.1. INTRODUCTION

When a concentration gradient exists within a fluid consisting of two or more components, there is a tendency for each constituent to flow in such a direction as to reduce the concentration gradient; this process is known as mass transfer. In a still fluid, or in a fluid flowing under streamline conditions in a direction at right angles to the concentration gradient, the transfer is effected as a result of the random motion of the molecules. In a turbulent fluid this mechanism is supplemented by transference of material by eddy currents.

Mass transfer can take place in either a gas phase or a liquid phase, or in both simultaneously. When a liquid evaporates into a still gas, vapour is transferred from the surface to the bulk of the gas as a result of the concentration gradient; the process continues until the whole of the liquid has evaporated or until the gas is saturated and the concentration gradient reduced to zero. In the absorption of a soluble gas from a mixture with an insoluble gas, mass transfer takes place from the bulk of the gas to the liquid surface and then into the bulk of the liquid. Neither the insoluble gas nor the solvent moves in the direction of mass transfer. In a distillation column, on the other hand, the less volatile component diffuses in the gas phase to the liquid surface and the more volatile material diffuses at an approximately equal molar rate in the opposite direction. In the liquid phase a similar process takes place, with the less volatile material diffusing away from the gas–liquid interface.

The rate of transfer of **A** in a mixture of two components **A** and **B** will therefore be determined not only by the rate of diffusion of **A** but also by the behaviour of **B**. The molar rate of transfer of **A**, per unit area, due to molecular motion is given by:

$$N_A = -D_{AB}\frac{dC_A}{dy} \tag{8.1}$$

where N_A is the molar rate of diffusion per unit area, D_{AB} the diffusivity of **A** in **B**, a physical property of the two vapours, C_A the molar concentration of **A**, and y the distance in the direction of diffusion.

The corresponding rate of diffusion of **B** is given by:

$$N_B = -D_{BA}\frac{dC_B}{dy} \tag{8.2}$$

where D_{BA} is the diffusivity of **B** in **A** and C_B is the molar concentration of **B**.

Equation 8.1 is often referred to as Fick's law. Fick[1] in 1855 derived by analogy with heat transfer that $\partial C_A/\partial t = D(\partial^2 C_A/\partial y^2)$ from which equation 8.1 is obtained when $\partial C_A/\partial t = 0$.

If the total pressure, and hence the total molar concentration, is everywhere constant, dC_A/dy and dC_B/dy must be equal and opposite, and therefore **A** and **B** tend to diffuse in opposite directions. In a distillation process where the two components have equal molar latent heats, condensation of one mole of the less volatile material releases just sufficient heat for the vaporisation of one mole of the more volatile component, and therefore *equimolecular counterdiffusion* takes place with two components diffusing at equal and opposite rates, as determined by Fick's law. In an absorption process there is net transfer of only one of the components although there is a concentration difference of the other. It is therefore necessary to study the two cases separately.

In many processes **B** will neither remain stationary nor will it diffuse at an equal and opposite molar rate to **A**. Exact calculations relating to this type of problem are difficult. An example of this nature is the mass transfer in a distillation column when the two components have unequal molar latent heats.

When the fluid is turbulent, eddy diffusion takes place in addition to molecular diffusion and the rate of diffusion is increased and

$$N_A = -(D_{AB} + E_D)(dC_A/dy) \tag{8.3}$$

where E_D is known as the eddy diffusivity. E_D will increase as the turbulence is increased and is more difficult to evaluate than the molecular diffusivity.

8.2. DIFFUSION IN THE GAS PHASE

8.2.1. Equimolecular Counterdiffusion

Suppose two vapours **A** and **B** are diffusing at equal and opposite rates and P_A and P_B are their partial pressures at any point in the system.

If **A** and **B** are ideal gases

$$P_A V = n_A \mathbf{R} T \text{ and } P_B V = n_B \mathbf{R} T \quad \text{(from equation 2.15)}$$

where n_A and n_B are the number of moles of **A** and **B** in a volume V so that:

$$P_A = \frac{n_A}{V} \mathbf{R} T = C_A \mathbf{R} T = \frac{c_A}{M_A} \mathbf{R} T \tag{8.4}$$

and

$$P_B = \frac{n_B}{V} \mathbf{R} T = C_B \mathbf{R} T = \frac{c_B}{M_B} \mathbf{R} T \tag{8.5}$$

where c_A and c_B are mass concentrations and M_A and M_B molecular weights. If the total pressure P is everywhere constant in the system, from equations 8.4 and 8.5:

$$P = P_A + P_B = RT(C_A + C_B) = \mathbf{R} T \left(\frac{c_A}{M_A} + \frac{c_B}{M_B} \right) \tag{8.6}$$

and

$$P = \mathbf{R} T C_T \tag{8.7}$$

where $C_T = C_A + C_B$ is the sum of the molar concentrations of **A** and **B**.

Thus

$$\frac{dP_A}{dy} = -\frac{dP_B}{dy} \tag{8.8}$$

$$\frac{dC_A}{dy} = -\frac{dC_B}{dy} \tag{8.9}$$

and

$$\frac{dc_A}{dy} = -\frac{dc_B}{dy} \frac{M_A}{M_B} \tag{8.10}$$

By substituting from equations 8.4 and 8.5 into equations 8.1 and 8.2, the mass transfer rates can be expressed in terms of partial pressure gradients rather than concentration gradients.

Thus

$$N_A = -\frac{D_{AB}}{RT} \frac{dP_A}{dy} \tag{8.11}$$

and

$$N_B = -\frac{D_{BA}}{\mathbf{R} T} \frac{dP_B}{dy} = +\frac{D_{BA}}{\mathbf{R} T} \frac{dP_A}{dy} \quad \text{(from equation 8.11)} \tag{8.12}$$

In equimolecular counterdiffusion, the total pressure at any point in the system will remain constant as a result of diffusion, and therefore equal numbers of molecules will diffuse in each direction, i.e.

$$N_A = - N_B$$

so that

$$D_{AB} = D_{BA} = D \text{ (say)}. \quad \text{(from equations 8.11 and 8.12)}$$

Thus

$$N_A = - D\frac{dC_A}{dy} = -\frac{D}{RT}\frac{dP_A}{dy} \tag{8.13}$$

and

$$N_B = + D\frac{dC_A}{dy} = +\frac{D}{RT}\frac{dP_A}{dy} \tag{8.14}$$

The above equations which give the transfer rate in terms of a point value of a concentration or partial pressure gradient can be integrated directly. Thus if the conditions at two different points in the system are denoted by the suffixes 1 and 2, integration of equation 8.12 gives:

$$N_A = - D\frac{(C_{A_2} - C_{A_1})}{y_2 - y_1} \tag{8.15}$$

or

$$N_A = -\frac{D(P_{A_2} - P_{A_1})}{RT(y_2 - y_1)} \tag{8.16}$$

8.2.2. Diffusion through a Stationary Gas

The rates of diffusion of **A** and **B** are given by equations 8.13 and 8.14:

$$N_A = -D\frac{dC_A}{dy} = -\frac{D}{RT}\frac{dP_A}{dy} \tag{8.17}$$

and

$$N_B = + D\frac{dC_A}{dy} = +\frac{D}{RT}\frac{dP_A}{dy} \tag{8.18}$$

If a surface is introduced on which **A** is absorbed but **B** is not absorbed (Fig. 8.1) a partial pressure gradient will be set up, causing **A** to diffuse towards, and **B** away from the surface. Imagine this process to continue for a short interval. **A** will be absorbed at the surface and **B** will tend to diffuse away and therefore a total pressure gradient will be produced causing a bulk motion of **A** and **B** towards the surface, in addition to the transfer by diffusion. Since there is no net motion of **B**, the bulk rate of flow must exactly balance its transfer by diffusion.

Thus the bulk rate of flow of **B** = $- N_B$

$$= - D(dC_A/dy)$$

The bulk flow of **B** is accompanied by a bulk flow of **A** and since the ratio of the number of molecules of **A** at a point to those of **B** is C_A/C_B:

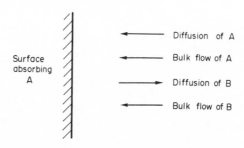

FIG. 8.1. Mass transfer through stationary gas.

$$\text{Bulk flow of } \mathbf{A} = -N_B(C_A/C_B)$$

$$= -D\frac{dC_A}{dy}\frac{C_A}{C_T - C_A} \tag{8.19}$$

The total rate of transfer of **A** is obtained by summing the transfers by diffusion and by bulk flow.

Thus, by adding equations 8.17 and 8.19 the total transfer N'_A is given by:

$$N'_A = -D\frac{dC_A}{dy} - D\frac{dC_A}{dy}\frac{C_A}{C_T - C_A}$$

$$= -D\frac{dC_A}{dy}\frac{C_T}{C_T - C_A} = -D\frac{C_T}{C_B}\frac{dC_A}{dy} \tag{8.20}$$

This relation is known as Stefan's law.[2]

Integration of equation 8.20 between two positions denoted by suffixes 1 and 2 gives:

$$N'_A = \frac{DC_T}{y_2 - y_1}\ln\frac{C_T - C_{A_2}}{C_T - C_{A_1}} \tag{8.21}$$

$$= -\frac{D}{y_2 - y_1}\frac{C_T}{(C_T - C_A)_m}(C_{A_2} - C_{A_1})$$

$$= -\frac{D}{y_2 - y_1}\frac{C_T}{C_{Bm}}(C_{A_2} - C_{A_1}) \tag{8.22}$$

$$= -\frac{D}{RT(y_2 - y_1)}\frac{P}{P_{Bm}}(P_{A_2} - P_{A_1}) \tag{8.23}$$

where suffix m denotes the logarithmic mean value of the quantity at the positions 1 and 2.

The bulk flow of the mixture gives rise to a bulk flow velocity u_F given by:

$$u_F C_A = -D\frac{C_A}{C_B}\frac{dC_A}{dy}$$

i.e.
$$u_F = -\frac{D}{C_B}\frac{dC_A}{dy} \tag{8.24}$$

Equation 8.21 can be simplified when the concentration of the diffusing component **A** is small. Under these conditions C_A is small compared with C_T. Equation 8.21 gives:

$$N'_A = \frac{DC_T}{y_2 - y_1}\ln\left[1 - \left(\frac{C_{A_2} - C_{A_1}}{C_T - C_{A_1}}\right)\right]$$

$$= \frac{DC_T}{y_2 - y_1}\left[-\left(\frac{C_{A_2} - C_{A_1}}{C_T - C_{A_1}}\right) - \frac{1}{2}\left(\frac{C_{A_2} - C_{A_1}}{C_T - C_{A_1}}\right)^2 - \cdots\right]$$

For small values of C_A, this reduces to:

$$N'_A \approx -\frac{D}{y_2 - y_1}(C_{A_2} - C_{A_1}) \tag{8.25}$$

This is identical to equation 8.15 for equimolecular counterdiffusion. Thus, the effects of bulk flow can be neglected at low concentrations.

8.2.3. Comparison of Rates of Mass Transfer in Equimolecular Counterdiffusion and in Diffusion through a Stationary Gas

For equimolecular counterdiffusion, from equation 8.15:

$$N_A = -D\frac{C_{A_2} - C_{A_1}}{y_2 - y_1} = h_D(C_{A_1} - C_{A_2}) \qquad (8.26)$$

$$= k_G'(P_{A_1} - P_{A_2}) \qquad (8.27)$$

where $h_D[=D/(y_2-y_1)]$ and $k_G'\{=D/[RT(y_2-y_1)]\}$ are mass transfer coefficients for equimolecular counterdiffusion.

For diffusion through a stationary gas **B** from equation 8.22:

$$N_A' = -D\frac{C_{A_2} - C_{A_1}}{y_2 - y_1}\frac{C_T}{C_{Bm}} = h_D(C_{A_1} - C_{A_2}) \qquad (8.28)$$

$$= k_G(P_{A_1} - P_{A_2}) \qquad (8.29)$$

where
$$h_D\left(=\frac{D}{y_2-y_1}\frac{C_T}{C_{Bm}}\right) \quad \text{and} \quad k_G\left(=\frac{D}{RT(y_2-y_1)}\frac{C_T}{C_{Bm}}\right)$$

are mass transfer coefficients for diffusion through a stationary gas.

Thus
$$\frac{N_A'}{N_A} = \frac{C_T}{C_{Bm}} = \frac{k_G}{k_G'} \qquad (8.30)$$

Where C_A is small compared with C_T, it is seen by comparing equations 8.15 and 8.25 that the rate of transfer of **A** is the same for equimolecular counterdiffusion as it is for diffusion through a stationary gas.

The coefficients h_D, k_G, k_G' should be compared with the film coefficients for heat transfer already discussed in Chapter 7.

8.2.4. Maxwell's Law of Diffusion

Maxwell[3] postulated that the partial pressure gradient in the direction of diffusion for a constituent of a two-component gaseous mixture was proportional to:

(a) the relative velocity of the molecules in the direction of diffusion, and
(b) the product of the molar concentrations of the components.

Thus
$$-\frac{dP_A}{dy} = FC_AC_B(u_A - u_B) \qquad (8.31)$$

$$\therefore \qquad -\frac{dC_A}{dy} = \frac{FC_AC_B}{RT}(u_A - u_B) \qquad (8.32)$$

where u_A and u_B are the mean molecular velocities of **A** and **B** respectively in the direction of mass transfer and F is a coefficient.

Equimolecular Counterdiffusion

We have
$$N_A = u_AC_A \qquad (8.33)$$

$$N_B = u_BC_B \qquad (8.34)$$

$$N_A = -N_B \quad \text{(by definition of equimolecular counterdiffusion)}$$

Hence
$$u_A = N_A/C_A \qquad (8.35)$$

and
$$u_B = -N_A/C_B \qquad (8.36)$$

Substituting in equation 8.32:

$$-\frac{dC_A}{dy} = \frac{FC_AC_B}{RT}\left(\frac{N_A}{C_A} + \frac{N_A}{C_B}\right)$$

$$= \frac{FN_A}{RT}(C_B + C_A) \tag{8.37}$$

Thus

$$N_A = -\frac{RT}{FC_T}\frac{dC_A}{dy} \tag{8.38}$$

Then, by comparison with Fick's law:

$$D = \frac{RT}{FC_T} \tag{8.39}$$

or

$$F = \frac{RT}{DC_T} \tag{8.40}$$

Transfer of A Through Stationary B

By analogy with equation 8.33:

$$N_A' = u_A C_A$$

and

$$u_B = 0$$

so that

$$-\frac{dC_A}{dy} = \frac{FC_AC_B}{RT}\frac{N_A'}{C_A}$$

or

$$N_A' = -\frac{RT}{FC_T}\frac{C_T}{C_B}\frac{dC_A}{dy} \tag{8.41}$$

Substituting from equation 8.40:

$$N_A' = -D\frac{C_T}{C_B}\frac{dC_A}{dy} \tag{equation 8.20}$$

Thus, Stefan's law can also be derived from Maxwell's law of diffusion.

Multicomponent Mass Transfer

The above argument can be applied to the diffusion of a constituent of a multicomponent gas. Consider the transfer of component A through a stationary gas consisting of components B, C, . . . Suppose that the total partial pressure gradient can be regarded as being made up of series of terms each representing the contribution of the individual component gases. Then, from equation 8.32:

$$-\frac{dC_A}{dy} = \frac{F_{AB}C_AC_BN_A'}{RT}\frac{N_A'}{C_A} + \frac{F_{AC}C_AC_CN_A'}{RT}\frac{N_A'}{C_A} + \cdots$$

$$= \frac{N_A'}{RT}(F_{AB}C_B + F_{AC}C_C + \cdots)$$

From equation 8.40, writing:

$$F_{AB} = \frac{RT}{D_{AB}C_T}, \quad \text{etc.}$$

where D_{AB} is the diffusivity of A in B, etc.:

$$-\frac{dC_A}{dy} = \frac{N_A'}{C_T}\left(\frac{C_B}{D_{AB}} + \frac{C_C}{D_{AC}} + \cdots\right)$$

$$\therefore \quad N_A' = - \frac{C_T}{\dfrac{C_B}{D_{AB}} + \dfrac{C_C}{D_{AC}} + \cdots} \frac{dC_A}{dy}$$

$$= - \frac{1}{\dfrac{C_B}{C_T - C_A} \dfrac{1}{D_{AB}} + \dfrac{C_C}{C_T - C_A} \dfrac{1}{D_{AC}} + \cdots} \frac{C_T}{C - C_A} \frac{dC_A}{dy}$$

$$= - \frac{1}{\dfrac{y_B'}{D_{AB}} + \dfrac{y_C'}{D_{AC}} + \cdots} \frac{C_T}{C_T - C_A} \frac{dC_A}{dy} \qquad (8.42)$$

where y_B' is the mole fraction of **B** in the stationary components of the gas, etc.

By comparing equation 8.42 with Stefan's law (equation 8.20) the effective diffusivity of **A** in the mixture (D') is given by

$$\frac{1}{D'} = \frac{y_B'}{D_{AB}} + \frac{y_C'}{D_{AC}} + \cdots \qquad (8.43)$$

8.2.5. Diffusivities of Various Vapours

Diffusivities of a number of vapours in air are given in Tables 8.1[4] and 8.2,[4] in which the ratio of the kinematic viscosity of a dilute mixture to the diffusivity $\mu/\rho D$ is also listed. This ratio is known as the Schmidt number and will be discussed further in Chapter 10.

Diffusivities of vapours are most conveniently determined by the method developed by Winkelmann.[5] Liquid is allowed to evaporate in a vertical glass tube over the top of which a stream of vapour-free gas is passed, sufficiently rapidly for the vapour pressure to be maintained almost at zero (Fig. 8.2). If the apparatus is maintained at a steady temperature, there will be no eddy currents in the vertical tube and mass transfer will take place from the surface by molecular diffusion alone. The rate of evaporation can be followed by the rate of fall of the liquid surface, and since the concentration gradient is known, the diffusivity can then be calculated.

TABLE 8.1. *Diffusion coefficients of gases and vapours in air at* 298 K *and atmospheric pressure*

Substance	D (m²/s × 10⁶)	$\dfrac{\mu}{\rho D}$	Substance	D (m²/s × 10⁶)	$\dfrac{\mu}{\rho D}$
Ammonia	23·6	0·66	Valeric acid	6·7	2·31
Carbon dioxide	16·4	0·94	i-Caproic acid	6·0	2·58
Hydrogen	41·0	0·22	Diethyl amine	10·5	1·47
Oxygen	20·6	0·75	Butyl amine	10·1	1·53
Water	25·6	0·60	Aniline	7·2	2·14
Carbon disulphide	10·7	1·45	Chlorobenzene	7·3	2·12
Ethyl ether	9·3	1·66	Chlorotoluene	6·5	2·38
Methanol	15·9	0·97	Propyl bromide	10·5	1·47
Ethanol	11·9	1·30	Propyl iodide	9·6	1·61
Propanol	10·0	1.55	Benzene	8·8	1.76
Butanol	9·0	1·72	Toluene	8·4	1.84
Pentanol	7·0	2.21	Xylene	7·1	2·18
Hexanol	5·9	2·60	Ethyl benzene	7·7	2·01
Formic acid	15·9	0·97	Propyl benzene	5·9	2·62
Acetic acid	13·3	1·16	Diphenyl	6·8	2·28
Propionic acid	9·9	1·56	n-Octane	6·0	2·58
i-Butyric acid	8·1	1·91	Mesitylene	6·7	2·31

Note: the group ($\mu/\rho D$) in the above table is evaluated for mixtures composed largely of air.

TABLE 8.2. *Diffusion coefficients of organic esters in air at 298 K*

Number of carbon atoms	D $(m^2/s \times 10^6)$	$\mu/\rho D$
2*	11·7	1·32
3†	9·7	1·59
4‡	8·6	1·79
5	7·8	1·97
6	6·9	2·23
7	6·5	2·37
8	5·7	2·70
9	4·9	3·14

Note. The group $(\mu/\rho D)$ in the above table is evaluated for mixtures composed largely of air.
  * Methyl formate.
  † Ethyl formate, methyl acetate.
  ‡ n-Propyl formate, i-propyl formate, ethyl acetate, methyl propionate.

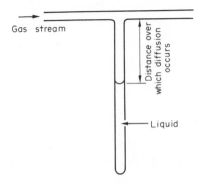

FIG. 8.2. Determination of diffusivities of vapours.

Example 8.1. The diffusivity of the vapour of a volatile liquid in air can be conveniently determined by Winkelmann's method in which liquid is contained in a narrow diameter vertical tube, maintained at a constant temperature, and an air stream is passed over the top of the tube sufficiently rapidly to ensure that the partial pressure of the vapour there remains approximately zero. On the assumption that the vapour is transferred from the surface of the liquid to the air stream by molecular diffusion, calculate the diffusivity of carbon tetrachloride vapour in air at 321 K and atmospheric pressure from the following experimental data:

Time from commencement of experiment			Liquid level (mm)
h	m	(ks)	
0	0	0·0	0·0
0	26	1·6	2·5
3	5	11·1	12·9
7	36	27·4	23·2
22	16	80·2	43·9
32	38	117·5	54·7
46	50	168·6	67·0
55	25	199·7	73·8
80	22	289·3	90·3
106	25	383·1	104·8

The vapour pressure of carbon tetrachloride at 321 K is 37·6 kN/m² and the density of the liquid is 1540 kg/m³. Take the kilogram molecular volume as 22·4 m³.

Solution. From equation 8.28 the rate of mass transfer is given by:

$$N'_A = D(C_A/L)(C_T/C_{Bm})$$

where C_A is the saturation concentration at the interface and L is the effective distance through which mass transfer is taking place. Considering the evaporation of the liquid:

$$N'_A = (\rho_L/M)\, dL/dt$$

where ρ_L is the density of the liquid.

Thus: $(\rho_L/M)\, dL/dt = D(C_A/L)(C_T/C_{Bm})$

Integrating and putting $L = L_0$ at $t = 0$

$$L^2 - L_0^2 = (2MD/\rho_L)(C_A C_T/C_{Bm})t$$

L_0 will not be measured accurately nor is the effective distance for diffusion, L, at time t. Accurate values of $(L - L_0)$ are available, hence:

$$(L - L_0)(L - L_0 + 2L_0) = (2MD/\rho_L)(C_A C_T/C_{Bm})t$$

or $t/(L - L_0) = (\rho_L/2MD)(C_{Bm}/C_A C_T)(L - L_0) + (\rho_L C_{Bm}/MDC_A C_T)L_0$

If s is the slope of a graph of $t/(L - L_0)$ against $(L - L_0)$:

$$s = (\rho_L C_{Bm})/(2MDC_A C_T) \quad \text{or} \quad D = (\rho_L C_{Bm})/(2MC_A C_T s)$$

From a plot of $t/(L - L_0)$ and $(L - L_0)$ as in Fig. 8.3:

$$s = 0·0310 \text{ ks/mm}^2 \quad \text{or} \quad 3·1 \times 10^{-7} \text{ s/m}^2$$
$$C_T = (1/22·4)(273/321) = 0·0380 \text{ kmol/m}^3$$
$$M = 154 \text{ kg/kmol}$$
$$C_A = (37·6/101·3) \times 0·0380 = 0·0141 \text{ kmol/m}^3$$
$$\rho_L = 1540 \text{ kg/m}^3$$

$$C_{B1} = 0·0380 \text{ kmol/m}^3, \qquad C_{B2} = [(101·3 - 37·6)/101·3] \times 0·0380 = 0·0238 \text{ kmol/m}^3$$

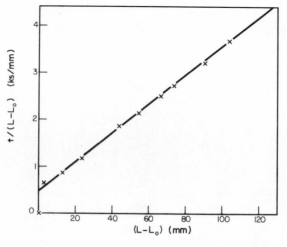

FIG. 8.3. $t/(L - L_0)$ versus $(L - L_0)$ for Example 8.1.

$$\therefore \quad C_{Bm} = (0\cdot0380 - 0\cdot0238)/\ln(0\cdot0380/0\cdot0238) = 0\cdot0303 \text{ kmol/m}^3$$

$$\therefore \quad D = (1540 \times 0\cdot0303)/(2 \times 154 \times 0\cdot0141 \times 0\cdot0380 \times 3\cdot1 \times 10^{-7})$$

$$= \underline{9\cdot12 \times 10^{-6} \text{ m}^2/\text{s}}$$

In many cases, the diffusivity D for the transfer of one gas in another may not be known and experimental determination may not be practicable. Many attempts have been made to express D in terms of other physical properties and the following empirical equation of Maxwell's, modified by Gilliland,[6] gives satisfactory agreement with the observed figures:

$$D = \frac{4\cdot3 T^{1\cdot5}\sqrt{(1/M_A) + (1/M_B)}}{P(V_A^{1/3} + V_B^{1/3})^2} \qquad (8.44)$$

where D is the diffusivity in cm^2/s, T the absolute temperature (K), M_A, M_B are the molecular weights of **A** and **B**, P is the total pressure in N/m^2, and V_A, V_B are the molecular volumes of **A** and **B**. The molecular volume is the volume in m^3 of 1 kmol of the material in the form of liquid at its boiling point, and is a measure of the volume occupied by the molecules themselves. It may not always be known, but an approximate value can be obtained by application of Kopp's law of additive volumes. Kopp gave a particular value for the equivalent atomic volume of each element. When the atomic volumes (Table 8.3) of the elements of the molecule in question are added in the appropriate proportions, the equivalent molecular volume is obtained approximately. There are certain exceptions to this rule, and additions or subtractions have to be made if the elements are combined in particular manners.

It will be noted from equation 8.44 that the diffusivity of a vapour is inversely proportional to the pressure and varies with the absolute temperature raised to the 1·5 power.

TABLE 8.3. *Atomic volumes m^3/kmol*[7]

Air	0·0299	Oxygen, double bonded	0·0074
Antimony	0·0242	Coupled to two other elements:	
Arsenic	0·0305	in aldehydes and ketones	0·0074
Bismuth	0·0480	in methyl esters	0·0091
Bromine	0·0270	in ethyl esters	0·0099
Carbon	0·0148	in higher esters and ethers	0·0110
Chlorine, terminal, as in R-Cl	0·0216	in acids	0·0120
medial, as in R-CHCl-R'	0·0246	in union with S, P, N	0·0083
Chromium	0·0274	Phosphorus	0·0270
Fluorine	0·0087	Silicon	0·0320
Germanium	0·0345	Sulphur	0·0256
Hydrogen	0·0037	Tin	0·0423
Nitrogen, in primary amines	0·0105	Titanium	0·0357
in secondary amines	0·0120	Vanadium	0·0320
		Zinc	0·0204

Notes
For three-membered ring, as in ethylene oxide, deduct 0·0060.
For four-membered ring, as in cyclobutane,   deduct 0·0085.
For five-membered ring, as in furane,   deduct 0·0115.
For six-membered ring, as in benzene, pyridine, deduct 0·0150.
For anthracene ring formation,   deduct 0·0475.

Example 8.2. Ammonia is absorbed in water from a mixture with air using a column operating at 101·3 kN/m^2 and 295 K. The resistance to transfer can be regarded as lying entirely within the gas phase. At a point in the column, the partial pressure of the ammonia is 7·0 kN/m^2. The back pressure at the water interface is negligible and the resistance to transfer can be regarded as lying in a stationary gas film 1 mm thick. If the diffusivity of ammonia in air is $2.36 \times 10^{-5} \, \text{m}^2/\text{s}$, what is the transfer rate per unit area at that point in the column? How would the rate of transfer be affected if the ammonia air mixture were compressed to double the pressure.

Solution. Concentration of ammonia in the gas

$$= (1/22.4)(101.3/101.3)(273/295)(7.0/101.3) = 0.00285 \, \text{kmol/m}^3$$

$$C_T/C_{Bm} = 101.3 \ln (101.3/94.3)/(101.3 - 94.3) = 1.036$$

From equation 8.22:

$$N_A' = -[D/(y_2 - y_1)](C_T/C_{Bm})(C_{A2} - C_{A1})$$
$$= -(2.36 \times 10^{-5}/1 \times 10^{-3}) \times 1.036 \times 0.00285$$
$$= 6.97 \times 10^{-5} \, \text{kmol/m}^2 \, \text{s}$$

If the pressure is increased to $202.6 \, \text{kN/m}^2$, the driving force is doubled, C_T/C_{Bm} is essentially unaltered, and the diffusivity, being inversely proportional to the pressure (equation 8.44), is halved. The mass transfer rate therefore remains the same.

8.3. DIFFUSION IN THE LIQUID PHASE

The rate of diffusion of a material in the liquid phase is represented by the same basic equation as for gas phase diffusion, i.e.

$$N_A = -D_L (dC_A/dy) \tag{8.45}$$

where D_L is the liquid phase diffusivity.

The integrated form of the equation for equimolecular counterdiffusion of a material, **A** is then of the same form as for gaseous diffusion, viz.

$$N_A = -D_L \frac{C_{A_2} - C_{A_1}}{y_2 - y_1} \tag{8.46}$$

where C_{A_1}, C_{A_2} are the molar concentrations of **A** at two points, y_1 and y_2.

For diffusion through a stagnant liquid **B**:

$$N_A' = -D_L \frac{C_{A_2} - C_{A_1}}{y_2 - y_1} \frac{C_T}{C_{Bm}} \tag{8.47}$$

where C_{A_1}, C_{A_2} are the molar concentrations of **A** at two points, $y_2 - y_1$ is the equivalent thickness of the liquid film through which diffusion is taking place; C_T is the total molar concentration, which may not remain constant throughout the whole of the fluid as was the case for diffusion in the gas phase; and C_{Bm} the logarithmic mean of the molar concentrations of **B** on each side of the liquid film.

Values of D_L for diffusion of various materials in water are given in Table 8.4.[4]

An empirical equation from which liquid phase diffusivities in dilute solutions can be calculated approximately is:

$$D_L = \frac{7.7 \times 10^{-16} T}{\mu (V^{1/3} - V_0^{1/3})} \tag{8.48}$$

where D_L is the diffusivity in m^2/s, T the temperature in K, μ the viscosity in Ns/m^2, $V_0 = 0.008$, 0.0149 and 0.0228 for diffusion in dilute solutions in water, methanol, and benzene respectively in m^3/kmol, and **V** the molecular volume, as in equation 8.44.

Equation 8.48 does not apply to electrolytes nor to concentrated solutions. Reid and Sherwood[8] discuss diffusion in electrolytes. Little information is available on diffusivities in

TABLE 8.4. *Diffusion coefficients in liquids at 293 K*

Solute	Solvent	D $(m^2/s \times 10^9)$	$(\mu/\rho D)*$
O_2	Water	1·80	558
CO_2	Water	1·50	670
N_2O	Water	1·51	665
NH_3	Water	1·76	570
Cl_2	Water	1·22	824
Br_2	Water	1·2	840
H_2	Water	5·13	196
N_2	Water	1·64	613
HCl	Water	2·64	381
H_2S	Water	1·41	712
H_2SO_4	Water	1·73	580
HNO_3	Water	2·6	390
Acetylene	Water	1·56	645
Acetic acid	Water	0·88	1140
Methanol	Water	1·28	785
Ethanol	Water	1·00	1005
Propanol	Water	0·87	1150
Butanol	Water	0·77	1310
Allyl alcohol	Water	0·93	1080
Phenol	Water	0·84	1200
Glycerol	Water	0·72	1400
Pyrogallol	Water	0·70	1440
Hydroquinone	Water	0·77	1300
Urea	Water	1·06	946
Resorcinol	Water	0·80	1260
Urethane	Water	0·92	1090
Lactose	Water	0·43	2340
Maltose	Water	0·43	2340
Glucose	Water	0·60	—
Mannitol	Water	0·58	1730
Raffinose	Water	0·37	2720
Sucrose	Water	0·45	2230
Sodium chloride	Water	1·35	745
Sodium hydroxide	Water	1·51	665
CO_2	Ethanol	3·4	445
Phenol	Ethanol	0·8	1900
Chloroform	Ethanol	1·23	1230
Phenol	Benzene	1·54	479
Chloroform	Benzene	2.11	350
Acetic acid	Benzene	1·92	384
Ethylene dichloride	Benzene	2·45	301

*Based on $\mu/\rho = 1·005 \times 10^{-6}$ m²/s for water, $7·37 \times 10^{-7}$ for benzene, and $1·511 \times 10^{-6}$ for ethanol, all at 293 K; applies only for dilute solutions.

concentrated solutions but it appears that, for ideal mixtures, the product μD_L is a linear function of the molar concentration.

8.4. UNSTEADY STATE MASS TRANSFER

8.4.1. Equimolecular Counterdiffusion

In the mass transfer problems considered so far, the process has been steady state in character. In many chemical engineering applications of mass transfer the process is time dependent. It is therefore necessary to develop the appropriate equations.

For conditions of equimolecular counterdiffusion the equation for unidirectional mass transfer can be obtained in the same manner as the corresponding equation for heat transfer, i.e.

$$\frac{\partial C_A}{\partial t} = D\frac{\partial^2 C_A}{\partial y^2} \qquad \text{(See equation 7.25)} \quad (8.49)$$

The equation for the three-dimensional transfer is:

$$\frac{\partial C_A}{\partial t} = D\left(\frac{\partial^2 C_A}{\partial x^2} + \frac{\partial^2 C_A}{\partial y^2} + \frac{\partial^2 C_A}{\partial z^2}\right) \qquad \text{(See equation 7.24)} \quad (8.50)$$

The solution to equation 8.49 can be obtained by the Laplace transform method already described in connection with unsteady state thermal conduction. The form of solution will be dependent on the boundary conditions which must be specified before the solution can be obtained.

For the case of a fluid, semi-infinite in extent, and initially with a constant concentration of A, C_{Ao}, through the whole depth, suppose that the free surface concentration is suddenly changed (at $t = 0$) to some new constant value C_{Ai} and is maintained at that value. This problem will have important applications later, where the *penetration* theory is discussed.

The boundary conditions applying to equation 8.49 are now as follows:

$$t = 0 \qquad 0 < y < \infty \qquad C_A = C_{Ao}$$
$$t > 0 \qquad y = 0 \qquad C_A = C_{Ai}$$
$$t > 0 \qquad y = \infty \qquad C_A = C_{Ao}$$

It is useful to define a new variable $C' = C_A - C_{Ao}$, where C' denotes the excess in concentration above the initial uniform value C_{Ao}. Substituting into equation 8.49:

$$\frac{\partial C'}{\partial t} = D\frac{\partial^2 C'}{\partial y^2} \tag{8.51}$$

By definition, the Laplace transform $\bar{C}'$ of C' is given by:

$$\bar{C}' = \int_0^\infty e^{-pt} C' \, dt \tag{8.52}$$

Then:

$$\overline{\frac{\partial C'}{\partial t}} = \int_0^\infty e^{-pt} \frac{\partial C'}{\partial t} \, dt$$

$$= [e^{-pt} C']_0^\infty + p\int_0^\infty e^{-pt} C' \, dt$$

$$= p\bar{C}' \tag{8.53}$$

Since the Laplace transform operation is independent of y:

$$\overline{\frac{\partial^2 C'}{\partial y^2}} = \frac{\partial^2 \bar{C}'}{\partial y^2} \tag{8.54}$$

Thus, taking Laplace transforms of both sides of equation 8.51:

$$p\bar{C}' = D\frac{\partial^2 \bar{C}'}{\partial y^2}$$

$$\therefore \qquad \frac{\partial^2 \bar{C}'}{\partial y^2} - \frac{p}{D}\bar{C}' = 0$$

Thus

$$\bar{C}' = B_1 e^{\sqrt{(p/D)}y} + B_2 e^{-\sqrt{(p/D)}y} \tag{8.55}$$

When $\qquad y = 0, \quad C_A = C_{Ai}, \quad \text{and} \quad C' = C_{Ai} - C_{Ao}$

and when $\qquad y = \infty, \quad C_A = C_{Ao}, \quad \text{and} \quad C' = 0$

Hence $\qquad\qquad\qquad B_1 = 0$

and $\qquad\qquad\qquad \bar{C}' = B_2\, e^{-\sqrt{(p/D)}\,y}$ (8.56)

Now $\qquad\qquad B_2 = \int_0^\infty (C_{Ai} - C_{Ao})\, e^{-pt}\, dt$

$$= \frac{1}{p}(C_{Ai} - C_{Ao})$$

Thus $\qquad\qquad \bar{C}' = \frac{1}{p}(C_{Ai} - C_{Ao})\, e^{-\sqrt{(p/D)}\,y}$ (8.57)

Taking the inverse of the transform (see Volume 3,[8]):

$$C' = C_A - C_{Ao} = (C_{Ai} - C_{Ao})\,\operatorname{erfc}\frac{y}{2\sqrt{Dt}}$$

or $\qquad\qquad \dfrac{C_A - C_{Ao}}{C_{Ai} - C_{Ao}} = \operatorname{erfc}\dfrac{y}{2\sqrt{Dt}}$ (8.58)

This expression gives concentration C_A as a function of position y and of time t. The concentration gradient is then obtained by differentiation with respect to y.

Thus $\qquad \dfrac{1}{C_{Ai} - C_{Ao}}\dfrac{\partial C_A}{\partial y} = \dfrac{\partial}{\partial y}\left[\dfrac{2}{\sqrt{\pi}}\int_{(y/2\sqrt{Dt})}^{\infty} e^{-y^2/4Dt}\, d\!\left(\dfrac{y}{2\sqrt{Dt}}\right)\right]$

$\therefore \qquad \dfrac{\partial C_A}{\partial y} = (C_{Ai} - C_{Ao})\dfrac{2}{\sqrt{\pi}}\dfrac{1}{2\sqrt{Dt}}(e^{-y^2/4Dt})$

$$= -(C_{Ai} - C_{Ao})\dfrac{1}{\sqrt{\pi Dt}}e^{-y^2/4Dt}$$ (8.59)

The mass transfer rate at the surface is then given by:

$$(N_A)_{y=0} = -D\left(\frac{\partial C_A}{\partial y}\right)_{y=0}$$

$$= (C_{Ai} - C_{Ao})\sqrt{\frac{D}{\pi t}}$$ (8.60)

This equation which is of basic importance in the *penetration theory* of mass transfer across a phase boundary eventually leads to equation 8.76 for the mass transfer rate.

8.4.2. Mass Transfer with Bulk Flow

It will be remembered that the mass transfer rate in a steady-state absorption process, in which mass transfer is taking place through a second gas which is not undergoing transfer, is greater than that for equimolecular counterdiffusion because of the effects of bulk flow (compare with equations 8.1 and 8.20). Again, in an unsteady state process bulk flow will give rise to an increase in the mass transfer rate though in this case the transfer of the insoluble component will be zero only at the gas–liquid interface. Elsewhere, transfer will be necessary in order to enable its concentration to change with time. This problem is important in a process of gas absorption where the concentration of the soluble gas is high and where there has either been insufficient time for steady concentration gradients to be established or surface replacement is being effected at intervals by the action of, say, the packing elements in a column.

Taking the interface to correspond with $y = 0$, the molar transfer rate of **A** and **B** through a unit plane a distance y from the interface can be expressed as follows:

	Diffusional flux	Flux due to bulk flow	Total flux	
A	$-D\dfrac{\partial C_A}{\partial y}$	$u_F C_A$	$N'_A = -D\dfrac{\partial C_A}{\partial y} + u_F C_A$	(8.61)
B	$-D\dfrac{\partial C_B}{\partial y}$	$u_F C_B$	$N'_B = -D\dfrac{\partial C_B}{\partial y} + u_F C_B$	(8.62)

where u_F is the bulk flow velocity (see equation 8.24).

A material balance over an element of unit area situated between the y and $y + dy$ planes then gives:

$$\frac{\partial C_A}{\partial t} = -\frac{\partial}{\partial y}\left(-D\frac{\partial C_A}{\partial y} + u_F C_A\right) = D\frac{\partial^2 C_A}{\partial y^2} - \frac{\partial}{\partial y}(u_F C_A) \tag{8.63}$$

and

$$\frac{\partial C_B}{\partial t} = -\frac{\partial}{\partial y}\left(-D\frac{\partial C_B}{\partial y} + u_F C_B\right) = -D\frac{\partial^2 C_A}{\partial y^2} - \frac{\partial}{\partial y}[u_F(C_T - C_A)] \tag{8.64}$$

Now since $C_A + C_B = C_T$, the constant total molar concentration:

$$\frac{\partial C_A}{\partial t} + \frac{\partial C_B}{\partial t} = 0 \tag{8.65}$$

Adding equations 8.63 and 8.64:
$$\frac{\partial}{\partial y}(u_F C_T) = 0$$

i.e.
$$C_T \frac{\partial u_F}{\partial y} = 0$$

i.e.
$$u_F \neq f(y) \tag{8.66}$$

Now the value of u_F is obtained from the boundary condition:

$$(N'_B)_{y=0} = 0 \tag{8.67}$$

Thus, from equation 8.64:
$$-D\left(\frac{\partial C_B}{\partial y}\right)_{y=0} + u_F(C_B)_{y=0} = 0$$

$\therefore$
$$u_F = \frac{D\left(\dfrac{\partial C_B}{\partial y}\right)_{y=0}}{(C_B)_{y=0}} = \frac{-D}{(C_T - C_A)_{y=0}}\left(\frac{\partial C_A}{\partial y}\right)_{y=0} \tag{8.68}$$

Thus
$$N'_A = -D\frac{\partial C_A}{\partial y} - \frac{DC_A}{(C_T - C_A)_{y=0}}\left(\frac{\partial C_A}{\partial y}\right)_{y=0}$$

$$= -D\left[\frac{\partial C_A}{\partial y} + \frac{C_A}{(C_T - C_A)_{y=0}}\left(\frac{\partial C_A}{\partial y}\right)_{y=0}\right] \tag{8.69}$$

and
$$\frac{\partial C_A}{\partial t} = D\frac{\partial^2 C_A}{\partial y^2} + \frac{D}{(C_T - C_A)_{y=0}}\left(\frac{\partial C_A}{\partial y}\right)_{y=0}\frac{\partial C_A}{\partial y}$$

i.e.
$$\frac{\partial C_A}{\partial t} = D\left[\frac{\partial^2 C_A}{\partial y^2} + \frac{1}{(C_T - C_A)_{y=0}}\left(\frac{\partial C_A}{\partial y}\right)_{y=0}\frac{\partial C_A}{\partial y}\right] \tag{8.70}$$

Thus, at the interface, $y = 0$:

$$(N_A')_{y=0} = -D\left(\frac{C_T}{C_T - C_A}\frac{\partial C_A}{\partial y}\right)_{y=0} \tag{8.71}$$

and

$$\left(\frac{\partial C_A}{\partial t}\right)_{y=0} = D\left[\left(\frac{\partial^2 C_A}{\partial y^2}\right)_{y=0} + \frac{1}{(C_T - C_A)_{y=0}}\left(\frac{\partial C_A}{\partial y}\right)^2_{y=0}\right] \tag{8.72}$$

The solution of equation 8.70 is more difficult than that obtained previously (8.49) for equimolecular counterdiffusion. It has been solved by Arnold[10] for evaporation from a surface.

8.5. MASS TRANSFER ACROSS A PHASE BOUNDARY

The theoretical treatment which has been developed in the preceding sections has related to mass transfer within a single phase in which no discontinuities exist. In most of the important applications of mass transfer, however, material is transferred across a phase boundary. Thus, in distillation a vapour and liquid are brought into contact in the fractionating column and the more volatile material is transferred from the liquid to the vapour while the less volatile constituent is transferred in the opposite direction; this is an example of equimolecular counterdiffusion. In gas absorption the soluble gas diffuses to the surface, dissolves in the liquid, and then passes into the bulk of the liquid; in this case, the carrier gas is not transferred. In both of these examples, one phase is a liquid and the other a gas. However, in liquid–liquid extraction, a solute is transferred from one liquid solvent to another across a phase boundary, and in the dissolution of a crystal the solute is transferred from a solid to a liquid.

Each of the above processes is characterised by a transference of material across an interface. Because no material accumulates there, the rate of transfer on each side of the interface must be the same, and therefore the concentration gradients automatically adjust themselves so that they are proportional to the resistance to transfer in the particular phase. Furthermore, if there is no resistance to transfer at the interface, the concentrations on each side will be related to each other by the phase equilibrium relationship. The existence or otherwise of a resistance to transfer at the phase boundary is the subject of conflicting opinions.[11, 12] It appears likely that it is not high, except in the case of crystallisation, and in the subsequent treatment equilibrium between the phases will be assumed to exist at the interface.

The mass transfer rate between two fluid phases will depend on the physical properties of the two phases, the concentration difference, the interfacial area, and the degree of turbulence. Mass transfer equipment is therefore designed to give a large area of contact between the phases and to promote turbulence in each of the fluids. In the majority of plants, the two phases flow continuously in a countercurrent manner. In a steady state process, therefore, though the composition of each element of fluid is changing as it passes through the equipment, conditions at any given point do not change with time. In most industrial equipment, the flow pattern is so complex that it is not capable of expression in mathematical terms and the interfacial area is not known precisely.

A number of mechanisms have been suggested to represent conditions in the region of the phase boundary. The earliest of these is the *two-film* theory propounded by Whitman[13] in 1923. He suggested that the resistance to transfer in each phase could be regarded as lying in a thin film close to the interface. The transfer across these films is regarded as a steady state process of molecular diffusion following equations of the type of equation 8.46. The turbulence in the bulk fluid is considered to die out at the interface of the films. In 1935 Higbie[14] suggested that the transfer process was largely attributable to fresh material being brought by the eddies to the interface, where a process of unsteady state transfer took place for a fixed period at the freshly exposed surface. This theory is generally known as the penetration theory. Danckwerts[15] has since suggested a modification of this theory and considers that the material brought to the surface will remain there for varying periods of time. He later discusses the random age distribution of such elements from which the transfer is by an unsteady state process to the second phase. Subsequently, Toor and Marchello[16] have proposed a more general theory, the film-penetration theory, and have shown that each of the earlier theories is a particular limiting case of their own. A number of other theoretical treatments have also been given, including those of Bakowski[17] and Kishinevskij.[18] The two-film theory and the penetration theory will now be considered, followed by an examination of the film-penetration theory.

8.5.1. The Two-film Theory

The *two-film* theory of Whitman[13] was the first serious attempt to represent conditions occurring when material is transferred from one fluid stream to another. Although it does not closely reproduce the conditions in most practical equipment, it gives expressions which can be applied to the experimental data which are generally available, and for that reason it is still extensively used.

In this theory, it is assumed that turbulence dies out at the interface and that a laminar layer exists in each of the two fluids. Outside the laminar layer, turbulent eddies supplement the action caused by the random movement of the molecules, and the resistance to transfer becomes progressively smaller. For equimolecular counterdiffusion the concentration gradient is therefore linear close to the interface, and gradually becomes less at greater distances as shown in Fig. 8.4 by the full lines ABC and DEF. The basis of the theory is the assumption

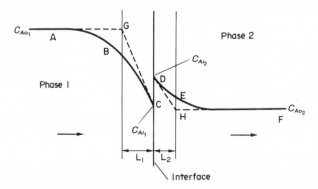

FIG. 8.4. Two-film theory.

that the zones in which the resistance to transfer lies can be replaced by two hypothetical layers, one on each side of the interface, in which the transfer is entirely by molecular diffusion. The concentration gradient is therefore linear in each of these layers and zero outside. The broken lines AGC and DHF indicate the hypothetical concentration distributions, and the thicknesses of the two films are L_1 and L_2. Equilibrium is assumed to exist at the interface and therefore the relative positions of the points C and D are determined by the equilibrium relation between the phases.

The mass transfer is treated as a steady state process and therefore the theory can be applied only if the time taken for the concentration gradients to become established is very small compared with the time of transfer, or if the capacity of the films is negligible.

From equation 8.15 the rate of transfer per unit area in terms of the two-film theory for equimolecular counterdiffusion is given for the first phase as:

$$N_A = \frac{D_1}{L_1}(C_{Ao_1} - C_{Ai_1}) = k_1'(C_{Ao_1} - C_{Ai_1}) \qquad (8.73)$$

where L_1 is the thickness of the film, C_{Ao_1} the molar concentration outside the film, and C_{Ai_1} the molar concentration at the interface.

For the second phase, with the same notation,

$$N_A = \frac{D_2}{L_2}(C_{Ai_2} - C_{Ao_2}) = k_2'(C_{Ai_2} - C_{Ao_2}) \qquad (8.74)$$

Because material does not accumulate at the interface, the two rates of transfer must be the same and:

$$\frac{k_1'}{k_2'} = \frac{C_{Ai_2} - C_{Ao_2}}{C_{Ao_1} - C_{Ai_1}} \qquad (8.75)$$

The flow conditions are too complex for the film thicknesses to be evaluated, but they are progressively decreased as the turbulence of the fluid is increased.

The theory can be applied equally well to the diffusion process which occurs in gas absorption, where there is no net transference of the carrier gas. In this case, the concentration gradient in the film will not be linear, and the mass transfer rate will be greater by a factor of C_T/C_{Bm}.

8.5.2. The Penetration Theory

The penetration theory was propounded in 1935 by Higbie[14] who was investigating whether or not a resistance to transfer existed at the interface when a pure gas was absorbed in a liquid. He carried out his experiments by allowing a slug-like bubble of carbon dioxide to rise through a vertical column of water in a 3 mm diameter glass tube. As the bubble rose, the displaced liquid ran back as a thin film between the bubble and the tube. Higbie assumed that each element of surface in this liquid was exposed to the gas for the time taken for the gas bubble to pass it; that is for the time given by the quotient of the bubble length and its velocity. He supposed that during this short period, which varied between 0·01 and 0·1 s in his experiments, absorption took place as the result of unsteady state molecular diffusion into the liquid; for the purposes of calculation he regarded the liquid as infinite in depth because the time of exposure was so short.

The way in which the concentration gradient builds up as a result of exposing a liquid—initially pure—to the action of a soluble gas is shown in Fig. 8.5 which is based on Higbie's calculations. The percentage saturation of the liquid is plotted against distance from the surface for a number of exposure times in arbitrary units. Initially only the surface layer contains solute and the concentration changes abruptly from 100 per cent to 0 per cent at the surface. For progressively longer exposure times the concentration profile develops as shown, until after an infinite time the whole of the liquid becomes saturated. The shape of the profiles is such that at any time the effective depth of liquid which contains an appreciable concentration of solute can be specified, and hence the theory is referred to as the penetration theory. If this depth of penetration is less than the total depth of liquid, no significant error is introduced by assuming that the total depth is infinite.

The work of Higbie laid the basis of the penetration theory in which it is assumed that the eddies in the fluid bring an element of fluid to the interface where it is exposed to the second phase for a definite interval of time, after which the surface element is mixed with the bulk again. Thus, fluid whose initial composition corresponds with that of the bulk fluid remote

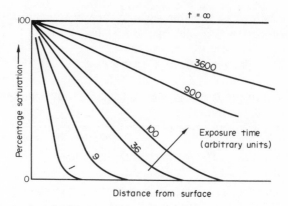

FIG. 8.5. Penetration of solute into a solvent.

from the interface is suddenly exposed to the second phase. It is assumed that equilibrium is immediately attained by the surface layers and that a process of unsteady state molecular diffusion then occurs and that the element is remixed after a fixed interval of time. In the calculation, the depth of the liquid element is assumed to be infinite and this is justifiable if the time of exposure is sufficiently short for penetration to be confined to the surface layers. Throughout, the existence of velocity gradients within the fluids is ignored and the fluid at all depths is assumed to be moving at the same rate as the interface.

Diffusion of solute A away from the interface (Y-direction) is thus given by equation 8.49:

$$\frac{\partial C_A}{\partial t} = D \frac{\partial^2 C_A}{\partial y^2} \qquad \text{(equation 8.49)}$$

for conditions of equimolecular counterdiffusion, or for an absorption process when the concentrations of diffusing materials are low. The following boundary conditions apply:

$t = 0 \quad 0 < y < \infty \quad C_A = C_{Ao}$ where C_{Ao} is the concentration in the body of the phase.

$t > 0 \quad y = 0 \quad\quad C_A = C_{Ai}$ the equilibrium value at the interface.

$t > 0 \quad y = \infty \quad\quad C_A = C_{Ao}$

Equation 8.49, with the same boundary conditions, was solved earlier to give the mass transfer rate at the interface as:

$$N_A = (C_{Ai} - C_{Ao})\sqrt{D/\pi t} \qquad \text{(equation 8.60)}$$

Equation 8.60 gives the instantaneous transfer rate when the surface element under consideration has an age t. If the element is exposed for a time t_e, the average rate of transfer is given by:

$$(N_A)_{av} = (C_{Ai} - C_{Ao}) \sqrt{\frac{D}{\pi}} \frac{1}{t_e} \int_0^{t_e} \frac{dt}{\sqrt{t}}$$

$$= 2(C_{Ai} - C_{Ao})\sqrt{D/\pi t_e} \qquad (8.76)$$

Thus, the shorter the time of exposure the greater is the rate of mass transfer. No precise value can be assigned to t_e in any industrial equipment, but its value will clearly become less as the degree of agitation of the fluid is increased.

If it is assumed that each element resides for the same time interval t_e in the surface, equation 8.76 gives the overall mean rate of transfer. It will be noted that the rate is a linear function of the driving force expressed as a concentration difference, as in the two-film theory, but that it is proportional to the diffusivity raised to the power of 0·5 instead of unity.

Example 8.3. In an experimental wetted wall column, pure carbon dioxide is absorbed in water. The mass transfer rate is calculated using the penetration theory, application of which is limited by the fact that the concentration should not reach more than 1 per cent of the saturation value at a depth below the surface at which the velocity is 95 per cent of the surface velocity. What is the maximum length of column to which the theory can be applied if the flowrate of water is $3 \text{ cm}^3/\text{s}$ per cm of perimeter?

$$\text{Viscosity of water} \qquad\qquad\qquad = 10^{-3} \text{ N s/m}^2$$

$$\text{Diffusivity of carbon dioxide in water} = 1 \cdot 5 \times 10^{-9} \text{ m}^2/\text{s}$$

Solution. For flow of a vertical film of fluid the mean velocity of flow is governed by equation 7.99:

$$u_m = \rho g s^2 / 3\mu$$

where s is the thickness of the film.

Flowrate per unit perimeter $= \rho g s^3 / 3\mu = 3 \times 10^{-4} \text{ m}^2/\text{s}$

Thus
$$s = \left(\frac{3 \times 10^{-4} \times 10^{-3} \times 3}{1000 \times 9 \cdot 81} \right)^{1/3}$$

$$= 4 \cdot 51 \times 10^{-4} \text{ m}$$

The velocity u_y at a distance y' from the column wall is given by equation 7.98 (using y' in place of y) as:

$$u_y = \rho g (s y' - \tfrac{1}{2} y'^2)/\mu$$

The free surface velocity u_s is given by putting $y' = s$:

$$u_s = \rho g s^2 / 2\mu$$

$$\therefore \qquad u_y / u_s = 2 \left(\frac{y'}{s} \right) - \left(\frac{y'}{s} \right)^2 = 1 - \left(1 - \frac{y'}{s} \right)^2$$

When $u_y / u_s = 0 \cdot 95$ (velocity is 95 per cent of surface velocity)

$$1 - y'/s = 0 \cdot 224$$

$$\therefore \qquad y = s - y' = 1 \cdot 010 \times 10^{-4} \text{ m (distance below free surface)}$$

Equation 8.58 gives the relationship between concentration C_A, time and depth as:

$$(C_A - C_{Ao})/(C_{Ai} - C_{Ao}) = \text{erfc} \, (y/2\sqrt{Dt})$$

The time at which concentration reaches $0 \cdot 01$ of saturation value at a depth of $1 \cdot 010 \times 10^{-4}$ m is given by:

$$0 \cdot 01 = \text{erfc} \, (1 \cdot 010 \times 10^{-4}/2\sqrt{1 \cdot 5 \times 10^{-9} t})$$

Thus
$$\text{erf} \, (1 \cdot 305/\sqrt{t}) = 0 \cdot 99$$

Using tables of error function (Volume 3–Appendix):

$$1 \cdot 305/\sqrt{t} = 1 \cdot 822$$

$$t = 0 \cdot 51 s$$

Surface velocity
$$u_s = \rho g s^2 / 2\mu$$

$$= [1000 \times 9 \cdot 81 \times (4 \cdot 51 \times 10^{-4})^2]/2 \times 10^{-3}$$

$$= 1 \text{ m/s}$$

Maximum length of column $\qquad = 1 \times 0 \cdot 51 = \underline{\underline{0 \cdot 51 \text{ m}}}$

Random Surface Renewal

Danckwerts[15] suggested that each element of surface would not be exposed for the same time, but that a random distribution of ages would exist. He assumed that the probability of any element of surface becoming destroyed and mixed with the bulk of the fluid was independent of the age of the element. On this basis he calculated the age distribution of the surface elements as follows:

Suppose that the rate of production of fresh surface per unit total area of surface is s, and that s is independent of the age of the element in question.

The area of surface of age between t and $t + dt$ will be a function of t and may be written as $f(t)dt$. This will be equal to the area passing in time dt from the age range $[(t - dt)$ to $t]$ to the age range $[t$ to $(t + dt)]$. Further, this in turn will be equal to the area in the age group $[(t - dt)$ to $t]$, less that replaced by fresh surface in time dt:

i.e.
$$f(t)dt = f(t - dt)dt - [f(t - dt)dt]s \, dt \tag{8.77}$$

$$\therefore \qquad \frac{f(t) - f(t - dt)}{dt} = - s f(t - dt)$$

$$\therefore \qquad f'(t) + s f(t) = 0 \tag{8.78}$$

$$\therefore \qquad e^{st} f(t) = \text{constant}$$

$$\therefore \qquad f(t) = \text{constant } e^{-st}$$

Now the total area of surface considered was unity:

$$\therefore \qquad \int_{0}^{\infty} f(t) \, dt = 1 \tag{8.79}$$

$$\therefore \qquad \text{constant} \times \frac{1}{s} = 1$$

$$\therefore \qquad f(t) = s \, e^{-st} \tag{8.80}$$

Thus the age distribution of the surface is of an exponential form. Now from equation 8.60 the mass transfer rate at unit area of surface of age t is given by:

$$N_A = (C_{Ai} - C_{Ao})\sqrt{D/\pi t} \qquad \text{(equation 8.60)}$$

Thus, the overall rate of transfer per unit area when the surface is renewed in a random manner is:

$$N_A = (C_{Ai} - C_{Ao}) \int_{t=0}^{t=\infty} \sqrt{\frac{D}{\pi t}} \, s \, e^{-st} \, dt$$

$$= (C_{Ai} - C_{Ao})s \sqrt{\frac{D}{\pi}} \int_{0}^{\infty} t^{-1/2} \, e^{-st} \, dt \tag{8.81}$$

Put $st = \beta^2$. Then $s \, dt = 2\beta \, d\beta$

$$\therefore \qquad N_A = (C_{Ai} - C_{Ao})s \sqrt{\frac{D}{\pi}} \frac{2}{s^{1/2}} \int_{0}^{\infty} e^{-\beta^2} \, d\beta$$

$$= (C_{Ai} - C_{Ao})\sqrt{Ds} \tag{8.82}$$

Equation 8.82 might be expected to underestimate the mass transfer rate because, in any practical equipment, there will be a finite upper limit to the age of any surface element. However, the proportion of the surface in the older age group is very small and the overall rate is largely unaffected. It will be seen that the mass transfer rate is again proportional to the concentration difference and to the square root of the diffusivity. The numerical value of s is difficult to estimate, but will clearly increase as the fluid becomes more turbulent. In a packed column, s will be of the same order as the velocity of the liquid flowing over the packing, divided by the length of the packing.

Varying Interface Composition

In the preceding sections the penetration theory was used to calculate the rate of mass transfer across an interface for conditions where the concentration C_{Ai} of solute A in the interfacial layers ($y = 0$) remained constant throughout the process. When there is no resistance to mass transfer in the other phase, for instance when this consists of pure solute A, there will be no concentration gradient in that phase and the composition at the interface will therefore at all times be the same as the bulk composition. Since the composition of the interfacial layers of the *penetration* phase is determined by the phase equilibrium relationship, it, too, will remain constant and the conditions necessary for the penetration theory to apply will hold. If, however, the other phase offers a significant resistance to transfer this condition will not, in general, be fulfilled.

As an example, suppose that in phase 1 there is a constant finite resistance to mass transfer which can in effect be represented as a resistance in a laminar film, and in phase 2 the penetration model is applicable. Immediately after surface renewal has taken place, the mass transfer resistance in phase 2 will be negligible and therefore the whole of the concentration driving force will lie across the film in phase 1. The interface compositions will therefore correspond to the bulk value in phase 2 (the penetration phase). As the time of exposure increases, the resistance to mass transfer in phase 2 will progressively increase and an increasing proportion of the total driving force will lie across this phase. Thus the interface composition, initially determined by the bulk composition in phase 2 (the penetration phase) will progressively approach the bulk composition in phase 1 as the time of exposure increases.

The mass transfer rate per unit area across the film at any time t is given by:

$$N_A = -\frac{D_f}{L}(C'_{Ai} - C'_{Ao}) \tag{8.83}$$

where D_f is the diffusivity in the film of thickness L, and C'_{Ai} and C'_{Ao} are the concentrations of A at the interface and in the bulk.

The capacity of the film will be assumed to be small so that the hold-up of solute is negligible. If Henry's law is applicable, the interface concentration in the second (penetration) phase is given by:

$$C_{Ai} = \frac{1}{\mathcal{H}} C'_{Ai} \tag{8.84}$$

Here C_A is used to denote concentration in the penetration phase. Thus, by substitution, the interface composition C_{Ai} is obtained in terms of the mass transfer rate N_A.

However, N_A must also be given by applying Fick's law to the interfacial layers of phase 2 (the penetration phase).

Thus $$N_A = -D\left(\frac{\partial C_A}{\partial y}\right)_{y=0} \quad \text{(from equation 8.1)} \tag{8.85}$$

Combining equations 8.83, 8.84, and 8.85:

$$C_{Ai} = \frac{C'_{Ao}}{\mathcal{H}} + \frac{DL}{D_f \mathcal{H}}\left(\frac{\partial C_A}{\partial y}\right)_{y=0} \tag{8.86}$$

Then, using this value of C_{Ai} (in place of the constant value used previously) in the penetration theory, the instantaneous mass transfer rate per unit area at time t can be shown to be given by:

$$N_A = (C'_{Ao} - \mathcal{H}C_{Ao}) \frac{D_f}{L} \left(e^{[(D_f^2 \mathcal{H}^2)/(DL^2)]t} \operatorname{erfc} \sqrt{\frac{D_f^2 \mathcal{H}^2 t}{DL^2}} \right) \tag{8.87}$$

Average rates of mass transfer can then be obtained, as previously, by using either the Higbie or the Danckwerts model for surface renewal.

Penetration Model with Laminar Film at Interface

Harriott[19] suggested that, as a result of the effects of interfacial tension, the layers of fluid in the immediate vicinity of the interface would frequently be unaffected by the mixing process postulated in the penetration theory. There would then be a thin laminar layer unaffected by the mixing process and offering a constant resistance to mass transfer. The overall resistance can then be calculated in a manner similar to that used in the previous section where the total resistance to transfer was made up of two components—a film resistance in one phase and a penetration model resistance in the other. In equation 8.87, it will be necessary to put the Henry's law constant equal to unity and the diffusivity D_f in the film equal to that in the remainder of the fluid D. The driving force will be $C_{Ai} - C_{Ao}$ in place of $C'_{Ao} - \mathcal{H}C_{Ao}$. Thus the mass transfer rate at time t is given by:

$$N_A = (C_{Ai} - C_{Ao}) \frac{D}{L} \left(e^{Dt/L^2} \operatorname{erfc} \sqrt{\frac{Dt}{L^2}} \right) \tag{8.88}$$

The average transfer rate according to the Higbie model for surface age distribution then becomes:

$$(N_A)_{av} = (C_{Ai} - C_{Ao}) \left[\frac{L}{t_e} (e^{Dt_e/L^2} \operatorname{erfc} \sqrt{Dt_e/L^2} - 1) + 2\sqrt{\frac{D}{\pi t_e}} \right] \tag{8.89}$$

Using the Danckwerts model:

$$(N_A)_{av} = (C_{Ai} - C_{Ao}) \frac{D}{L} \left(1 + \sqrt{\frac{D}{L^2 s}} \right)^{-1} \tag{8.90}$$

8.5.3. The Film-penetration Theory

A theory which incorporates some of the principles of both the two-film theory and the penetration theory has subsequently been suggested by Toor and Marchello.[16] The whole of the resistance to transfer is regarded as lying within a laminar film at the interface, as in the two-film theory, but the mass transfer is regarded as an unsteady state process. It is assumed that fresh surface is formed at intervals from fluid which is brought from the bulk of the fluid to the interface by the action of the eddy currents. Mass transfer then takes place as in the penetration theory, with the exception that the resistance is confined to the finite film, and material which traverses the film is immediately completely mixed with the bulk of the fluid. For short times of exposure, when none of the diffusing material has reached the far side of the layer, the process is identical to that postulated in the penetration theory. For prolonged periods of exposure when a steady concentration gradient has developed, conditions are similar to those considered in the two-film theory.

The mass transfer process is again governed by equation 8.49, but the third boundary condition is applied at $y = L$, the film thickness, and not at $y = \infty$.

Thus
$$\overline{C'} = B_1 e^{\sqrt{(p/D)}y} + B_2 e^{-\sqrt{(p/D)}y} \qquad \text{(equation 8.55)}$$

$$t > 0 \quad y = 0 \quad C_A = C_{Ai} \quad C' = C_{Ai} - C_{Ao} \quad \overline{C'} = (1/p)C'_i$$

$$t > 0 \quad y = L \quad C_A = C_{Ao} \quad C' = 0 \quad \overline{C'} = 0$$

Thus
$$0 = B_1 e^{\sqrt{(p/D)}L} + B_2 e^{-\sqrt{(p/D)}L}$$

$$\therefore \quad \frac{C'_i}{p} = B_1 + B_2$$

$$\therefore \quad B_1 = -B_2 e^{-2\sqrt{(p/D)}L}$$

$$\therefore \quad B_2 = \frac{C'_i}{p}(1 - e^{-2\sqrt{(p/D)}L})^{-1}$$

$$B_1 = -\frac{C'_i}{p} e^{-2\sqrt{(p/D)}L}(1 - e^{-2\sqrt{(p/D)}L})^{-1}$$

$$\therefore \quad \overline{C'} = \frac{C'_i}{p}(1 - e^{-2\sqrt{(p/D)}L})^{-1}(e^{-\sqrt{(p/D)}y} - e^{-\sqrt{(p/D)}(2L-y)}) \qquad (8.91)$$

Since there is no inverse of equation 8.91 in its present form, it is necessary to expand using the binomial theorem to give:

$$\overline{C'} = \frac{C'_i}{p}(e^{-\sqrt{(p/D)}y} - e^{-\sqrt{(p/D)}(2L-y)})\sum_{n=0}^{n=\infty} e^{-2n\sqrt{(p/D)}L}$$

$$= C'_i \left[\sum_{n=0}^{n=\infty}\frac{1}{p} e^{-\sqrt{(p/D)}(2nL+y)} - \sum_{n=0}^{n=\infty}\frac{1}{p} e^{-\sqrt{(p/D)}\{2(n+1)L-y\}}\right] \qquad (8.92)$$

On inversion of equation 8.92:

$$\frac{C_A - C_{Ao}}{C_{Ai} - C_{Ao}} = \frac{C'}{C'_i} = \sum_{n=0}^{n=\infty}\operatorname{erfc}\frac{(2nL+y)}{2\sqrt{Dt}} - \sum_{n=0}^{n=\infty}\operatorname{erfc}\frac{2(n+1)L-y}{2\sqrt{Dt}} \qquad (8.93)$$

Differentiating with respect to y:

$$\frac{1}{C_{Ai} - C_{Ao}}\frac{\partial C_A}{\partial y} = \sum_{n=0}^{n=\infty} -\frac{2}{\sqrt{\pi}}\frac{1}{2\sqrt{Dt}} e^{-(2nL+y)^2/(4Dt)} - \sum_{n=0}^{n=\infty}\frac{2}{\sqrt{\pi}}\frac{1}{2\sqrt{Dt}} e^{-[2(n+1)L-y]^2/4(Dt)}$$

At free surface, $y = 0$ and:

$$\frac{1}{C_{Ai} - C_{Ao}}\left(\frac{\partial C_A}{\partial y}\right)_{y=0} = -\frac{1}{\sqrt{\pi Dt}}\left(\sum_{n=0}^{n=\infty} e^{-(n^2L^2)/(Dt)} + \sum_{n=0}^{n=\infty} e^{-[(n+1)^2L^2]/(Dt)}\right)$$

$$= -\frac{1}{\sqrt{\pi Dt}}\left(1 + 2\sum_{n=1}^{n=\infty} e^{-(n^2L^2)/(Dt)}\right)$$

Mass transfer rate across interface per unit area is given by:

$$N_A = -D\left(\frac{\partial C_A}{\partial y}\right)_{y=0}$$

$$= (C_{Ai} - C_{Ao})\sqrt{\frac{D}{\pi t}}\left(1 + 2\sum_{n=1}^{n=\infty} e^{-(n^2L^2)/(Dt)}\right) \qquad (8.94)$$

Equation 8.94 converges rapidly for high values of L^2/Dt. For low values of L^2/Dt it is convenient to use an alternative form by using the identity:[20]

$$\sqrt{\frac{\alpha}{\pi}}\left(1 + 2\sum_{n=1}^{n=\infty} e^{-n^2\alpha}\right) \equiv 1 + 2\sum_{n=1}^{n=\infty} e^{-(n^2\pi^2)/\alpha} \qquad (8.95)$$

Taking $\alpha = L^2/Dt$,

$$N_A = (C_{Ai} - C_{Ao})\frac{D}{L}\left(1 + 2\sum_{n=1}^{n=\infty} e^{-(n^2\pi^2 Dt)/(L^2)}\right) \tag{8.96}$$

It will be noted that equation 8.94 and 8.96 become identical in form and in convergence when $L^2/Dt = \pi$.

Then

$$N_A = (C_{Ai} - C_{Ao})\frac{D}{L}\left(1 + 2\sum_{n=1}^{n=\infty} e^{-n^2\pi}\right)$$

$$= (C_{Ai} - C_{Ao})\frac{D}{L}[1 + 2(e^{-\pi} + e^{-9\pi} + \cdots)]$$

$$= (C_{Ai} - C_{Ao})\frac{D}{L}(1 + 0{\cdot}0864 + 6{\cdot}92 \times 10^{-6} + 1{\cdot}03 \times 10^{-12} + \cdots)$$

Thus, provided the rate of convergence is not worse than that for $L^2/Dt = \pi$, all terms other than the first in the series may be neglected. Now equation 8.94 will converge more rapidly than this for $L^2/Dt > \pi$, and equation 8.96 will converge more rapidly for $L^2/Dt < \pi$.

Thus:

$$\pi \leqslant \frac{L^2}{Dt} < \infty \qquad N_A = (C_{Ai} - C_{Ao})\sqrt{\frac{D}{\pi t}}(1 + 2\,e^{-L^2/Dt}) \tag{8.97}$$

$$0 < \frac{L^2}{Dt} \leqslant \pi \qquad N_A = (C_{Ai} - C_{Ao})\frac{D}{L}(1 + 2\,e^{-(\pi^2 Dt)/L^2}) \tag{8.98}$$

Now it will be noted that the second terms in equations 8.97 and 8.98 never exceeds 8·64 per cent of the first term. Thus, with an error not exceeding 8·64 per cent,

$$\pi \leqslant \frac{L^2}{Dt} < \infty \qquad N_A = (C_{Ai} - C_{Ao})\sqrt{\frac{D}{\pi t}} \tag{8.99}$$

$$0 < \frac{L^2}{Dt} \leqslant \pi \qquad N_A = (C_{Ai} - C_{Ao})\frac{D}{L} \tag{8.100}$$

Thus either the penetration theory of the film theory (equation 8.99 or 8.100) respectively can be used to describe the mass transfer process. The error will not exceed about 10 per cent provided that the appropriate equation is used, equation 8.99 for $L^2/Dt > \pi$ and equation 8.100 for $L^2/Dt < \pi$. Equation 8.100 will frequently apply quite closely in a wetted-wall column or in a packed tower with large packings. Equation 8.99 will be likely to apply when one of the phases is dispersed in the form of droplets, as in a spray tower, or in a packed tower with small packing elements.

Equations 8.97 and 8.98 give the point value of N_A at time t. The average values of N_A can then be obtained by applying the age distribution functions obtained by Higbie and by Danckwerts, respectively, as in the earlier section on the penetration theory.

8.5.4. Other Theories of Mass Transfer

Kishinevskij[18] has developed a model for mass transfer across an interface in which molecular diffusion is assumed to play no part. He considers that fresh material is continuously brought to the interface as a result of turbulence within the fluid and that, after exposure to the second phase, the fluid element attains equilibrium with it and then becomes mixed again with the bulk of the phase. The model thus presupposes surface renewal without penetration by diffusion and therefore the effect of diffusivity should not be important. No reliable experimental results are available to test the theory adequately.

Bakowski[17] has considered the rate of transfer from a turbulent gas stream to a liquid interface. He postulates that the transfer rate will be a function of the number of molecules of the diffusing component in the gas layer in contact with the liquid. The absorption of molecules by the liquid is assumed to be rapid so that the rate of replacement of the component in the surface layer is the rate determining factor. Hence the velocity of the gas stream over the surface directly influences the mass transfer rate in that it affects transfer to the surface layer.

8.5.5. Mass Transfer Coefficients

On the basis of each of the theories discussed, the rate of mass transfer in the absence of bulk flow is directly proportional to the driving force, expressed as a molar concentration difference, and, therefore:

$$N_A = k'(C_{Ai} - C_{Ao})\tag{8.101}$$

where k' is a mass transfer coefficient (see equation 8.73). In the two-film theory, k' is directly proportional to the diffusivity and inversely proportional to the film thickness. According to the penetration theory it is proportional to the square root of the diffusivity and, when all surface elements are exposed for an equal time, it is inversely proportional to the square root of time of exposure; when random surface renewal is assumed, it is proportional to the square root of the rate of renewal. In the film-penetration theory, the mass transfer coefficient is a complex function of the diffusivity, the film thickness, and either the time of exposure or the rate of renewal of surface.

In most cases, the value of the transfer coefficient cannot be calculated from the first principles. However, the way in which the coefficient will vary as operating conditions are altered can frequently be predicted by using the theory which is most nearly applicable to the problem in question.

The penetration and film-penetration theories have been developed for conditions of equimolecular counterdiffusion only; the equations are too complex to solve explicitly for transfer through a stationary carrier gas. For gas absorption, therefore, they apply only when the concentration of diffusing material is low. On the other hand, in the two-film theory the additional contribution to the mass transfer which is caused by bulk flow is easily calculated and k' is equal to $(D/L)(C_T/C_{Bm})$ instead of D/L.

In a process where mass transfer takes place across a phase boundary, the same type of theoretical approach can be applied to each of the phases, though it does not follow that the same theory is best applied to both phases. For example, the film model might be applicable to one phase and the penetration model to the other. This problem has already been discussed earlier.

When the film theory is applicable to each phase (the two-film theory), the process is steady state throughout and the interface composition does not then vary with time. For this case the two film coefficients can readily be combined. Because material does not accumulate at the interface, the mass transfer rate on each side of the phase boundary will be the same and for two phases it follows that:

$$N_A = k_1'(C_{Ao1} - C_{Ai1}) = k_2'(C_{Ai2} - C_{Ao2})\tag{8.102}$$

If there is no resistance to transfer at the interface C_{Ai1} and C_{Ai2} will be corresponding values in the phase–equilibrium relationship.

Usually, the values of the concentration at the interface are not known and the mass transfer coefficient is considered for the overall process. Overall transfer coefficients are then defined by the relations:

$$N_A = K_1(C_{Ao1} - C_{Ae1}) = K_2(C_{Ae2} - C_{Ao2})\tag{8.103}$$

where C_{Ae1} is the concentration in phase 1 in equilibrium with C_{Ao2} in phase 2, and C_{Ae2} is the concentration in phase 2 in equilibrium with C_{Ao1} in phase 1. If the equilibrium relationship is linear:

$$\mathcal{H} = \frac{C_{Ai1}}{C_{Ai2}} = \frac{C_{Ae1}}{C_{Ao2}} = \frac{C_{Ao1}}{C_{Ae2}} \tag{8.104}$$

where $\mathcal{H}$ is a proportionality constant.

The relationships between the various transfer coefficients are obtained as follows:
From equations 8.102 and 8.103:

$$\frac{1}{K_1} = \frac{1}{k_1'}\frac{C_{Ae1}-C_{Ao1}}{C_{Ai1}-C_{Ao1}} = \frac{1}{k_1'}\frac{C_{Ae1}-C_{Ai1}}{C_{Ai1}-C_{Ao1}} + \frac{1}{k_1'}\frac{C_{Ai1}-C_{Ao1}}{C_{Ai1}-C_{Ao1}}$$

But

$$\frac{1}{k_1'} = \frac{1}{k_2'}\frac{C_{Ai1}-C_{Ao1}}{C_{Ao2}-C_{Ai2}} \quad \text{(from equation 8.102)}$$

$$\therefore \quad \frac{1}{K_1} = \frac{1}{k_1'} + \frac{1}{k_2'}\frac{C_{Ae1}-C_{Ai1}}{C_{Ai1}-C_{Ao1}}\frac{C_{Ai1}-C_{Ao1}}{C_{Ao2}-C_{Ai2}}$$

But from equation 8.104:

$$\frac{C_{Ae1}-C_{Ai1}}{C_{Ao2}-C_{Ai2}} = \mathcal{H}$$

$$\therefore \quad \frac{1}{K_1} = \frac{1}{k_1'} + \frac{\mathcal{H}}{k_2'} \tag{8.105}$$

Similarly:

$$\frac{1}{K_2} = \frac{1}{\mathcal{H}k_1'} + \frac{1}{k_2'} \tag{8.106}$$

and hence:

$$\frac{1}{K_1} = \frac{\mathcal{H}}{K_2} \tag{8.107}$$

It follows, therefore, that when k_1' is large compared with k_2', K_2 and k_2' are approximately equal, and, when k_2' is large compared with k_1', K_1 and k_1' are almost equal.

The above relations between the various coefficients are valid provided that the transfer rate is linearly related to the driving force and that the equilibrium relationship is a straight line. They are therefore applicable for the two-film theory, and for any instant of time for the penetration and film-penetration theories. In general, application to time-averaged coefficients obtained from the penetration and film-penetration theories is not permissible because the condition at the interface will be time-dependent unless all of the resistance lies in one of the phases.

8.5.6. Countercurrent Mass Transfer and Transfer Units

Mass transfer processes involving two fluid streams are frequently carried out continuously using either countercurrent or cocurrent flow in a column apparatus. The former is the more usual, but cocurrent flow has advantages in special cases. Three common examples of processes of countercurrent mass transfer are as follows:

1. *In a packed distillation column.* Here a vapour stream is rising against the downward flow of a liquid reflux, and a state of dynamic equilibrium is set up in a steady state process. The more volatile constituent is transferred under the action of a concentration gradient from the liquid to the interface where it evaporates and then diffuses into the vapour stream. The less volatile component is transferred in the opposite direction, and if the molar latent heats of the components are the same, equimolecular counterdiffusion takes place.

2. *In a packed absorption column.* The flow pattern is similar to that in a packed distillation column but the vapour stream is replaced by a carrier gas together with a solute gas. The solute diffuses through the gas phase to the liquid surface where it dissolves and is

then transferred to the bulk of the liquid. In this case there is no mass transfer in the opposite direction and the transfer rate is supplemented by bulk flow.

3. *In a liquid–liquid extraction column.* Here the process is similar to that occurring in an absorption column with the exception that both streams are liquid, and the lighter liquid rises through the denser one.

The mass transfer relations do differ in a number of important respects in distillation, gas absorption and liquid–liquid extraction. Thus in distillation, equimolecular counterdiffusion frequently takes place and the molar rate of flow of the two phases remains approximately constant throughout the whole height of the column. On the other hand, in gas absorption the mass transfer process is increased as a result of bulk flow and, when the solute gas is present in high concentrations, the molar rate of flow at the top of the column will be less than that at the bottom. For low concentrations of the solute gas however, conditions are similar in distillation and gas absorption. Consideration will now be given to mass transfer in a column under these conditions.

In Fig. 8.6 the conditions existing in a column during the steady state operation of a countercurrent process are illustrated. The molar rates of flow of the two streams are G_1 and G_2, which are assumed constant over the whole column. Suffixes 1 and 2 will be used to denote the two phases, and t and b to denote the top and bottom of the column.

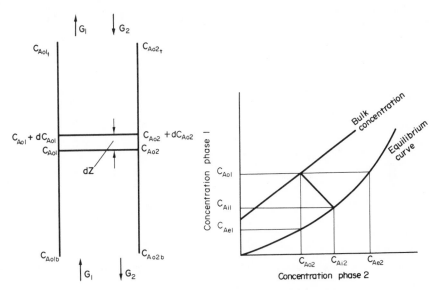

FIG. 8.6. Countercurrent mass transfer in a column.

The height of the column is Z, its total cross-sectional area is S, and a is the interfacial area between the two phases per unit volume of column. Then the rate of transfer of the diffusing component in a height dZ of column:

$$= G_1 \frac{1}{C_T} dC_{Ao1} = k_1'(C_{Ai1} - C_{Ao1}) Sa \, dZ \tag{8.108}$$

giving

$$\frac{dC_{Ao1}/dZ}{C_{Ai1} - C_{Ao1}} = \frac{k_1' a S C_T}{G_1}$$

$$= k_1' a C_T / G_1' \tag{8.109}$$

where G_1' is the molar rate of flow per unit cross-section of column. The exact interfacial area cannot normally be determined independently of the transfer coefficient, and therefore values of the product $k_1'a$ are usually quoted for any particular system.

The left-hand side of equation 8.109 is the rate of change of concentration with height for unit driving force, and is therefore a measure of the efficiency of the column. Thus a high value of $(k_1'aC_T)/G_1'$ will be obtained in a column of high efficiency. The reciprocal of this quantity is $G_1'/(k_1'aC_T)$ which has linear dimensions and is known as the *height of the transfer unit* $\mathbf{H}_1$ (HTU).

If equation 8.109 is rearranged and integrated:

$$\int_{C_{Ao1^b}}^{C_{Ao1_t}} \frac{\mathrm{d}C_{Ao1}}{C_{Ai1} - C_{Ao1}} = \frac{k_1'aC_T}{G_1'} Z = \frac{Z}{G_1'/(k_1'aC_T)} \tag{8.110}$$

The right-hand side of equation 8.110 is the height of the column divided by the HTU and is known as the *number of transfer units* $\mathbf{N}_1$. It is obtained by evaluating the integral on the left-hand side of the equation.

Therefore:

$$\mathbf{N}_1 = Z/\mathbf{H}_1 \tag{8.111}$$

Equation 8.108 could have been written in terms of the film coefficient for the second phase (k_2') or either of the overall transfer coefficients (K_1, K_2). Transfer units based on either film coefficient or overall coefficient can therefore be defined. The following equations are analogous to equation 8.111:

The number of transfer units based on phase 2, $\qquad \mathbf{N}_2 = Z/\mathbf{H}_2 \tag{8.112}$

Number of overall transfer units based on phase 1, $\qquad \mathbf{N}_{o1} = Z/\mathbf{H}_{o1} \tag{8.113}$

Number of overall transfer units based on phase 2, $\qquad \mathbf{N}_{o2} = Z/\mathbf{H}_{o2} \tag{8.114}$

In this notation the introduction of o into the suffix indicates an overall transfer unit. The values of $\mathbf{H}_1, \mathbf{H}_2, \mathbf{H}_{o1}, \mathbf{H}_{o2}$ are of the following form:

$$\mathbf{H}_1 = G_1'/(k_1'aC_T) \tag{8.115}$$

If one phase is a gas, as in gas absorption, it is often more convenient to express concentrations as partial pressures so that:

$$\mathbf{H}_1 = \mathbf{H}_G = G_1'/(k_GaP) \tag{8.116}$$

where, in SI units, G_1' can be expressed in $\mathrm{kmol/m^2\,s}$, k_G in $\mathrm{kmol/[m^2\,s(kN/m^2)]}$, P in $\mathrm{kN/m^2}$, and a in $\mathrm{m^2/m^3}$.

The overall values of the HTU can be expressed in terms of the film values by using equation 8.105 which gives the relation between the coefficients.

$$\frac{1}{K_1} = \frac{1}{k_1'} + \frac{\mathscr{H}}{k_2'} \tag{equation 8.105}$$

Substituting for the coefficients in terms of the HTU:

$$\mathbf{H}_{o1} \frac{aC_T}{G_1'} = \mathbf{H}_1 \frac{aC_T}{G_1'} + \frac{\mathscr{H}aC_T}{G_2'} \mathbf{H}_2 \tag{8.117}$$

i.e.

$$\mathbf{H}_{o1} = \mathbf{H}_1 + \mathscr{H} \frac{G_1'}{G_2'} \mathbf{H}_2 \tag{8.118}$$

Similarly:

$$\mathbf{H}_{o2} = \mathbf{H}_2 + \frac{1}{\mathscr{H}} \frac{G_2'}{G_1'} \mathbf{H}_1 \tag{8.119}$$

The advantage of using the transfer unit in preference to the transfer coefficient is that the former remains much more nearly constant as flow conditions are altered. This is particularly important in problems of gas absorption where the concentration of the solute gas is high and the flow pattern changes in the column because of the change in the total rate of flow of gas at different sections. In most cases the coefficient is proportional to the flowrate raised to a power slightly less than unity and therefore the HTU is substantially constant.

As already seen, for equimolecular counterdiffusion, the film transfer coefficients, and hence the corresponding HTUs, can be expressed in terms of the physical properties of the system and the assumed film thickness or exposure time, using the two-film, the penetration, or the film-penetration theories. For conditions where bulk flow is important, however, the transfer rate of constituent A is increased by the factor C_T/C_{Bm} and the diffusion equations can be solved only on the basis of the two-film theory. In the design of equipment it is usual to work in terms of transfer coefficients or HTUs and not to endeavour to evaluate them in terms of properties of the system.

8.5.7. Other Applications of Theories of Interphase Transfer

The two-film, penetration, and film-penetration theories can be applied satisfactorily to problems involving mass transfer between a fluid and the surface of a solid. In this case the basic assumption that turbulence dies out at the interface is fully justified. The layer of fluid immediately in contact with the solid is in laminar flow and constitutes the laminar sub-layer. However, in the vicinity of a solid surface, appreciable velocity gradients exist within the fluid and the calculation of transfer rates become complex if they are taken into account.

The preceding treatment has related to processes of mass transfer. When a temperature difference exists between two phases, heat will be transferred from the hotter fluid to the cooler, and that proportion of the heat which is transferred by forced convection can be calculated by the application of the theories discussed here. Furthermore, if there is a change of phase at the interface, as in a humidifying or water cooling tower, the heat transfer rate in the two phases will not, in general, be the same.

Example 8.4. Ammonia is absorbed at $101 \cdot 3 \, kN/m^2$ from an ammonia–air stream by passing it up a vertical tube, down which dilute sulphuric acid is flowing. The following laboratory data are available:

$$\text{Length of tube} \quad = 825 \, mm$$
$$\text{Diameter of tube} = 15 \, mm$$

Partial pressures of ammonia:

$$\text{at inlet} \; = 7 \cdot 5 \, kN/m^2; \quad \text{at outlet} = 2 \cdot 0 \, kN/m^2$$
$$\text{Air rate} = 2 \times 10^{-5} \, kmol/s$$

What is the transfer coefficient K_G?

Solution. Driving force at inlet $\quad = 7 \cdot 5 \, kN/m^2$

Driving force at outlet $= 2 \cdot 0 \, kN/m^2$

Mean driving force $\quad = (7 \cdot 5 - 2 \cdot 0)/\ln (7 \cdot 5/2 \cdot 0) = 4 \cdot 2 \, kN/m^2$

Ammonia absorbed $\quad = 2 \times 10^{-5}[(7 \cdot 5/93 \cdot 8) - (2 \cdot 0/99 \cdot 3)] = 1 \cdot 120 \times 10^{-6} \, kmol/s$

Wetted surface $\quad = \pi \times 0 \cdot 015 \times 0 \cdot 825 = 0 \cdot 0388 \, m^2$

Hence $K_G = 1 \cdot 120 \times 10^{-6}/(0 \cdot 0388 \times 4 \cdot 2) = \underline{\underline{6 \cdot 87 \times 10^{-6} \, kmol/[m^2 \, s \, (kN/m^2)]}}$

8.6. MASS TRANSFER TO A SPHERE

In the problems considered previously, the transfer has been taking place in a single direction of a rectangular coordinate system. In many applications of mass transfer, one of the fluids is injected as approximately spherical droplets into a second immiscible fluid, and transfer of the solute occurs as the droplet passes through the continuous medium.

Consider the problem of a sphere of pure liquid of radius b which is suddenly immersed in a gas, the whole of the mass transfer resistance lying within the liquid. On the assumption that mass transfer is governed by Fick's law and that angular symmetry exists:

At radius r within sphere, mass transfer rate $= -(4\pi r^2)D(\partial C_A/\partial r)$

Change in mass transfer rate over a distance dr

$$= \frac{\partial}{\partial r}\left(-4\pi r^2 D\frac{\partial C_A}{\partial r}\right)dr$$

Then material balance over a shell gives:

$$-\frac{\partial}{\partial r}\left(-4\pi r^2 D\frac{\partial C_A}{\partial r}\right)dr = \frac{\partial C_A}{\partial t}(4\pi r^2\,dr)$$

$$\therefore \qquad r^2\frac{\partial C_A}{\partial t} = D\frac{\partial}{\partial r}\left(r^2\frac{\partial C_A}{\partial r}\right) \qquad\qquad (8.120)$$

This equation may be solved by taking Laplace transforms.

The concentration gradient at the surface of the sphere $r = b$ is then found to be given by:

$$\left(\frac{\partial C_A}{\partial r}\right)_{r=b} = \frac{-C_{Ai}}{b} + \frac{C_{Ai}}{\sqrt{\pi Dt}}\left(1 + 2\sum_{n=1}^{n=\infty}e^{-(n^2b^2)/Dt}\right) \qquad (8.121)$$

The mass transfer rate at surface of sphere at time t

$$= 4\pi b^2(-D)\left(\frac{\partial C_A}{\partial r}\right)_{r=b}$$

$$= 4\pi bDC_{Ai}\left[1 - \frac{b}{\sqrt{\pi Dt}}\left(1 + 2\sum_{n=1}^{n=\infty}e^{-(n^2b^2)/Dt}\right)\right] \qquad (8.122)$$

The total mass transfer during the passage of a drop is therefore obtained by integration of equation 8.122 over the time of exposure.

8.7. MASS TRANSFER AND CHEMICAL REACTION

In many applications of mass transfer the solute will react with the medium as in the case, for example, of the absorption of carbon dioxide in an alkaline solution. The mass transfer rate will then decrease in the direction of diffusion as a result of the reaction. Consider the unidirectional molecular diffusion of a component A through a distance dy over unit area. Then, neglecting the effects of bulk flow, a material balance for a reaction of order n gives:

$$\frac{\partial C_A}{\partial t}\,dy = -\frac{\partial}{\partial y}\left(-D\frac{\partial C_A}{\partial y}\right)dy - kC_A{}^n\,dy$$

i.e.

$$\frac{\partial C_A}{\partial t} = D\frac{\partial^2 C_A}{\partial y^2} - kC_A{}^n \qquad\qquad (8.123)$$

where k is the reaction rate constant. This equation has no analytical solution for the general case. For a steady-state process:

$$D\frac{\partial^2 C_A}{\partial y^2} - kC_A{}^n = 0 \tag{8.124}$$

Equation 8.124 may be integrated using the appropriate boundary conditions. The most important case is for a first-order reaction. Putting $n = 1$ in equation 8.124:

$$D\frac{\partial^2 C_A}{\partial y^2} - kC_A = 0 \tag{8.125}$$

The solution of equation 8.125 is:

$$C_A = B_1' e^{\sqrt{(k/D)}y} + B_2' e^{-\sqrt{(k/D)}y} \tag{8.126}$$

B_1' and B_2' must then be evaluated using the appropriate boundary conditions.

Example 8.5. In a gas absorption process, the solute gas A diffuses into a solvent liquid with which it reacts. The mass transfer is one of steady state unidirectional molecular diffusion and the concentration of A is always sufficiently small for bulk flow to be negligible. Under these conditions the reaction is first order with respect to the solute A.

At a depth l below the liquid surface, the concentration of A has fallen to one-half of the value at the surface. What is the ratio of the mass transfer rate at this depth l to the rate at the surface? Calculate the numerical value of the ratio when $l\sqrt{k/D} = 0.693$, where D is the molecular diffusivity and k the first-order rate constant.

Solution. The above process is described by equation 8.126:

$$C_A = B_1' e^{\sqrt{(k/D)}y} + B_2' e^{-\sqrt{(k/D)}y} \tag{equation 8.126}$$

If C_{Ai} is the surface concentration ($y = 0$):

$$C_{Ai} = B_1' + B_2'$$

At $y = l$, $C_A = C_{Ai}/2$

$$\therefore \qquad C_{Ai}/2 = B_1' e^{\sqrt{(k/D)}l} + B_2' e^{-\sqrt{(k/D)}l}$$

Solving for B_1' and B_2':

$$B_1' = \frac{C_{Ai}}{2}(1 - 2 e^{-\sqrt{(k/D)}l})(e^{\sqrt{(k/D)}l} - e^{-\sqrt{(k/D)}l})^{-1}$$

$$B_2' = -\frac{C_{Ai}}{2}(1 - 2 e^{\sqrt{(k/D)}l})(e^{\sqrt{(k/D)}l} - e^{-\sqrt{(k/D)}l})^{-1}$$

$$\frac{(N_A)_{y=l}}{(N_A)_{y=0}} = \frac{-D(dC_A/dy)_{y=l}}{-D(dC_A/dy)_{y=0}} = \frac{(dC_A/dy)_{y=l}}{(dC_A/dy)_{y=0}}$$

$$\frac{dC_A}{dy} = \sqrt{\frac{k}{D}}(B_1' e^{\sqrt{(k/D)}y} - B_2' e^{-\sqrt{(k/D)}y})$$

$$\frac{(N_A)_{y=l}}{(N_A)_{y=0}} = \frac{B_1' e^{\sqrt{(k/D)}l} - B_2' e^{-\sqrt{(k/D)}l}}{B_1' - B_2'}$$

$$= \frac{(1 - 2 e^{-\sqrt{(k/D)}l}) e^{\sqrt{(k/D)}l} + (1 - 2 e^{\sqrt{(k/D)}l}) e^{-\sqrt{(k/D)}l}}{(1 - 2 e^{-\sqrt{(k/D)}l}) + (1 - 2 e^{\sqrt{(k/D)}l})}$$

$$= \frac{e^{\sqrt{(k/D)}l} + e^{-\sqrt{(k/D)}l} - 4}{2(1 - e^{-\sqrt{(k/D)}l} - e^{\sqrt{(k/D)}l})}$$

When
$$l\sqrt{\frac{k}{D}} = 0.693, \quad e^{\sqrt{(k/D)l}} = 2, \quad e^{-\sqrt{(k/D)l}} = 0.5$$

$$\frac{(N_A)_{y=l}}{(N_A)_{y=0}} = \frac{2 + \frac{1}{2} - 4}{2(1 - 2 - \frac{1}{2})} = \underline{\underline{0.5}}$$

8.8. PRACTICAL STUDIES OF MASS TRANSFER

8.8.1. The j-Factor of Chilton and Colburn

Results of experimental studies of mass transfer can be conveniently represented by means of the j-factor, originally developed by Chilton and Colburn[21, 22] for representing data on heat transfer between a turbulent fluid and the wall of a pipe. From equation 7.53:

$$Nu = 0.023 \, Re^{0.8} \, Pr^{0.33} \qquad \text{(equation 7.53)}$$

where the viscosity is measured at the mean film temperature, and Nu, Re, and Pr denote the Nusselt, Reynolds, and Prandtl numbers.

If both sides are divided by the product $Re \, Pr$:

$$\frac{Nu}{Re \, Pr} = St = \frac{h}{C_p \rho u} = 0.023 \, Re^{-0.2} \, Pr^{-0.67} \qquad (8.127)$$

where St is the Stanton number

or
$$St \, Pr^{0.67} = j_h = 0.023 \, Re^{-0.2} \qquad (8.128)$$

Chilton and Colburn found that a plot of j_h against Re gave approximately the same curve as the friction chart for flow through tubes.

By analogy with the derivation given above for heat transfer, Chilton and Colburn[21, 22] have derived a factor for mass transfer j_d which they have expressed as:

$$j_d = \frac{h_D}{u} \frac{C_{Bm}}{C_T} \left(\frac{\mu}{\rho D}\right)^{0.67} \qquad (8.129)$$

Here $\mu/\rho D$ is the Schmidt number (Sc) which corresponds to the Prandtl number in heat transfer (see Chapter 10).

The factor C_{Bm}/C_T, the logarithmic mean of the concentration of the inert component **B** divided by the total concentration, is introduced because the concentration of **B** may alter substantially and $h_D C_{Bm}$ has been found to be more nearly constant than h_D. It has already been seen that j_h can be related to the Reynolds group and the friction term and been seen that the factor j_d obtained in experiments on mass transfer will now be used to find the experimental relation between j_d and j_h.

Mass Transfer Inside Vertical Tubes

Several workers have measured the rate of transfer from a liquid flowing down the inside wall of a tube to a gas passing upwards. Gilliland and Sherwood[23] have vaporised a number of liquids including water, toluene, aniline and propyl, amyl and butyl alcohols into an air stream flowing up the tube. They worked with a small tube 25 mm diameter (d) and 450 mm long, fitted with calming sections at the top and bottom, and varied the pressure from 14 to about 300 kN/m^2. The data were plotted logarithmically as:

$$\frac{h_D d}{D} \frac{C_{Bm}}{C_T} \quad \text{against} \quad Re$$

when straight lines are obtained. Introducing the Schmidt group, they were able to bring the points for all the liquids on to the same line, the equation of which was:

$$\frac{h_D}{u}\frac{C_{Bm}}{C_T}\left(\frac{\mu}{\rho D}\right)^{0.56} = 0.023\,Re^{-0.17} \qquad (8.130)$$

The index of the Schmidt group is much less than the index of 0·67 for the Prandtl group as obtained with heat transfer, but the range of values of Sc used was very small and insufficient to confirm this. Sherwood and Pigford[7] have shown that if the data of Gilliland and Sherwood and others are plotted with the Schmidt group raised to the power 0·67, as shown in Fig. 8.7, a reasonably good correlation is obtained. Although the points are rather more scattered than with heat transfer it is reasonable to assume that both j_d and j_h are approximately equal to $R/\rho u^2$. Equation 8.130 applies in the absence of ripples which can be responsible for a very much increased rate of mass transfer. For instance, Kafesjian, Planck and Gerhard[41] suggest that the constant of 0·023 in equation 8.130 must be replaced by $0.00814(4\Gamma/\mu)^{0.15}$, where Γ is the mass rate of flow per unit perimeter.

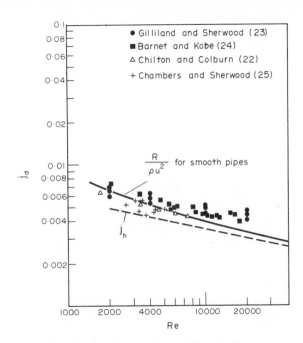

FIG. 8.7. Mass transfer in wetted wall columns.

Mass Transfer to Plane Surfaces

The rate of vaporisation of liquids into an air stream passing over the liquid surface under conditions of forced convection has been measured by several investigators. In addition, some experiments have been carried out in the absence of an airstream, under what have been termed still air conditions. Most of these experiments have been carried out in some form of wind tunnel where the rate of flow of the air could be controlled and its temperature and humidity measured. In such experiments it is important to keep the liquid level constant and to avoid any form of turbulence promoter on the leading edge of the pan.

Hinchley and Himus[26] measured the rate of vaporisation of water from heated rectangular pans fitting flush with the floor of a wind tunnel 457 mm wide by 229 mm high. They were able to show that the rate of vaporisation was proportional to the pressure difference $P_s - P_w$, where P_s is the vapour pressure of the water and P_w the partial pressure of the water in the air.

They presented their results in an empirical equation,

$$W = \text{constant} \, (P_s - P_w) \qquad (8.131)$$

where the constant varies with the geometry of the pan and the velocity.

The importance of this early work lies in the fact that the pressure difference $P_s - P_w$ was shown to be the driving force in the process. More systematic work in more elaborate equipment has been done by Powell and Griffiths,[27] Wade,[28] and by Pasquill.[29] Wade, working with a small pan of 88 mm side in a tunnel, vaporised a variety of organic liquids including acetone, benzene, and trichlorethylene, all at atmospheric pressure. Powell and Griffiths used a stretched canvas over rectangular pans in such a way that the canvas was always wet and gave the effect of a water surface. For all these liquids the relationships between the rate of vaporisation and both the partial pressure difference and the rate of flow of the air were the same. Powell and Griffiths found that the rate of vaporisation per unit area of the surface decreased with increase in the down-wind length L of the wet surface and that for rectangular pans the rate of vaporisation was proportional to $L^{0.77}$. On this basis Sherwood[3] has plotted the results in terms of the Reynolds group Re_x, with the length of the pan as the characteristic dimension. Figure 8.8, taken from his work, shows j_d plotted against

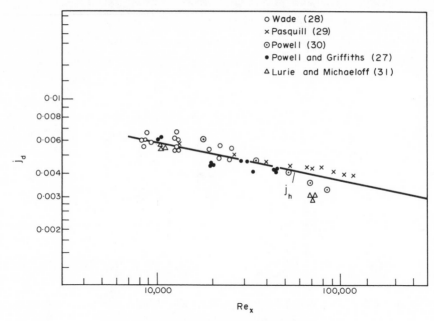

FIG. 8.8. Evaporation from plane surfaces.

Re_x for a number of liquids. Although the individual points do not lie too well on a smooth curve it is reasonable to write $j_d \approx j_h$. It should be noted that the Schmidt group was varied only over a small range and that almost equally good results would be obtained by omitting it from the correlation. The inclusion of the term C_{Bm} is not satisfactory, since the results for a number of workers, including those of Wade on many liquids, show that the vaporisation is proportional to $P_s - P_w$ even when this term amounts to $50 \, \text{kN/m}^2$ ($= 0.5 \, \text{bar}$).

The true influence of the Schmidt group was not established until Linton and Sherwood[32] measured the rates of solution of benzoic acid, cinnamic acid, and β-naphthol in water. With these relatively insoluble materials the Schmidt group had values from 1000 to 3000 and the results confirmed the index of 0.67 for this group.

Since j_h and j_d are approximately equal, the values of mass transfer coefficients can be calculated from the corresponding values of the heat transfer coefficients.

Now $\dfrac{h_D}{u}\dfrac{C_{Bm}}{C_T}Sc^{0.67}=\dfrac{h}{C_p\rho u}Pr^{0.67}$ (from equations 8.128 and 8.129):

so that
$$h_D=\frac{h}{C_p\rho}\frac{C_T}{C_{Bm}}\left(\frac{Pr}{Sc}\right)^{0.67}\qquad(8.132)$$

Difficulties arise in the evaluation of mass transfer coefficients since the composition of the material varies with distance from the surface, and the physical properties therefore are not constant and some mean value must be used.

The results quoted above give reasonably good support to the treatment of heat, mass, and momentum transfer by the j-factor, but it is important to remember that in all cases considered the drag is almost entirely in the form of skin friction (i.e. viscous drag at the surface). As soon as an attempt is made to apply the relation to cases where form drag (i.e. additional drag caused by the eddies set up as a result of the fluid impinging on an obstruction) is important, such as beds of granular solids or evaporation from cylinders or spheres, the j-factor and the friction factor are found no longer to be equal. This problem will receive further consideration in Volume 2. Sherwood[33, 34] carried out experiments where the form drag was large compared with the skin friction which he calculated approximately by subtracting the form drag from the total drag force. He obtained reasonable agreement between the corresponding value of $R/\rho u^2$ and j_h and j_d.

Gamson et al.[35] have successfully used the j-factor method to correlate their experimental results for heat and mass transfer between a bed of granular solids and a gas stream.

Pratt[36] has examined the effect of using artificially roughened surfaces and of introducing "turbulence promoters", which increase the amount of form drag. He found that the values of $R/\rho u^2$ and the heat and mass transfer coefficients were a minimum for smooth surfaces and all three quantities increased as the surface roughness was increased. $R/\rho u^2$ increased far more rapidly than either of the other two quantities however, and the heat and mass transfer coefficients were found to reach a limiting value whereas $R/\rho u^2$ could be increased almost indefinitely. Pratt has suggested that these limiting values are reached when the velocity gradient at the surface corresponds with that in the turbulent part of the fluid; i.e. at a condition where the buffer layer ceases to exist (see Chapter 9).

Mass Transfer to Single Spheres

Mass transfer from single drops to still air follows the law for molecular diffusion. Thus, for radial diffusion into a large expanse of stationary medium where the partial pressure falls off to zero over an infinite distance, then by analogy with the corresponding heat transfer equation 7.22:

$$Sh=\frac{h_D d}{D}=2.0\qquad(8.133)$$

where Sh is the Sherwood number.

For conditions of forced convection, Frössling[37] studied the evaporation of drops of nitrobenzene, aniline and water, and of spheres of naphthalene into an air stream. The drops were mainly small and of the order of 1 mm diameter. Powell[30] measured the evaporation of water from the surfaces of wet spheres up to 150 mm diameter and from spheres of ice.

The experimental results of Frössling could be represented by the equation:

$$Sh=\frac{h_D d}{D}=2.0(1+0.276\,Re^{0.5}\,Sc^{0.33})\qquad(8.134)$$

Pigford[7] found that the effect of the Schmidt group was also influenced by the Reynolds group and that the available data were fairly well correlated as shown in Fig. 8.9, in which $(h_D d)/D$ is plotted against $Re\,Sc^{0.67}$.

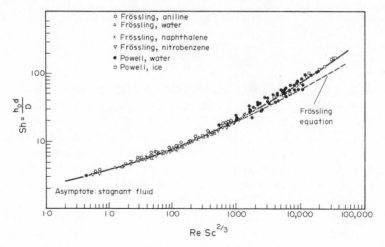

FIG. 8.9. Mass transfer to single spheres.

Garner and Keey[38, 39] dissolved pelleted spheres of organic acids in water in a low-speed water tunnel at particle Reynolds numbers between 2·3 and 255 and compared their results with other data available at Reynolds numbers up to 900. Natural convection was found to exert some influence at Reynolds numbers up to 750. At Reynolds numbers greater than 250, the results were correlated by equation 8.135:

$$Sh = 0 \cdot 94 \, Re^{0 \cdot 5} \, Sc^{0 \cdot 33} \tag{8.135}$$

They also studied mass transfer under conditions of natural convection.

Rowe et al.[40] have reviewed the literature on heat and mass transfer between spherical particles and a fluid. For heat transfer, the results are generally well represented by equation 7.76:

$$Nu = 2 + \beta'' \, Pr^{0 \cdot 33} \, Re^{0 \cdot 5} \quad (0 \cdot 4 < \beta'' < 0 \cdot 8) \tag{8.136}$$

For mass transfer:

$$Sh = \alpha + \beta' \, Sc^{0 \cdot 33} \, Re^{0 \cdot 5} \quad (0 \cdot 3 < \beta' < 1 \cdot 0) \tag{8.137}$$

The constant α appears to be a function of the Grashof number, but approaches a value of about 2 as the Grashof number approaches zero.

In an experimental investigation[40] they confirmed that equations 8.136 and 8.137 can be used to represent the results obtained for transfer from both air and water to spheres. The constants β', β'' varied from 0·68 to 0·79.

8.9.   FURTHER READING

BIRD, R. B., STEWART, W. E. and LIGHTFOOT, E. N.: Transport Phenomena (John Wiley, 1960).
ECKERT, E. R. G. and DRAKE, R. M.: Analysis of Heat and Mass Transfer (McGraw-Hill, 1972).
BENNETT, C. O. and MYERS, J. E.: Momentum, Heat, and Mass Transfer (McGraw-Hill, 1962).
TREYBAL, R. E.: Mass-Transfer Operations, 2nd edn. (McGraw-Hill, 1968).
DANCKWERTS, P. V.: Gas–liquid Reactors (McGraw-Hill, 1970).
SHERWOOD, T. K.: Chemical Engineering Education (Fall, 1974), 204, A review of the development of mass transfer theory.
SHERWOOD, T. K., PIGFORD, D. L. and WILKE, C. R.: Mass Transfer (McGraw-Hill, 1975).

8.10. REFERENCES

1. FICK, A.: *Ann. Phys.* **94** (1855) 59. Ueber Diffusion.
2. STEFAN, J.: *Wiener Akad. Wissensch.* **68** (1873) 385; **79** (1879) 169; **98** (1889) 1418; *Ann. Physik* **41** (1890) 723. Versuche über die Verdampfung.
3. MAXWELL, J. C.: *Phil. Trans. Roy. Soc.* **157** (1867) 49. The dynamical theory of gases.
4. PERRY, J. H.: *Chemical Engineers' Handbook* (McGraw-Hill, 1950).
5. WINKELMANN, A.: *Ann. Physik.* **22** (1884) 1, 152. Ueber die Diffusion von Gasen und Dämpfen.
6. GILLILAND, E. R.: *Ind. Eng. Chem.* **26** (1934) 681. Diffusion coefficients in gaseous systems.
7. SHERWOOD, T. K. and PIGFORD, R. L.: *Absorption and Extraction* (McGraw-Hill, 1952).
8. REID, R. C. and SHERWOOD, T. K.: *The Properties of Gases and Liquids* (McGraw-Hill, 1958).
9. COULSON, J. M. and RICHARDSON, J. F.: *Chemical Engineering Vol. 3.* (Pergamon Press Oxford, 1970).
10. ARNOLD, J. H.: *Trans. Am. Inst. Chem. Eng.* **40** (1944) 361. Studies in diffusion: III, Steady-state vaporization and absorption.
11. GOODRIDGE, F. and BRICKNELL, D. J.: *Trans. Inst. Chem. Eng.* **40** (1962) 54. Interfacial resistance in the carbon dioxide–water system.
12. ROBB, D.: University of Durham M.Sc. thesis (1963). Studies in mass transfer across a gas–liquid interface.
13. WHITMAN, W. G.: *Chem. and Met. Eng.* **29** (1923) 147. The two-film theory of absorption.
14. HIGBIE, R.: *Trans. Am. Inst. Chem. Eng.* **31** (1935) 365. The rate of absorption of pure gas into a still liquid during short periods of exposure.
15. DANCKWERTS, P. V.: *Ind. Eng. Chem.* **43** (1951) 1460. Significance of liquid film coefficients in gas absorption.
16. TOOR, H. L. and MARCHELLO, J. M.: *A.I.Ch.E. Jl* **4** (1958) 97. Film-penetration model for mass and heat transfer.
17. BAKOWSKI, S.: *Trans. I. Chem. E.* **32** (1954) S.37. A new approach to the problem of mass transfer in the gas phase.
18. KISHINEVSKIJ, M. K.: *Zhur. Priklad. Khim.* **24** (1951) 542; *J. Appl. Chem. U.S.S.R.* **24** 593. The kinetics of absorption under intense mixing.
19. HARRIOTT, P.: *Chem. Eng. Sci.* **17** (1962) 149. A random eddy modification of the penetration theory.
20. DWIGHT, H. B.: *Tables of Integrals and Other Mathematical Data* (The Macmillan Company, 1957).
21. COLBURN, A. P.: *Trans. Am. Inst. Chem. Eng.* **29** (1933) 174. A method of correlating forced convection heat transfer data and a comparison with fluid friction.
22. CHILTON, T. H. and COLBURN, A. P.: *Ind. Eng. Chem.* **26** (1934) 1183. Mass transfer (absorption) coefficients—production from data on heat transfer and fluid friction.
23. GILLILAND, E. R. and SHERWOOD, T. K.: *Ind. Eng. Chem.* **26** (1934) 516. Diffusion of vapours into air streams.
24. BARNET, W. I. and KOBE, K. A.: *Ind. Eng. Chem.* **33** (1941) 436. Heat and vapour transfer in a wetted-wall column.
25. CHAMBERS, F. S. and SHERWOOD, T. K.: *Ind. Eng. Chem.* **29** (1937) 579. *Trans. Am. Inst. Chem. Eng.* **33** (1937) 579. Absorption of nitrogen dioxide by aqueous solutions.
26. HINCHLEY, J. W. and HIMUS, G. W.: *Trans. Inst. Chem. Eng.* **2** (1924) 57. Evaporation in currents of air.
27. POWELL, R. W. and GRIFFITHS, E.: *Trans. Inst. Chem. Eng.* **13** (1935) 175. The evaporation of water from plane and cylindrical surfaces.
28. WADE, S. H.: *Trans. Inst. Chem. Eng.* **20** (1942) 1. Evaporation of liquids in currents of air.
29. PASQUILL, F.: *Proc. Roy. Soc.* A **182** (1943) 75. Evaporation from a plane free liquid surface into a turbulent air stream.
30. POWELL, R. W.: *Trans. Inst. Chem. Eng.* **18** (1940) 36. Further experiments on the evaporation of water from saturated surfaces.
31. LURIE, M. and MICHAILOFF, M.: *Ind. Eng. Chem.* **28** (1936) 345. Evaporation from free water surfaces.
32. LINTON, W. H. and SHERWOOD, T. K.: *Chem. Eng. Prog.* **46** (1950) 258. Mass transfer from solid shapes to water in streamline and turbulent flow.
33. SHERWOOD, T. K.: *Trans. Am. Inst. Chem. Eng.* **36** (1940) 817. Mass transfer and friction in turbulent flow.
34. SHERWOOD, T. K.: *Ind. Eng. Chem.* **42** (1950) 2077. Heat transfer, mass transfer, and fluid friction.
35. GAMSON, B. W., THODOS, G. and HOUGEN, O. A.: *Trans Am. Inst. Chem. Eng.* **39** (1943) 1. Heat mass and momentum transfer in the flow of gases through granular solids.
36. PRATT, H. R. C.: *Trans. Inst. Chem. Eng.* **28** (1950) 77. The application of turbulent flow theory to transfer processes in tubes containing turbulence promoters and packings.
37. FRÖSSLING, N.: *Gerlands Beitr. Geophys.* **52** (1938) 170. Über die Verdunstung fallender Tropfen.
38. GARNER, F. H. and KEEY, R. B.: *Chem. Eng. Sci.* **9** (1958) 119. Mass transfer from single solid spheres—I. Transfer at low Reynolds numbers.
39. GARNER, F. H. and KEEY, R. B.: *Chem. Eng. Sci.* **9** (1959) 218. Mass transfer from single solid spheres—II. Transfer in free convection.
40. ROWE, P. N., CLAXTON, K. T., and LEWIS, J. B.: *Trans. Inst. Chem. Eng.* **41** (1965) T14. Heat and mass transfer from a single sphere in an extensive flowing fluid.
41. KAFESJIAN, R., PLANK, C. A. and GERHARD, E. R.: *A. I. Ch. E. Jl* **7** (1961) 463. Liquid flow and gas phase mass transfer in wetted-wall towers.

8.11. NOMENCLATURE

		Units in SI system	Dimensions in $\mathbf{M, L, T,}$ θ
a	Interfacial area per unit volume	m²/m³	L^{-1}
B_1, B_2	Integration constants	kmol s/m³	$ML^{-3}T$
B'_1, B'_2	Integration constants	kmol/m³	ML^{-3}
b	Radius of sphere	m	L
C	Molar concentration	kmol/m³	ML^{-3}
C_A, C_B	Molar concentration of $\mathbf{A, B}$	kmol/m³	ML^{-3}
C_p	Specific heat at constant pressure	J/kg K	$L^2T^{-2}\theta^{-1}$
C_T	Total molar concentration	kmol/m³	ML^{-3}
C_{Bm}	Logarithmic mean value of C_B	kmol/m³	ML^{-3}
C'	$C_A - C_{Ao}$	kmol/m³	ML^{-3}
C'_A, C'_B	Molar concentration of $\mathbf{A, B}$ in film	kmol/m³	ML^{-3}
$\bar{C}'$	Laplace transform of C'	kmol s/m³	$ML^{-3}T$
c	Mass concentration	kg/m³	ML^{-3}
D	Diffusivity	m²/s	L^2T^{-1}
D_f	Diffusivity of fluid in film	m²/s	L^2T^{-1}
D_L	Liquid phase diffusivity	m²/s	L^2T^{-1}
D_{AB}	Diffusivity of $\mathbf{A}$ in $\mathbf{B}$	m²/s	L^2T^{-1}
D_{BA}	Diffusivity of $\mathbf{B}$ in $\mathbf{A}$	m²/s	L^2T^{-1}
D'	Effective diffusivity in multicomponent system	m²/s	L^2T^{-1}
d	Pipe diameter	m	L
E_D	Eddy diffusivity	m²/s	L^2T^{-1}
F	Coefficient in Maxwell's law of diffusion	m³/kmol s	$M^{-1}L^3T^{-1}$
G	Molar flow of stream	kmol/s	MT^{-1}
G'	Molar flow per unit area	kmol/m² s	$ML^{-2}T^{-1}$
$\mathscr{H}$	Ratio of equilibrium values of concentrations in two phases	—	$MT^{-3}\theta^{-1}$
$\mathbf{H}$	Height of transfer unit	m	L
h	Heat transfer coefficient	W/m² K	$MT^{-3}\theta^{-1}$
h_D	Mass transfer coefficient	m/s	LT^{-1}
j_d	"j-factor" for mass transfer	—	—
j_h	"j-factor" for heat transfer	—	—
K	Overall mass transfer coefficient	m/s	LT^{-1}
k	Mass transfer coefficient	m/s	LT^{-1}
k'	Mass transfer coefficient	m/s	LT^{-1}
k_G	Mass transfer coefficient for transfer through stationary fluid	s/m	$L^{-1}T$
k'_G	Mass transfer coefficient for equimolecular counterdiffusion	s/m	$L^{-1}T$
k	Reaction rate constant	*	*
L	Length of surface, or film thickness	m	L
M	Molecular weight	—	—
N	Molar rate of diffusion per unit area	kmol/m² s	$ML^{-2}T^{-1}$
N'	Total molar rate of transfer per unit area	kmol/m² s	$ML^{-2}T^{-1}$
$\mathbf{N}$	Number of transfer units	—	—
n	Number of moles of gas	kmol	M
n	Order of reaction, or number of term in series	—	—
P	Total pressure	N/m²	$ML^{-1}T^{-2}$
P_A, P_B	Partial pressure of $\mathbf{A, B}$	N/m²	$ML^{-1}T^{-2}$
P_s	Vapour pressure of water	N/m²	$ML^{-1}T^{-2}$
P_w	Partial pressure of water in gas stream	N/m²	$ML^{-1}T^{-2}$
P_{As}, P_{Bs}	Partial pressure of $\mathbf{A, B}$ outside boundary layer	N/m²	$ML^{-1}T^{-2}$
P_{Bm}	Logarithmic mean value of P_B	N/m²	$ML^{-1}T^{-2}$
p	Parameter in Laplace Transform	1/s	T^{-1}
R	Shear stress acting on surface	N/m²	$ML^{-1}T^{-2}$
$\mathbf{R}$	Universal gas constant	J/kmol K	$L^2T^{-2}\theta^{-1}$
r	Radius within sphere	m	L
S	Cross-sectional area of flow	m²	L^2
s	Rate of production of fresh surface per unit area	1/s	T^{-1}
T	Absolute temperature	K	θ
t	Time	s	T
t_e	Time of exposure of surface element	s	T
u	Mean velocity	m/s	LT^{-1}
u_A, u_B	Mean molecular velocity in direction of transfer	m/s	LT^{-1}
u_F	Velocity due to bulk flow	m/s	LT^{-1}
V	Volume	m³	L^3

*Dimensions depend on order of reaction.

		Units in SI system	Dimensions in $\mathbf{M}, \mathbf{L}, \mathbf{T}, \theta$
$\mathbf{V}$	Molecular volume	m³/kmol	$\mathbf{M^{-1}L^3}$
$\mathbf{V_0}$	Correction term in equation 8.48	m³/kmol	$\mathbf{M^{-1}L^3}$
W	Mass rate of evaporation	kg/s	$\mathbf{MT^{-1}}$
x	Distance from leading edge of surface or in X-direction	m	$\mathbf{L}$
y	Distance from surface or in direction of diffusion	m	$\mathbf{L}$
y'	Mol fraction in stationary gas	—	—
Z	Height of column	m	$\mathbf{L}$
z	Distance in Z-direction	m	$\mathbf{L}$
μ	Viscosity of fluid	N s/m²	$\mathbf{ML^{-1}T^{-1}}$
ρ	Density of fluid	kg/m³	$\mathbf{ML^{-3}}$
Nu	Nusselt number hd/k	—	—
Re	Reynolds number $ud\rho/\mu$	—	—
Re_x	Reynolds number $u_s x\rho/\theta$	—	—
Pr	Prandtl number	—	—
Sc	Schmidt number $\mu/\rho D$	—	—
Sh	Sherwood number $h_D d/D$	—	—
St	Stanton number $h/C_p \rho u$	—	—

Suffixes

1	Phase 1
2	Phase 2
A	Component $\mathbf{A}$
B	Component $\mathbf{B}$
AB	Of $\mathbf{A}$ in $\mathbf{B}$
b	Bottom of column
e	Value in equilibrium with bulk of other phase
G	Gas phase
i	Interface value
o	Value in bulk of phase
t	Top of column
av	Average value

CHAPTER 9

The Boundary Layer and Pipe Flow

9.1. INTRODUCTION

When a fluid flows over a solid surface a velocity gradient is set up at right angles to the direction of flow because of the viscous forces acting within the fluid. The fluid in contact with the surface must necessarily be brought to rest since, otherwise, the velocity gradient and the shear stress at the surface would be infinite. The drag force resulting from the retardation of the fluid at the surface is transmitted throughout the whole of the fluid and therefore the velocity gradient also extends through the whole of the fluid. At progressively greater distances from the surface, however, the effect of the drag becomes smaller and, for all practical purposes, can be regarded as being confined to a region close to the surface and known as the *boundary layer*. The whole of the velocity gradient is thus assumed to lie within the boundary layer and, outside it, the velocity is assumed to remain constant.

The thickness of the boundary layer will be a function of the distance from the leading edge of the surface (Fig. 9.1). Since the viscous drag of the fluid can be transmitted only at a finite rate, the boundary layer thickness will be zero at the leading edge and gradually increase as the distance from the leading edge increases. In any plane at right angles to the direction of flow, however, the velocity within the boundary layer will vary from zero at the surface to u_s, the velocity of the undisturbed stream, at its outer edge. Where the boundary layer thickness is small, the flow is streamline and the velocity at any distance from the surface is a simple function of that distance. At a certain critical thickness, however, the flow changes from streamline to turbulent, except within a very thin layer near the surface, where it remains streamline: this thin layer is known as the *laminar sub-layer*. Between the laminar sub-layer and the turbulent portion of the boundary layer is a region in which the flow is neither streamline nor fully turbulent; this is known as the *buffer layer*.

The change from streamline to turbulent conditions in the boundary layer occurs at a certain critical distance from the leading edge. This distance depends on the shape of the leading edge and the roughness of the surface, and also on the velocity and properties of the fluid: thus with a rough surface or a blunt edge it is comparatively short. For a given surface the transition takes place at some critical value of the Reynolds group with respect to distance x from the leading edge. This Reynolds group will be denoted by the symbol Re_x and its critical value at a distance x_c from the leading edge by the symbol Re_{x_e}; Re_{x_e} is of the order of 10^5.

The transition from streamline to turbulent flow in the neighbourhood of a surface has been studied by F. N. M. Brown[1] with the aid of a three-dimensional smoke tunnel with transparent faces. Photographs were taken using short-interval flash lamps giving effective exposure times of about 20 μs. Figure 9.2 refers to the flow over an aerofoil and shows the

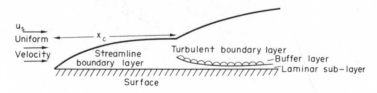

FIG. 9.1. Development of boundary layer.

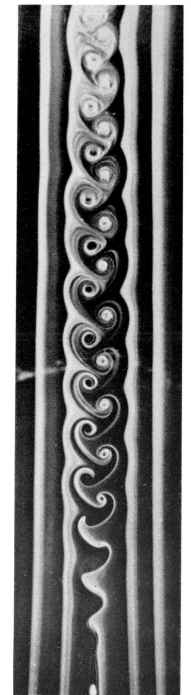

Fig. 9.2. Formation of vortices in flow over an aerofoil.

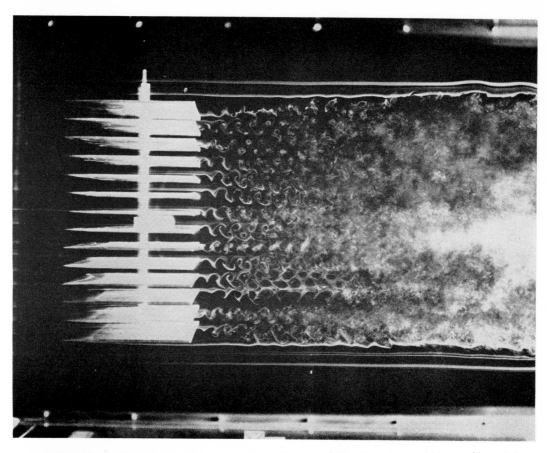

FIG. 9.3. Wake produced by cascade of flat plates with sharpened leading edges and blunt trailing edges.

manner in which vortices are created. Figure 9.3 shows the wake of a cascade of flat plates with sharpened leading edges and blunt trailing edges.

A knowledge of conditions in the boundary layer is necessary in order to calculate the resultant drag force of the surface on the fluid and in the calculation of heat and mass transfer coefficients. Expressions will now be derived for the thickness and velocity of flow at any point in the boundary layer.

9.2. THE MOMENTUM EQUATION

A fluid of density ρ and viscosity μ flows over a plane surface and the velocity of flow outside the boundary layer is u_s. A boundary layer of thickness δ forms near the surface, and at a height y above the surface the velocity of the fluid is reduced to a value u_x.

Consider the equilibrium of an element of fluid bounded by the planes 1–2 and 3–4 at distances x and $x + dx$ from the leading edge; the surface and the plane 2–4 parallel to the surface and at a distance l from it; and by two planes parallel to the plane of the paper and unit distance apart. The distance l is greater than the boundary layer thickness δ (Fig. 9.4).

Consider the velocities and forces in the X-direction.

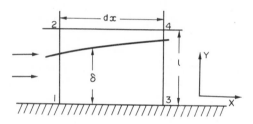

FIG. 9.4. Element of boundary layer.

At plane 1–2, mass rate of flow through a strip of thickness dy at distance y from the surface

$$= \rho u_x \, 1 \, dy$$

Total flow through plane 1–2

$$= \int_0^l \rho u_x \, dy$$

Rate of transfer of momentum through the elementary strip

$$= \rho u_x \, 1 \, dy \, u_x = \rho u_x^2 \, dy$$

Total rate of transfer of momentum through plane 1–2

$$= \int_0^l \rho u_x^2 \, dy$$

In passing from plane 1–2 to plane 3–4, the mass flow changes by:

$$\frac{\partial}{\partial x} \left(\int_0^l \rho u_x \, dy \right) dx$$

and the rate of transfer of momentum changes by:

$$\frac{\partial}{\partial x}\left(\int_0^l \rho u_x^2 \, dy\right) dx$$

A mass flow of fluid equal to the difference between the flows at planes 3–4 and 1–2 must therefore occur through plane 2–4.

Since plane 2–4 lies outside the boundary layer, this entering fluid must have a velocity u_s in the X-direction. Because the fluid in the boundary layer is being retarded, there will be a smaller flow at plane 3–4 than at 1–2, and hence the amount of fluid entering through plane 2–4 is negative i.e. fluid actually leaves the element.

Thus the rate of transfer of momentum through plane 2–4 into the element

$$= u_s \frac{\partial}{\partial x}\left(\int_0^l \rho u_x \, dy\right) dx$$

The net rate of change of momentum in the X-direction on the element must be equal to the momentum added from outside together with the net force acting on it.

The forces in the X-direction acting on the element of fluid are:

(1) A shear force resulting from the shear stress R_0 acting at the surface. This is a retarding force and therefore R_0 is negative.
(2) The force produced as a result of a difference in pressure dP between the planes 3–4 and 1–2. However, if the velocity outside the boundary layer remains constant, there can be no pressure gradient in the X-direction (i.e. $\partial P/\partial x = 0$) and therefore will be zero.

Thus the net force acting $= R_0 \, dx$

Hence

$$\frac{\partial}{\partial x}\left(\int_0^l \rho u_x^2 \, dy\right) dx = u_s \frac{\partial}{\partial x}\left(\int_0^l \rho u_x \, dy\right) dx + R_0 \, dx$$

Since u_s does not vary with x:

$$\frac{\partial}{\partial x}\int_0^l \rho(u_s - u_x)u_x \, dy = -R_0 \tag{9.1}$$

This expression, known as the *momentum equation*, can be integrated if the relations between u_x and y and between P and x are known. The pressure drop in the X-direction will be negligible if the velocity of the main stream remains constant at u_s, and for an incompressible fluid or for a compressible fluid where the pressure changes are small compared with the total pressure, the expression can therefore be written:

$$\rho \frac{\partial}{\partial x}\int_0^l (u_s - u_x)u_x \, dy = -R_0 \tag{9.2}$$

It will be noted that no assumptions have been made concerning the nature of the flow within the boundary layer and therefore this relation is applicable to both the streamline and the turbulent regions. The relation between u_x and y will now be derived for streamline and turbulent flow over a plane surface and equation 9.2 will be integrated.

9.3. THE STREAMLINE PORTION OF THE BOUNDARY LAYER

In the streamline boundary layer the only forces acting within the fluid are pure viscous forces and no transfer of momentum takes place by eddy motion.

Assume that the relation between u_x and y can be expressed approximately by:

$$u_x = u_0 + a'y + b'y^2 + c'y^3 \tag{9.3}$$

The coefficients a', b', c', and u_0 can be evaluated because the boundary conditions which the relation must satisfy are known (Fig. 9.5).

The fluid in contact with the surface is at rest and therefore u_0 must be zero. Furthermore, all the fluid close to the surface is moving at very low velocity and therefore any changes in its momentum as it flows parallel to the surface must be extremely small. Consequently the net shear force acting on an element of fluid near the surface is negligible, the retarding force at its lower boundary being balanced by the accelerating force at its upper boundary. Thus the shear stress R_0 in the fluid near the surface approaches a constant value.

Since $R_0 = -\mu(du_x/dy)_{y=0}$, du_x/dy must also be constant at small values of y and:

$$(\partial^2 u_x/\partial y^2)_{y=0} = 0$$

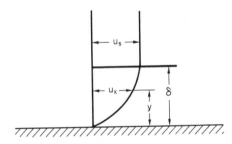

FIG. 9.5. Velocity distribution in streamline boundary layer.

At the distant edge of the boundary layer it is assumed that the velocity is equal to the main stream velocity.

Thus, when $y = \delta$: $u_x = u_s$ and $\partial u_x/\partial y = 0$

and equation 9.3 becomes $u_x = a'y + b'y^2 + c'y^3$

$$\partial u_x/\partial y = a' + 2b'y + 3c'y^2$$

and $\partial^2 u_x/\partial y^2 = 2b' + 6c'y$

At $y = 0$: $\partial^2 u_x/\partial y^2 = 0$

Thus $b' = 0$

At $y = \delta$: $u_x = a'\delta + c'\delta^3 = u_s$

and $\partial u_x/\partial y = a' + 3c'\delta^2 = 0$

Thus $a' = -3c'\delta^2$

Hence $c' = -u_s/2\delta^3$ and $a' = 3u_s/2\delta$

The equation for the velocity profile is therefore:

$$u_x = \frac{3u_s}{2}\frac{y}{\delta} - \frac{u_s}{2}\left(\frac{y}{\delta}\right)^3 \tag{9.4}$$

i.e.
$$\frac{u_x}{u_s} = \frac{3}{2}\left(\frac{y}{\delta}\right) - \frac{1}{2}\left(\frac{y}{\delta}\right)^3 \tag{9.5}$$

This relation applies over the range $0 < y < \delta$.

When $y > \delta$:
$$u_x = u_s \tag{9.6}$$

The integral in the momentum equation (9.2) can now be evaluated for the streamline boundary layer by considering the ranges $0 < y < \delta$ and $\delta < y < l$ separately.

Thus

$$\int_0^l (u_s - u_x)u_x \, dy = \int_0^\delta u_s^2\left(1 - \frac{3}{2}\frac{y}{\delta} + \frac{y^3}{2\delta^3}\right)\left(\frac{3}{2}\frac{y}{\delta} - \frac{y^3}{2\delta^3}\right) dy + \int_\delta^l (u_s - u_s)u_s \, dy$$

$$= u_s^2\int_0^\delta \left(\frac{3}{2}\frac{y}{\delta} - \frac{9}{4}\frac{y^2}{\delta^2} - \frac{1}{2}\frac{y^3}{\delta^3} + \frac{3}{2}\frac{y^4}{\delta^4} - \frac{1}{4}\frac{y^6}{\delta^6}\right) dy$$

$$= u_s^2\delta(\tfrac{3}{4} - \tfrac{3}{4} - \tfrac{1}{8} + \tfrac{3}{10} - \tfrac{1}{28})$$

$$= 39u_s^2\delta/280 \tag{9.7}$$

Now
$$R_0 = -\mu\left(\frac{\partial u_x}{\partial y}\right)_{y=0} = -\frac{3}{2}\mu\frac{u_s}{\delta} \tag{9.8}$$

Substitution from equations 9.7 and 9.8 in equation 9.2, gives:

$$\rho\frac{\partial}{\partial x}\left(\frac{39}{280}\delta u_s^2\right) = \frac{3}{2}\mu\frac{u_s}{\delta}$$

$\therefore$
$$\delta \, d\delta = \frac{140}{13}\frac{\mu}{\rho}\frac{1}{u_s}\,dx$$

$\therefore$
$$\frac{\delta^2}{2} = \frac{140}{13}\frac{\mu x}{\rho u_s} \quad \text{(since } \delta = 0 \text{ when } x = 0) \tag{9.9}$$

Thus
$$\delta = 4{\cdot}64\sqrt{\frac{\mu x}{\rho u_s}} \tag{9.10}$$

and
$$\frac{\delta}{x} = 4{\cdot}64\sqrt{\frac{\mu}{x\rho u_s}} = 4{\cdot}64 Re_x^{-1/2} \tag{9.11}$$

This relation for the thickness of the boundary layer has been obtained on the assumption that the velocity profile can be described by a polynomial of the form of equation 9.3 and that the main stream velocity is reached at a distance δ from the surface, whereas, in fact, the stream velocity is approached asymptotically. Whereas equation 9.4 gives the velocity u_x accurately as a function of y, it does not provide a means of calculating accurately the distance from the surface at which u_x has a particular value, when u_x is near u_s. The thickness of the boundary layer as calculated is therefore a function of the particular approximate relation which is taken to represent the velocity profile. This difficulty can be overcome by introducing a new term, the *displacement thickness* δ^*.

When a viscous fluid flows over a surface it is retarded and the overall flowrate is therefore reduced. A non-viscous fluid, however, would not be retarded and therefore a boundary layer would not form. The displacement thickness δ^* is defined as the distance the surface would have to be moved in the Y-direction in order to obtain the same rate of flow with this non-viscous fluid as would be obtained with the viscous fluid.

Mass rate of flow of a frictionless fluid between $y = \delta*$ and $y = \infty$

$$= \rho \int_{\delta*}^{\infty} u_s \, dy$$

Mass rate of flow of the real fluid between $y = 0$ and $y = \infty$

$$= \rho \int_{0}^{\infty} u_x \, dy$$

Then, by definition of the displacement thickness:

$$\rho \int_{\delta*}^{\infty} u_s \, dy = \rho \int_{0}^{\infty} u_x \, dy$$

i.e.

$$\int_{0}^{\infty} u_s \, dy - \int_{0}^{\delta*} u_s \, dy = \int_{0}^{\infty} u_x \, dy$$

giving

$$\delta* = \int_{0}^{\infty} \left(1 - \frac{u_x}{u_s} \right) dy \tag{9.12}$$

$$= \int_{0}^{\delta} \left(1 - \frac{3}{2} \frac{y}{\delta} + \frac{1}{2} \frac{y^3}{\delta^3} \right) dy$$

(since equation 9.5 applies only in the limits $0 < y < \delta$ and outside this region $u_x = u_s$ and the integral is zero):

i.e.

$$\delta* = \delta(1 - \tfrac{3}{4} + \tfrac{1}{8})$$
$$= 0 \cdot 375\delta \tag{9.13}$$

9.3.1. Shear Stress at the Surface

The shear stress in the fluid at the surface:

$$R_0 = -\frac{3}{2} \mu \frac{u_s}{\delta} \quad \text{(from equation 9.8)}$$

$$= -\frac{3}{2} \mu u_s \frac{1}{x} \frac{1}{4 \cdot 64} \sqrt{\frac{x \rho u_s}{\mu}}$$

$$= -0 \cdot 323 \rho u_s^2 \sqrt{\frac{\mu}{x \rho u_s}} = -0 \cdot 323 \rho u_s^2 \, Re_x^{-1/2}$$

The shear stress R acting on the surface itself is equal to $-R_0$.

Thus

$$R/\rho u_s^2 = 0 \cdot 323 \, Re_x^{-1/2} \tag{9.14}$$

This gives the point value of R and $R/\rho u_s^2$ at $x = x$. In order to calculate the total frictional force acting at the surface, it is necessary to multiply the average value of R between $x = 0$ and $x = x$ by the area of the surface.

The average value of $R/\rho u_s^2$ denoted by the symbol $(R/\rho u_s^2)_m$ is given by:

$$\left(\frac{R}{\rho u_s^2}\right)_m x = \int_0^x \frac{R}{\rho u_s^2}\, dx$$

$$= \int_0^x 0{\cdot}323 \sqrt{\frac{\mu}{x\rho u_s}}\, dx \quad \text{(from equation 9.14)}$$

$$= 0{\cdot}646 x \sqrt{\frac{\mu}{x\rho u_s}}$$

$$\therefore \quad \left(\frac{R}{\rho u_s^2}\right)_m = 0{\cdot}646 \sqrt{\frac{\mu}{x\rho u_s}}$$

$$= 0{\cdot}646\, Re_x^{-0{\cdot}5} \approx 0{\cdot}65\, Re_x^{-0{\cdot}5} \tag{9.15}$$

9.4. THE TURBULENT PORTION OF THE BOUNDARY LAYER

A useful simplified treatment of the flow conditions within the turbulent boundary layer is obtained by neglecting the existence of the buffer layer (Fig. 9.6). It is assumed that there is a region close to the surface, the laminar sublayer, in which momentum transfer is by molecular motion alone; outside this region the viscous forces are assumed to be negligible so that transfer is effected entirely by eddy motion. The method is based on the assumption that the shear stress at a plane surface can be calculated from the simple power law developed by Blasius, already referred to in Chapter 3.

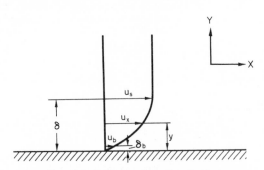

FIG. 9.6. Turbulent boundary layer.

Blasius[2] has given an approximate expression for the shear stress at a plane smooth surface over which a fluid is flowing with a velocity u_s, for conditions where $Re_x < 10^7$. His equation is as follows:

$$\frac{R}{\rho u_s^2} = 0{\cdot}0228 \left(\frac{\mu}{u_s \delta \rho}\right)^{0{\cdot}25} \tag{9.16}$$

Thus the shear stress is expressed as a function of the boundary layer thickness δ and it is therefore implicitly assumed that a certain velocity profile exists in the fluid. As a first assumption, assume that a simple power relation exists between the velocity and the distance from the surface in the boundary layer,

i.e. $(u_x/u_s) = (y/\delta)^f$ \tag{9.17}

Hence $\quad R = 0.0228\rho u_s^2(\mu/u_s\delta\rho)^{0.25}$

$$= 0.0228\rho^{0.75}\mu^{0.25}\delta^{-0.25}u_s^{1.75}$$

$$= 0.0228\rho^{0.75}\mu^{0.25}\delta^{-0.25}u_x^{1.75}(\delta/y)^{1.75f} \quad \text{(from equation 9.17)}$$

$$= 0.0228\rho^{0.75}\mu^{0.25}u_x^{1.75}y^{-1.75f}\delta^{1.75f-0.25} \tag{9.18}$$

If the velocity profile is the same for all stream velocities, the shear stress must be defined by specifying the velocity u_x at any distance y from the surface. The boundary layer thickness (determined by the velocity profile) is no longer an independent variable so that the index of δ in equation 9.18 must be zero, i.e.

$$1.75f - 0.25 = 0$$

and

$$f = 1/7$$

Thus

$$\frac{u_x}{u_s} = \left(\frac{y}{\delta}\right)^{1/7} \tag{9.19}$$

This is sometimes known as the Prandtl seventh power law.

In the analysis, no assumption has been made concerning the shape of the surface and this relation therefore applies to cylindrical as well as plane surfaces.

Differentiating equation 9.19 with respect to y:

$$\frac{\partial u_x}{\partial y} = \frac{1}{7}u_s\delta^{-1/7}y^{-6/7} \tag{9.20}$$

This relation gives an infinite velocity gradient at the surface and a finite velocity gradient at the outer edge of the boundary layer. This is in contradiction to the conditions which have been seen to exist in the stream. However, little error will be introduced by using this relation for the whole of the boundary layer in the momentum equation, since the velocities and hence the momenta near the surface are very low and it gives the correct value of the velocity at the edge of the boundary layer. Accepting equation 9.19 for the limits $0 < y < \delta$, the integral in equation 9.2 becomes:

$$\int_0^l (u_s - u_x)u_x \, dy = u_s^2\left\{\int_0^\delta\left[1 - \left(\frac{y}{\delta}\right)^{1/7}\right]\left(\frac{y}{\delta}\right)^{1/7} dy\right\} + \int_\delta^l (u_s - u_s)u_s \, dy$$

$$= u_s^2\int_0^\delta\left[\left(\frac{y}{\delta}\right)^{1/7} - \left(\frac{y}{\delta}\right)^{2/7}\right] dy$$

$$= u_s^2\delta\left(\frac{7}{8} - \frac{7}{9}\right)$$

$$= \frac{7}{72}u_s^2\delta \tag{9.21}$$

Now, from the Blasius equation:

$$-R_0 = R = 0.0228\rho u_s^2\left(\frac{\mu}{u_s\delta\rho}\right)^{1/4} \quad \text{(from equation 9.16)}$$

Substituting from equations 9.16 and 9.21 in equation 9.2:

$$\rho\frac{\partial}{\partial x}\frac{7}{72}u_s^2\delta = 0.0228\rho u_s^2\left(\frac{\mu}{u_s\delta\rho}\right)^{1/4}$$

i.e.

$$\delta^{1/4} \, d\delta = 0.235\left(\frac{\mu}{u_s\rho}\right)^{1/4} dx$$

i.e.
$$\frac{4}{5}\delta^{5/4} = 0.235x\left(\frac{\mu}{u_s\rho}\right)^{1/4}$$

This assumes that $\delta = 0$ when $x = 0$, i.e. that the turbulent boundary layer extends to the leading edge of the surface. An error is introduced by this assumption, but it is small except where the surface is only slightly longer than the critical distance x_c.

Thus
$$\delta = 0.376x^{0.8}(\mu/u_s\rho)^{0.2} \tag{9.22}$$
$$= 0.376x(\mu/u_s\rho x)^{0.2} \tag{9.23}$$
$$\delta/x = 0.376\,Re_x^{-0.2} \tag{9.24}$$
i.e.

Now the displacement thickness δ^* is given by equation 9.12:

$$\delta^* = \int_0^\infty \left(1 - \frac{u_x}{u_s}\right)dy \qquad \text{(equation 9.12)}$$

$$= \int_0^\delta \left(1 - \left(\frac{y}{\delta}\right)^{1/7}\right)dy \quad \text{(approximately)}$$

$$= \delta/8 \tag{9.25}$$

As already explained δ^* is independent of the particular approximation used for the velocity profile.

It is seen from equations 9.11 and 9.23 that at any given value of x, the rate of increase of thickness of the boundary layer is greater in the turbulent region than in the streamline region. At the transition point between the two types of flow, there is a small critical region in which the flow is unstable.

9.4.1. The Laminar Sub-layer

Suppose that at $x = x$ the laminar sub-layer is of thickness δ_b and that the total thickness of the boundary layer is δ.

It has already been explained that the shear stress and hence the velocity gradient are almost constant near the surface. Since the laminar sub-layer is very thin, the velocity gradient within it can therefore be taken as constant.

Thus the shear stress in the fluid at the surface

$$R_0 = -\mu\left(\frac{\partial u_x}{\partial y}\right)_{y=0} = -\mu\frac{u_x}{y}, \quad \text{where } y < \delta_b$$

Equating this to the value obtained from equation 9.16:

$$0.0228\rho u_s^2\left(\frac{\mu}{us\delta\rho}\right)^{1/4} = \mu\frac{u_x}{y}$$

so that
$$u_x = 0.0228\rho u_s^2\frac{1}{\mu}\left(\frac{\mu}{u_s\delta\rho}\right)^{1/4}y$$

If the velocity at the edge of the laminar sub-layer is u_b, i.e. if $u_x = u_b$ when $y = \delta_b$:

$$u_b = 0.0228\rho u_s^2\frac{1}{\mu}\left(\frac{\mu}{u_s\delta\rho}\right)^{1/4}\delta_b$$

$$= 0.0228\frac{\rho u_s^2}{\mu}\frac{\mu}{u_s\delta\rho}\delta_b\left(\frac{\mu}{u_s\delta\rho}\right)^{-3/4}$$

Thus
$$\frac{\delta_b}{\delta} = \frac{1}{0.0228}\left(\frac{u_b}{u_s}\right)\left(\frac{\mu}{u_s\delta\rho}\right)^{3/4} \tag{9.26}$$

Now the velocity at the inner edge of the turbulent region must also be given by the equation for the velocity distribution in the turbulent region.

Hence
$$\left(\frac{\delta_b}{\delta}\right)^{1/7} = \frac{u_b}{u_s} \quad \text{(from equation 9.19)}$$

Thus
$$\left(\frac{u_b}{u_s}\right)^7 = \frac{1}{0 \cdot 0228}\left(\frac{u_b}{u_s}\right)\left(\frac{\mu}{u_s\delta\rho}\right)^{3/4} \quad \text{(from equation 9.26)}$$

i.e.
$$\frac{u_b}{u_s} = 1 \cdot 87\left(\frac{\mu}{u_s\delta\rho}\right)^{1/8}$$
$$= 1 \cdot 87 \, Re \, \delta^{-1/8} \tag{9.27}$$

Now since
$$\delta = 0 \cdot 376 x^{0 \cdot 8}\left(\frac{\mu}{u_s\rho}\right)^{0 \cdot 2}, \tag{equation 9.22}$$
$$\frac{u_b}{u_s} = 1 \cdot 87\left(\frac{u_s\rho}{\mu} 0 \cdot 376 \frac{x^{0 \cdot 8}}{u_s^{0 \cdot 2}} \frac{\mu^{0 \cdot 2}}{\rho^{0 \cdot 2}}\right)^{-1/8}$$
$$= \frac{1 \cdot 87}{0 \cdot 376^{1/8}}\left(\frac{u_s^{0 \cdot 8} x^{0 \cdot 8} \rho^{0 \cdot 8}}{\mu^{0 \cdot 8}}\right)^{-1/8}$$
$$= 2 \cdot 11 \, Re_x^{-0 \cdot 1} \approx 2 \cdot 1 \, Re_x^{-0 \cdot 1} \tag{9.28}$$

The thickness of the laminar sub-layer is given by:

$$\frac{\delta_b}{\delta} = \left(\frac{u_b}{u_s}\right)^7 = \frac{190}{Re_x^{0 \cdot 7}} \quad \text{(from equations 9.19 and 9.28)}$$

i.e.
$$\frac{\delta_b}{x} = \frac{190}{Re_x^{0 \cdot 7}}\frac{0 \cdot 376}{Re_x^{0 \cdot 2}} \quad \text{(from equation 9.24)}$$
$$= 71 \cdot 5 \, Re_x^{-0 \cdot 9} \tag{9.29}$$

Thus $\delta_b \propto x^{0 \cdot 1}$, i.e. it increases very slowly as x increases. Further, $\delta_b \propto u_s^{-0 \cdot 9}$ and therefore decreases rapidly as the velocity is increased, and heat and mass transfer coefficients are therefore considerably influenced by the velocity.

The shear stress at the surface, at a distance x from the leading edge, is given by:

$$R_0 = -\mu(u_b/\delta_b)$$

Then, since $R_0 = -R$:

$$R = \mu \, 2 \cdot 11 u_s \, Re_x^{-0 \cdot 1} \frac{1}{x}\frac{1}{71 \cdot 5} Re_x^{0 \cdot 9} \quad \text{(from equations 9.28 and 9.29)}$$

$$= 0 \cdot 0296 \, Re_x^{0 \cdot 8} \frac{\mu u_s}{x}$$

$$= 0 \cdot 0296 \rho u_s^2 Re_x^{-0 \cdot 2} \approx 0 \cdot 03 \rho u_s^2 \, Re_x^{-0 \cdot 2} \tag{9.30}$$

i.e.
$$(R/\rho u_s^2) = 0 \cdot 0296 \, Re_x^{-0 \cdot 2} \tag{9.31}$$

or approximately
$$(R/\rho u_s^2) = 0 \cdot 03 \, Re_x^{-0 \cdot 2} \tag{9.32}$$

The mean value of $R/\rho u_s^2$ over the range $x = 0$ to $x = x$ is given by:

$$\left(\frac{R}{\rho u_s^2}\right)_m x = \int_0^x \left(\frac{R}{\rho u_s^2}\right) dx$$

$$= \int_0^x 0.0296 \left(\frac{\mu}{u_s x \rho}\right)^{0.2} dx$$

$$= 0.0296 \left(\frac{\mu}{u_s x \rho}\right)^{0.2} \frac{x}{0.8}$$

i.e.
$$\left(\frac{R}{\rho u_s^2}\right)_m = 0.037\, Re_x^{-0.2} \qquad\qquad (9.33)$$

The total shear force acting on the surface is found by adding the forces acting in the streamline $(x < x_c)$ and turbulent $(x > x_c)$ regions. This can be done provided the critical value Re_{x_c}, is known.

In the streamline region: $(R/\rho u_s^2)_m = 0.646\, Re_x^{-0.5}$ (equation 9.15)

In the turbulent region: $(R/\rho u_s^2)_m = 0.037\, Re_x^{-0.2}$ (equation 9.33)

In calculating the mean value of $(R/\rho u_s^2)_m$ in the turbulent region, it was assumed that the turbulent boundary layer extended to the leading edge. A more accurate value for the mean value of $(R/\rho u_s^2)_m$ over the whole surface can be obtained by using the expression for streamline conditions over the range from $x = 0$ to $x = x_c$ (where x_c is the critical distance from the leading edge) and the expression for turbulent conditions in the range $x = x_c$ to $x = x$.

Mean value of $(R/\rho u_s^2)_m$

$$= \frac{1}{x}(0.646\, Re_{x_c}^{-0.5} x_c + 0.037\, Re_x^{-0.2} x - 0.037\, Re_{x_c}^{-0.2} x_c)$$

$$= 0.646\, Re_{x_c}^{-0.5} \frac{Re_{x_c}}{Re_x} + 0.037\, Re_x^{-0.2} - 0.037\, Re_{x_c}^{-0.2} \frac{Re_{x_c}}{Re_x}$$

$$= 0.037\, Re_x^{-0.2} + Re_x^{-1}(0.646 Re_{x_c}^{0.5} - 0.037\, Re_{x_c}^{0.8}) \qquad\qquad (9.34)$$

Example 9.1. Water flows at a velocity of 1 m/s over a plane surface 0·6 m wide and 1 m long. Calculate the total drag force acting on the surface if the transition from streamline to turbulent flow in the boundary layer occurs when the Reynolds group $Re_x = 10^5$.

Solution. Taking $\mu = 1$ mN s/m$^2 = 10^{-3}$ Ns/m^2,

at the far end of the surface, $Re_x = 1 \times 1 \times 10^3/10^{-3} = 10^6$

Mean value of $R/\rho u_s^2$ from equation 9.34

$$= 0.037(10^6)^{-0.2} + (10^6)^{-1}[0.646(10^5)^{0.5} - 0.037(10^5)^{0.8}]$$

$$= 0.00214$$

Total drag force $= (R/\rho u_s^2)(\rho u_s^2)$ (area of surface)

$$= 0.00214 \times 1000 \times 1^2 \times 1 \times 0.6$$

$$= \underline{\underline{1.28\, N}}$$

Example 9.2. Calculate the thickness of the boundary layer at a distance of 150 mm from the leading edge of a surface over which oil, of viscosity 0·05 N s/m² and density 1000 kg/m³ flows with a velocity of 0·3 m/s. What is the displacement thickness of the boundary layer?

Solution. $Re_x = 0.150 \times 0.3 \times 1000/0.05 = 900$

$$\delta/x = 4.64/Re_x^{0.5} \quad \text{for streamline flow (from equation 9.11)}$$
$$= 4.64/900^{0.5} = 0.1545$$

Hence $\qquad\qquad \delta = 0.1545 \times 0.150 = 0.0232 \text{ m} = \underline{\underline{23.2 \text{ mm}}}$

Displacement thickness $= 0.375 \times 23.2 = \underline{\underline{8.7 \text{ mm}}}$ (from equation 9.13)

9.5. BOUNDARY LAYER THEORY APPLIED TO PIPE FLOW

9.5.1. Entry Conditions

When a fluid flowing with a uniform velocity enters a pipe, a boundary layer forms at the walls and gradually thickens as the distance from the entry point increases. Since the fluid in the boundary layer is retarded and the total flow remains constant, the fluid in the central stream will be accelerated. At a certain distance from the inlet, the boundary layers, which have formed in contact with the walls, join at the axis of the pipe, and, from that point onwards, occupy the whole cross-section and consequently remain of constant thickness. *Fully developed flow* then exists. If the boundary layers are still streamline when fully developed flow commences, the flow in the pipe remains streamline. On the other hand, if the boundary layers are already turbulent, turbulent flow will persist (Fig. 9.7).

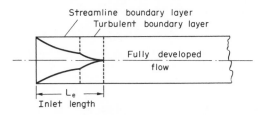

FIG. 9.7. Conditions at entry to pipe.

An approximate experimental expression for the inlet length L_e is:

$$L_e/d = 0.0288 \, Re \qquad\qquad (9.35)$$

where d is the diameter of the pipe and Re is the Reynolds group with respect to pipe diameter, and based on the mean velocity of flow in the pipe. This expression is only approximate, and is inaccurate for Reynolds numbers in the region of 2500 because the boundary layer thickness increases very rapidly in this region. An average value of L_e at a Reynolds number of 2500 is 50–100d. The inlet length is somewhat arbitrary as steady conditions in the pipe are approached asymptotically—the boundary layer thickness being a function of the assumed velocity profile.

At the inlet to the pipe the velocity across the whole section will be constant. At some distance from the inlet, the velocity at the pipe axis will have increased and will reach a maximum value when the boundary layer joins. After this point the whole velocity distribution, and hence the velocity at the axis, will remain constant. Since the velocity of the fluid at the axis is increased, the kinetic energy per unit mass increases, and therefore there must be a corresponding fall in the pressure energy.

Under streamline conditions, the velocity at the axis u_s will increase from a value u at the inlet to a value $2u$ where fully developed flow exists (Fig. 9.8).

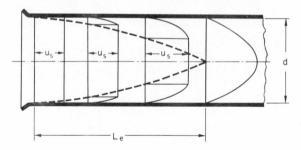

FIG. 9.8. Development of laminar velocity profile at entry to pipe.

Thus kinetic energy per unit mass of the fluid at the axis inlet $= u^2/2$.

Corresponding kinetic energy at end of inlet length $= (1/2)(2u)^2 = 2u^2$.

Increase in kinetic energy per unit mass $= 3u^2/2$.

Thus fall in pressure due to increase of velocity of fluid $= 3\rho u^2/2$.

If the flow in the pipe is turbulent, the velocity at the axis increases from u to only about $u/0.82$.

Under these conditions, fall in pressure

$$= \tfrac{1}{2}\rho u^2 \left(\frac{1}{0.82^2} - 1\right)$$

$$= \rho u^2/4 \tag{9.36}$$

The existence of a *vena contracta* near the entry to the pipe has been neglected here.

9.5.2. Application of Boundary Layer Theory

The velocity distribution and frictional resistance have already been calculated from purely theoretical considerations for the streamline flow of a fluid in a pipe. The boundary layer theory can now be applied in order to calculate, approximately, the conditions when the fluid is turbulent. For this purpose it is assumed that the boundary layer expressions can be applied to flow over a cylindrical surface and that the flow conditions in the region of fully developed flow are the same as those when the boundary layers first join. The thickness of the boundary layer is thus taken to be equal to the radius of the pipe and the velocity at the outer edge of the boundary layer is assumed to be the velocity at the axis.

The velocity of the fluid can be assumed to obey the Prandtl one-seventh power law (equation 9.19). If the boundary layer thickness δ is replaced by the pipe radius r this becomes:

$$u_x = u_s(y/r)^{1/7} \tag{9.37}$$

The relation between the mean velocity and the velocity at the axis has already been derived using this expression in Chapter 3. The mean velocity u was found to be 0.82 times the velocity u_s at the axis, but in this calculation the thickness of the laminar sub-layer was neglected and the Prandtl velocity distribution assumed to apply over the whole cross-section. The result therefore is strictly applicable only at very high Reynolds numbers where the thickness of the laminar sub-layer is very small. At lower Reynolds numbers the mean velocity will be rather less than 0·82 times the velocity at the axis.

The expressions for the shear stress at the walls, the thickness of the laminar sub-layer, and the velocity at the outer edge of the laminar sub-layer can be applied to the turbulent flow of a fluid in a pipe. It is convenient to express these relations in terms of the mean velocity in the pipe, the pipe diameter, and the Reynolds group with respect to the mean velocity and diameter.

The shear stress at the walls is given by the Blasius equation (9.16):

$$\frac{R}{\rho u_s^2} = 0.0228 \left(\frac{\mu}{u_s r \rho}\right)^{1/4}$$

Writing $u = 0.82u_s$ and $d = 2r$:

$$\frac{R}{\rho u^2} = 0.0384 \left(\frac{\mu}{udp}\right)^{1/4} = 0.0384\, Re^{-1/4} \tag{9.38}$$

This equation is more usually written:

$$R/\rho u^2 = 0.0396\, Re^{-1/4} \tag{9.39}$$

The constant of 0·0396 corresponds to $u = 0.81u_s$. It will be appreciated that $u/u_s = 0.82$ only when the laminar sub-layer is so thin that the one-seventh power law can be regarded as applying over the whole cross-section of the pipe.

Equation 9.39 is applicable for Reynolds numbers up to 10^5.

The velocity at the edge of the laminar sub-layer is given by:

$$\frac{u_b}{u_s} = 1.87 \left(\frac{\mu}{u_s r \rho}\right)^{1/8} \tag{equation 9.27}$$

which becomes:

$$\frac{u_b}{u} = 2.49 \left(\frac{\mu}{udp}\right)^{1/8}$$

$$= 2.49\, Re^{-1/8} \tag{9.40}$$

and

$$u_b/u_s = 2.0\, Re^{-1/8} \tag{9.41}$$

The thickness of the laminar sub-layer is given by:

$$\frac{\delta_b}{r} = \left(\frac{u_b}{u_s}\right)^7 \quad \text{(from equation 9.19)}$$

$$= 1.87^7 \left(\frac{\mu}{u_s r \rho}\right)^{7/8} \quad \text{(from equation 9.27)}$$

Thus

$$\frac{\delta_b}{d} = 62 \left(\frac{\mu}{udp}\right)^{7/8}$$

$$= 62\, Re^{-7/8} \tag{9.42}$$

Thus the thickness of the laminar sub-layer is approximately inversely proportional to the Reynolds number and hence to the velocity.

Example 9.3. Calculate the thickness of the laminar sub-layer when benzene flows through a pipe 50 mm in diameter at 2000 cm³/s. What is the velocity of the benzene at the edge of the laminar sub-layer? Assume that fully developed flow exists within the pipe and for benzene, $\rho = 870$ kg/m² and $\mu = 0.7$ mN s/m².

Solution. $2000 \text{ cm}^3/\text{s} = 2000 \times 10^{-6} \times 870$

$$= 1 \cdot 74 \text{ kg/s}$$

Reynolds number $= 4G/\mu\pi D = 4 \times 1 \cdot 74/(0 \cdot 7 \times 10^{-3} \pi \times 0 \cdot 050)$

$$= 63,290$$

From equation 9.42: $\delta_b/d = 62 \, Re^{-\frac{7}{8}}$

$\therefore$ $\delta_b = 62 \times 0 \cdot 050/63290^{\frac{7}{8}}$

$$= 1 \cdot 95 \times 10^{-4} \text{ m}$$

or $\underline{0 \cdot 195 \text{ mm}}$

Mean velocity $= 1 \cdot 74/[870 \times (\pi/4)0 \cdot 050^2]$

$$= 1 \cdot 018 \text{ m/s}$$

From equation 9.40: $u_b/u = 2 \cdot 49/Re^{\frac{1}{8}}$

$\therefore$ $u_b = (2 \cdot 49 \times 1 \cdot 018)/63290^{\frac{1}{8}}$

$$= \underline{0 \cdot 637 \text{ m/s}}$$

9.6.   THE DEVELOPMENT OF EXPRESSIONS FOR VELOCITY PROFILE AND SHEAR STRESS IN TURBULENT FLOW

9.6.1.   The Application of the Prandtl Concept of Mixing Length

Prandtl[3] introduced the concept of a *mixing length* in order to obtain an expression for the velocity distribution in turbulent flow: a simplified form of his analysis is given below. It must be remembered that the high frictional losses in turbulent flow are attributable to the exchange of momentum between elements of fluid which move from one plane to another in a fluid stream. Slow-moving elements entering a faster-moving fluid exert a drag and fast elements entering a slow-moving fluid speed it up; thus the existence of velocity gradients is responsible for the shear stress between the layers.

The curve in Fig. 9.9 shows the velocity profile in two dimensions of a stream moving towards the right. Suppose an element of fluid to move from position 1 at a distance y above the surface to a position 2 a distance λ_E beyond section 1. Then if the mean velocity at position 1 is u_x, at position 2 it is $u_x + \lambda_E (du_x/dy)$ (approximately).

Since the flow is turbulent there will be superimposed on these velocities fluctuating velocities u'_x in the X-direction and u'_y in the Y-direction, where the average values of u'_x and u'_y (i.e. $\bar{u}'_x$, $\bar{u}'_y$) are zero. It is further assumed that these fluctuating velocities are of the same order of magnitude as the difference in velocities at sections 1 and 2:

i.e. $$u'_x \approx \lambda_E \frac{du_x}{dy} \tag{9.43}$$

Then the shear stress between sections 1 and 2

$$= \text{rate of change of momentum:}$$

i.e. $$-R_y = \rho u'_y \lambda_E \frac{du_x}{dy} \tag{9.44}$$

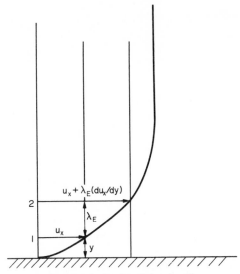

FIG. 9.9. Prandtl mixing length.

Prandtl takes $u'_x \triangleq u'_y$, so that:

$$- R_y = \rho \left(\lambda_E \frac{du_x}{dy} \right)^2 \tag{9.45}$$

The negative sign must prefix the shear stress term because momentum is being transferred *towards* the surface, i.e. in the negative sense.

Considering the shear stress round a cylindrical surface of radius s with its axis a distance r from the wall:

$$R_s = \frac{dP}{dx} \frac{s}{2}$$

and at the wall:

$$R_0 = \frac{dP}{dx} \frac{r}{2}$$

where dP/dx is the pressure gradient in the direction of flow.

Thus

$$- R_s = - R_0 \frac{s}{r} = \rho \left(\lambda_E \frac{du_x}{dy} \right)^2 \tag{9.46}$$

Noting that if y is the distance from the wall:

$$\frac{s}{r} = 1 - \frac{y}{r}$$

and writing $R(= - R_0)$ as the shear stress acting on the pipe surface:

$$R \left(1 - \frac{y}{r} \right) = \rho \left(\lambda_E \frac{du_x}{dy} \right)^2$$

i.e.

$$\lambda_E \frac{du_x}{dy} = \sqrt{\frac{R}{\rho}} \sqrt{1 - \frac{y}{r}} \tag{9.47}$$

Now $\sqrt{R/\rho} = \sqrt{\text{stress/density}}$ and has dimensions of velocity and is denoted by u^* and is known as the friction or shearing stress velocity.

It is further assumed that the roughness of the surface and the viscosity of the fluid do not influence λ_E. However, λ_E increases with the distance from the wall, and Prandtl assumed that $\lambda_E = Ky$. Experiments by Nikuradse[4,5] gave a value of 0·4 for K.

Near the wall $1 - (y/r)$ is approximately equal to unity so that:

$$Ky(du_x/dy) = u^*$$

Giving, on integration:

$$u_x = \frac{u^*}{K} \ln y + B \qquad\qquad (9.48)$$

When $y = r$:

$$u_x = u_{max} = u_s$$

Thus

$$u_s = \frac{u^*}{K} \ln r + B$$

or

$$u_x - u_s = \frac{u^*}{K} \ln \frac{y}{r}.$$

Putting $k = 0.4$ gives the Prandtl *velocity defect law* for velocity distribution as:

$$\frac{u_s - u_x}{u^*} = 2.5 \ln \frac{r}{y} \qquad\qquad (9.49)$$

This relationship, which is shown in Fig. 9.10, is applicable to both smooth and rough pipes.

Furthermore, substituting $K = 0.4$ into equation 9.48 gives:

$$\frac{u_x}{u^*} = 2.5 \ln y + \frac{B}{u^*}$$

which may be written as

$$\frac{u_x}{u^*} = 2.5 \ln \frac{yu^*\rho}{\mu} + B' \qquad\qquad (9.50)$$

u_x/u^* is a dimensionless derivative of u_x and can be denoted by u^+; $yu^*\rho/\mu$ is a dimensionless derivative of y and can be denoted by y^+.

Then

$$u^+ = 2.5 \ln y^+ + B' \qquad\qquad (9.51)$$

In the preceding analysis the transference of momentum has been attributed entirely to the turbulent eddies in the fluid, and the equations are therefore applicable under those conditions where the flow is fully turbulent so that transfer by molecular movement can be neglected in comparison with eddy transfer.

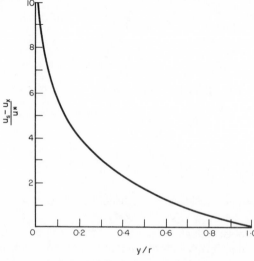

FIG. 9.10. Velocity defect law.

9.6.2. Universal Velocity Profile

A general expression for the velocity distribution in a flowing fluid may be obtained if note is taken of the fact that the flow in the vicinity of a surface or tube wall may conveniently be divided into three regions for turbulent flow.

The Fully Turbulent Core

In this region the contribution of the eddies to momentum transfer far exceeds that of the individual molecules, and thus equation 9.50 is applicable. If u^+ (to a linear scale) is plotted against y^+ (to a logarithmic scale) for flow over a smooth surface, a straight line of slope 2·5 is obtained, as shown in Fig. 9.11, for values of y^+ exceeding about 30. By extrapolation to $y^+ = 1$ ($\ln y^+ = 0$), the constant C' is found to have a value of about 5·5. For rough surfaces the slope of the line obtained is still 2·5, as for the smooth surface, but the intercept C' has a value which decreases with increase in the roughness e of the surface.

The Laminar Sub-layer

Close to the surface, the turbulent eddies die out and the transfer of momentum is effected solely by the action of the molecules, and therefore a region of viscous shear exists. The region in which this occurs is known as the laminar sub-layer. Because it is thin, the velocity gradient is approximately linear and can therefore be written as $u_x/y = u_b/\delta_b$, where δ_b is the thickness of the layer and u_b is the fluid velocity at its outer edge. Thus the shear stress is given by:

$$R = \mu \frac{u_b}{\delta_b} = \mu \frac{u_x}{y}$$

Then

$$u^{*2} = \frac{R}{\rho} = \frac{\mu u_x}{\rho y}$$

$$\therefore \quad \frac{u_x}{u^*} = \frac{yu^*\rho}{\mu} \tag{9.52}$$

or

$$u^+ = y^+ \tag{9.53}$$

This relationship holds reasonably well, as can be seen from Fig. 9.11, for values of y^+ up to about 5. It applies to both rough and smooth surfaces.

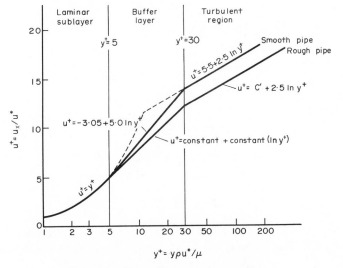

FIG. 9.11. The universal velocity profile.

The Transitional or Buffer Layer

In the region between the laminar sub-layer and the turbulent zone, the contributions of the molecules and of the eddies to momentum transfer are comparable. In Fig. 9.11 the data are represented by a short transitional curve.

Dimensionless Velocity-Distance Relations for Pipe Flow

Figure 9.11 is often known as the *universal velocity profile* which is a representation of the data for the complete velocity profile.

For a smooth surface:

$$0 < y^+ < 5 \qquad u^+ = y^+ \tag{equation 9.53}$$

$$5 < y^+ < 30 \qquad u^+ = 5 \cdot 0 \ln y^+ - 3 \cdot 05 \tag{9.54}$$

$$y^+ > 30 \qquad u^+ = 2 \cdot 5 \ln y^+ + 5 \cdot 5 \tag{9.55}$$

The corresponding curves for rough and smooth surfaces are shown schematically in Fig. 9.11.

For flow in a pipe the quantities u^*, u^+, and y^+ can be expressed in terms of the Reynolds number ($Re = ud\rho/\mu$) and the friction factor ($\phi = R/\rho u^2$).

Thus

$$u^* = \sqrt{\frac{R}{\rho}} = \sqrt{\frac{R}{\rho u^2}}\, u = \phi^{1/2} u \tag{9.56}$$

$$u^+ = \frac{u_x}{u^*} = \frac{u_x}{u}\, \phi^{-1/2} \tag{9.57}$$

$$y^+ = \frac{yu^*\rho}{\mu} = \frac{y}{d}\frac{u^*}{u}\frac{ud\rho}{\mu} = \frac{y}{d}\, \phi^{1/2}\, Re \tag{9.58}$$

The thickness of the laminar sub-layer (δ_b) and the velocity at its outer edge (u_b) are found by substituting in equation 9.52. Thus, since $u^+ = y^+ = 5$:

$$\delta_b u^* \rho / \mu = 5$$

$$\therefore \qquad \frac{\delta_b}{d} = \frac{5\mu}{u^* \rho d} = 5\frac{\mu}{ud\rho}\frac{u}{u^*} = 5\, Re^{-1}\, \phi^{-1/2} \tag{9.59}$$

and

$$u_x/u^* = 5$$

$$\therefore \qquad u_x/u = 5\phi^{1/2} \tag{9.60}$$

Using the Blasius equation (3.9), $\phi = 0 \cdot 04\, Re^{-1/4}$:

$$u_x/u = Re^{-1/8} \tag{9.61}$$

Velocity Gradients

From equations 9.53 to 9.55:

$$0 < y^+ < 5 \qquad \frac{du^+}{dy^+} = 1 \tag{9.62}$$

$$5 < y^+ < 30 \qquad \frac{du^+}{dy^+} = \frac{5 \cdot 0}{y^+} \tag{9.63}$$

$$y^+ > 30 \qquad \frac{du^+}{dy^+} = \frac{2 \cdot 5}{y^+} \tag{9.64}$$

Equations 9.62 and 9.64 apply to both rough and smooth surfaces; equation 9.63 applies to smooth surfaces only. The velocity gradient can then be calculated since:

$$\frac{du_x}{dy} = \frac{R}{\mu}\frac{du^+}{dy^+} \tag{9.65}$$

Thus
$$0 < y^+ < 5 \qquad \frac{du_x}{dy} = \frac{R}{\mu} \tag{9.66}$$

$$5 < y^+ < 30 \qquad \frac{du_x}{dy} = \frac{5 \cdot 0}{y^+} \frac{R}{\mu}$$

$$= 5 \cdot 0 \phi^{1/2} \frac{u}{y} \tag{9.67}$$

$$y^+ > 30 \qquad \frac{du_x}{dy} = 2 \cdot 5 \phi^{1/2} \frac{u}{y} \tag{9.68}$$

It should be noted that equation 9.68 must not be used in the centre of the pipe as it will then give excessive values for the velocity gradient. The reason for this is that $1 - (y/r)$ no longer approaches unity in this region.

9.6.3. Eddy Kinematic Viscosity

When a fluid is in streamline motion, momentum transfer and shear stresses result from the random movement of the molecules. The relation between shear stress and rate of shear for Newtonian behaviour is then given by:

$$R_y = -\mu \frac{du_x}{dy} = -\frac{\mu}{\rho} \frac{d(\rho u_x)}{dy} \tag{9.69}$$

In turbulent flow the eddies are responsible for most of the transfer of momentum and the shear stress can be related to the shear rate by the following equation:

$$R_y = -E\rho \frac{du_x}{dy} = -E \frac{d(\rho u_x)}{dy} \tag{9.70}$$

When both mechanisms make a significant contribution to momentum transfer; as for example in the buffer zone:

$$R_y = -(\mu + E\rho) \frac{du_x}{dy} = -\left(\frac{\mu}{\rho} + E\right) \frac{d(\rho u_x)}{dy} \tag{9.71}$$

The *kinematic viscosity* μ/ρ is a physical property of the fluid, but the *eddy kinematic viscosity* E is a function of the flow pattern, in addition to the physical properties of the fluid. By comparing equations 9.44, 9.45, and 9.70, it will be seen that

$$E = u'_y \lambda_E = \lambda_E^2 \frac{du_x}{dy} \tag{9.72}$$

In general, E will vary with distance y from a boundary surface.

9.6.4. Distribution of Shear Stress and Eddy Kinematic Viscosity

The distribution of shear stress over the cross-section of a pipe is simply obtained by considering a force balance, firstly over the whole cross-section of the pipe, and, secondly, over a central core. In a pipe of length L over which the pressure difference is ΔP:

$$-\Delta P \pi r^2 = -R_0 2 \pi r L$$

and
$$-\Delta P \pi (r - y)^2 = -R_y 2 \pi (r - y) L$$

Dividing:

$$\frac{R_y}{R_0} = 1 - \frac{y}{r} \tag{9.73}$$

It is now possible to determine the variation of the eddy kinematic viscosity over the cross-section of a smooth pipe. In the turbulent core, where $E \gg \mu/\rho$:

$$R_y = -E\rho \frac{\partial u_x}{\partial y} \qquad \text{(equation 9.70)}$$

Writing $R_0 = -R$ and substituting from equations 9.68 and 9.73:

$$E = \frac{R}{\rho}\left(1 - \frac{y}{r}\right)\frac{1}{2 \cdot 5}\phi^{-1/2}\frac{y}{u}$$

$$= 0 \cdot 4 \phi^{1/2} uy\left(1 - \frac{y}{r}\right) \qquad (9.74)$$

In the buffer zone:

$$R_y = -(\mu + E\rho)\frac{\partial u_x}{\partial y} \qquad \text{(equation 9.71)}$$

Then, substituting from equations 9.67 and 9.73:

$$E = \frac{R}{\rho}\left(1 - \frac{y}{r}\right)\frac{1}{5}\phi^{-1/2}\frac{y}{u} - \frac{\mu}{\rho}$$

$$= 0 \cdot 2\phi^{1/2} uy\left(1 - \frac{y}{r}\right) - \frac{\mu}{\rho} \qquad (9.75)$$

9.6.5. Friction Factor for a Smooth Pipe

Equation 9.55 can be used in order to calculate the friction factor $\phi = R/\rho u^2$ for the turbulent flow of fluid in a pipe. It is first necessary to obtain an expression for the mean velocity of the fluid from the relation:

$$u = \frac{\displaystyle\int_0^r [2\pi(r - y)\,\mathrm{d}yu_x]}{\pi r^2}$$

$$= 2\int_0^1 u_x\left(1 - \frac{y}{r}\right)\mathrm{d}\left(\frac{y}{r}\right) \qquad (9.76)$$

Now the velocity at the pipe axis u_s is obtained by putting $y = r$ into equation 9.55.

Thus $$u_s = u^*\left(2 \cdot 5 \ln\frac{r\rho u^*}{\mu} + 5 \cdot 5\right) \qquad (9.77)$$

Substituting for u_x in equation 9.76:

$$u = 2\int_0^1 \left(u_s - 2 \cdot 5 u^* \ln\frac{r}{y}\right)\left(1 - \frac{y}{r}\right)\mathrm{d}\left(\frac{y}{r}\right)$$

i.e. $$\frac{u}{u_s} = 2\int_0^1 \left(1 + 2 \cdot 5\frac{u^*}{u_s}\ln\frac{y}{r}\right)\left(1 - \frac{y}{r}\right)\mathrm{d}\left(\frac{y}{r}\right)$$

$$= 2\left\langle \frac{y}{r} - \frac{1}{2}\left(\frac{y}{r}\right)^2 + 2 \cdot 5\frac{u^*}{u_s}\left\{\left(\ln\frac{y}{r}\right)\left[\frac{y}{r} - \frac{1}{2}\left(\frac{y}{r}\right)^2\right] - \int\left(\frac{y}{r}\right)^{-1}\left[\frac{y}{r} - \frac{1}{2}\left(\frac{y}{r}\right)^2\right]\mathrm{d}\left(\frac{y}{r}\right)\right\}\right\rangle_0^1$$

$$= 2\left\langle \frac{y}{r} - \frac{1}{2}\left(\frac{y}{r}\right)^2 + 2 \cdot 5\frac{u^*}{u_s}\left\{\left(\ln\frac{y}{r}\right)\left[\frac{y}{r} - \frac{1}{2}\left(\frac{y}{r}\right)^2\right] - \left(\frac{y}{r}\right) + \frac{1}{4}\left(\frac{y}{r}\right)^2\right\}\right\rangle_0^1$$

$$= 2\left[\frac{1}{2} + 2 \cdot 5\frac{u^*}{u_s}\left(-\frac{3}{4}\right)\right]$$

$$= 1 - 3 \cdot 75\frac{u^*}{u_s} \qquad (9.78)$$

Substituting into equation 9.77:

$$u + 3 \cdot 75u^* = u^* \left\{ 2 \cdot 5 \ln \left[\left(\frac{d\rho u}{\mu} \right) \left(\frac{r}{d} \right) \left(\frac{u^*}{u} \right) \right] + 5 \cdot 5 \right\}$$

i.e.
$$u/u^* = 2 \cdot 5 \ln \{(Re)(u^*/u)\}$$

Now
$$\phi = R/\rho u^2 = (u^*/u)^2$$

∴
$$\phi^{-1/2} = 2 \cdot 5 \ln [(Re)\phi^{1/2}] \qquad (9.79)$$

The experimental results for ϕ as a function of Re closely follow equation 9.79 modified by a correction term of $0 \cdot 3$ to give:

$$\phi^{-1/2} = 2 \cdot 5 \ln [(Re)\phi^{1/2}] + 0 \cdot 3 \qquad (9.80)$$

Equation 9.80 is identical with equation 3.10.

Example 9.4. Air flows through a smooth circular duct of internal diameter 250 mm at an average velocity of 15 m/s. Calculate the fluid velocity at points 50 mm and 5 mm from the wall. What will be the thickness of the laminar sub-layer if this extends to $u^+ = y^+ = 5$? The density and viscosity of air may be taken as $1 \cdot 10$ kg/m³ and 20×10^{-6} N s/m² respectively.

Solution. Reynolds number: $Re = 0 \cdot 250 \times 15 \times 1 \cdot 10/(20 \times 10^{-6}) = 2 \cdot 06 \times 10^5$

Hence, from Fig. 3.7: $(R/\rho u^2) = 0 \cdot 0018$

$$u_s = u/0 \cdot 82 = 15/0 \cdot 82 = 18 \cdot 3 \text{ m/s}$$
$$u^* = u\sqrt{(R/\rho u^2)} = 15\sqrt{0 \cdot 0018} = 0 \cdot 636 \text{ m/s}$$

At 50 mm from the wall: $y/r = 0 \cdot 050/0 \cdot 125 = 0 \cdot 40$

Hence, from equation 9.49:
$$u_x = u_s + 2 \cdot 5u^* \ln (y/r)$$
$$= 18 \cdot 3 + 2 \cdot 5 \times 0 \cdot 636 \ln 0 \cdot 4$$
$$= \underline{\underline{16 \cdot 8 \text{ m/s}}}$$

At 5 mm from the wall: $y/r = 0 \cdot 005/0 \cdot 125 = 0 \cdot 04$

Hence:
$$u_x = 18 \cdot 3 + 2 \cdot 5 \times 0 \cdot 636 \ln 0 \cdot 04$$
$$= \underline{\underline{13 \cdot 2 \text{ m/s}}}$$

The thickness of the laminar sub-layer is given by equation 9.59:

$$\delta_b = 5d/[Re\sqrt{(R/\rho u^2)}]$$
$$= 5 \times 0 \cdot 250/[2 \cdot 06 \times 10^5\sqrt{(0 \cdot 0018)}]$$
$$= 1 \cdot 43 \times 10^{-4} \text{ m}$$

or
$$\underline{\underline{0 \cdot 143 \text{ mm}}}$$

9.7.   EFFECT OF SURFACE ROUGHNESS ON SHEAR STRESS[4, 5]

Experiments have been carried out on artificially roughened surfaces in order to determine the effect of obstructions of various heights. Experimentally, it has been shown that the shear force is not affected by the presence of an obstruction of height e unless:

$$u_e e \rho / \mu > 40 \qquad (9.81)$$

where u_e is the velocity of the fluid at a distance e above the surface.

If the obstruction lies entirely within the laminar sub-layer the velocity u_e is given by:

$$R = \mu (\partial u_x / \partial y)_{y=0}$$
$$= \mu (u_e / e), \quad \text{approximately}$$

Now the shearing stress velocity:

$$u^* = \sqrt{\frac{R}{\rho}} = \sqrt{\frac{\mu u_e}{\rho e}}$$

so that

$$u_e = \frac{\rho e}{\mu} u^{*2}$$

Thus

$$\frac{u_e e \rho}{\mu} = \frac{e \rho}{\mu} u^{*2} \frac{e \rho}{\mu} = \left(\frac{e \rho u^*}{\mu}\right)^2 \qquad (9.82)$$

$e \rho u^* / \mu$ is known as the *roughness Reynolds number*, Re_r.

For the flow of a fluid in a pipe:

$$u^* = \sqrt{R/\rho} = u \phi^{1/2}$$

where u is the mean velocity over the whole cross-section. Re_r will now be expressed in terms of the three dimensionless groups used in the friction chart.

Thus

$$Re_r = e \rho u^* / \mu$$
$$= (d \rho u / \mu)(e / d) \phi^{1/2}$$
$$= Re(e/d) \phi^{1/2} \qquad (9.83)$$

The shear stress should then be unaffected by the obstruction if $Re_r < \sqrt{40}$, i.e. if $Re_r < 6 \cdot 5$ (from equations 9.75 and 9.76).

However, if the surface has a number of closely spaced obstructions, all of the same height e, the shear stress is affected when $Re_r >$ about 3. If the obstructions are of varying heights, with e as the arithmetic mean, the shear stress is increased if $Re_r >$ about 0·3, because the effect of one relatively large obstruction is greater than that of several small ones.

For hydrodynamically smooth pipes, through which fluid is flowing under turbulent conditions the shear stress is given approximately by the Blasius equation:

$$\phi = R/\rho u^2 \propto Re^{-1/4}$$

so that

$$R \propto u^{1 \cdot 75}$$

and is independent of the roughness.

For smooth pipes, the frictional drag at the surface is known as *skin friction*. With rough pipes, however, an additional drag known as *form drag* results from the eddy currents caused by impact of the fluid on the obstructions and, when the surface is very rough, it becomes large compared with the skin friction. Since form drag involves dissipation of kinetic energy, the losses are proportional to the square of the velocity of the fluid, so that $R \propto u^2$. This applies when $Re_r > 50$.

Thus, when

$$Re_r < 0 \cdot 3, \quad R \propto u^{1 \cdot 75}$$

when

$$Re_r > 50, \quad R \propto u^2$$

and when

$$0 \cdot 3 < Re_r < 50, \quad R \propto u^w \quad \text{where } 1 \cdot 75 < w < 2.$$

When the thickness of the laminar sub-layer is large compared with the height of the obstructions, the pipe behaves as a smooth pipe (when $e < \delta_b/3$). Since the thickness of the laminar sub-layer decreases as the Reynolds number is increased, a surface which is hydrodynamically smooth at low Reynolds numbers may behave as a rough surface at higher values. This explains the shapes of the curves obtained for ϕ plotted against Reynolds number (Fig. 3.7). The curves, for all but the roughest of pipes, follow the curve for the smooth pipe at low Reynolds numbers and then diverge at higher values. The greater the roughness of the surface, the lower is the Reynolds number at which the curve starts to diverge. At high Reynolds numbers, the curves for rough pipes become parallel to the Reynolds number axis, indicating that skin friction is negligible and $R \propto u^2$. Under these conditions, the shear stress can be calculated from equation 3.11.

Nikuradse's data for rough pipes gives:

$$u^+ = 2 \cdot 5 \ln (y/e) + 8 \cdot 5 \tag{9.84}$$

9.8. THE BOUNDARY LAYER FOR HEAT TRANSFER

It has been shown that, when a fluid flows over a solid surface, a velocity gradient is set up in the boundary layer. If there is a difference in temperature between the surface and the fluid, heat transfer will take place and a temperature gradient will form in a similar manner to the velocity gradient. The whole of the temperature gradient can be considered to exist within a layer of fluid close to the surface, termed the *thermal boundary layer*. The idea of a boundary layer is only hypothetical, since the temperature gradient extends to an indefinite depth in the fluid, but it provides a useful basis for heat transfer calculations. The thermal boundary layer does not necessarily correspond with the velocity boundary layer, which is not affected by the heat transfer, unless the physical properties of the fluid are appreciably altered.

When heat transfer is taking place between the surface and the element of fluid referred to in Section 9.2, conditions are as shown in Fig. 9.12, in which the distance l is greater than either of the boundary layer thicknesses.

The heat transferred through an element of the plane 1–2, of thickness dy, and at a distance y from the surface

$$= C_p \rho \theta u_x \, \mathrm{d}y$$

where C_p is the specific heat of the fluid at constant pressure, ρ the density of the fluid, and θ and u_x are the temperature and velocity at a distance y from the surface.

The total rate of transfer of heat through the plane, 1–2,

$$= C_p \rho \int_0^l \theta u_x \, \mathrm{d}y$$

if it is assumed that the physical properties of the fluid are independent of temperature.

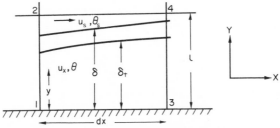

FIG. 9.12. The thermal boundary layer.

In the distance dx this heat flow changes by an amount:

$$C_p\rho \frac{\partial}{\partial x} \left(\int_0^l \theta u_x \, dy \right) dx$$

It has already been shown that there is a mass rate of flow of fluid through plane, 2–4, into the element equal to $\rho \frac{\partial}{\partial x} \left(\int_0^l u_x \, dy \right) dx$ for an incompressible fluid.

Since the plane, 2–4, lies outside the boundary layers, the heat introduced into the element through the plane

$$= C_p\rho\theta_s \frac{\partial}{\partial x} \left(\int_0^l u_x \, dy \right) dx$$

where θ_s is the temperature outside the thermal boundary layer.

The heat transferred by thermal conduction into the element through plane, 1–3,

$$= - k \, dx (\partial\theta/\partial y)_{y=0}$$

If the temperature θ_s of the main stream is unchanged, a heat balance on the element gives:

$$C_p\rho \left(\frac{\partial}{\partial x} \int_0^l \theta u_x \, dy \right) dx = C_p\rho\theta_s \left(\frac{\partial}{\partial x} \int_0^l u_x \, dy \right) dx - k \left(\frac{\partial\theta}{\partial y} \right)_{y=0} dx$$

i.e.

$$\frac{\partial}{\partial x} \int_0^l u_x (\theta_s - \theta) \, dy = D_H \left(\frac{\partial\theta}{\partial y} \right)_{y=0} \tag{9.85}$$

where $D_H = k/C_p\rho$, the thermal diffusivity of the fluid.

The relations between u_x and y have already been obtained for both streamline and turbulent flow. A relation between θ and y for streamline conditions in the boundary layer will now be derived, but it is not possible to define the conditions in the turbulent boundary layer sufficiently precisely to derive a similar expression for that case.

9.8.1. Heat Transfer for Streamline Flow over a Plane Surface

Consider the flow of fluid over a plane surface, which is heated at distances greater than x_0 from the leading edge. The velocity boundary layer will therefore start at the leading edge and the thermal boundary layer at a distance x_0 from it. Let the temperature of the heated portion of the plate remain constant and take this as the datum temperature (Fig. 9.13). Assume that the temperature at a distance y from the surface can be represented by the expression:

$$\theta = a_0 y + b_0 y^2 + c_0 y^3 \tag{9.86}$$

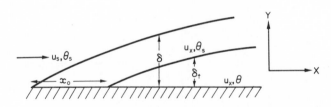

FIG. 9.13. Thermal boundary layer—streamline flow.

Since the fluid in contact with the surface is at rest, heat flow in the vicinity of the surface must be by pure thermal conduction. Thus heat transferred per unit area and unit time q

$$= -k(\partial\theta/\partial y)_{y=0}$$

$\therefore$ $\quad (\partial\theta/\partial y)_{y=0} = $ a constant $\quad$ and $\quad (\partial^2\theta/\partial y^2)_{y=0} = 0$

At the outer edge of the thermal boundary layer, the temperature is θ_s and the temperature gradient $(\partial\theta/\partial y) = 0$.

Thus the conditions for the thermal boundary layer, with respect to temperature, are the same as those for the velocity boundary layer with respect to velocity. Then, if the thickness of the thermal boundary layer is δ_t, the temperature distribution is given by:

$$\frac{\theta}{\theta_s} = \frac{3}{2}\frac{y}{\delta_t} - \frac{1}{2}\left(\frac{y}{\delta_t}\right)^3 \tag{9.87}$$

(cf. equation 9.5)

and

$$\left(\frac{\partial\theta}{\partial y}\right)_{y=0} = \frac{3\theta}{2\delta_t} \tag{9.88}$$

It will now be assumed that the velocity boundary layer is everywhere thicker than the thermal boundary layer, so that $\delta > \delta_t$ (Fig. 9.13). Thus the velocity distribution everywhere within the thermal boundary layer is given by equation 9.5.

The integral in equation 9.85 clearly has a finite value within the thermal boundary layer but is zero outside it. When the expression for the temperature distribution in the boundary layer is inserted therefore, the upper limit of integration must be altered from l to δ_t.

Thus $\displaystyle\int_0^l (\theta_s - \theta)u_x \, \mathrm{d}y$

$$= \theta_s u_s \int_0^{\delta_t} \left[1 - \frac{3}{2}\frac{y}{\delta_t} + \frac{1}{2}\left(\frac{y}{\delta_t}\right)^3\right]\left[\frac{3y}{2\delta} - \frac{1}{2}\left(\frac{y}{\delta}\right)^3\right]\mathrm{d}y$$

$$= \theta_s u_s \left[\frac{3}{4}\frac{\delta_t^2}{\delta} - \frac{3}{4}\frac{\delta_t^2}{\delta} - \frac{1}{8}\frac{\delta_t^4}{\delta^3} + \frac{3}{20}\left(\frac{\delta_t^2}{\delta} + \frac{\delta_t^4}{\delta^3}\right) - \frac{1}{28}\frac{\delta_t^4}{\delta^3}\right]$$

$$= \theta_s u_s \left(\frac{3}{20}\frac{\delta_t^2}{\delta} - \frac{3}{280}\frac{\delta_t^4}{\delta^3}\right)$$

$$= \theta_s u_s \delta\left(\frac{3}{20}\sigma^2 - \frac{3}{280}\sigma^4\right) \tag{9.89}$$

where $\sigma = \delta_t/\delta$.

Since $\delta_t < \delta$, the second term is small compared with the first and:

$$\int_0^l (\theta_s - \theta)u_x \, \mathrm{d}y = \frac{3}{20}\theta_s u_s \delta\sigma^2 \tag{9.90}$$

Substituting from equations 9.88 and 9.90 in equation 9.85:

$$\frac{\partial}{\partial x}\left(\frac{3}{20}\theta_s u_s \delta\sigma^2\right) = D_H\frac{3\theta_s}{2\delta_t} = D_H\frac{3\theta_s}{2\delta\sigma}$$

$\therefore$

$$\frac{1}{10}u_s\delta\sigma\frac{\partial}{\partial x}\delta\sigma^2 = D_H$$

$\therefore$

$$\frac{1}{10}u_s\left(\delta\sigma^3\frac{\partial\delta}{\partial x} + 2\delta^2\sigma^2\frac{\partial\sigma}{\partial x}\right) = D_H \tag{9.91}$$

It has already been shown that:

$$\delta^2 = \frac{280\mu x}{13\rho u_s} = 21\cdot5\,\frac{\mu x}{\rho u_s}$$

(equation 9.9)

Thus

$$\delta\frac{\partial\delta}{\partial x} = \frac{140}{13}\frac{\mu}{\rho u_s}$$

Substituting in equation 9.91:

$$\frac{u_s}{10}\frac{\mu}{\rho u_s}\left(\frac{140}{13}\sigma^3 + \frac{560}{13}x\sigma^2\frac{\partial\sigma}{\partial x}\right) = D_H$$

$$\therefore \qquad \frac{14}{13}\frac{\mu}{\rho D_H}\left(\sigma^3 + 4x\sigma^2\frac{\partial\sigma}{\partial x}\right) = 1$$

$$\therefore \qquad \sigma^3 + \frac{4x}{3}\frac{\partial\sigma^3}{\partial x} = \frac{13}{14}Pr^{-1}$$

(where the Prandtl number $Pr = C_p\mu/k = \mu/\rho D_H$)

$$\therefore \qquad \frac{3}{4}x^{-1}\sigma^3 + \frac{\partial\sigma^3}{\partial x} = \frac{13}{14}Pr^{-1}\frac{3}{4}x^{-1}$$

$$\therefore \qquad \frac{3}{4}x^{-1/4}\sigma^3 + x^{3/4}\frac{\partial\sigma^3}{\partial x} = \frac{13}{14}Pr^{-1}\frac{3}{4}x^{-1/4}$$

Integrating:

$$x^{3/4}\sigma^3 = \frac{13}{14}Pr^{-1}x^{3/4} + \text{constant}$$

i.e.

$$\sigma^3 = \frac{13}{14}Pr^{-1} + \text{constant } x^{-3/4}$$

Now, when $x = x_0$,

$$\sigma = 0,$$

so that

$$\text{constant} = -\frac{13}{14}Pr^{-1}x_0^{3/4}$$

Hence

$$\sigma^3 = \frac{13}{14}Pr^{-1}\left[1 - \left(\frac{x_0}{x}\right)^{3/4}\right]$$

and

$$\sigma = 0\cdot98\,Pr^{-1/3}\left[1 - \left(\frac{x_0}{x}\right)^{3/4}\right]^{1/2}$$

$$\approx Pr^{-1/3}\left[1 - \left(\frac{x_0}{x}\right)^{3/4}\right]^{1/2}$$

(9.92)

If the whole length of the plate is heated, $x_0 = 0$ and:

$$\sigma \approx Pr^{-1/2}$$

(9.93)

In the above derivation, it has been assumed that $\sigma < 1$.

For all liquids other than molten metals, $Pr > 1$ and hence $\sigma < 1$ (from equation 9.93). For gases, $Pr \nleq 0\cdot6$, so that $\sigma \ngtr 1\cdot18$.

Thus only a small error is introduced when the above expression is applied to gases. The only serious deviations occur for molten metals, which have very low Prandtl numbers.

If h is the heat transfer coefficient:

$$q = -h\theta_s$$

Thus

$$-h\theta_s = -k(\partial\theta/\partial y)_{y=0}$$

i.e.

$$h = \frac{k}{\theta_s}\frac{3}{2}\frac{\theta_s}{\delta_t} \quad \text{(from equation 9.88)}$$

$$= 3k/2\delta_t = 3k/2\delta\sigma$$

(9.94)

Substituting for δ from equation 9.10, and σ from equation 9.92:

$$h = \frac{3k}{2} \frac{1}{4\cdot64} \sqrt{\frac{\rho u_s}{\mu x}} \frac{Pr^{1/3}}{[1-(x_0/x)^{3/4}]^{1/3}}$$

i.e.
$$\frac{hx}{k} = 0\cdot323 \, Pr^{1/3} \, Re_x^{1/2} \frac{1}{[1-(x_0/x)^{3/4}]^{1/3}} \tag{9.95}$$

If the surface is heated over its entire length, so that $x_0 = 0$:

$$Nu_x = hx/k = 0\cdot323 \, Pr^{1/3} \, Re_x^{1/2} \tag{9.96}$$

It is seen that the heat transfer coefficient has an infinite value at the leading edge, where the thickness of the thermal boundary layer is zero, and that it decreases progressively as the boundary layer thickens. Equation 9.96 gives the point value of the heat transfer coefficient at a distance x from the leading edge. The mean value between $x = 0$ and $x = x$ is given by:

$$h_m = x^{-1} \int_0^x h \, dx$$

$$= x^{-1} \int_0^x \psi x^{-1/2} \, dx$$

where ψ is not a function of x

$$= x^{-1}(2\psi x^{1/2})_0^x = 2h$$

Thus the mean value of the heat transfer coefficient between $x = 0$ and $x = x$ is equal to twice the point value at $x = x$. The mean value of the Nusselt group is given by:

$$(Nu_x)_m = 0\cdot65 \, Pr^{1/2} \, Re_x^{1/2} \tag{9.97}$$

9.8.2. Streamline Flow of a Fluid through a Heated Tube

Equation 9.97 relates to heat transfer between a fluid in streamline flow and a plane surface for conditions where the thickness of the boundary layer is not limited. For the common problem of heat transfer between a fluid and a tube wall, the boundary layers are limited in thickness to the radius of the pipe and, furthermore, the effective area for heat flow decreases with distance from the surface. The problem can conveniently be divided into two parts. Firstly, heat transfer in the entry length in which the boundary layers are developing, and, secondly, heat transfer under conditions of fully developed flow.

For the region of fully developed flow (Fig. 9.14) the rate of flow of heat Q through a

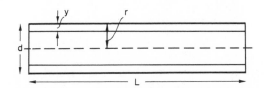

FIG. 9.14. Flow through a heated tube.

cylindrical surface in the fluid at a distance y from the walls is given by:

$$Q = -k2\pi L(r-y)\frac{d\theta}{dy}$$

Thus

$$\frac{d\theta}{dy} = -\frac{Q}{k2\pi L}(r-y)^{-1} \quad \text{and} \quad \left(\frac{d\theta}{dy}\right)_{y=0} = -\frac{Q}{2\pi kLr}$$

$$\frac{d^2\theta}{dy^2} = -\frac{Q}{k2\pi L}(r-y)^{-2} \quad \text{and} \quad \left(\frac{d^2\theta}{dy^2}\right)_{y=0} = -\frac{Q}{2\pi kLr^2}$$

Thus

$$\left(\frac{d^2\theta}{dy^2}\right)_{y=0} = r^{-1}\left(\frac{d\theta}{dy}\right)_{y=0} \tag{9.98}$$

Assume that the temperature of the walls remains constant at the datum temperature and that the temperature at any distance y from the walls is given by:

$$\theta = a_0 y + b_0 y^2 + c_0 y^3 \tag{9.99}$$

Thus

$$\frac{d\theta}{dy} = a_0 + 2b_0 y + 3c_0 y^2 \quad \text{and} \quad \left(\frac{d\theta}{dy}\right)_{y=0} = a_0$$

$$\frac{d^2\theta}{dy^2} = 2b_0 + 6c_0 y \quad \text{and} \quad \left(\frac{d^2\theta}{dy^2}\right)_{y=0} = 2b_0$$

Thus

$$2b_0 = a_0/r \quad \text{(from equation 9.98)}$$

i.e.

$$b_0 = a_0/2r$$

Now let the temperature of the fluid at the axis of the pipe be θ_s. The temperature gradient at the axis, from symmetry, must be zero.

Thus

$$0 = a_0 + 2r(a_0/2r) + 3c_0 r^2$$

$$\therefore \quad c_0 = -2a_0/3r^2$$

and

$$\theta_s = a_0 r + r^2(a_0/2r) + r^3(-2a_0/3r^2)$$

$$= 5a_0 r/6$$

$$\therefore \quad a_0 = 6\theta_s/5r$$

$$b_0 = 3\theta_s/5r^2$$

and

$$c_0 = -4\theta_s/5r^3$$

Thus

$$\frac{\theta}{\theta_s} = \frac{6}{5}\frac{y}{r} + \frac{3}{5}\left(\frac{y}{r}\right)^2 - \frac{4}{5}\left(\frac{y}{r}\right)^3 \tag{9.100}$$

Thus the rate of heat transfer per unit area at the wall:

$$q = -k(d\theta/dy)_{y=0}$$

$$= -6k\theta_s/5r \tag{9.101}$$

In general, the temperature θ_s at the axis is not known, and the heat transfer coefficient is related to the temperature difference between the walls and the bulk fluid. The bulk temperature of the fluid is defined as the ratio of the heat content to the heat capacity of the fluid flowing at any section. Thus the bulk temperature θ_B is given by:

$$\theta_B = \frac{\int_0^r C_p \rho \theta u_x 2\pi (r-y)\,dy}{\int_0^r C_p \rho u_x 2\pi (r-y)\,dy}$$

$$= \frac{\int_0^r \theta u_x (r-y)\,dy}{\int_0^r u_x (r-y)\,dy} \tag{9.102}$$

Now from Poiseuille's law (equation 3.30):

$$u_x = \frac{-\Delta P}{4\mu L}[r^2 - (r-y)^2] = \frac{-\Delta P}{4\mu L}(2ry - y^2)$$

Hence

$$u_s = \frac{-\Delta P}{4\mu L}r^2$$

where u_s is the velocity at the pipe axis, and:

$$\frac{u_x}{u_s} = \frac{2y}{r} - \left(\frac{y}{r}\right)^2$$

Thus

$$\int_0^r u_x(r-y)\,dy = r^2 u_s \int_0^1 \left[2\frac{y}{r} - \left(\frac{y}{r}\right)^2\right]\left(1 - \frac{y}{r}\right)\,d\left(\frac{y}{r}\right)$$

$$= r^2 u_s \int_0^1 \left[2\left(\frac{y}{r}\right) - 3\left(\frac{y}{r}\right)^2 + \left(\frac{y}{r}\right)^3\right]\,d\left(\frac{y}{r}\right)$$

$$= r^2 u_s / 4 \tag{9.103}$$

Since

$$\frac{\theta}{\theta_s} = \frac{6}{5}\frac{y}{r} + \frac{3}{5}\left(\frac{y}{r}\right)^2 - \frac{4}{5}\left(\frac{y}{r}\right)^3 \tag{equation 9.100}$$

$$\int_0^r \theta u_x(r-y)\,dy$$

$$= r^2 u_s \theta_s \int_0^1 \left[\frac{6}{5}\frac{y}{r} + \frac{3}{5}\left(\frac{y}{r}\right)^2 - \frac{4}{5}\left(\frac{y}{r}\right)^3\right]\left[2\left(\frac{y}{r}\right) - 3\left(\frac{y}{r}\right)^2 + \left(\frac{y}{r}\right)^3\right]\,d\left(\frac{y}{r}\right)$$

$$= r^2 u_s \theta_s \int_0^1 \left[\frac{12}{5}\left(\frac{y}{r}\right)^2 - \frac{12}{5}\left(\frac{y}{r}\right)^3 - \frac{11}{5}\left(\frac{y}{r}\right)^4 + 3\left(\frac{y}{r}\right)^5 - \frac{4}{5}\left(\frac{y}{r}\right)^6\right]\,d\left(\frac{y}{r}\right)$$

$$= r^2 u_s \theta_s \left(\frac{4}{5} - \frac{3}{5} - \frac{11}{25} + \frac{1}{2} - \frac{4}{35}\right)$$

$$= 51 r^2 u_s \theta_s / 350 \tag{9.104}$$

Substituting from equations 9.103 and 9.104 in equation 9.102:

$$\theta_B = \frac{51 r^2 u_s \theta_s / 350}{r^2 u_s / 4}$$

$$= 102 \theta_s / 175 = 0.583 \theta_s \qquad (9.105)$$

Then, the heat transfer coefficient h is given by:

$$h = -q / \theta_B$$

where q is the rate of heat transfer per unit area of tube.

Thus
$$h = \frac{6 k \theta_s / 5 r}{0.583 \theta_s} \quad \text{(from equations 9.101 and 9.105)}$$

$$= 2.06 k / r = 4.1 \, k / d$$

$$\therefore \qquad Nu = hd / k = 4.1 \qquad (9.106)$$

This expression is applicable only to the region of fully developed flow. The heat transfer coefficient for the inlet length can be calculated approximately, using the expressions that have been derived for the flow over a plane surface. It should be borne in mind that it has been assumed throughout that the physical properties of the fluid are not appreciably dependent on temperature and therefore the expressions will not be expected to hold accurately if the temperature differences are large and if the properties vary widely with temperature. An empirical equation (7.60) has already been given in Chapter 7 to cover the whole of the tube.

9.9. BOUNDARY LAYER FOR MASS TRANSFER

If a concentration gradient exists within a fluid flowing over a surface, mass transfer will take place. Again the whole of the resistance to transfer can be regarded as lying within a third boundary layer in the vicinity of the surface. If the concentration gradients, and hence the mass transfer rates, are small, it can be shown that the velocity and thermal boundary layers are unaffected.[6] The basic equation for mass transfer is then:

$$\frac{\partial}{\partial x} \int_0^l (C_{As} - C_A) u_x \, dy = D \left(\frac{\partial C_A}{\partial y} \right)_{y=0} \qquad (9.107)$$

where C_A and C_{As} are the molar concentrations of A at a distance y from the surface and outside the boundary layer respectively; and l is a distance at right angles to the surface which is greater than the thickness of any of the three boundary layers. Equation 9.107 is obtained in exactly the same manner as equation 9.85 for heat transfer.

The integral in equation 9.107 can be obtained by substituting an approximate expression for the concentration profile, only for conditions of streamline flow and where the mass transfer rate is so low that it does not disturb the velocity and temperature gradients. Under these conditions, the solution corresponds with that for the heat transfer problem (equation 9.96) and the point value of the mass transfer coefficient is given by:

$$Sh_x = h_D x / D = 0.323 Sc^{1/3} Re_x^{1/2} \qquad (9.108)$$

The mean value of the coefficient between $x = 0$ and $x = x$ is then given by:

$$(Sh_x)_m = 0.65 \, Sc^{1/3} Re_x^{1/2} \qquad (9.109)$$

In equations 9.108 and 9.109 Sh_x and $(Sh_x)_m$ represent the point and mean values respectively of the Sherwood numbers.

For fully developed streamline flow in a pipe, by analogy with equation 9.106:

$$Sh = h_D d/D = 4 \cdot 1 \qquad (9.110)$$

No satisfactory solution of the integral in equation 9.107 is available for turbulent flow.

9.10. FURTHER READING

WHITE, F. M.: *Viscous Fluid Flow* (McGraw-Hill, 1974).

9.11. REFERENCES

1. BROWN, F. N. M.: *Proc. Midwest Conf. Fluid Mechanics* (Sept. 1959) 331. The organized boundary layer.
2. BLASIUS, H.: *Forsch. Ver. deut. Ing.* **131** (1913). Das Ähnlichkeitsgesetz bei Reibungsvorgängen in Flüssigkeiten.
3. PRANDTL, L.: *Z. Ver. deut. Ing.* **77** (1933) 105. Neuere Ergebnisse der Turbulenzforschung.
4. NIKURADSE, J.: *Forsch. Ver. deut. Ing.* **356** (1932). Gesetzmässigkeiten der turbulenten Strömung in glatten Rohren.
5. NIKURADSE, J.: *Forsch. Ver. deut. Ing.* **361** (1933). Strömungsgesetze in rauhen Rohren.
6. ECKERT, E. R. G. and DRAKE, R. M.: *Analysis of Heat and Mass Transfer* (McGraw Hill, 1972).

9.12. NOMENCLATURE

		Units in SI System	Dimensions in M, L, T, θ
A	Area of surface	m²	L^2
a_0	Coefficient of y	K/m	$L^{-1}\theta$
a'	Coefficient of y	1/s	T^{-1}
B	Integration constant	m/s	LT^{-1}
B'	B/u^*	—	—
b	Ratio of θ_b to θ_s	—	—
b_0	Coefficient of y^2	K/m²	$L^{-2}\theta$
b'	Coefficient of y^2	1/sm	$L^{-1}T^{-1}$
C_A	Molar concentration of A	kmol/m³	ML^{-3}
C_{AS}	Molar concentration of A outside boundary layer	kmol/m³	ML^{-3}
C_p	Specific heat at constant pressure	J/kg K	$L^2T^{-2}\theta^{-1}$ or —
c_0	Coefficient of y^3	K/m³	$L^{-3}\theta$
c'	Coefficient of y^3	1/s²m²	$L^{-2}T^{-1}$
D	Gas phase diffusivity	m²/s	L^2T^{-1}
D_H	Thermal diffusivity	m²/s	L^2T^{-1}
d	Pipe diameter	m	L
E	Eddy kinematic viscosity	m²/s	L^2T^{-1}
e	Surface roughness	m	L
f	Index	—	—
h	Heat transfer coefficient	W/m²K	$MT^{-3}\theta^{-1}$
h_D	Mass transfer coefficient	kmol/(m²)(s)(kmol/m³)	LT^{-1}
h_m	Mean value of heat transfer coefficient	W/m²K	$MT^{-3}\theta^{-1}$
K	Ratio of mixing length to distance from surface	—	—
k	Thermal conductivity	W/m K	$MLT^{-3}\theta^{-1}$
L	Length of pipe or surface	—	L
L_e	Inlet length of pipe	m	L
l	Thickness of element of fluid	m	L
P	Total pressure	N/m²	$ML^{-1}T^{-2}$
Q	Rate of transfer of heat	W	ML^2T^{-3}
q	Rate of transfer of heat per unit area at walls	W/m²	MT^{-3}
q_y	Rate of transfer of heat per unit area at $y = y$	W/m²	MT^{-3}
R	Shear stress acting on surface	N/m²	$ML^{-1}T^{-2}$
R_0	Shear stress acting on fluid at surface	N/m²	$ML^{-1}T^{-2}$

		Units in SI System	Dimensions in M, L, T, θ
R_y	Shear stress in fluid at $y = y$	N/m²	$\mathbf{ML^{-1}T^{-2}}$
r	Radius of pipe	m	$\mathbf{L}$
s	Radius of cylindrical surface	m	$\mathbf{L}$
T	Absolute temperature	K	$\boldsymbol{\theta}$
t	Time	s	$\mathbf{T}$
u	Mean velocity	m/s	$\mathbf{LT^{-1}}$
u_b	Velocity at edge of laminar sub-layer	m/s	$\mathbf{LT^{-1}}$
u_E	Mean velocity in eddy	m/s	$\mathbf{LT^{-1}}$
u_e	Velocity at distance e from surface	m/s	$\mathbf{LT^{-1}}$
u_0	Velocity of fluid at surface	m/s	$\mathbf{LT^{-1}}$
u_s	Velocity of fluid outside boundary layer, or at pipe axis	m/s	$\mathbf{LT^{-1}}$
u_x	Velocity in X-direction at $y = y$	m/s	$\mathbf{LT^{-1}}$
u^+	Ratio of u_y to $u*$	—	—
$u*$	Shearing stress velocity, $\sqrt{R/\rho}$	m/s	$\mathbf{LT^{-1}}$
x	Distance from leading edge of surface in X-direction	m	$\mathbf{L}$
x_c	Value of x at which flow becomes turbulent	m	$\mathbf{L}$
x_0	Unheated length of surface	m	$\mathbf{L}$
y	Distance from surface	m	$\mathbf{L}$
y^+	Ratio of y to $\mu/\rho u*$	—	—
δ	Thickness of boundary layer	m	$\mathbf{L}$
δ_b	Thickness of laminar sub-layer	m	$\mathbf{L}$
δ_t	Thickness of thermal boundary layer	m	$\mathbf{L}$
$\delta*$	Displacement thickness of boundary layer	m	$\mathbf{L}$
λ_E	Mixing length	m	$\mathbf{L}$
μ	Viscosity of fluid	Ns/m²	$\mathbf{ML^{-1}T^{-1}}$
ρ	Density of fluid	kg/m³	$\mathbf{ML^{-3}}$
σ	Ratio of δ_t to δ	—	—
θ	Temperature at $y = y$	K	$\boldsymbol{\theta}$
θ_B	Bulk temperature of fluid	K	$\boldsymbol{\theta}$
θ_s	Temperature outside boundary layer, or at pipe axis	K	$\boldsymbol{\theta}$
ϕ	Friction factor $R/\rho u^2$	—	—
Nu	Nusselt number hd/k	—	—
Nu_x	Nusselt number hx/k	—	—
Re	Reynolds number $ud\rho/\mu$	—	—
Re_r	Roughness Reynolds number $u_s e\rho/\mu$	—	—
Re_x	Reynolds number $u_s x\rho/\mu$	—	—
Re_{x_c}	Reynolds number $u_s x_c\rho/\mu$	—	—
Re_δ	Reynolds number $u_s\delta\rho/\mu$	—	—
Pr	Prandtl number $C_p\mu/k$	—	—
Sc	Schmidt number $\mu/\rho D$	—	—
Sh_x	Sherwood number $h_D x/D$	—	—

CHAPTER 10

Momentum, Heat, and Mass Transfer

10.1. INTRODUCTION

In the previous chapters the stresses arising from relative motion within a fluid, the transfer of heat by conduction and convection, and the mechanism of mass transfer have all been discussed. These three major processes of momentum, heat, and mass transfer have, however, been regarded as independent problems.

In most of the unit operations encountered in the chemical and petroleum industries, one or more of the processes of momentum, heat, and mass transfer is involved. Thus, in the flow of a fluid under adiabatic conditions through a bed of granular particles, a pressure gradient is set up in the direction of flow and a velocity gradient develops approximately perpendicularly to the direction of motion in each fluid stream; momentum transfer takes place between the fluid elements which are moving at different velocities. If there is a temperature difference between the fluid and the pipe wall or the particles, heat transfer will take place as well, and the convective component of the heat transfer will be directly affected by the flow pattern of the fluid. Here, then, is an example of a process of simultaneous momentum and heat transfer in which the same fundamental mechanism is affecting both processes. Fractional distillation and gas absorption are frequently carried out in a packed column in which the gas or vapour stream rises countercurrently to a liquid. The function of the packing in this case is to provide a large interfacial area between the phases and to promote turbulence within the fluids. In a very turbulent fluid the rates of transfer per unit area of both momentum and mass are high; and as the pressure rises the rates of transfer of both momentum and mass increase together. In some cases, momentum, heat, and mass transfer all occur simultaneously as, for example, in a water-cooling tower (see Chapter 11), where transfer of sensible heat and evaporation both take place from the surface of the water droplets. It will now be shown not only that the process of momentum, heat, and mass transfer are physically related, but also that quantitative relations between them can be developed.

Another form of interaction between the transfer processes is responsible for the phenomenon of *thermal diffusion* in which a component in a mixture moves under the action of a temperature gradient. Although these are important applications of thermal diffusion, the magnitude of the effect is usually small relative to that arising from concentration gradients.

When a fluid is flowing under streamline conditions over a surface, a forward component of velocity is superimposed on the random distribution of velocities of the molecules, and movement at right angles to the surface occurs solely as a result of the random motion of the molecules. Thus if two adjacent layers of fluid are moving at different velocities, there will be a tendency for the faster moving layer to be retarded and the slower moving layer to be accelerated by virtue of the continuous passage of molecules in each direction. There will therefore be a net transfer of momentum from the fast- to the slow-moving stream. Similarly, the molecular motion will tend to reduce any temperature gradient or any concentration gradient if the fluid consists of a mixture of two or more components. At the boundary the effects of the molecular transfer are balanced by the drag forces at the surface.

If the motion of the fluid is turbulent, the transfer of fluid by eddy motion is superimposed on the molecular transfer process. In this case, the rate of transfer to the surface will be a function of the degree of turbulence. When the fluid is highly turbulent, the rate of transfer by molecular motion will be negligible compared with that by eddy motion. For small degrees of turbulence the two may be of the same order.

It was shown in the previous chapter that when a fluid flows under turbulent conditions over a surface, the flow can conveniently be divided into three regions:

(1) At the surface, the laminar sub-layer, in which the only motion at right angles to the surface is due to molecular diffusion.
(2) Next, the buffer layer, in which molecular diffusion and eddy motion are of comparable magnitude.
(3) Finally, over the greater part of the fluid, the turbulent region in which eddy motion is large compared with molecular diffusion.

In addition to momentum, both heat and mass can be transferred either by molecular diffusion alone or by molecular diffusion combined with eddy diffusion. Because the effects of eddy diffusion are generally far greater than those of the molecular diffusion, the main resistance to transfer will lie in the regions where only molecular diffusion is occurring. Thus the main resistance to the flow of heat or mass to a surface lies within the laminar sub-layer. It has been shown in Chapter 9 that the thickness of the laminar sub-layer is almost inversely proportional to the Reynolds number for fully developed turbulent flow in a pipe. Thus the heat and mass transfer coefficients are much higher at high Reynolds numbers.

10.2. TRANSFER BY MOLECULAR DIFFUSION

10.2.1. Momentum Transfer

When the flow characteristics of the fluid are *Newtonian*, the shear stress R_y in a fluid is proportional to the velocity gradient and to the viscosity.

Thus
$$R_y = -\mu \frac{du_x}{dy} = -\frac{\mu}{\rho} \frac{d(\rho u_x)}{dy} \qquad \text{(cf. equation 3.21)} \tag{10.1}$$

where u_x is the velocity of the fluid parallel to the surface at a distance y from it.

The shear stress R_y within the fluid, at a distance y from the boundary surface, is a measure of the rate of transfer of momentum per unit area at right angles to the surface.

Since (ρu_x) is the momentum per unit volume of the fluid, the rate of transfer of momentum per unit area is proportional to the gradient in the Y-direction of the momentum per unit volume. The negative sign indicates that momentum is transferred from the fast- to the slow-moving fluid and the shear stress acts in such a direction as to oppose the motion of the fluid.

10.2.2. Heat Transfer

From the definition of thermal conductivity, the heat transferred per unit time through a unit area at a distance y from the surface is given by:

$$q_y = -k \frac{d\theta}{dy} = -\frac{k}{C_p \rho} \frac{d(C_p \rho \theta)}{dy} \tag{10.2}$$

where C_p is the specific heat of the fluid at constant pressure, θ the temperature, and k the thermal conductivity.

The term $(C_p \rho \theta)$ represents the heat content per unit volume of fluid and therefore the flow of heat is proportional to the gradient in the Y-direction of the heat content per unit volume. The proportionality constant $k/C_p \rho$ is called the thermal diffusivity D_H.

10.2.3. Mass Transfer

It has been shown in Chapter 8, from Fick's Law of diffusion, that the rate of diffusion of a constituent **A** in a mixture is proportional to its concentration gradient.

Thus, from equation 10.1:

$$N_A = -D\frac{dC_A}{dy} \tag{10.3}$$

where N_A is the molar rate of diffusion of constituent **A** per unit area, C_A the molar concentration of constituent **A** and D the diffusivity.

The essential similarity between the three processes is that the rates of transfer of momentum, heat, and mass are all proportional to the concentration gradients of these quantities. In the case of gases the proportionality constants μ/ρ, D_H, and D, all of which have the dimensions length^2/time, all have a physical significance. For liquids the constants cannot be interpreted in a similar manner. The viscosity, thermal conductivity, and diffusivity of a gas will now be considered.

10.2.4. Viscosity

Consider the flow of a gas parallel to a solid surface and the movement of molecules at right angles to this direction through a plane a–a of unit area, parallel to the surface and sufficiently close to it to be within the laminar sublayer (Fig. 10.1). During an interval of time dt, molecules with an average velocity $i_1 u_m$ in the Y-direction will pass through the plane (where u_m is the root mean square velocity and i_1 is some fraction of it, depending on the actual distribution of velocities).

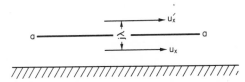

FIG. 10.1. Transfer of momentum near a surface.

If all these molecules can be considered as having the same component of velocity in the Y-direction, molecules from a volume $i_1 u_m\, dt$ will pass through the plane in time dt.

If N is the numerical concentration of molecules close to the surface, the number of molecules passing $= i_1 u_m\, \mathrm{N}\, dt$.

Thus the rate of passage of molecules $= i_1 u_m \mathrm{N}$.

These molecules have a mean velocity u_x (say) in the x-direction.

Thus the rate at which momentum is transferred across the plane away from the surface

$$= i_1 \mathrm{N} u_m m u_x$$

where m is the mass of each molecule.

By similar reasoning there must be an equivalent stream of molecules also passing through the plane in the opposite direction; otherwise there would be a resultant flow perpendicular to the surface.

Let this other stream of molecules have originated at a distance $j\lambda$ from the previous ones, and let the mean component of their velocities in the X-direction be u'_x (where λ is the mean free path of the molecules and j is some fraction of the order of unity).

The net rate of transfer of momentum away from the surface

$$= i_1 \mathrm{N} u_m m (u_x - u'_x)$$

The gradient of the velocity with respect to the Y-direction

$$= du_x/dy = (u_x' - u_x)/j\lambda$$

since λ is small.

Thus the rate of transfer of momentum per unit area which can be written as:

$$R_y = -i_1 N u_m m j\lambda \, (du_x/dy)$$

$$= -i_1 j\rho u_m \lambda \, (du_x/dy) \tag{10.4}$$

(since $Nm = \rho$, the density of the fluid).

But $R_y = -\mu(du_x/dy)$ (from equation 10.1)

$\therefore$

$$\mu/\rho = i_1 j u_m \lambda \tag{10.5}$$

The value of the product $i_1 j$ has been variously given by different workers, from statistical treatment of the velocities of the molecules;[1] a value of 0·5 will be taken.

Thus $$\mu/\rho = u_m \lambda /2 \tag{10.6}$$

It is now possible to give a physical interpretation to the Reynolds number, which is complementary to that given in Chapter 3.

$$Re = \frac{ud\rho}{\mu} = ud\frac{2}{u_m\lambda} = 2\frac{u}{u_m}\frac{d}{\lambda} \tag{10.7}$$

i.e. it is proportional to the product of the ratio of the flow velocity to the molecular velocity and the ratio of the characteristic linear dimension of the system to the mean free path of the molecules.

From the kinetic theory,[1] $u_m = \sqrt{(8RT/\pi M)}$ and is independent of pressure, and $\rho\lambda$ is a constant.

Thus the viscosity of a gas would be expected to be a function of temperature but not of pressure.

10.2.5.   Thermal Conductivity

Consider now the case where there is a temperature gradient in the Y-direction.

The rate of passage of molecules through the unit plane $a - a = i_2 u_m N$ (where i_2 is some fraction of the order of unity). Now if the temperature difference between two planes situated a distance $j\lambda$ apart is $\theta - \theta'$, the net heat transferred as one molecule passes in one direction and another molecule passes in the opposite direction $= c_m(\theta - \theta')$, where c_m is the heat capacity per molecule.

The net rate of heat transfer per unit area $= i_2 u_m N c_m(\theta - \theta')$.

The temperature gradient $d\theta/dy = (\theta' - \theta)/j\lambda$ since λ is small.

Thus the net rate of heat transfer per unit area

$$= -i_2 j u_m N c_m \lambda \, (d\theta/dy)$$

$$= -i_2 j u_m C_v \rho \lambda \, (d\theta/dy) \tag{10.8}$$

(since $Nc_m = \rho C_v$, the specific heat per unit volume of fluid)

$$= -\bar{k}(d\theta/du) \quad \text{(from equation 10.2)}$$

Thus the thermal diffusivity:

$$k/C_p\rho = i_2 j u_m \lambda \, (C_v/C_p) \tag{10.9}$$

From statistical calculations[1] the value of $i_2 j$ has been given as $(9\gamma - 5)/8$ (where $\gamma = C_p/C_v$, the ratio of the specific heat at constant pressure to the specific heat at constant volume).

Thus
$$k/C_p\rho = u_m\lambda(9\gamma - 5)/8\gamma \tag{10.10}$$

The Prandtl number Pr is defined as the ratio of the kinematic viscosity to the thermal diffusivity.

Thus
$$Pr = \frac{\mu/\rho}{k/C_p\rho} = \frac{C_p\mu}{k} = \frac{\frac{1}{2}u_m\lambda}{u_m\lambda(9\lambda - 5)/8\gamma}$$
$$= 4\gamma/(9\gamma - 5) \tag{10.11}$$

Values of Pr calculated from equation 10.11 are in close agreement with practical figures.

10.2.6. Diffusivity

Consider the diffusion, in the Y-direction, of one constituent **A** of a mixture across the plane $a - a$.

If the numerical concentration is N_A on one side of the plane and N'_A on the other side at a distance of $j\lambda$:

Net rate of passage of molecules per unit area

$$= i_3 u_m (N_A - N'_A)$$

where i_3 is an appropriate fraction of the order of unity.

Rate of mass transfer per unit area

$$= i_3 u_m (N_A - N'_A)m$$

Concentration gradient of **A** in the Y-direction

$$= dC_A/dy = (N'_A - N_A)m/j\lambda$$

Thus rate of mass transfer per unit area

$$= -i_3 j\lambda u_m (dC_A/dy) \tag{10.12}$$
$$= -D(dC_A/dy) \quad \text{(from equation 10.3)}$$

Thus
$$D = i_3 j u_m \lambda \tag{10.13}$$

There is, however, no satisfactory evaluation of the product $i_3 j$.

The ratio of the kinematic viscosity to the diffusivity is the Schmidt number, Sc, i.e.

$$Sc = (\mu/\rho)/D = \mu/\rho D \tag{10.14}$$

It is thus seen that the kinematic viscosity, the thermal diffusivity, and the diffusivity for mass transfer are all proportional to the product of the mean free path and the root mean square velocity of the molecules, and that the expressions for the transfer of momentum, heat, and mass are of the same form.

For liquids the same qualitative forms of relationships exist, but it is not possible to express the physical properties of the liquids in terms of molecular velocities and distances.

10.3. EDDY TRANSFER

In steady-state turbulent flow the velocity and direction of flow at a point will be subject to rapid fluctuations but the average values will, over a period, be constant. Thus, at any instant, the velocity at a point can be regarded as composed of two components—the mean velocity and the temporary deviation. By definition the mean values of the fluctuating components are zero. In the case where the magnitude of the fluctuating component in each of the three principal directions is the same, the turbulence is said to be isotropic. If the direction of flow over a surface is taken as one of the principal directions, the average velocity in each of the other principal directions will be zero, but there will exist a fluctuating component equal to that in the direction of flow if the turbulence is isotropic.

The fluctuating velocity components are responsible for momentum transfer within the fluid and therefore give rise to shear stresses. If the fluid is flowing across a surface the rate at which fluid will be transferred across a plane of unit area parallel to the surface as a result of a fluctuating velocity component u_{Ey} perpendicular to the surface is ρu_{Ey}. If the fluctuating component in the direction of flow is u_{Ex}, the momentum transferred in time dt which is attributable to these fluctuations is $\rho u_{Ey} u_{Ex} \, dt$. Now although the average values of u_{Ey} and u_{Ex} individually are zero, the products such as $u_{Ey} u_{Ex}$ will not, in general, be zero. Thus equating the shear stress in the fluid to the rate of transference of momentum per unit area and noting that R_y is negative:

$$-R_y = \rho u_{Ex} u_{Ey}$$

i.e.

$$-R_y/\rho = u_{Ex} u_{Ey} \tag{10.15}$$

Turbulence in a fluid is characterised by two quantities—the *intensity of turbulence* and the *scale of turbulence*. The time average of the fluctuating components is necessarily zero, but the time average of the root mean square values $\sqrt{\overline{u_E^2}}$ will have a finite value. The intensity of turbulence is defined as the ratio $\sqrt{\frac{1}{3}(\overline{u_{Ex}^2} + \overline{u_{Ey}^2} + \overline{u_{Ez}^2})}/u_x$. In the case of isotropic turbulence this reduces to $\sqrt{\overline{u_{Ex}^2}}/u_x$, where u_E, the mean eddy velocity component, is the same in all directions. Equation 10.15 then becomes:

$$-R_y/\rho = u_E^2 \tag{10.16}$$

The scale of turbulence is related to the size of the eddies. Thus, at two points close together in a turbulent fluid there will be a good correlation between the values of the instantaneous velocities because they will usually refer to the same eddy. On the other hand, if the points are far apart the correlation will be negligible or zero. There will be a gradual change in the correlation coefficient with distance between the reference points, but there is no definite boundary between consecutive eddies as they tend to merge into one another.

In the previous chapter, the concept of a *mixing length* λ_E was discussed. For turbulent motion, Prandtl[2] and Taylor[4] have defined λ_E as the mean distance which a fluid in an eddy travels at right angles to the direction of flow before it can be considered to have lost its identity and to be assimilated by the fluid at this new position. The mixing length is analogous to the mean free path for molecular diffusion. By using the same reasoning as was used for molecular diffusion it was shown that the eddy kinematic viscosity E which controls momentum transfer in a turbulent fluid is proportional to the product of the mixing length λ_E and the eddy velocity.

Then, for isotropic turbulence:

$$E \propto \lambda_E u_E \tag{10.17}$$

Taking u_E as proportional to the product of the velocity gradient and the mixing length (see equation 9.43):

$$u_E \propto \lambda_E |du_x/dy| \tag{10.18}$$

where $|du_x/dy|$ denotes the positive value of the velocity gradient. Then from equations 10.17 and 10.18:

$$E \propto \lambda_E^2 |du_x/dy| \tag{10.19}$$

Putting the proportionality constant arbitrarily equal to unity:

$$E = \lambda_E^2 |du_x/dy| \tag{10.20}$$

This involves a slight redefinition of the mixing length and the rate of momentum transfer due to eddy motion is given by:

$$R_y = -E\, d(\rho u_x)/dy$$

Substituting for E from equation 10.20:

$$R_y = -\lambda_E^2\, d(\rho u_x)/dy\, |du_x/dy| \tag{10.21}$$

In a turbulent fluid the eddies will be responsible for heat transfer if a temperature gradient exists and for mass transfer if there is a gradient in the concentration of one of the components of a mixture. Since the mechanism of transference of momentum, heat, and mass by the eddies is similar, the eddy thermal diffusivity E_H and the eddy diffusivity E_D will both be proportional to the product of the mixing length and the mean eddy velocity (cf. equation 10.17).

Thus
$$E_H \propto \lambda_E u_E \tag{10.22}$$

and
$$E_D \propto \lambda_E u_E \tag{10.23}$$

The ratio E/E_H is known as the *turbulent Prandtl number* and the ratio E/E_D as the *turbulent Schmidt number*. In practice, E, E_H, and E_D can be taken as equal.

Thus
$$E_H = \lambda_E^2 \left|\frac{du_x}{dy}\right| \tag{10.24}$$

and
$$E_D = \lambda_E^2 \left|\frac{du_x}{dy}\right| \tag{10.25}$$

Thus for the eddy transfer of heat:

$$q_y = -E_H \frac{d(C_p\rho\theta)}{dy} = -\lambda_E^2 \left|\frac{du_x}{dy}\right| \frac{d(C_p\rho\theta)}{dy} \tag{10.26}$$

and similarly for mass transfer:

$$N_A = -E_D \frac{dC_A}{dy} = -\lambda_E^2 \left|\frac{du_x}{dy}\right| \frac{dC_A}{dy} \tag{10.27}$$

In the neighbourhood of a surface du_x/dy will be positive and thus:

$$R_y = -\lambda_E^2 \frac{du_x}{dy} \frac{d(\rho u_x)}{dy} \tag{10.28}$$

$$q_y = -\lambda_E^2 \frac{du_x}{dy} \frac{d(C_p\rho\theta)}{dy} \tag{10.29}$$

and
$$N_A = -\lambda_E^2 \frac{du_x}{dy} \frac{dC_A}{dy} \tag{10.30}$$

For conditions of constant density, equation 10.21 gives:

$$\sqrt{\frac{-R_y}{\rho}} = \lambda_E \frac{du_x}{dy} \tag{10.31}$$

When molecular and eddy transport both contribute significantly:

$$R_y = -\left(\frac{\mu}{\rho} + E\right)\frac{d(\rho u_x)}{dy} \qquad = -\mu\frac{du_x}{dy} - \lambda_E^2 \rho\left(\frac{du_x}{dy}\right)^2 \tag{10.32}$$

$$q_y = -\left(\frac{k}{C_p\rho} + E_H\right)\frac{d(C_p\rho\theta)}{dy} = -k\frac{d\theta}{dy} - \lambda_E^2 \rho C_p\left(\frac{du_x}{dy}\right)\left(\frac{d\theta}{dy}\right) \tag{10.33}$$

$$N_A = -(D + E_D)\frac{dC_A}{dy} \qquad = -D\frac{dC_A}{dy} - \lambda_E^2 \left(\frac{du_x}{dy}\right)\left(\frac{dC_A}{dy}\right) \tag{10.34}$$

Whereas the kinematic viscosity μ/ρ, the thermal diffusivity $k/C_p\rho$, and the diffusivity D are physical properties of the system and can therefore be taken as constant provided that physical conditions do not vary appreciably, the eddy coefficients E, E_H, and E_D will be affected by the flow pattern and will vary throughout the fluid. Each of the eddy coefficients is proportional to the square of the mixing length. The mixing length will normally increase with distance from a surface and the eddy coefficients will therefore increase rapidly with position. It is therefore necessary to assume some approximate relation between λ_E and distance y from the surface before the above equations can be solved. In Chapter 9 the equations for momentum transfer were integrated on the assumption that λ_E was directly proportional to y.

The relations may be summarised as follows:

	Molecular processes only	Molecular and eddy transfer together	Eddy transfer predominating
Momentum transfer	$R_y = -\dfrac{\mu}{\rho}\dfrac{d(\rho u_x)}{dy}$	$R_y = -\left(\dfrac{\mu}{\rho} + E\right)\dfrac{d(\rho u_x)}{dy}$	$R_y = -E\dfrac{d(\rho u_x)}{dy}$
Heat transfer	$q_y = -\dfrac{k}{C_p\rho}\dfrac{d(C_p\rho\theta)}{dy}$	$q_y = -\left(\dfrac{k}{C_p\rho} + E_H\right)\dfrac{d(C_p\rho\theta)}{dy}$	$q_y = -E_H\dfrac{d(C_p\rho\theta)}{dy}$
Mass transfer	$N_A = -D\dfrac{dC_A}{dy}$	$N_A = -(D + E_D)\dfrac{dC_A}{dy}$	$N_A = -E_D\dfrac{dC_A}{dy}$

$$\text{where } E \approx E_H \approx E_D \approx \lambda_E^2 |du_x/dy|$$

10.4. REYNOLDS ANALOGY

The simple concept of the Reynolds analogy has already been referred to in Chapter 8 during the discussion of models for representing the mass transfer process in the neighbourhood of an interface. The analogy was first suggested by Reynolds[3] to relate heat transfer rates to shear stress, but it is also applicable to mass transfer. It is assumed that elements of fluids are brought from remote regions to the surface by the action of the turbulent eddies; the elements do not undergo any mixing with the intermediate fluid through which they pass, and they instantaneously reach equilibrium on contact with the interfacial layers. An equal volume of fluid is, at the same time, displaced in the reverse direction. Thus in a flowing fluid there is a transference of momentum and a simultaneous transfer of heat if there is a temperature gradient, and of mass transfer if there is a concentration gradient. The turbulent fluid is assumed to have direct access to the surface and the existence of a buffer layer and laminar sub-layer is neglected. Modification of the model has been made by Taylor and Prandtl to take account of the laminar sub-layer. Subsequently, the effect of the buffer layer has been incorporated by applying the *universal velocity profile*.

Consider the equilibrium set up when an element of fluid moves from a region at high temperature, lying outside the boundary layer, to a solid surface at a lower temperature if no mixing with the intermediate fluid takes place. Turbulence is therefore assumed to persist right up to the surface. The relationship between the rates of transfer of momentum and heat can then be deduced as follows (Fig. 10.2).

Consider the fluid to be flowing in a direction parallel to the surface (X-direction) and for momentum and heat transfer to be taking place in a direction at right angles to the surface (Y-direction positive away from surface).

Suppose a mass **M** of fluid situated at a distance from the surface to be moving with a velocity u_s in the X-direction. If this element moves to the surface where the velocity is zero, it will give up its momentum Mu_s in time t, say. If the temperature difference between the

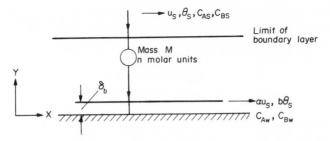

FIG. 10.2. The Reynolds analogy—momentum, heat, and mass transfer.

element and the surface is θ_s and C_p is the specific heat of the fluid, the heat transferred to the surface will be $MC_p\theta_s$. Over a surface of area A the rate of heat transfer is given by:

$$MC_p\theta_s/t = -qA \tag{10.35}$$

where $-q$ is the heat transferred to the surface per unit area per unit time (NB—the negative sign has been introduced as the positive direction is away from the surface).

If the shear stress at the surface is R_0, the shearing force over the area A will equal the rate of change in momentum, i.e.

$$Mu_s/t = -R_0A \tag{10.36}$$

Thus
$$C_p\theta_s/u_s = q/R_0 \tag{10.37}$$

The shear stress R_0 in the fluid at the walls will be equal and opposite to the shear stress R acting on the walls themselves. Thus, writing $R = -R_0$ and h as the heat transfer coefficient between the fluid and the surface:

$$-q/\theta_s = h = -R_0C_p/u_s = RC_p/u_s$$

or
$$h/C_p\rho u_s = R/\rho u_s^2 \tag{10.38}$$

The dimensionless group $(h/C_p\rho u_s)$, is the Stanton Group, St.

In this analysis, no allowance has been made for the variations in physical properties of the fluid with temperature.

10.4.1. Reynolds Analogy Applied to Mass Transfer

If a concentration gradient, rather than a temperature gradient, exists within the fluid, the movement of the element of fluid will give rise to a net transfer of one of the components. In this case it is necessary to consider two important cases: firstly, the process in which the second constituent of the mixture undergoes no net transfer, such as that encountered in gas absorption; and, secondly, a process of equimolecular counterdiffusion in which equal molar quantities of material are absorbed and liberated at the surface.

Mass Transfer with Bulk Flow

Consider the movement of an element of fluid consisting of n molar units of a mixture of two constituents **A** and **B** from a region outside the boundary layer, where the molecular concentrations are C_{As} and C_{Bs}, to the surface where the corresponding concentrations are C_{Aw} and C_{Bw}. The transfer is effected in a time t and takes place at an area A of surface.

There is no net transference of the component **B**. When n molar units of material are transferred from outside the boundary layer to the surface:

$$\text{Transfer of } \mathbf{A} \text{ towards surface} = n(C_{As}/C_T)$$
$$\text{Transfer of } \mathbf{B} \text{ towards surface} = n(C_{Bs}/C_T)$$

In this case the molar rate of transfer of **B** away from the surface is equal to the transfer towards the surface.

$\therefore$ Transfer of **B** away from surface $= n(C_{Bs}/C_T)$

Associated transfer of **A** away from surface

$$= n(C_{Bs}/C_T)(C_{Aw}/C_{Bw})$$

Thus the net transfer of **A** towards the surface

$$-N'_A At = n\left(\frac{C_{As}}{C_T} - \frac{C_{Bs}}{C_T}\frac{C_{Aw}}{C_{Bw}}\right)$$

$$= n\left(\frac{C_T C_{As} - C_{Aw}C_{As} - C_{Aw}C_T + C_{Aw}C_{As}}{C_T(C_T - C_{Aw})}\right)$$

$$= n(C_{As} - C_{Aw})/C_{Bw}$$

It is assumed that the total molar concentration is everywhere constant. Thus the rate of transfer per unit area and unit time is given by:

$$-N'_A = n(C_{As} - C_{Aw})/C_{Bw}At \qquad (10.39)$$

The net transfer of momentum per unit time

$$= -R_0 A = (nu_s\rho/C_T)/t \qquad (10.40)$$

ρ is taken as the mean mass density of the fluid.

$\therefore$ $$-R_0 = n\rho u_s/C_T At \qquad (10.41)$$

Dividing equations 10.39 and 10.41:

$$\frac{N'_A}{R_0} = \frac{C_{As} - C_{Aw}}{\rho u_s}\frac{C_T}{C_{Bw}} \qquad (10.42)$$

Writing $R_0 = -R$ and defining the mass transfer coefficient by the relation:

$$\frac{N'_A}{C_{As} - C_{Aw}} = -h_D \qquad (10.43)$$

$$\frac{h_D}{u_s}\frac{C_{Bw}}{C_T} = \frac{R}{\rho u_s^2} \qquad (10.44)$$

Thus there is a direct proportionality between the momentum transfer and that portion of the mass transfer which is *not* attributable to bulk flow. The dimensionless group h_D/u_s is analogous to the Stanton group for heat transfer.

If the concentration of the non-diffusing component **B** is small, and in cases of equimolecular counterdiffusion, equation 10.44 reduces to:

$$h_D/u_s = R/\rho u_s^2 \qquad (10.45)$$

10.4.2. Analogy between Heat and Mass Transfer

The relation between momentum and heat transfer is given by equation 10.38:

$$h/C_p\rho u_s = R/\rho u_s^2$$

The relation between momentum and mass transfer is given by equation 10.45 for equimolecular counterdiffusion and by equation 10.44 when the second constituent is not transferred.

$$h_D/u_s = R/\rho u_s^2 \qquad \text{(equation 10.45)}$$

$$(h_D/u_s)(C_{Bw}/C_T) = R/\rho u_s^2 \qquad \text{(equation 10.44)}$$

Therefore for equimolecular counterdiffusion:

$$h_D = h/C_p\rho \tag{10.46}$$

and for conditions where bulk flow occurs:

$$h_D(C_{Bw}/C_T) = h/C_p\rho \tag{10.47}$$

Equations 10.46 and 10.47 permit the calculation of the mass transfer coefficient for conditions where the heat transfer coefficient is known. Equation 10.47 is sometimes known as the *Lewis relation*. It reduces to equation 10.46 at low concentrations of diffusing component.

It has been seen in Chapter 3 that for the flow through a tube the friction factor $R/\rho u^2$ is related approximately to the Reynolds group by the Blasius equation, $R/\rho u^2 \propto Re^{-0.25}$, so that the relation between heat transfer and the Reynolds group is obtained as:

$$h/C_p\rho u \propto Re^{-0.25} \tag{10.48}$$

For mass transfer under conditions of equimolecular counterdiffusion:

$$h_D/u \propto Re^{-0.25} \tag{10.49}$$

For diffusion through a stationary gas:

$$(h_D/u)(C_{Bw}/C_T) \propto Re^{-0.25} \tag{10.50}$$

(since the mean velocity u is proportional to the velocity u_s remote from the surface).

10.4.3 Modified Reynolds Analogy for Heat Transfer and Mass Transfer

The original Reynolds analogy involves a number of simplifying assumptions which are justifiable only in a limited range of conditions. Thus it was assumed that fluid was transferred from outside the boundary layer to the surface without mixing with the intervening fluid, that it was brought to rest at the surface, and that thermal equilibrium was established. Various modifications have been made to this simple theory to take account of the existence of the laminar sub-layer and the buffer layer close to the surface.

Taylor[4] and Prandtl[5, 6] allowed for the existence of the laminar sub-layer but ignored the existence of the buffer layer in their treatment and assumed that the simple Reynolds analogy was applicable to the transfer of heat and momentum from the main stream to the edge of the laminar sub-layer of thickness δ_b. Transfer through the laminar sub-layer was then presumed to be attributable solely to molecular motion.

Let αu_s and $b\theta_s$ be the velocity and temperature, respectively, at the edge of the laminar sub-layer. Applying the Reynolds analogy (equation 10.37) for transfer across the turbulent region:

$$\frac{q}{R_0} = \frac{C_p(\theta_s - b\theta_s)}{u_s - \alpha u_s} \tag{10.51}$$

The rate of transfer of heat by conduction through the laminar sub-layer from a surface of area A is given by:

$$qA = -kb\theta_s A/\delta_b \tag{10.52}$$

The rate of transfer of momentum is equal to the shearing force and therefore:

$$R_0 A = -\mu \alpha u_s A / \delta_b = -RA \tag{10.53}$$

Dividing equations 10.52 and 10.53:

$$q / R_0 = kb\theta_s / \mu \alpha u_s \tag{10.54}$$

Thus from equations 10.51 and 10.54:

$$\frac{kb\theta_s}{\mu \alpha u_s} = \frac{C_p(1-b)\theta_s}{(1-\alpha)u_s}$$

$$\therefore \quad Pr(1-b)/b = (1-\alpha)/\alpha$$

$$\therefore \quad \frac{b}{\alpha} = \frac{1}{\alpha + (1-\alpha)Pr^{-1}}$$

Substituting in equation 10.54:

$$\frac{q}{R_0} = \frac{C_p\theta_s}{u_s}\frac{1}{1+\alpha(Pr-1)} = \frac{h\theta_s}{R}$$

i.e.

$$St = \frac{h}{C_p\rho u_s} = \frac{R/\rho u_s^2}{1+\alpha(Pr-1)} \tag{10.55}$$

The quantity α, which is the ratio of the velocity at the edge of the laminar sub-layer to the stream velocity, was evaluated in Chapter 9 in terms of the Reynolds number for flow over the surface. For flow over a plane surface, from Chapter 9:

$$\alpha = 2 \cdot 1 Re_x^{-0 \cdot 1} \qquad \text{(equation 9.28)}$$

where Re_x is the Reynolds number $u_s x \rho / \mu$, x being the distance from the leading edge of the surface.

For flow through a pipe of diameter d (see Chapter 9):

$$\alpha = 2 \cdot 0 Re^{-1/8} \qquad \text{(equation 9.41)}$$

where Re is the Reynolds number $ud\rho / \mu$.

For mass transfer to a surface, a similar relation to equation 10.55 can be derived for equimolecular counterdiffusion except that the Prandtl number is replaced by the Schmidt number. It follows that:

$$\frac{h_D}{u_s} = \frac{R/\rho u_s^2}{1+\alpha(Sc-1)} \tag{10.56}$$

where Sc is the Schmidt number ($\mu / \rho D$). It is possible also to derive an expression to take account of bulk flow, but many simplifying assumptions must be made and the final result is not very useful.

By taking account of the existence of the laminar sub-layer, correction factors are introduced into the simple Reynolds analogy.

For heat transfer, the factor is $[1 + \alpha(Pr - 1)]$ and for mass transfer it is $[1 + \alpha(Sc - 1)]$.

There are two sets of conditions under which the correction factor approaches unity:

(i) For gases both the Prandtl and Schmidt groups are approximately unity, and therefore the simple Reynolds analogy is closely followed. Furthermore, the Lewis relation which is based on the simple analogy would be expected to hold closely for gases. The Lewis relation will also hold for any system for which the Prandtl and Schmidt numbers are equal. In this case, the correction factors for heat and mass transfer will be approximately equal.

(ii) When the fluid is highly turbulent, the laminar sub-layer will become very thin and the velocity at the edge of the laminar sub-layer will be small. In these circumstances again, the correction factor will approach unity.

10.4.4. Use of Universal Velocity Profile[7] to Calculate Heat Transfer Coefficients

In the Taylor–Prandtl modification of the theory of heat transfer to a turbulent fluid, it was assumed that the heat passed directly from the turbulent fluid to the laminar sub-layer and the existence of the buffer layer was neglected. It was therefore possible to apply the simple theory for the boundary layer in order to calculate the heat transfer. In most cases, the results so obtained are sufficiently accurate, but errors become significant when the relations are used to calculate heat transfer to liquids of high viscosities. A more accurate expression can be obtained if the temperature difference across the buffer layer is taken into account. The exact conditions in the buffer layer are difficult to define and any mathematical treatment of the problem involves a number of assumptions. However, the conditions close to the surface over which fluid is flowing can be calculated approximately using the *universal velocity profile*.

The method is based on a calculation of the temperature difference across the laminar sub-layer, the buffer layer, and the turbulent region in turn. This is carried out on the assumption that the laminar sub-layer extends from $y^+ = 0$ to $y^+ = 5$, the buffer layer from $y^+ = 5$ to $y^+ = 30$, and the turbulent region for y^+ greater than 30. The temperature difference across the turbulent region is obtained by direct application of the Reynolds analogy, with the velocity at the edge calculated by substituting $y^+ = 30$ in equation 9.55. For the buffer layer it is shown that the eddy kinematic viscosity and the eddy thermal diffusivity are approximately equal, so that the temperature gradient can be expressed in terms of the eddy viscosity. The total temperature drop is then obtained by integration between the limits of 5 and 30 for y^+. For the laminar sub-layer a constant temperature gradient is assumed and the thickness of the layer is obtained from the fact that y^+ extends from 0 to 5.

Laminar Sub-Layer

$$0 < y^+ < 5$$

Heat transfer is by thermal conduction so that, on the assumption of a linear temperature gradient:

$$- q_0 = k\theta_s/y_s$$

Now

$$y^+ = yu^*\rho/\mu \quad \text{(from equation 9.58)}$$

so that

$$y_s = 5\mu/u^*\rho$$

∴

$$\theta_s = -\frac{q_0}{k}\frac{5\mu}{u^*\rho} \tag{10.57}$$

Buffer Layer

For momentum transfer:

$$- R_0 = (\mu + E\rho)\frac{du_x}{dy} \quad \text{(from equation 10.32)}$$

For a smooth surface, $\dfrac{du^+}{dy^+} = \dfrac{5\cdot0}{y^+}$ (from equation 9.63)

∴

$$\frac{du_x}{dy} = \frac{5\cdot0\rho u^{*2}}{\mu y^+}$$

Thus
$$E = -\frac{R_0 \mu y^+}{5 \cdot 0 \rho^2 u^{*2}} - \frac{\mu}{\rho}$$

$$= \frac{\mu y^+}{5\rho} - \frac{\mu}{\rho} \quad \left(\text{since } \frac{-R_0}{\rho} = u^{*2}\right)$$

$$= \frac{\mu}{\rho}\left(\frac{y^+}{5} - 1\right) \tag{10.58}$$

For heat transfer:
$$-q_0 = (k + E_H C_P \rho)\frac{d\theta}{dy} \quad \text{(from equation 10.33)}$$

Assuming that $E \approx E_H$:
$$-q_0 = \left[k + C_P \rho \frac{\mu}{\rho}\left(\frac{y^+}{5} - 1\right)\right]\frac{d\theta}{dy}$$

$$= \left[k + C_P \mu \left(\frac{y^+}{5} - 1\right)\right]\frac{d\theta}{dy^+}\frac{u^*\rho}{\mu}$$

$$\therefore \quad \frac{d\theta}{dy^+} = \frac{-\dfrac{5q_0}{(C_P \rho u^*)}}{\dfrac{5k}{C_P \mu} - 5 + y^+}$$

$$\theta\big|_5^{30} = -\frac{5q_0}{C_P \rho u^*}\{\ln[y^+ + 5(Pr^{-1} - 1)]\}\big|_5^{30}$$

$$= -\frac{5q_0}{C_P \rho u^*}\ln(5\,Pr + 1). \tag{10.59}$$

Turbulent Zone

$$y^+ > 30$$

From the Reynolds analogy:

$$\frac{-q_0}{-R_0} = \frac{C_P(\theta_s - \theta_{30})}{u_s - u_{30}} \tag{equation 10.33}$$

Velocity at $y^+ = 30$ is given by:
$$u^+ = 5 \cdot 0 \ln y^+ - 3 \cdot 05 \tag{equation 9.54}$$

$$\therefore \quad u_{30} = u^*(5 \cdot 0 \ln 30 - 3 \cdot 05)$$

and
$$\theta_s - \theta_{30} = -\frac{q_0}{C_P \rho u^{*2}}(u_s - 5 \cdot 0 u^* \ln 30 + 3 \cdot 05 u^*) \tag{10.60}$$

Overall temperature difference θ_s is obtained by addition of equations 10.53, 10.55, and 10.56.

Thus $\theta_s = -\dfrac{q_0}{C_P \rho u^{*2}}\left[\dfrac{5 C_P \mu}{k}u^* + 5u^* \ln(5\,Pr + 1) + u_s - 5 \cdot 0 u^* \ln 30 + 3 \cdot 05 u^*\right]$

$$= -\frac{q_0}{C_P \rho u^{*2}}\left[5\,Pr\,u^* + 5u^* \ln\left(\frac{Pr}{6} + \frac{1}{30}\right) + u_s + 3 \cdot 05 u^*\right] \tag{10.61}$$

i.e. $\dfrac{u^{*2}}{u_s^2} = -\dfrac{q_0}{C_P \rho u_s \theta_s}\left\{1 + 5\dfrac{u^*}{u_s}\left[Pr + \ln\left(\dfrac{Pr}{6} + \dfrac{1}{30}\right) + \ln 5 - 1\right]\right\}$ (since $5 \ln 5 = 8 \cdot 05$)

$$\therefore \quad \frac{R}{\rho u_s^2} = \frac{h}{C_P \rho u_s}\left\{1 + 5\sqrt{\frac{R}{\rho u_s^2}}\left[(Pr - 1) + \ln\left(\frac{5}{6}Pr + \frac{1}{6}\right)\right]\right\} \tag{10.62}$$

where
$$\frac{h}{C_P \rho u_s} = St \quad \text{(the Stanton number)}$$

A similar expression can also be derived for mass transfer in the absence of bulk flow:

$$\frac{R}{\rho u_s^2} = \frac{h_0}{u_s}\left\{1 + 5\sqrt{\frac{R}{\rho u_s^2}}\left[(Sc - 1) + \ln\left(\tfrac{5}{6}Sc + \tfrac{1}{6}\right)\right]\right\}$$

10.4.5. Flow over a Plane Surface

The simple Reynolds analogy gives a relation between the friction factor $R/\rho u_s^2$ and the Stanton number for heat transfer:

$$R/\rho u_s^2 = h/C_p\rho u_s \qquad\qquad \text{(equation 10.38)}$$

This equation can be used for calculating the point value of the heat transfer coefficient by substituting for $R/\rho u_s^2$ in terms of the Reynolds group Re_x using equation 9.32:

$$R/\rho u_s^2 = 0.03\, Re_x^{-0.2} \qquad\qquad \text{(equation 9.32)}$$

$$\therefore \qquad St = h/C_p\rho u_s = 0.03\, Re_x^{-0.2} \qquad\qquad \text{(10.63)}$$

Equation 10.63 gives the point value of the heat transfer coefficient. If the whole surface is effective for heat transfer, the mean value is given by:

$$St = h/C_p\rho u_s = \frac{1}{x}\int_0^x 0.03\, Re_x^{-0.2}\, dx$$

$$= 0.037\, Re_x^{-0.2} \qquad\qquad \text{(10.64)}$$

The above equations take no account of the existence of the laminar sub-layer and therefore give unduly high values for the transfer coefficient, especially with liquids. The effect of the laminar sub-layer is allowed for by using the Taylor–Prandtl modification:

$$St = \frac{h}{C_p\rho u_s} = \frac{R/\rho u_s^2}{1 + \alpha(Pr - 1)} \qquad\qquad \text{(equation 10.55)}$$

where

$$\alpha = u_b/u_s = 2.1\, Re_x^{-0.1} \qquad\qquad \text{(equation 9.28)}$$

Thus

$$St = Nu_x\, Re_x^{-1}\, Pr^{-1} = \frac{0.03\, Re_x^{-0.2}}{1 + 2.1\, Re_x^{-0.1}(Pr - 1)} \qquad\qquad \text{(10.65)}$$

This expression will give the point value of the Stanton number and hence of the heat transfer coefficient. The mean value over the whole surface is obtained by integration. No general expression for the mean coefficient can be obtained and a graphical or numerical integration must be carried out after the insertion of the appropriate values of the constants.

Similarly, substitution may be made from equation 9.32 into equation 10.62 to give the point values of the Stanton number and the heat transfer coefficient, thus:

$$St = h/C_p\rho u_s = \frac{0.030\, Re_x^{-0.2}}{1 + 0.87\, Re_x^{-0.1}[(Pr - 1) + \ln(\tfrac{5}{6}Pr + \tfrac{1}{6})]} \qquad\qquad \text{(10.66)}$$

Mean values may be obtained by graphical integration.

The same procedure may be used for obtaining relationships for mass transfer coefficients, for equimolecular counterdiffusion or where the concentration of the non-diffusing constituent is small:

$$R/\rho u_s^2 = h_D/u_s \qquad\qquad \text{(equation 10.45)}$$

For flow over a plane surface, substitution from equation 9.32 gives:

$$h_D/u_s = 0.03\, Re_x^{-0.2} \qquad\qquad \text{(10.67)}$$

Equation 10.67 gives the point value of h_D. The mean value over the surface is obtained in the same manner as equation 10.64 as:

$$h_D/u_s = 0.037\,Re_x^{-0.2} \qquad\qquad (10.68)$$

For mass transfer through a stationary second component:

$$R/\rho u_s^2 = (h_D/u_s)(C_{Bw}/C_T) \qquad\qquad (\text{equation } 10.44)$$

The correction factor C_{Bw}/C_T must then be introduced into equations 10.67 and 10.68.

The above equations are applicable only when the Schmidt number Sc is very close to unity or where the velocity of flow is so high that the resistance of the laminar sub-layer is small. The resistance of the laminar sub-layer can be taken into account, however, for equimolecular counterdiffusion or for low concentration gradients by using equation 10.56:

$$\frac{h_D}{u_s} = \frac{R/\rho u_s^2}{1+\alpha(Sc-1)} \qquad\qquad (\text{equation } 10.56)$$

Substitution for $R/\rho u_s^2$ and α using equations 9.32 and 9.28 gives:

$$h_D/u_s = Sh_x\,Re_x^{-1}\,Sc^{-1} = \frac{0.03\,Re_x^{-0.2}}{1+2.1\,Re_x^{-0.1}(Sc-1)} \qquad\qquad (10.69)$$

10.4.6. Flow in a Pipe

For the inlet length of a pipe in which the boundary layers are forming, the equations in the previous section will give an approximate value for the heat transfer coefficient. It should be remembered, however, that the flow in the boundary layer at the entrance to the pipe will, in general, be streamline and the point of transition to turbulent flow is not easily defined. The results therefore are, at best, approximate.

In fully developed flow, equations 10.38 and 10.55 can be used, but it is preferable to work in terms of the mean velocity of flow and the ordinary pipe Reynolds number Re. Furthermore, the heat transfer coefficient is generally expressed in terms of a driving force equal to the difference between the bulk fluid temperature and the wall temperature. If the fluid is highly turbulent, however, the bulk temperature will be quite close to the temperature θ_s at the axis.

Into equations 10.37 and 10.55, equation 3.56 may then be substituted:

$$u = 0.82u_s \qquad\qquad (\text{equation } 3.56)$$
$$u_b/u = 2.49\,Re^{-1/8} \qquad\qquad (\text{equation } 9.40)$$
$$R/\rho u^2 = 0.0396\,Re^{-1/4} \qquad\qquad (\text{equation } 9.39)$$

Firstly, using the simple Reynolds analogy (equation 10.38):

$$\begin{aligned}
h/C_p\rho u &= (h/C_p\rho u_s)(u_s/u)\\
&= (R/\rho u_s^2)(u_s/u)\\
&= (R/\rho u^2)(u/u_s)\\
&= 0.032\,Re^{-1/4} \qquad\qquad (10.70)
\end{aligned}$$

Then, using the Taylor–Prandtl modification (equation 10.55):

$$\begin{aligned}
\frac{h}{C_p\rho u} &= \frac{0.032\,Re^{-1/4}}{1+(u_b/u)(u/u_s)(Pr-1)}\\
&= \frac{0.032\,Re^{-1/4}}{1+2.0\,Re^{-1/8}(Pr-1)} \qquad\qquad (10.71)
\end{aligned}$$

Finally, using equation 10.66:

$$St = \frac{h}{C_p \rho u} = \frac{0 \cdot 82(R/\rho u^2)}{1 + 0 \cdot 82 \sqrt{(R/\rho u^2)} \, 5[(Pr - 1) + \ln(\frac{5}{6} Pr + \frac{1}{6})]}$$

$$= \frac{0 \cdot 032 \, Re^{-1/4}}{1 + 0 \cdot 82 \, Re^{-1/8}[(Pr - 1) + \ln(\frac{5}{6} Pr + \frac{1}{6})]} \tag{10.72}$$

For mass transfer, the equations corresponding to equations 10.68 and 10.69 are obtained in the same way as the analogous heat transfer equations.

Thus, using the simple Reynolds analogy for equimolecular counterdiffusion:

$$h_D/u = 0 \cdot 032 \, Re^{-1/4} \tag{10.73}$$

and for diffusion through a stationary gas.

$$(h_D/u)(C_{Bw}/C_T) = 0 \cdot 032 \, Re^{-1/4} \tag{10.74}$$

Using the Taylor–Prandtl form for equimolecular counterdiffusion or low concentration gradients:

$$\frac{h_D}{u} = \frac{0 \cdot 032 \, Re^{-1/4}}{1 + 2 \cdot 0 \, Re^{-1/8}(Sc - 1)} \tag{10.75}$$

Example 10.1. Water flows at 0·50 m/s through a 20 mm tube lined with β-naphthol. What is the mass transfer coefficient if the Schmidt number is 2330?

Solution. Reynolds number: $\qquad Re = 0 \cdot 020 \times 0 \cdot 50 \times 1000/1 \times 10^{-3} = 10,000$

From equation 10.75:

$$h_D/u = 0 \cdot 032 \, Re^{-\frac{1}{4}}[1 + 2 \cdot 0(Sc - 1) \, Re^{-\frac{1}{8}}]^{-1}$$

$$\therefore \qquad h_D = 0 \cdot 032 \times 0 \cdot 50 \times 0 \cdot 1[1 + 2(2329)0 \cdot 316]^{-1}$$

$$= \underline{1 \cdot 085 \times 10^{-6} \, \text{kmol/m}^2 \, \text{s} \, (\text{kmol/m}^3)}$$

Example 10.2. Calculate the rise in temperature of water which is passed at 3·5 m/s through a smooth 25 mm diameter pipe, 6 m long. The water enters at 300 K and the tube wall may be assumed constant at 330 K. The following methods may be used:

(a) the simple Reynolds analogy (equation 10.70);
(b) the Taylor–Prandtl modification (equation 10.71);
(c) the universal velocity profile (equation 10.72);
(d) $Nu = 0 \cdot 023 \, Re^{0 \cdot 8} \, Pr^{0 \cdot 33}$ (equation 7.53).

Solution. Taking the fluid properties at 310 K and assuming that fully developed flow exists, an approximate solution will be obtained neglecting the variation of properties with temperature.

$$Re = 0 \cdot 025 \times 3 \cdot 5 \times 1000/0 \cdot 7 \times 10^{-3} = 1 \cdot 25 \times 10^5$$

$$Pr = 4 \cdot 18 \times 10^3 \times 0 \cdot 7 \times 10^{-3}/0 \cdot 65 = 4 \cdot 50$$

(a) *Reynolds analogy*

$$h/C_p \rho u = 0 \cdot 032 \, Re^{-0 \cdot 25} \qquad \text{(equation 10.70)}$$

$$\therefore \qquad h = 4 \cdot 18 \times 1000 \times 1000 \times 3 \cdot 5 \times 0 \cdot 032/(1 \cdot 25 \times 10^5)^{-0 \cdot 25}$$

$$= 24,902 \, \text{W/m}^2 \, \text{K} \quad \text{or} \quad \underline{24 \cdot 9 \, \text{kW/m}^2 \, \text{K}}$$

Heat transferred per unit time in length dL of pipe $= h\pi 0{\cdot}025\,dL\,(330 - \theta)$ kW, where θ is the temperature at a distance L m from the inlet.

Rate of increase of heat content of fluid $= (\pi/4)0{\cdot}025^2 \times 3{\cdot}5 \times 1000 \times 4{\cdot}18\,d\theta$ kW

Outlet temperature θ' is then given by:

$$\int_{300}^{\theta'} \frac{d\theta}{(330 - \theta)} = 0{\cdot}0109h \int_{0}^{6} dL$$

where h is in kW/m^{2}K.

$\therefore$ $\qquad\qquad \log_{10}(330 - \theta') = \log_{10} 30 - 0{\cdot}0654\,h/2{\cdot}303 = 1{\cdot}477 - 0{\cdot}0283h$

In this case $\qquad\qquad h = 24{\cdot}9\,\text{kW/m}^2\,\text{K}$

$\therefore$ $\qquad\qquad \log_{10}(330 - \theta') = 1{\cdot}477 - 0{\cdot}705 = 0{\cdot}772$

$\therefore$ $\qquad\qquad\qquad \underline{\theta' = 324{\cdot}1\,\text{K}}$

(b) *Taylor–Prandtl equation*

$$h/C_p\rho u = 0{\cdot}032\,Re^{-\frac{1}{4}}[1 + 2{\cdot}0\,Re^{-\frac{1}{8}}(Pr - 1)]^{-1} \qquad\qquad \text{(equation 10.71)}$$

$\therefore$ $\qquad\qquad h = 24{\cdot}9/(1 + 2{\cdot}0 \times 3{\cdot}5/4{\cdot}34)$

$\qquad\qquad\quad = \underline{9{\cdot}53\,\text{kW/m}^2\,\text{K}}$

$\therefore$ $\qquad\qquad \log_{10}(330 - \theta') = 1{\cdot}477 - (0{\cdot}0283 \times 9{\cdot}53) = 1{\cdot}207$

$\therefore$ $\qquad\qquad\qquad \underline{\theta' = 313{\cdot}9\,\text{K}}$

(c) *Universal Velocity Profile equation*

$$h/C_p\rho u = 0{\cdot}032\,Re^{-\frac{1}{4}}\{1 + 0{\cdot}82\,Re^{-\frac{1}{8}}[(Pr - 1) + \ln(0{\cdot}83\,Pr + 0{\cdot}17)]\}^{-1}$$

$$\text{(equation 10.72)}$$

$\qquad\qquad\quad = 24{\cdot}9/[1 + (0{\cdot}82/4{\cdot}34)(3{\cdot}5 + 2{\cdot}303 \times 0{\cdot}591)]$

$\qquad\qquad\quad = \underline{12{\cdot}98\,\text{kW/m}^2\,\text{K}}$

$\therefore$ $\qquad\qquad \log_{10}(330 - \theta') = 1{\cdot}477 - (0{\cdot}0283 \times 12{\cdot}98) = 1{\cdot}110$

$\therefore$ $\qquad\qquad\qquad \underline{\theta' = 317{\cdot}1\,\text{K}}$

(d) $Nu = 0{\cdot}023\,Re^{0{\cdot}8}\,Pr^{0{\cdot}33}$ $\qquad\qquad\qquad\qquad\qquad$ (equation 7.53)

$\qquad\qquad h = (0{\cdot}023 \times 0{\cdot}65/0{\cdot}0250)(1{\cdot}25 \times 10^5)^{0{\cdot}8}(4{\cdot}50)^{0{\cdot}33}$

$\qquad\qquad\quad = 0{\cdot}596 \times 1{\cdot}195 \times 10^4 \times 1{\cdot}64$

$\qquad\qquad\quad = 1{\cdot}168 \times 10^4\,\text{W/m}^2\,\text{K}$ $\quad$ or $\quad$ $\underline{11{\cdot}68\,\text{kW/m}^2\,\text{K}}$

$\therefore$ $\qquad\qquad \log_{10}(330 - \theta') = 1{\cdot}477 - (0{\cdot}0283 \times 11{\cdot}68) = 1{\cdot}147$

$\therefore$ $\qquad\qquad\qquad \underline{\theta' = 316{\cdot}0\,\text{K}}$

Comparing the results:

	$h\,(\text{kW/m}^2\,\text{K})$	$\theta'\,(\text{K})$
(a)	24·9	324·1
(b)	9·5	313·9
(c)	13·0	317·1
(d)	11·7	316·0

it is seen that the simple Reynolds analogy is far from accurate in calculating heat transfer to a liquid.

Example 10.3. The same tube is maintained at 350 K and air is passed through it at 3·5 m/s, the initial temperature of the air being 290 K. What is the outlet temperature of the air for the four cases used in Example 10.2?

Solution. Taking the physical properties of air at 310 K and assuming that fully developed flow exists in the pipe:

$$Re = 0 \cdot 0250 \times 3 \cdot 5 \times (29/22 \cdot 4)(273/310)/(0 \cdot 018 \times 10^{-3}) = 5535$$
$$Pr = 1 \cdot 003 \times 1000 \times 0 \cdot 018 \times 10^{-3}/0 \cdot 024 = 0 \cdot 75$$

The heat transfer coefficients and final temperatures are then calculated as in Example 10.2

	h (W/m² K)	θ' (K)
(a)	15·5	348·1
(b)	18·3	349·0
(c)	17·9	348·9
(d)	21·2	349·4

In this case the result obtained using the Reynolds analogy agrees much more closely with the other three methods.

10.5. FURTHER READING

BENNETT, C. O. and MYERS, J. E.: Momentum, Heat and Mass Transfer (McGraw Hill, 1962).

10.6. REFERENCES

1. JEANS, J. H.: Kinetic Theory of Gases (Cambridge, 1940).
2. PRANDTL, L.: Z. Angew Math. u. Mech. 5 (1925) 136. Untersuchungen zur ausgebildeten Turbulenz.
3. REYNOLDS, O.: Proc. Manchester Lit. Phil. Soc. 14 (1874) 7. On the extent and action of the heating surface for steam boilers.
4. TAYLOR, G. I.: N.A.C.A. Rep. and Mem. No. 272 (1916) 423. Conditions at the surface of a hot body exposed to the wind.
5. PRANDTL, L.: Physik. Z. 11 (1910) 1072. Eine Beziehung zwischen Wärmeaustausch und Strömungswiderstand der Flüssigkeiten.
6. PRANDTL, L.: Physik. Z. 29 (1928) 487. Bemerkung über den Wärmeübergang im Rohr.
7. MARTINELLI, R. C.: Trans. Am. Soc. Mech. Eng. 69 (1947) 947. Heat transfer to molten metals.

10.7. NOMENCLATURE

		Units in SI system	Dimensions in M, L, T, θ
A	Area of surface	m²	L^2
b	Ratio of θ_b to θ_s	—	—
C	Molar concentration	kmol/m³	ML^{-3}
C_A, C_B	Molar concentration of A, B	kmol/m³	ML^{-3}
C_{As}, C_{Bs}	Molar concentration of A, B outside boundary layer	kmol/m³	ML^{-3}
C_{Aw}, C_{Bw}	Molar concentration of A, B at wall	kmol/m³	ML^{-3}
C_p	Specific heat at constant pressure	J/kg K	$L^2T^{-2}\theta^{-1}$

		Units in SI system	Dimensions in $\mathbf{M, L, T}, \theta$
C_T	Total molar concentration	kmol/m³	$\mathbf{ML^{-3}}$
C_v	Specific heat at constant volume	J/kg K	$\mathbf{L^2T^{-2}\theta^{-1}}$
c_m	Heat capacity of one molecule	J/K	$\mathbf{ML^2T^{-2}\theta^{-1}}$
D	Diffusivity	m²/s	$\mathbf{L^2T^{-1}}$
D_H	Thermal diffusivity	m²/s	$\mathbf{L^2T^{-1}}$
d	Pipe or particle diameter	m	$\mathbf{L}$
E	Eddy kinematic viscosity	m²/s	$\mathbf{L^2T^{-1}}$
E_D	Eddy diffusivity	m²/s	$\mathbf{L^2T^{-1}}$
E_H	Eddy thermal diffusivity	m²/s	$\mathbf{L^2T^{-1}}$
h	Heat transfer coefficient	W/m²K	$\mathbf{MT^{-3}\theta^{-1}}$
h_D	Mass transfer coefficient	kmol/(kmol/m³)(m²)(s)	$\mathbf{LT^{-1}}$
i	Fraction of root mean square velocity of molecules	—	—
j	Fraction of mean free path of molecules	—	—
k	Thermal conductivity	W/mK	$\mathbf{MLT^{-3}\theta^{-1}}$
M	Molecular weight	kg/kmol	—
$\mathbf{M}$	Mass of fluid	kg	$\mathbf{M}$
m	Mass of gas molecule	kg	$\mathbf{M}$
N	Molar rate of diffusion per unit area	kmol/m²s	$\mathbf{ML^{-2}T^{-1}}$
N'	Total molar rate of transfer per unit area	kmol/m²s	$\mathbf{ML^{-2}T^{-1}}$
$\mathbf{N}$	Number of molecules per unit volume	1/m³	$\mathbf{L^{-3}}$
n	Number of molar units	kmol	$\mathbf{M}$
q	Rate of transfer of heat per unit area at walls	W/m²	$\mathbf{MT^{-3}}$
q_y	Rate of transfer of heat per unit area at $y = y$	W/m²	$\mathbf{MT^{-3}}$
R	Shear stress acting on surface	N/m²	$\mathbf{ML^{-1}T^{-2}}$
R_0	Shear stress acting on fluid at surface	N/m²	$\mathbf{ML^{-1}T^{-2}}$
R_y	Shear stress in fluid at $y = y$	N/m²	$\mathbf{ML^{-1}T^{-2}}$
$\mathbf{R}$	Universal gas constant	8·314 kJ/kmol K	$\mathbf{L^2T^{-2}\theta^{-1}}$
T	Absolute temperature	K	$\boldsymbol{\theta}$
t	Time	s	$\mathbf{T}$
u	Mean velocity	m/s	$\mathbf{LT^{-1}}$
u_b	Velocity at edge of laminar sub-layer	m/s	$\mathbf{LT^{-1}}$
u_E	Mean velocity in eddy	m/s	$\mathbf{LT^{-1}}$
u_{Ex}	Mean component of eddy velocity in X-direction	m/s	$\mathbf{LT^{-1}}$
u_{Ey}	Mean component of eddy velocity in Y-direction	m/s	$\mathbf{LT^{-1}}$
u_m	Root mean square velocity of molecules	m/s	$\mathbf{LT^{-1}}$
u_s	Velocity of fluid outside boundary layer, or at pipe axis	m/s	$\mathbf{LT^{-1}}$
u_x	Velocity in X-direction at $y = y$	m/s	$\mathbf{LT^{-1}}$
u'_x	Velocity in X-direction at $y = y + j\lambda$	m/s	$\mathbf{LT^{-1}}$
u^+	Ratio of u_y to u^*	—	—
u^*	Shearing stress velocity, $\sqrt{R/\rho}$	m/s	$\mathbf{LT^{-1}}$
y	Distance from surface	m	$\mathbf{L}$
y^+	Ratio of y to $\mu/\rho u^*$	—	—
α	Ratio of u_b to u_s	—	—
γ	Ratio of C_v to C_v	—	—
δ	Thickness of boundary layer	m	$\mathbf{L}$
δ_b	Thickness of laminar sub-layer	m	$\mathbf{L}$
λ	Mean free path of molecules	m	$\mathbf{L}$
λ_E	Mixing length	m	$\mathbf{L}$
μ	Viscosity of fluid	Ns/m²	$\mathbf{ML^{-1}T^{-1}}$
ρ	Density of fluid	kg/m³	$\mathbf{ML^{-3}}$
θ	Temperature at $y = y$	K	$\boldsymbol{\theta}$
θ_s	Temperature outside boundary layer, or at pipe axis	K	$\boldsymbol{\theta}$
θ'	Temperature at $y = y + j\lambda$	K	$\boldsymbol{\theta}$
Nu	Nusselt number hd/k	—	—
Nu_x	Nusselt number hx/k	—	—
Re	Reynolds number $ud\rho/\mu$	—	—
Re_x	Reynolds number $u_s x\rho/\mu$	—	—
Pr	Prandtl number $C_p\mu/k$	—	—
Sc	Schmidt number $\mu/\rho D$	—	—
Sh	Sherwood number $h_D d/D$	—	—
St	Stanton number $h/C_p\rho u$ or $h/C_p\rho u_s$	—	—

CHAPTER 11

Humidification and Water Cooling

11.1. INTRODUCTION

In the processing of materials it is often necessary either to increase the amount of vapour present in a gas stream, an operation known as *humidification,* or to reduce the vapour present, a process referred to as *dehumidification.* In humidification, the vapour content can be increased by passing the gas over a liquid which then evaporates into the gas stream. This transfer into the main stream takes place by diffusion, and at the interface simultaneous transfer of heat and mass takes place according to the relations developed in previous chapters. In the reverse operation, that is dehumidification, partial condensation must be effected and the condensed vapour removed.

The most widespread application of humidification and dehumidification operations involves the air–water system, and a discussion of this forms the greater part of the present chapter. Although the drying of wet solids is an example of a humidification operation, the reduction of the moisture content of the solids is the main aim and the humidification of the air stream is a secondary effect. Much of the present chapter is, however, of vital significance in any drying operation. Air conditioning and gas drying also involve humidification and dehumidification operations. For example, moisture must be removed from wet chlorine so that the gas can be handled in steel equipment which would otherwise be severely corroded. Similarly, the gases used in the manufacture of sulphuric acid must be dried or dehumidified before they enter the converters, and this is achieved by passing the gas through a dehydrating agent such as sulphuric acid, in essence an absorption operation, or by an alternative dehumidification process discussed later.

In order that hot condenser water may be returned to the plant, it is normally cooled in contact with an air stream. The equipment usually takes the form of a tower in which the hot water is run in at the top and allowed to flow downwards over a packing against a counter-current flow of air which enters at the bottom of the cooling tower. The design of such towers forms an important part of the present chapter, though at the outset it is necessary to consider basic definitions of the various quantities involved in humidification, in particular *wet-bulb* and *adiabatic saturation temperatures,* and the way in which humidity data can be presented on charts and graphs. While the present discussion is devoted to the very important air–water system, which is in some ways unique, the same principles can be applied to other liquids and gases, and this point is covered in a final section.

11.2. HUMIDIFICATION TERMS

11.2.1. Definitions

The more important terms used in relation to humidification are defined as follows:

Humidity ($\mathscr{H}$)   mass of vapour associated with unit mass of dry gas

Humidity of saturated   humidity of the gas when it is saturated with vapour at a given
   gas ($\mathscr{H}_0$)   temperature

Percentage humidity $100\,\mathscr{H}/\mathscr{H}_0$

Humid heat (s)   heat required to raise unit mass of dry gas and its associated vapour
         through unit temperature difference at constant pressure.

That is:

$$s = C_a + \mathscr{H}C_w$$

where C_a and C_w are the specific heat capacities of the gas and the vapour, respectively.
For the air–water system, the humid heat is approximately:

$$s = 1{\cdot}00 + 1{\cdot}9\mathscr{H} \text{ kJ/kg K}$$

Humid volume   volume occupied by unit mass of dry gas and its associated vapour

Saturated volume   humid volume of saturated gas

Dew point   temperature at which the gas is saturated with vapour. If a gas is cooled, the dew point is the temperature at which condensation will first occur.

Percentage relative humidity $100 \times \dfrac{\text{partial pressure of vapour in gas}}{\text{partial pressure of vapour in saturated gas}}$

The relationship between the partial pressure of the vapour and the humidity of a gas may be derived as follows. In unit volume of gas:

$$\text{mass of vapour} = P_w M_w / \mathbf{R}T$$

and mass of non-condensable gas $= (P - P_w)M_A/\mathbf{R}T$

The humidity is therefore given by:

$$\mathscr{H} = [P_w/(P - P_w)](M_w/M_A) \tag{11.1}$$

and the humidity of the saturated gas is:

$$\mathscr{H}_0 = [P_{w_0}/(P - P_{w_0})](M_w/M_A) \tag{11.2}$$

where, P_w is the partial pressure of vapour in the gas, P_{w_0} the partial pressure of vapour in the saturated gas at the same temperature, M_A the mean molecular weight of the dry gas, M_w the molecular weight of the vapour, P the total pressure, $\mathbf{R}$ the gas constant ($8{\cdot}314$ kJ/kmol K in SI units), and T the absolute temperature.

For the air–water system, P_w is frequently small compared with P and hence, substituting the molecular weights:

$$\mathscr{H} = 18P_w/29P$$

The relationship between the percentage humidity of a gas and the percentage relative humidity may be derived as follows:

The percentage humidity, by definition $= 100\mathscr{H}/\mathscr{H}_0$

Substituting from equations 11.1 and 11.2 and simplifying:

Percentage humidity $= [(P - P_{w_0})/(P - P_w)](100P_w/P_{w_0})$

$$= [(P - P_{w_0})/(P - P_w)] \text{ [percentage relative humidity]} \tag{11.3}$$

When $(P - P_{w_0})/(P - P_w) \approx 1$, the percentage relative humidity and the percentage humidity are equal. This condition is approached when the partial pressure of the vapour is only a small proportion of the total pressure or when the gas is almost saturated, that is as $P_w \to P_{w_0}$.

Example 11.1. In a process in which benzene is used as a solvent, it is evaporated into dry nitrogen. At 297 K and 101.3 kN/m², the resulting mixture has a percentage relative humidity of

60. It is required to recover 80 per cent of the benzene present by cooling to 283 K and compressing to a suitable pressure. What should this pressure be?

Vapour pressure of benzene:

$$\text{at } 297\ K = 12\cdot2\ kN/m^2 \qquad \text{at } 283\ K = 6\cdot0\ kN/m^2$$

Solution. From the definition of percentage relative humidity (RH):

$$P_w = P_{w_0}(\text{RH}/100)$$

At 297 K: $\qquad\qquad\qquad P_w = (12\cdot2 \times 60/100) = 7\cdot32\ kN/m^2$

In the benzene–nitrogen mixture:

mass of benzene $= P_w M_w / RT = (7\cdot32 \times 78)/(8\cdot314 \times 297) = 0\cdot231\ kg$

mass of nitrogen $= (P - P_w)M_A/RT = (101\cdot3 - 7\cdot32)28/(8\cdot314 \times 297) = 1\cdot066\ kg$

Hence the humidity is: $\qquad \mathscr{H} = (0\cdot231/1\cdot066) = 0\cdot217\ kg/kg$

In order to recover 80 per cent of the benzene, the humidity must be reduced to 20 per cent of the initial value. As the vapour will be in contact with liquid benzene, the nitrogen will be saturated with benzene vapour and hence at 283 K:

$$\mathscr{H}_0 = (0\cdot217 \times 20/100) = 0\cdot0433\ kg/kg$$

Thus in equation 11.2:

$$0\cdot0433 = [6\cdot0/(P - 6\cdot0)](78/28)$$

from which $\qquad\qquad\qquad \underline{P = 392\ kN/m^2}$

Example 11.2. In a vessel at $101\cdot3\ kN/m^2$ and 300 K, the percentage relative humidity of the water vapour in the air is 25. If the partial pressure of water vapour when air is saturated with vapour at 300 K is $3\cdot6\ kN/m^2$, calculate:

(a) the partial pressure of the water vapour in the vessel;
(b) the specific volumes of the air and water vapour;
(c) the humidity of the air and humid volume; and
(d) the percentage humidity.

Solution. (a) From Example 11.1:

$$P_w = P_{w_0}(\text{RH}/100) = (3\cdot6 \times 25/100) = \underline{0\cdot9\ kN/m^2}$$

(b) In 1 m³ of air:

mass of water vapour $= 0\cdot9 \times 18/(8\cdot314 \times 300) = 0\cdot0064\ kg$

mass of air $\qquad = (101\cdot3 - 0\cdot9)29/(8\cdot314 \times 300) = 1\cdot1673\ kg$

Hence: specific volume of water vapour $= 1/0\cdot0064 = \underline{156\cdot3\ m^3/kg}$

specific volume of air $\qquad = 1/1\cdot1673 = \underline{0\cdot857\ m^3/kg}$

(c) Humidity: $\qquad\qquad \mathscr{H} = 0\cdot0064/1\cdot1673 = \underline{0\cdot00548\ kg/kg}$

(Using the approximate relationship:

$$\mathscr{H} = 18 \times 0\cdot9/(29 \times 101\cdot3) = 0\cdot00551\ kg/kg)$$

In 1 kg air vapour mixture, there is $0\cdot0055/1\cdot0055 = 0\cdot0054\ kg$ water and $(1 - 0\cdot0054) = 0\cdot9946\ kg$ air.

∴ $\qquad\qquad$ Humid volume $= (0\cdot9946 \times 0\cdot857) + (0\cdot0054 \times 156\cdot3) = \underline{1\cdot696\ m^3/kg}$

(d) From equation 11.3:

$$\text{percentage humidity} = [(101 \cdot 3 - 3 \cdot 6)/(101 \cdot 3 - 0 \cdot 9)]25$$
$$= \underline{\underline{24.3 \text{ per cent}}}$$

11.2.2. Wet-bulb Temperature

When a stream of unsaturated gas is passed over the surface of a liquid, the humidity of the gas is increased due to evaporation of the liquid. The temperature of the liquid falls below that of the gas and heat is transferred from the gas to the liquid. At equilibrium the rate of heat transfer from the gas just balances that required to vaporise the liquid and the liquid is said to be at the *wet-bulb temperature*. The rate at which this temperature is reached depends on the initial temperatures and the rate of flow of gas past the liquid surface. With a small area of contact between the gas and the liquid and a high gas flowrate, the temperature and the humidity of the gas stream remain virtually unchanged.

The rate of transfer of heat from the gas to the liquid can be written as:

$$Q = hA(\theta - \theta_w) \tag{11.4}$$

where Q is the heat flow, h the coefficient of heat transfer, A the area for transfer, and θ and θ_w are the temperatures of the gas and liquid phases.

The liquid evaporating into the gas is transferred by diffusion from the interface to the gas stream as a result of a concentration difference $(c_0 - c)$, where c_0 is the concentration of the vapour at the surface (mass per unit volume) and c is the concentration in the gas stream. The rate of evaporation is then given by:

$$W = h_D A(c_0 - c) = (h_D A)(M_w/\mathbf{R}T)(P_{w0} - P_w) \tag{11.5}$$

where h_D is the coefficient of mass transfer (see equation 8.28).

The partial pressures of the vapour, P_w and P_{w0}, can be expressed in terms of the corresponding humidities $\mathcal{H}$ and $\mathcal{H}_w$ by equations 11.1 and 11.2.

If P_w and P_{w0} are small compared with P, $(P - P_w)$ and $(P - P_{w0})$ can be replaced by a mean partial pressure of the gas P_A and:

$$W = [h_D A(\mathcal{H}_w - \mathcal{H})M_w/\mathbf{R}T]P_A M_A/M_w$$
$$= h_D A \rho_A(\mathcal{H}_w - \mathcal{H}) \tag{11.6}$$

where ρ_A is the density of the gas at the partial pressure P_A.

The heat transfer required to maintain this rate of evaporation is:

$$Q = h_D A \rho_A(\mathcal{H}_w - \mathcal{H})\lambda \tag{11.7}$$

where λ is the latent heat of vaporisation of the liquid.

Thus, equating 11.4 and 11.7:

$$(\mathcal{H} - \mathcal{H}_w) = -(h/h_D \rho_A \lambda)(\theta - \theta_w) \tag{11.8}$$

Both h and h_D are dependent on the equivalent gas film thickness, and thus any decrease in the thickness, as a result of increasing the gas velocity for example, increases both h and h_D. At normal temperatures, (h/h_D) is virtually independent of the gas velocity provided this is greater than about 5 m/s. Under these conditions the heat transfer by convection from the gas stream is large compared with that from the surroundings by radiation and conduction.

The wet-bulb temperature θ_w depends only on the temperature and the humidity of the gas and values normally quoted are determined for comparatively high gas velocities such that the condition of the gas does not change appreciably as a result of being brought into contact with the liquid and the ratio (h/h_D) has reached a constant value. For the air–water system, the ratio $(h/h_D\rho_A)$ is about $1\cdot0$ kJ/kg K and varies from $1\cdot5$ to $2\cdot0$ kJ/kg K for organic liquids.

Example 11.3. Moist air at 310 K has a wet-bulb temperature of 300 K. If the latent heat of vaporisation of water at 300 K is 2440 kJ/kg, estimate the humidity of the air and the percentage relative humidity. The total pressure is 105 kN/m² and the vapour pressure of water vapour at 300 K is $3\cdot60$ kN/m² and $6\cdot33$ kN/m² at 310 K.

Solution. The humidity of air saturated at the wet-bulb temperature is given by equation 11.2:

$$\mathcal{H}_w = [P_{wo}/(P - P_{wo})](M_w/M_A)$$
$$= [3\cdot6/(105\cdot0 - 3\cdot6)](18/29) = 0\cdot0220 \text{ kg/kg}$$

Therefore, taking $(h/h_D\rho_A)$ as $1\cdot0$ kJ/kg K, in equation 11.8:

$$(0\cdot0220 - \mathcal{H}) = (1\cdot0/2440)(310 - 300)$$

or
$$\underline{\underline{\mathcal{H} = 0\cdot018 \text{ kg/kg}}}$$

At 310 K, $\qquad\qquad\qquad P_{wo} = 6\cdot33 \text{ kN/m}^2$

In equation 11.2: $\qquad\qquad 0\cdot0780 = [18P_w/(105\cdot0 - P_w)29]$

∴ $\qquad\qquad\qquad\qquad P_w = 2\cdot959 \text{ kN/m}^2$

and percentage relative humidity $\quad = (100 \times 2\cdot959/6\cdot33) = \underline{\underline{46\cdot7 \text{ per cent}}}$

11.2.3. Adiabatic Saturation Temperature

In the system just considered, neither the humidity nor the temperature of the gas is appreciably changed. If the gas is passed over the liquid at such a rate that the time of contact is sufficient for equilibrium to be established, the gas will become saturated and both phases will be brought to the same temperature. In a thermally insulated system, the total sensible heat falls by an amount equal to the latent heat of the liquid evaporated. As a result of continued passage of the gas, the temperature of the liquid gradually approaches an equilibrium value which is known as the *adiabatic saturation temperature*.

These conditions are achieved in an infinitely tall thermally insulated humidification column through which gas of a given initial temperature and humidity flows countercurrently to the liquid under conditions where the gas is completely saturated at the top of the column. If the liquid is continuously circulated round the column, and if any fresh liquid which is added is at the same temperature as the circulating liquid, the temperature of the liquid at the top and bottom of the column, and of the gas at the top, approach the adiabatic saturation temperature. Temperature and humidity differences are a maximum at the bottom and zero at the top, and therefore the rates of transfer of heat and mass decrease progressively from the bottom to the top of the tower. This is illustrated in Fig. 11.1.

Making a heat balance over the column it is seen that the heat of vaporisation of the liquid must come from the sensible heat in the gas. The temperature of the gas falls from θ to the adiabatic saturation temperature θ_s, and its humidity increases from $\mathcal{H}$ to $\mathcal{H}_s$ (the saturation value at θ_s). Then working to a basis of unit mass of dry gas:

$$(\theta - \theta_s)s = (\mathcal{H}_s - \mathcal{H})\lambda$$

or
$$(\mathcal{H} - \mathcal{H}_s) = -(s/\lambda)(\theta - \theta_s) \qquad\qquad (11.9)$$

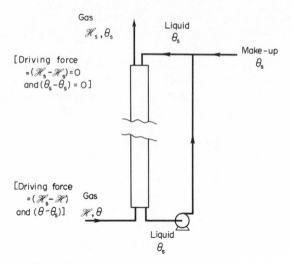

FIG. 11.1. Adiabatic saturation temperature θ_s.

where s is the humid heat of the gas and λ the latent heat of vaporisation at θ_s. s is almost constant for small changes in $\mathcal{H}$.

Equation 11.9 indicates an approximately linear relationship between humidity and temperature for all mixtures of gas and vapour having the same adiabatic saturation temperature θ_s. A curve of humidity versus temperature for gases with a given adiabatic saturation temperature is known as an *adiabatic cooling line*. For a range of adiabatic saturation temperatures, a family of curves, approximating to straight lines of slopes equal to $-(s/\lambda)$, is obtained. These lines are not exactly straight and parallel because of variations in λ and s.

Comparing equations 11.8 and 11.9, it is seen that the adiabatic saturation temperature is equal to the wet-bulb temperature when $s = h/h_D \rho_A$. This is the case for most water vapour systems and accurately so when $\mathcal{H} = 0.047$. The ratio $(h/h_D \rho_A s) = b$ is sometimes known as the psychrometric ratio and, as indicated, b is approximately unity for the air–water system. For most systems involving air and an organic liquid, $b = 1.3$–2.5 and the wet-bulb temperature is higher than the adiabatic saturation temperature. This was confirmed in 1932 by Sherwood and Comings[1] who worked with water, ethanol, n-propanol, n-butanol, benzene, toluene, carbon tetrachloride, and n-propyl acetate, and found that the wet-bulb temperature was always higher than the adiabatic saturation temperature except in the case of water.

In Chapter 9 it was shown that when the Schmidt and Prandtl numbers for a mixture of gas and vapour are approximately equal to unity, the *Lewis relation* applies, or:

$$h_D = h/C_p \rho$$

where C_p and ρ are the mean specific heat and density of the vapour phase.

Therefore
$$(h/h_D \rho_A) = C_p \rho/\rho_A \qquad (11.10)$$

where the humidity is relatively low, $C_p \approx s$ and $\rho \approx \rho_A$ and hence:

$$s \approx h/h_D \rho_A \qquad (11.11)$$

For systems containing vapour other than that of water, s is only approximately equal to $h/h_D \rho_A$ and the difference between the two quantities may be as high as 50 per cent.

If an unsaturated gas is brought into contact with a liquid which is at the adiabatic saturation temperature of the gas, a simultaneous transfer of heat and mass takes place. The

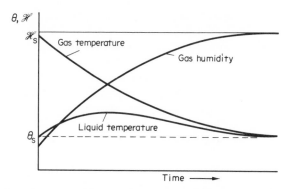

FIG. 11.2. Saturation of gas with liquid other than water at adiabatic saturation temperature.

temperature of the gas falls and its humidity increases (Fig. 11.2). The temperature of the liquid at any instant tends to change and approach the wet-bulb temperature corresponding to the particular condition of the gas at that moment. For a liquid other than water, the adiabatic saturation temperature is less than the wet-bulb temperature and therefore in the initial stages the temperature of the liquid rises. As the gas becomes humidified, however, its wet-bulb temperature falls and consequently the temperature to which the liquid is tending decreases as evaporation takes place. In due course, therefore, a point is reached where the liquid actually reaches the wet-bulb temperature of the gas in contact with it. It does not remain at this temperature, however, because the gas is not then completely saturated, and further humidification is accompanied by a continued lowering of the wet-bulb temperature. The temperature of the liquid therefore starts to fall and continues to fall until the gas is completely saturated. The liquid and gas are then both at the adiabatic saturation temperature.

The air–water system is unique, however, in that the Lewis relation holds quite accurately, so that the adiabatic saturation temperature is the same as the wet-bulb temperature. If, therefore, an unsaturated gas is brought into contact with water at the adiabatic saturation temperature of the gas, there is no tendency for the temperature of the water to change, and it remains in a condition of dynamic equilibrium through the whole of the humidification process (Fig. 11.3). In this case, the adiabatic cooling line represents the conditions of gases of constant wet-bulb temperatures as well as constant adiabatic saturation temperatures. The change in the condition of a gas as it is humidified with water vapour is therefore represented by the adiabatic cooling line and the intermediate conditions of the gas during the process are readily obtained. This is particularly useful because only partial humidification is normally obtained in practice.

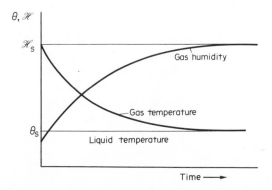

FIG. 11.3. Saturation of air with water at adiabatic saturation temperature.

11.3. HUMIDITY DATA FOR AIR–WATER SYSTEM

To facilitate calculations, various properties of the air–water system are plotted on a humidity chart. Such a chart is based on either the temperature or the enthalpy of the gas. The temperature–humidity chart is the more commonly used though the enthalpy–humidity chart is particularly useful for determining the effect of mixing two gases or of mixing a gas and a liquid. Each chart refers to a particular total pressure of the system. A humidity–temperature chart for the air–water system at atmospheric pressure is given in Fig. 11.4 and the corresponding humidity–enthalpy chart is given in Fig. 11.5.

11.3.1. Temperature–Humidity Chart

In Fig. 11.4 it will be seen that the following quantities are plotted against temperature:
(i) The *humidity* $\mathcal{H}$ for various values of the percentage relative humidity.

For saturated gas: $\mathcal{H}_0 = [P_{w0}/(P - P_{w0})](M_w/M_A)$ (equation 11.2)

From equations 11.1 and 11.3 for a gas of percentage relative humidity Z:

$$\mathcal{H} = \mathcal{H}_0\left(\frac{Z}{100}\right)\frac{(P - P_{w0})}{P - (ZP_{w0}/100)} \tag{11.12}$$

(ii) *The specific volume of dry gas.* This is a linear function of temperature.
(iii) *The saturated volume.* This increases more rapidly with temperature than the specific volume of dry gas because both the quantity and the specific volume of vapour increase with temperature. At a given temperature the humid volume varies linearly with humidity and hence the humid volume of unsaturated gas can be found by interpolation.
(iv) *The latent heat of vaporisation*
In addition, the *humid heat* is plotted as the abscissa in Fig. 11.4 with the humidity as the ordinate.

Adiabatic cooling lines are included in the diagram and, as already discussed, these have a slope of $-(s/\lambda)$ and they are slightly curved since s is a function of $\mathcal{H}$. On the chart they appear as straight lines, however, since the inclination of the axis has been correspondingly adjusted. Each adiabatic cooling line represents the composition of all gases whose adiabatic saturation temperature is given by its point of intersection with the 100 per cent relative humidity curve. For the air–water system, the adiabatic cooling lines represent conditions of constant wet-bulb temperature as well and, as previously mentioned, enable the change in composition of a gas to be followed as it is humidified by contact with water at the adiabatic saturation temperature of the gas.

Example 11.4. Air containing 0·005 kg water vapour per kg of dry air is heated to 325 K in a dryer and passed to the lower shelves. It leaves these shelves at 60 per cent relative humidity and is reheated to 325 K and passed over another set of shelves, again leaving at 60 per cent relative humidity. This is again repeated for the third and fourth sets of shelves, after which the air leaves the dryer. On the assumption that the material on each shelf has reached the wet-bulb temperature and that heat losses from the dryer can be neglected, determine:

(a) the temperature of the material on each tray;
(b) the amount of water removed in kg/s, if 5 m³/s moist air leaves the dryer;
(c) the temperature to which the inlet air would have to be raised to carry out the drying in a single stage.

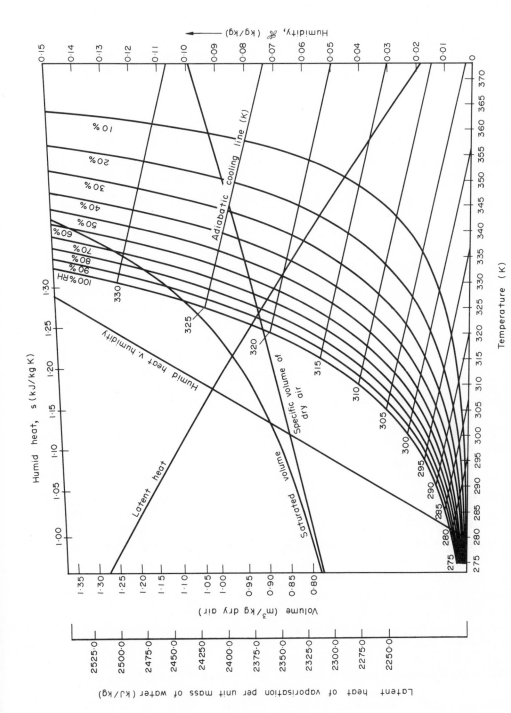

FIG. 11.4. Humidity–temperature diagram for air–water vapour system at atmospheric pressure.

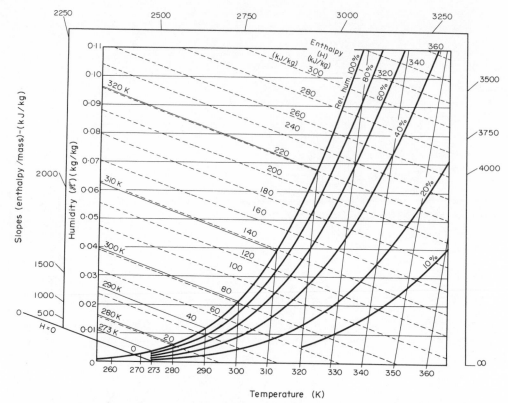

FIG. 11.5. Humidity–enthalpy diagram for air–water vapour system at atmospheric pressure.

Solution. For each of the four sets of shelves, the condition of the air is changed to 60 per cent relative humidity along an adiabatic cooling line.

Initial condition of air: $\qquad\qquad \theta = 325$ K, $\mathcal{H} = 0.005$ kg/kg

On humidifying to 60 per cent relative humidity:

$$\theta = 301 \text{ K}, \ \mathcal{H} = 0.015 \text{ kg/kg and } \theta_w = 296 \text{ K}$$

At the end of the second pass: $\quad \theta = 308$ K, $\mathcal{H} = 0.022$ kg/kg and $\theta_w = 301$ K

At the end of the third pass: $\qquad \theta = 312$ K, $\mathcal{H} = 0.027$ kg/kg and $\theta_w = 305$ K

At the end of the fourth pass: $\quad \theta = 315$ K, $\mathcal{H} = 0.032$ kg/kg and $\theta_w = 307$ K

Thus the temperatures of the material on each of the trays are:

$$\underline{296 \text{ K, } 301 \text{ K, } 305 \text{ K, and } 307 \text{ K}}$$

Total increase in humidity $\ = (0.032 - 0.005) = 0.027$ kg/kg

The air leaving the system is at 315 K and 60 per cent relative humidity.

From Fig. 11.4, specific volume of dry air $\qquad = 0.893$ m³/kg

$\qquad\qquad$ specific volume of saturated air

$\qquad\qquad\qquad$ (saturated volume) $\qquad\qquad = 0.968$ m³/kg

Therefore, by interpolation, the humid volume of air of 60 per cent relative humidity = 0·937 m³/kg

Mass of air passing through the dryer = (5/0·937) = 5·34 kg/s

Mass of water evaporated = (5·34 × 0·027) = <u>0·144 kg/s</u>

If the material is to be dried by air in a single pass, the air must be heated before entering the dryer such that its wet-bulb temperature is 307 K.

For air with a humidity of 0·005 kg/kg, this corresponds to a dry bulb temperature of <u>380 K</u>. The various steps in this calculation are shown in Fig. 11.6.

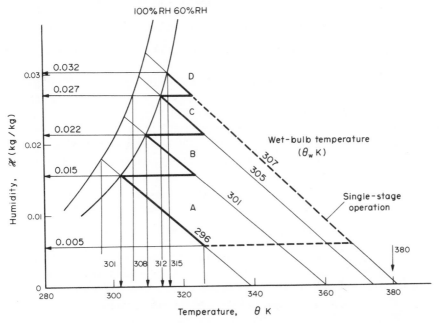

FIG. 11.6. Humidification stages for Example 11.4.

11.3.2. Enthalpy–Humidity Chart

In the calculation of enthalpies it is necessary to define some standard reference state at which the enthalpy is taken as zero. It is most convenient to take the melting point of the material constituting the vapour as the reference temperature and the liquid state of the material as its standard state.

If H is the enthalpy of the humid gas per unit mass of dry gas, H_a the enthalpy of the dry gas per unit mass, H_w the enthalpy of the vapour per unit mass, C_a the specific heat of the gas at constant pressure, C_w the specific heat of the vapour at constant pressure, θ the temperature of the humid gas, θ_0 the reference temperature, λ the latent heat of vaporisation of the liquid at θ_0 and $\mathcal{H}$ the humidity of the gas:

then for an unsaturated gas $\qquad H = H_a + H_w \mathcal{H}$ (11.13)

where $\qquad H_a = C_a (\theta - \theta_0)$ (11.14)

and $\qquad H_w = C_w (\theta - \theta_0) + \lambda$ (11.15)

Thus in equation 11.13: $\qquad H = (C_a + \mathcal{H} C_w)(\theta - \theta_0) + \mathcal{H}\lambda$ (11.16)
$$= (\theta - \theta_0)(s + \mathcal{H}\lambda)$$

If the gas contains more liquid or vapour than is required to saturate it at the temperature in question, either the gas will be supersaturated or the excess material will be present in the form of liquid or solid according to whether the temperature θ is greater or less than the reference temperature θ_0. The supersaturated condition is unstable and will not be considered further.

If the temperature θ is greater than θ_0 and if the humidity $\mathscr{H}$ is greater than the humidity $\mathscr{H}_0$ of saturated gas, the enthalpy H per unit mass of dry gas is given by:

$$H = C_a(\theta - \theta_0) + \mathscr{H}_0[C_w(\theta - \theta_0) + \lambda] + C_L(\mathscr{H} - \mathscr{H}_0)(\theta - \theta_0) \qquad (11.17)$$

where C_L is the specific heat of the liquid.

If the temperature θ, is less than θ_0, the corresponding enthalpy H is given by:

$$H = C_a(\theta - \theta_0) + \mathscr{H}_0[C_w(\theta - \theta_0) + \lambda] + (\mathscr{H} - \mathscr{H}_0)[C_s(\theta - \theta_0) + \lambda_f] \qquad (11.18)$$

where C_s is the specific heat of the solid and λ_f is the latent heat of freezing of the liquid, a negative quantity.

Equations 11.16 to 11.18 give the enthalpy in terms of the temperature and humidity of the humid gas for the three conditions: $\theta = \theta_0$, $\theta > \theta_0$, and $\theta < \theta_0$ respectively. Thus, given the percentage relative humidity and the temperature, the humidity may be obtained from Fig. 11.4, the enthalpy calculated from equations 11.16, 11.17 or 11.18 and plotted against the humidity, usually with enthalpy as the abscissa. Such a plot is shown in Fig. 11.7 for the air–water system, which includes the curves for 100 per cent relative humidity and for some lower value, say Z per cent.

Considering the nature of the isothermals for the three conditions dealt with previously, at constant temperature θ the relation between enthalpy and humidity for an unsaturated gas is:

$$H = \text{constant} + [C_w(\theta - \theta_0) + \lambda]\mathscr{H} \qquad (11.19)$$

Thus the isothermal is a straight line of slope $[C_w(\theta - \theta_0) + \lambda]$ with respect to the humidity axis. At the reference temperature θ_0, the slope is λ; at higher temperatures the slope is greater than λ, and at lower temperatures it is less than λ. Because the latent heat is normally large compared with the sensible heat, the slope of the isothermals remains positive down to very low temperatures. Since the humidity is plotted as the ordinate, the slope of the isothermal relative to the X-axis decreases with increase in temperature. When $\theta > \theta_0$ and $\mathscr{H} > \mathscr{H}_0$, the saturation humidity, the vapour phase consists of a saturated gas with liquid

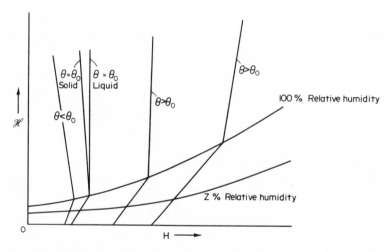

Fig. 11.7. Humidity–enthalpy diagram for air–water system—rectangular axes.

droplets in suspension. The relation between enthalpy and humidity at constant temperature θ is:

$$H = \text{constant} + C_L(\theta - \theta_0)\mathcal{H} \qquad (11.20)$$

The isothermal is therefore a straight line of slope $C_L(\theta - \theta_0)$. At the reference temperature θ_0, the slope is zero and the isothermal is parallel to the humidity axis. At higher temperatures the slope has a small positive value. When $\theta < \theta_0$ and $\mathcal{H} > \mathcal{H}_0$, solid particles are formed and the equation of the isothermal is:

$$H = \text{constant} + [C_s(\theta - \theta_0) + \lambda_f]\mathcal{H} \qquad (11.21)$$

This represents a straight line of slope $[C_s(\theta - \theta_0) + \lambda_f]$. Both $C_s(\theta - \theta_0)$ and λ_f are negative and therefore the slopes of all these isothermals are negative. When $\theta = \theta_0$, the slope is λ_f. In the supersaturated region therefore, there are two distinct isothermals at temperature θ_0; one corresponds to the condition where the excess vapour is present in the form of liquid droplets and the other to the condition where it is present as solid particles. The region between these isothermals represents conditions where a mixture of liquid and solid is present in the saturated gas at the temperature θ_0.

The shape of the humidity–enthalpy line for saturated air is such that the proportion of the total area of the diagram representing saturated, as opposed to supersaturated, air is small when rectangular axes are used. In order to enable greater accuracy to be obtained in the use of the diagram, oblique axes are normally used, as in Fig. 11.5, so that the isothermal for unsaturated gas at the reference temperature θ_0 is parallel to the humidity axis.

It should be noted that the curves of humidity plotted against either temperature or enthalpy have a discontinuity at the point corresponding to the freezing point of the humidifying material. Above the temperature θ_0 the lines are determined by the vapour–liquid equilibrium and below it by the vapour–solid equilibrium.

Two cases may be considered to illustrate the use of enthalpy–humidity charts. These are the mixing of two streams of humid gas and the addition of liquid or vapour to a gas.

Mixing of Two Streams of Humid Gas

Consider the mixing of two gases of humidities $\mathcal{H}_1$ and $\mathcal{H}_2$, temperatures θ_1 and θ_2, and enthalpies H_1 and H_2 to give a mixed gas of temperature θ, enthalpy H, and humidity $\mathcal{H}$. Let the masses of dry gas concerned be m_1, m_2, and m respectively.

Then taking a balance on the dry gas, vapour, and enthalpy:

$$m_1 + m_2 = m \qquad (11.22)$$

$$m_1\mathcal{H}_1 + m_2\mathcal{H}_2 = m\mathcal{H} \qquad (11.23)$$

and

$$m_1 H_1 + m_2 H_2 = mH \qquad (11.24)$$

Elimination of m gives:

$$m_1(\mathcal{H} - \mathcal{H}_1) = m_2(\mathcal{H}_2 - \mathcal{H}) \qquad (11.25)$$

and

$$m_1(H - H_1) = m_2(H_2 - H)$$

Dividing these two equations:

$$(\mathcal{H} - \mathcal{H}_1)/(H - H_1) = (\mathcal{H} - \mathcal{H}_2)/(H - H_2) \qquad (11.26)$$

The condition of the resultant gas is therefore represented by a point on the straight line joining $(\mathcal{H}_1, H_1)$ and $(\mathcal{H}_2, H_2)$. The humidity $\mathcal{H}$ is given, from equation 11.25, by:

$$(\mathcal{H} - \mathcal{H}_1)/(\mathcal{H}_2 - \mathcal{H}) = m_2/m_1 \qquad (11.27)$$

The gas formed by mixing two unsaturated gases may be either unsaturated, saturated, or supersaturated. The possibility of producing supersaturated gas arises because the 100 per cent relative humidity line on the humidity–enthalpy diagram is concave towards the humidity axis.

Example 11.5. In an air-conditioning system, 1 kg/s air at 350 K and 10 per cent relative humidity is mixed with 5 kg/s air at 300 K and 30 per cent relative humidity. What is the enthalpy, humidity, and temperature of the resultant stream?

Solution. From Fig. 11.4:

$$\text{at } \theta_1 = 350 \text{ K and RH} = 10 \text{ per cent;} \quad \mathscr{H}_1 = 0\cdot043 \text{ kg/kg}$$
$$\text{at } \theta_2 = 300 \text{ K and RH} = 30 \text{ per cent;} \quad \mathscr{H}_2 = 0\cdot0065 \text{ kg/kg}$$

Thus, in equation 11.23:

$$(1 \times 0\cdot043) + (5 \times 0\cdot0065) = (1 + 5)\mathscr{H}$$

and
$$\underline{\underline{\mathscr{H} = 0\cdot0125 \text{ kg/kg}}}$$

From Fig. 11.5:

$$\text{at } \theta_1 = 350 \text{ K and } \mathscr{H}_1 = 0\cdot043 \text{ kg/kg;} \quad H_1 = 192 \text{ kJ/kg}$$
$$\text{at } \theta_2 = 300 \text{ K and } \mathscr{H}_2 = 0\cdot0065 \text{ kg/kg;} \quad H_2 = 42 \text{ kJ/kg}$$

Thus, in equation 11.25:

$$1(H - 192) = 5(42 - H)$$

and
$$\underline{\underline{H = 67 \text{ kJ/kg}}}$$

From Fig. 11.5:

$$\text{at } H = 67 \text{ kJ/kg and } \mathscr{H} = 0\cdot0125 \text{ kg/kg}$$

$$\underline{\underline{\theta = 309 \text{ K}}}$$

The data used in this example are shown in Fig. 11.8.

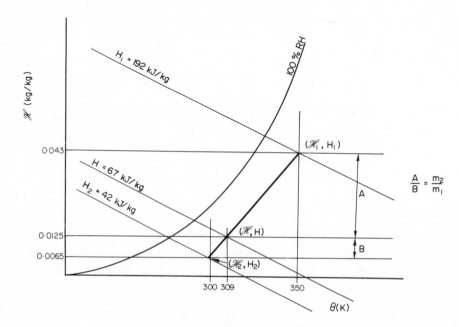

FIG. 11.8. Data used in Example 11.5.

Addition of Liquid or Vapour to a Gas

Let a mass m_3 of liquid or vapour of enthalpy H_3 be added to a gas of humidity $\mathcal{H}_1$ and enthalpy H_1 and containing a mass m_1 of dry gas. Then:

$$m_1(\mathcal{H} - \mathcal{H}_1) = m_3 \qquad (11.28)$$

$$m_1(H - H_1) = m_3 H_3 \qquad (11.29)$$

Thus $$(H - H_1)/(\mathcal{H} - \mathcal{H}_1) = H_3 \qquad (11.30)$$

where $\mathcal{H}$ and H are the humidity and enthalpy of the gas produced on mixing.

The composition and properties of the mixed stream are therefore represented by a point on the straight line of slope H_3, relative to the humidity axis which passes through the point $(H_1, \mathcal{H}_1)$. In Fig. 11.5 the edges of the plot are marked with points which, when joined to the origin, give a straight line of the slope indicated. Thus in using the chart, a line of slope H_3 is drawn through the origin and a parallel line drawn through the point $(H_1, \mathcal{H}_1)$. The point representing the final gas stream is then given from equation 11.28:

$$\mathcal{H} - \mathcal{H}_1 = m_3/m_1$$

It can be seen from Fig. 11.5 that for the air–water system a straight line, of slope equal to the enthalpy of dry saturated steam (2675 kJ/kg), is almost parallel to the isothermals, so that the addition of live steam has only a small effect on the temperature of the gas. The addition of water spray, even if the water is considerably above the temperature of the gas, results in a lowering of the temperature after the water is evaporated. This arises because the latent heat of vaporisation of the liquid constitutes the major part of the enthalpy of the vapour. Thus when steam is added it gives up a small amount of sensible heat to the gas, whereas when hot liquid is added a small amount of sensible heat is given up and a very much larger amount of latent heat is absorbed from the gas.

Example 11.6. 0.15 kg/s steam at atmospheric pressure and superheated to 400 K is bled into an air stream at 320 K and 20 per cent RH. What is the temperature, enthalpy, and humidity of the mixed stream if the air is flowing at 5 kg/s? How much steam would be required to provide an exit temperature of 330 K and what would be the humidity of this mixture?

Solution. Steam at atmospheric pressure is saturated at 373 K at which the latent heat

$$= 2257 \text{ kJ/kg}$$

Taking the specific heat of superheated steam as 2.0 kJ/kg K;

enthalpy of the steam: $H_3 = 4.18(373 - 273) + 2257 + 2.0(400 - 373)$

$$= 2729 \text{ kJ/kg}$$

From Fig. 11.4:

at $\theta_1 = 320$ K and 20 per cent RH; $\mathcal{H}_1 = 0.014$ kg/kg

In Fig. 11.5,

when $\theta_1 = 320$ K and $\mathcal{H}_1 = 0.014$ kg/kg; $H_1 = 84$ kJ/kg

The line joining the axis and slope $H_3 = 2729$ kJ/kg at the edge of the chart is now drawn in and a parallel line is drawn through $(H_1, \mathcal{H}_1)$.

Now $$\mathcal{H} - \mathcal{H}_1 = m_3/m_1 = 0.15/5 = 0.03 \text{ kg/kg}$$

and $$\mathcal{H} = 0.03 + 0.014 = 0.044 \text{ kg/kg}$$

At the intersection of $\mathcal{H} = 0.044$ kg/kg and the line through $(\mathcal{H}_1, H_1)$

$$\underline{H = 171 \text{ kJ/kg and } \theta = 325 \text{ K}}$$

When $\theta = 330$ K the intersection of this isotherm and the line through $(\mathcal{H}_1, H_1)$ gives an outlet stream in which $\mathcal{H} = 0.09$ kg/kg (70 per cent RH) and $H = 288$ kJ/kg.

Thus, in equation 11.28

$$m_3 = 5(0.09 - 0.014) = \underline{0.38 \text{ kg/s}}$$

The data used in this example are shown in Fig. 11.9.

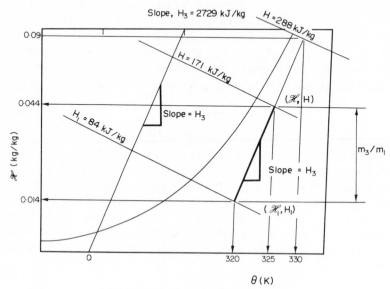

FIG. 11.9. Data used in Example 11.6.

11.4. DETERMINATION OF HUMIDITY

The most important methods for determining humidity are as follows:

(1) *Chemical methods.* A known volume of the gas is passed over a suitable absorbent, the increase in weight of which is measured. The efficiency of the process can be checked by arranging a number of vessels containing absorbent in series and ascertaining that the increase in weight in the last of these is negligible. The method is very accurate but is laborious. Satisfactory absorbents for water vapour are phosphorus pentoxide dispersed in pumice, and concentrated sulphuric acid.

(2) *Determination of the wet-bulb temperature.* Equation 11.8 gives the humidity of a gas in terms of its temperature, its wet-bulb temperature, and various physical properties of the gas and vapour. The wet-bulb temperature is normally determined as the temperature attained by the bulb of a thermometer which is covered with a piece of material which is maintained saturated with the liquid. The gas should be passed over the surface of the wet bulb at a high enough velocity (> 5 m/s) (a) for the condition of the gas stream not to be affected appreciably by the evaporation of liquid, (b) for the heat transfer by convection to be large compared with that by radiation and conduction from the surroundings, and (c) for the ratio of the coefficients of heat and mass transfer to have reached a constant value. The gas should be passed long enough for equilibrium to be attained and, for accurate work, the liquid should be cooled nearly to the wet-bulb temperature before it is applied to the material.

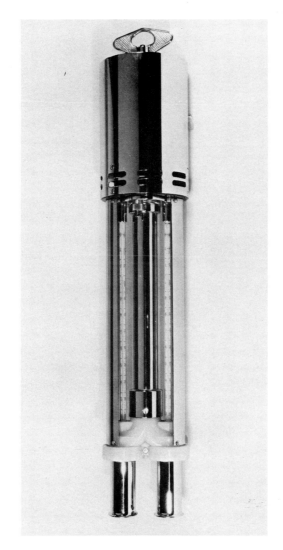

FIG. 11.10. Wet-bulb thermometer.

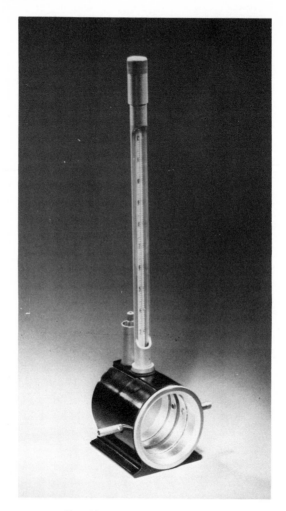

FIG. 11.11. Dew-point apparatus.

FIG. 11.12. Hair hygrometer.

FIG. 11.14. Water-cooling tower. View of spray distribution system.

The stream of gas over the liquid surface may be produced by a small fan or other similar means (Fig. 11.10). The crude forms of wet-bulb thermometer which make no provision for the rapid passage of gas cannot be used for accurate determinations of humidity.

(3) *Determination of the dew point.* The dew point is determined by cooling a highly polished surface (Fig. 11.11) in the gas and observing the highest temperature at which condensation takes place. The humidity of the gas is equal to the humidity of saturated gas at the dew-point.

(4) *Measurement of the change in length of a hair or fibre.* The length of a hair or fibre is influenced by the humidity of the surrounding atmosphere (Fig. 11.12). Many forms of apparatus for automatic recording of humidity depend on this property. The method has the disadvantage that the apparatus needs frequent calibration because the zero tends to shift. This difficulty is most serious when the instrument is used over a wide range of humidities.

(5) *Measurement of conductivity of a fibre.* If a fibre is impregnated with an electrolyte, such as lithium chloride, its electrical resistance will be governed by its moisture content, which in turn depends on the humidity of the atmosphere in which it is situated. In a lithium chloride cell, a skein of very fine fibres is wound on a plastic frame carrying the electrodes and the current flowing at a constant applied voltage gives a direct measure of the relative humidity.

Some of the more important methods of determining humidity have been outlined above, but reference should be made to standard works on psychrometry for details of the apparatus and experimental techniques.

11.5. HUMIDIFICATION AND DEHUMIDIFICATION

11.5.1. Methods of Increasing Humidity

The following methods may be used for increasing the humidity of a gas:

(1) Live steam may be added directly in the required quantity. It has been shown that this produces only a slight increase in the temperature, but the method is not generally favoured because any impurities that are present in the steam are added at the same time.

(2) Water may be sprayed into the gas at such a rate that, on complete vaporisation, it gives the required humidity. In this case, the temperature of the gas will fall as the latent heat of vaporisation must be supplied from the sensible heat of the gas and liquid.

(3) The gas may be mixed with a stream of gas of higher humidity. This method is frequently used in laboratory work when the humidity of a gas supplied to an apparatus is controlled by varying the proportions in which two gas streams are mixed.

(4) The gas may be brought into contact with water in such a way that only part of the liquid is evaporated. This is perhaps the most common method and will now be considered in more detail.

In order to obtain a high rate of humidification, the area of contact between the air and the water is made as large as possible by supplying the water in the form of a fine spray; alternatively, the interfacial area is increased by using a packed column. Evaporation occurs if the humidity at the surface is greater than that in the bulk of the air; that is, if the temperature of the water is above the dew point of the air.

When humidification is carried out in a packed column, the water which is not evaporated can be recirculated so as to reduce the requirements of fresh water. As a result of continued recirculation, the temperature of the water will approach the adiabatic saturation temperature of the air, and the air leaving the column will be cooled—in some cases to within 1 K of the temperature of the water. If the temperature of the air is to be maintained constant, or raised, the water must be heated.

Two methods of changing the humidity and temperature of a gas from $A(\theta_1, \mathcal{H}_1)$ to $B(\theta_2, \mathcal{H}_2)$ can be traced on the humidity chart as shown in Fig. 11.13. The first method consists of saturating the air by water artificially maintained at the dew point of air of humidity $\mathcal{H}_2$ (line AC) and then heating at constant humidity to θ_2 (line CB). In the second method, the air is heated (line AD) so that its adiabatic saturation temperature corresponds with the dew point of air of humidity $\mathcal{H}_2$. It is then saturated by water at the adiabatic saturation temperature (line DC) and heated at constant humidity to θ_2 (line CB). In this second method, an additional operation—the preliminary heating—is carried out on the air, but the water temperature automatically adjusts itself to the required value.

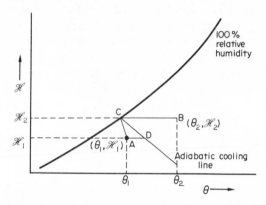

FIG. 11.13. Two methods of changing conditions of gas from $(\theta_1, \mathcal{H}_1)$ to $(\theta_2, \mathcal{H}_2)$.

Since complete humidification is not always obtained, an allowance must be made when designing air humidification cycles. For example, if only 95 per cent saturation is obtained the adiabatic cooling line should be followed only to the point corresponding to that degree of saturation, and therefore the gas must be heated to a slightly higher temperature before adiabatic cooling is commenced.

Example 11.7. Air at 300 K and 20 per cent RH is to be heated in two stages with intermediate saturation with water to 90 per cent RH so that the final stream is at 320 K and 20 per cent RH. What is the humidity of the exit stream and the conditions at the end of each stage?

Solution. At $\theta_1 = 300$ K and 20 per cent RH: $\mathcal{H}_1 = 0\cdot0045$ kg/kg, from Fig. 11.4

At $\theta_2 = 320$ K and 20 per cent RH: $\underline{\mathcal{H}_2 = 0\cdot0140 \text{ kg/kg}}$

When $\mathcal{H}_2 = 0\cdot0140$ kg/kg, air is saturated at 292 K and has a relative humidity of 90 per cent at 293 K.

The adiabatic cooling line corresponding to 293 K intersects with $\mathcal{H} = 0\cdot0045$ kg/kg at a temperature, $\theta = 318$ K.

Thus the stages are:

(i) Heat the air at $\mathcal{H} = 0\cdot0045$ from 300 to 318 K.

(ii) Saturate with water at an adiabatic saturation temperature of 293 K until 90 per cent RH is attained. At the end of this stage:

$$\mathcal{H} = 0\cdot0140 \text{ kg/kg} \quad \text{and} \quad \theta = 294\cdot5 \text{ K}$$

(iii) Heat at $\underline{\mathcal{H} = 0\cdot0140 \text{ kg/kg from } 294\cdot5 \text{ to } 320 \text{ K}}$

11.5.2. Dehumidification

Dehumidification of air can be effected by bringing it into contact with a cold surface, either liquid or solid. If the temperature of the surface is lower than the dew point of the gas, condensation takes place and the temperature of the gas falls. The temperature of the surface tends to rise because of the transfer of latent and sensible heat from the air. It would be expected that the air would cool at constant humidity until the dew point was reached, and that subsequent cooling would be accompanied by condensation. It is found, in practice, that this occurs only when the air is well mixed. Normally the temperature and humidity are reduced simultaneously throughout the whole of the process. The air in contact with the surface is cooled below its dew point, and condensation of vapour therefore occurs before the more distant air has time to cool. Where the gas stream is cooled by cold water, countercurrent flow should be employed because the temperature of the water and air are changing in opposite directions.

The humidity can be reduced by compressing air, allowing it to cool again to its original temperature, and draining off the water which has condensed. During compression, the partial pressure of the vapour is increased and condensation takes place as soon as it reaches the saturation value. Thus, if air is compressed to a high pressure, it becomes saturated with vapour, but the partial pressure is a small proportion of the total pressure. Compressed air from a cylinder therefore has a low humidity. Town gas is frequently compressed before it is circulated so as to prevent condensation in the mains.

Many large air-conditioning plants incorporate automatic control of the humidity and temperature of the issuing air. Temperature control is effected with the aid of a thermocouple or resistance thermometer, and humidity control by means of a thermocouple recording the difference between the wet- and dry-bulb temperatures.

11.6. WATER COOLING

11.6.1. Cooling Towers

Cooling of water can be carried out on a small scale either by allowing it to stand in an open pond or by the spray pond technique in which it is dispersed in spray form and then collected in a large, open pond. Cooling takes place both by the transference of sensible heat and by evaporative cooling as a result of which sensible heat in the water provides the latent heat of vaporisation.

On the large scale, air and water are brought into countercurrent contact in a cooling tower which may employ either natural draught or mechanical draught. The water flows down over a series of wooden slats which give a large interfacial area and promote turbulence in the liquid. The air is humidified and heated as it rises, while the water is cooled.

The natural draught cooling tower depends on the chimney effect produced by the presence in the tower of air and vapour of higher temperature and therefore of lower density than the surrounding atmosphere. Thus atmospheric conditions and the temperature and quantity of the water will exert a very important effect on the operation of the tower. Not only will these factors influence the quantity of air drawn through the tower, but they will also affect the velocities and flow patterns and hence the transfer coefficients between gas and liquid. One of the prime considerations in design therefore is to construct a tower in such a way that the resistance to air flow is low. Hence the packings and distributors must be arranged in open formation. The draught of a cooling tower at full load is usually only about 50 N/m^2,[2] and the air velocity in the region of 1·2–1·5 m/s, so that under the atmospheric conditions prevailing in this country the air usually leaves the tower in a saturated condition.

The density of the air stream at outlet is therefore determined by its temperature. Calculation of conditions within the tower is carried out in the manner described in the following pages. It is, however, necessary to work with a number of assumed air flowrates and to select the one which fits both the transfer conditions and the relationship between air rate and pressure difference in the tower.

The natural draught cooling tower consists of an empty shell constructed either of timber or ferroconcrete; the upper portion is empty and merely serves to increase the draught. The lower portion, amounting to about 10–12 per cent of the total height, is usually fitted with grids on to which the water is fed by means of distributors or sprays as shown in Fig. 11.14. The shells of cooling towers are now generally constructed in ferroconcrete in a shape corresponding approximately to a hyperboloid of revolution. The shape is chosen mainly for constructional reasons, but it does take account of the fact that the entering air will have a radial velocity component; the increase in cross-section towards the top causes a reduction in the outlet velocity and there is a small recovery of kinetic energy into pressure energy. Ferroconcrete towers are now made up to about 100–140 m in height.

The mechanical draught cooling tower may employ forced draught with the fan at the bottom, or induced draught with the fan driving the moist air out at the top. The air velocity can be increased appreciably above that in the natural draught tower, and a greater depth of packing can be used. The tower will extend only to the top of the packing unless atmospheric conditions are such that a chimney must be provided in order to prevent recirculation of the moist air. The danger of recirculation is considerably less with the induced-draught type because the air is expelled with a higher velocity. Mechanical draught towers are generally confined to small installations and to conditions where the water must be cooled to as low a temperature as possible. In some cases it is possible to cool the water to within 1 K of the wet-bulb temperature of the air. Although the initial cost of the tower is less, maintenance and operating costs are of course higher than in natural draught towers which are now used for all large installations.

In the cooling tower the temperature of the liquid falls and the temperature and humidity of the air rise, and its action is thus similar to that of an air humidifier. The limiting temperature to which the water can be cooled is the wet-bulb temperature corresponding to the condition of the air at inlet. The enthalpy of the air stream does not remain constant since the temperature of the liquid changes rapidly in the upper portion of the tower. Towards the bottom, however, the temperature of the liquid changes less rapidly because the temperature differences are smaller. At the top of the tower, the temperature falls from the bulk of the liquid to the interface and then again from the interface to the bulk of the gas. Thus the liquid is cooled by transfer of sensible heat and by evaporation at the surface. At the bottom of a tall tower, however, the temperature gradient in the liquid is in the same direction, though smaller, but the temperature gradient in the gas is in the opposite direction. Transfer of sensible heat to the interface therefore takes place from the bulk of the liquid and from the bulk of the gas, and all the cooling is caused by the evaporation at the interface. In most cases, about 80 per cent of the heat loss from the water is accounted for by evaporative cooling.

11.6.2. Height of Packing[3]

The height of a water-cooling tower can be determined by setting up a material balance on the water, an enthalpy balance, and rate equations for the transfer of heat in the liquid and gas and for mass transfer in the gas phase. There is no concentration gradient in the liquid and therefore there is no resistance to mass transfer in the liquid phase.

Consider the countercurrent flow of water and air in a tower of height z (Fig. 11.15). The mass rate of flow of air per unit cross-section G' is constant throughout the whole height of the tower and, because only a small proportion of the total supply of water is normally

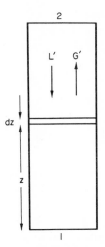

FIG. 11.15. Flow in water-cooling tower.

evaporated (1–5 per cent), the liquid rate per unit area L' can be taken as constant. The temperature, enthalpy, and humidity will be denoted by the symbols θ, H, and $\mathcal{H}$ respectively, suffixes G, L, 1, 2, and f being used to denote conditions in the gas and liquid, at the bottom and top of the column, and of the air in contact with the water.

The five basic equations for an incremental height of column, dz, are as follows:

(1) Water balance:

$$dL' = G' \, d\mathcal{H} \tag{11.31}$$

(2) Enthalpy balance:

$$G' \, dH_G = L' \, dH_L \tag{11.32}$$

since only a small proportion of the liquid is evaporated.

Now

$$H_G = s(\theta_G - \theta_0) + \lambda\mathcal{H} \tag{11.33}$$

and

$$H_L = C_L(\theta_L - \theta_0) \tag{11.34}$$

Thus

$$G' \, dH_G = L'C_L \, d\theta_L \tag{11.35}$$

and

$$dH_G = s \, d\theta_G + \lambda \, d\mathcal{H} \tag{11.36}$$

Integration of this expression over the whole height of the column, on the assumption that the physical properties of the materials do not change appreciably, gives:

$$G'(H_{G2} - H_{G1}) = L'C_L(\theta_{L2} - \theta_{L1}) \tag{11.37}$$

(3) Heat transfer from the body of the liquid to the interface:

$$h_L a \, dz(\theta_L - \theta_f) = L'C_L \, d\theta_L \tag{11.38}$$

where h_L is the heat transfer coefficient in the liquid phase and a is the interfacial area per unit volume of column. It will be assumed that the area for heat transfer is equal to that available for mass transfer, though it may be somewhat greater when the packing is not completely wetted.

Rearranging equation 11.38:

$$d\theta_L/(\theta_L - \theta_f) = h_L a \, dz/L'C_L \tag{11.39}$$

(4) Heat transfer from the interface to the bulk of the gas:

$$h_G a \, dz(\theta_f - \theta_G) = G's \, d\theta_G \tag{11.40}$$

where h_G is the heat transfer coefficient in the gas phase.

Rearranging:
$$\frac{d\theta_G}{\theta_f - \theta_G} = \frac{h_G a}{G' s} dz \qquad (11.41)$$

(5) Mass transfer from the interface to the gas:
$$h_D \rho a \, dz (\mathcal{H}_f - \mathcal{H}) = G' \, d\mathcal{H} \qquad (11.42)$$

where h_D is the mass transfer coefficient for the gas and ρ is the mean density of the air (see equation 11.6).

Rearranging:
$$\frac{d\mathcal{H}}{\mathcal{H}_f - \mathcal{H}} = \frac{h_D \rho a}{G'} dz \qquad (11.43)$$

These equations cannot be integrated directly since the conditions at the interface are not necessarily constant; nor can they be expressed directly in terms of the corresponding property in the bulk of the gas or liquid.

If the Lewis relation (equation 11.11) is applied, it is possible to obtain workable equations in terms of enthalpy instead of temperature and humidity. Thus, writing h_G as $h_D \rho s$, from equation 11.40:
$$G's \, d\theta_G = h_D \rho a \, dz (s\theta_f - s\theta_G) \qquad (11.44)$$

and from equation 11.42;
$$G'\lambda \, d\mathcal{H} = h_D \rho a \, dz (\lambda \mathcal{H}_f - \lambda \mathcal{H}) \qquad (11.45)$$

Adding these two equations gives:
$$G'(s \, d\theta_G + \lambda \, d\mathcal{H}) = h_D \rho a \, dz [(s\theta_f + \lambda \mathcal{H}_f) - (s\theta_G + \lambda \mathcal{H})]$$

$$G' \, dH_G = h_D \rho a \, dz (H_f - H_G) \quad \text{(from equation 11.33)} \qquad (11.46)$$

or:
$$\frac{dH_G}{H_f - H_G} = \frac{h_D \rho a}{G'} dz \qquad (11.47)$$

The use of an enthalpy driving force, as in equation 11.46, was first suggested by Merkel,[4] and the following development of the treatment was proposed by Mickley.[3]

Combination of equations 11.35, 11.38, and 11.46 gives:
$$\frac{H_G - H_f}{\theta_L - \theta_f} = -\frac{h_L}{h_D \rho} \qquad (11.48)$$

From equations 11.44 and 11.46:
$$\frac{H_G - H_f}{\theta_G - \theta_f} = \frac{dH_G}{d\theta_G} \qquad (11.49)$$

and from equations 11.44 and 11.42:
$$\frac{\mathcal{H} - \mathcal{H}_f}{\theta_G - \theta_f} = \frac{d\mathcal{H}}{d\theta_G} \qquad (11.50)$$

These equations will now be employed in the determination of the required height of a cooling tower for a given duty. The method consists of the graphical evaluation of the relation between the enthalpy of the body of gas and the enthalpy of the gas at the interface with the liquid. The required height of the tower is then obtained by integration of equation 11.47.

Suppose water is to be cooled at a mass rate L' per unit area from a temperature θ_{L2} to θ_{L1}. The air will be assumed to have a temperature θ_{G1}, a humidity $\mathcal{H}_1$, and an enthalpy H_{G1} (which can be calculated from the temperature and humidity), at the inlet point at the bottom of the tower, and its mass flow per unit area will be taken as G'. The change in the condition of the liquid and gas phases will now be followed on an enthalpy–temperature diagram (Fig. 11.16). The enthalpy–temperature curve PQ for saturated air is plotted either using calculated data or

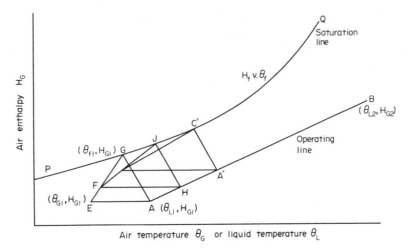

FIG. 11.16. Construction for height of water-cooling tower.

from the humidity chart (Fig. 11.4). The region below this line relates to unsaturated air and the region above it to supersaturated air. If it is assumed that the air in contact with the liquid surface is saturated with water vapour, this curve represents the relation between air enthalpy H_f and temperature θ_f at the interface.

The curve connecting air enthalpy and water temperature is now drawn using equation 11.37. This is known as the operating line and is a straight line of slope $(L'C_L/G')$, passing through the points $A(\theta_{L1}, H_{G1})$ and $B(\theta_{L2}, H_{G2})$. Since θ_{L1}, H_{G1} are specified, the procedure is to draw a line through (θ_{L1}, H_{G1}) of slope $L'C_L/G'$ and to produce it to a point whose abscissa is equal to θ_{L2}. This point B then corresponds to conditions at the top of the tower and the ordinate gives the enthalpy of the air leaving the column.

Equation 11.48 gives the relation between liquid temperature, air enthalpy, and conditions at the interface, for any position in the tower, and is represented by a family of straight lines of slope $-(h_L/h_D\rho)$. The line for the bottom of the column passes through the point $A(\theta_{L1}, H_{G1})$ and cuts the enthalpy–temperature curve for saturated air at the point C, representing conditions at the interface. The difference in ordinates of points A and C is the difference in the enthalpy of the air at the interface and that of the bulk air at the bottom of the column.

Similarly, line $A'C'$, parallel to AC, enables the difference in the enthalpies of the bulk air and the air at the interface to be determined at some other point in the column. The procedure can be repeated for a number of points and the value of $(H_f - H_G)$ obtained as a function of H_G for the whole tower.

$$\frac{dH_G}{H_f - H_G} = \frac{h_D a \rho}{G'} dz \qquad \text{(equation 11.47)}$$

on integration:

$$z = \int_1^2 dz = \frac{G'}{h_D \rho a} \int_1^2 \frac{dH_G}{H_f - H_G} \qquad (11.51)$$

assuming h_D to remain approximately constant.

Since $(H_f - H_G)$ is now known as a function of H_G, $1/(H_f - H_G)$ can be plotted against H_G and the integral evaluated between the required limits. The height of the tower is thus determined.

The integral in equation 11.51 cannot be evaluated by taking a logarithmic mean driving force because the saturation line PQ is far from linear. Carey and Williamson[5] have given a useful approximate method of evaluating the integral. They assume that the enthalpy difference $(H_f - H_G) = \Delta H$ varies in a parabolic manner. The three fixed points taken to define the parabola are at the bottom and top of the column (ΔH_1 and ΔH_2 respectively) and ΔH_m, the value at the mean water temperature in the column. The effective mean driving force is $f \Delta H_m$, where f is a factor for converting the driving force at the mean water temperature to the effective value. In Fig. 11.17, $(\Delta H_m / \Delta H_1)$ is plotted against $(\Delta H_m / \Delta H_2)$ and contours representing constant values of f are included.

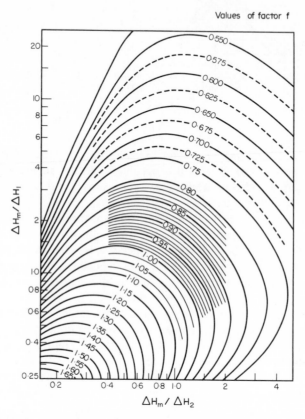

FIG. 11.17. Correction factor f for obtaining mean effective driving force in column.

Using the mean driving force, integration of equation 11.51 gives:

$$(H_{G2} - H_{G1})/f \Delta H_m = h_D a \rho z / G' \tag{11.52}$$

or:

$$z = (G'/h_D a \rho)(H_{G2} - H_{G1})/f \Delta H_m$$

11.6.3. Change in Air Condition

The change in the humidity and temperature of the air is obtained as follows. The enthalpy and temperature of the air are known only at the bottom of the tower, where fresh air is admitted. Here the condition of the air can be represented by a point E with coordinates (H_{G1}, θ_{G1}). Thus the line AE is parallel to the temperature axis.

Since

$$(H_G - H_f)/(\theta_G - \theta_f) = dH_G/d\theta_G \tag{equation 11.49}$$

the slope of the line EC is $(dH_G/d\theta_G)$ and represents the rate of change of air enthalpy with air temperature at the bottom of the column. If the gradient $(dH_G/d\theta_G)$ is taken as constant over a small section, the point F, on EC, will represent the condition of the gas at a small distance from the bottom. The corresponding liquid temperature is found by drawing through F a line parallel to the temperature axis. This cuts the operating line at some point H, which indicates the liquid temperature. The corresponding value of the temperature and enthalpy of the gas at the interface is then obtained by drawing a line through H, parallel to AC. This line then cuts the curve for saturated air at a point J, which represents the conditions of the gas at the interface. The rate of change of enthalpy with temperature for the gas is then given by the slope of the line FJ. Again, this slope can be considered to remain constant over a small height of the column, and the condition of the gas is thus determined for the next point in the tower. The procedure is then repeated until the curve representing the condition of the gas has been extended to a point whose ordinate is equal to the enthalpy of the gas at the top of the column. This point is obtained by drawing a straight line through B, parallel to the temperature axis. The final point on the line then represents the condition of the air which leaves the top of the water-cooling tower.

The size of the individual increments of height which are considered must be decided for the particular problem under consideration and will depend, primarily, on the rate of change of the gradient $(dH_G/d\theta_G)$. It should be noted that, for the gas to remain unsaturated throughout the whole of the tower, the line representing the condition of the gas must be below the curve for saturated gas. If at any point in the column, the air has become saturated, it is liable to become supersaturated as it passes further up the column and comes into contact with hotter liquid. It is difficult to define precisely what happens beyond this point as partial condensation may occur, giving rise to a mist. Under these conditions the preceding equations will no longer be applicable. However, an approximate solution is obtained by assuming that once the air stream becomes saturated it remains so during its subsequent contact with the water through the column.

11.6.4. Evaluation of Heat and Mass Transfer Coefficients

In general, coefficients of heat and mass transfer in the gas phase and the heat transfer coefficient for the liquid phase are not known. They can be determined, however, by carrying out tests in the laboratory or pilot scale using the same packing. If, for the air–water system, a small column is operated at steady water and air rates and the temperature of the water at the top and bottom and the initial and final temperatures and humidities of the air stream are noted, the operating line for the system is obtained. Assuming a value of the ratio $-(h_L/h_D\rho)$, for the slope of the tie-lines AC, etc., the graphical construction is carried out, starting with the conditions at the bottom of the tower. The condition of the gas at the top of the tower is thus calculated and compared with the measured value. If the difference is significant, another value of $-(h_L/h_D\rho)$ is assumed and the procedure repeated. Now that the slope of the tie line is known, the value of the integral of $dH_G/(H_f - H_G)$ over the whole column can be calculated. Since the height of the column is known, the product $h_D a$ is found by solution of equation 11.47. $h_G a$ can then be calculated using the Lewis relation.

The values of the three transfer coefficients are therefore obtained at any given flow rates from a single experimental run. The effect of liquid and gas rate can be found by means of a series of similar experiments.

Several workers have measured heat and mass transfer coefficients in water-cooling towers and in humidifying towers. Thomas and Houston,[6] using a tower 2 m high and 0·3 m square in cross-section, fitted with wooden slats, give the following figures for heat and mass transfer coefficients for packed heights greater than 75 mm:

$$h_G a = 3 \cdot 0 L^{0 \cdot 26} G'^{0 \cdot 72} \tag{11.53}$$

$$h_L a = 1 \cdot 04 \times 10^4 L'^{0 \cdot 51} G'^{1 \cdot 00} \tag{11.54}$$

$$h_D a = 2 \cdot 95 L'^{0 \cdot 26} G'^{0 \cdot 72} \tag{11.55}$$

In the above equations, L' and G' are expressed in kg/s m^2, s in J/kg K, $h_G a$ and $h_L a$ in W/m^3 K, and $h_D a$ in s^{-1}. A comparison of the gas and liquid film coefficients can then be made for a number of gas and liquid rates; taking the humid heat s as $1 \cdot 17 \times 10^3$ J/kg K:

	$L' = G' = 0 \cdot 5$ kg/m^{2}s	$L' = G' = 1 \cdot 0$ kg/m^{2}s	$L' = G' = 2 \cdot 0$ kg/m^{2}s
$h_G a$	1780	3510	6915
$h_L a$	3650	10,400	29,600
$h_L a / h_G a$	2·05	2·96	4·28

Cribb[7] quotes values of the ratio h_L/h_G ranging from 2·4 to 8·5.

It will be seen that the liquid film coefficient is generally considerably higher than the gas film coefficient, but that it is not always safe to ignore the resistance to transfer in the liquid phase.

Lowe and Christie[8] used a 1·3 m square experimental column fitted with a number of different types of packing and measured heat and mass transfer coefficients and pressure drops. They showed that in most cases:

$$h_D a \propto L'^{1-n} G'^n \tag{11.56}$$

The index n was found to vary from about 0·4 to 0·8 according to the type of packing. It will be noted that when $n \approx 0 \cdot 75$, there is close agreement with the results of Thomas and Houston (equation 11.55).

The heat-transfer coefficient for the liquid is often large compared with that for the gas phase. As a first approximation, therefore, it can be assumed that the whole of the resistance to heat transfer lies within the gas phase and that the temperature at the water–air interface is equal to the temperature of the bulk of the liquid. Thus, everywhere in the tower, $\theta_f = \theta_L$. This simplifies the calculations, because the lines AC, HJ, etc., have a slope of $-\infty$, i.e. they become parallel to the enthalpy axis.

Some workers have attempted to base the design of humidifiers on the overall heat transfer coefficient between the liquid and gas phases. This treatment is not satisfactory since the quantities of heat transferred through the liquid and through the gas are not the same, as some of the heat is utilised in effecting evaporation at the interface. In fact, at the bottom of a tall tower, the transfer of heat in both the liquid and the gas phases may be towards the interface, as already mentioned. A further objection to the use of overall coefficients is that the Lewis relation can be applied only to the heat and mass transfer coefficients in the gas phase.

In the design of commercial units, nomographs are available which give a performance characteristic (KaV/L'), where K is a mass transfer coefficient (kg water/m^{2}s) and V is the active cooling volume (m^3/m^2 plan area), as a function of θ, θ_w and (L'/G').[9, 10] For a given duty (KaV/L') is calculated from:

$$(KaV/L') = \int_{\theta_1}^{\theta_2} d\theta/(H_f - H_G) \tag{11.57}$$

and then a suitable tower with this value of (KaV/L') is sought from performance curves.[11, 12] In normal applications the performance characteristic varies between 0·5–2·5.

Example 11.8. Water is to be cooled from 328 to 293 K by means of a counter-current air stream entering at 293 K with a relative humidity of 20 per cent. The flow of air is 0·68 m^3/m^2 s and the water throughput is 0.26 kg/m² s. The whole of the resistance to heat and mass transfer may be assumed to be in the gas phase and the product $(h_D a)$ may be taken as 0.2 (m/s)(m²/m³). What is the required height of packing and the condition of the exit air stream?

Solution. Assuming the latent heat of water at 273 K = 2495 kJ/kg

$$\text{specific heat of air} = 1.003 \text{ kJ/kg K}$$

$$\text{and specific heat of water vapour} = 2.006 \text{ kJ/kg K}$$

the enthalpy of the inlet air stream

$$H_{G1} = 1.003(193 - 272) + \mathscr{H}[2495 + 2.006(193 - 273)]$$

From Fig. 11.4:

at $\theta = 293$ K and 20 per cent RH, $\mathscr{H} = 0.003$ kg/kg, and hence

$$H_{G1} = (1.003 \times 20) + 0.003[2495 + (2.006 \times 20)]$$

$$= 27.67 \text{ kJ/kg}$$

In the inlet air, water vapour = 0.003 kg/kg dry air

or $(0.003/18)/(1/29) = 0.005$ kmol/kmol dry air

Thus flow of dry air $= (1 - 0.005)0.68 = 0.677$ m³/m²s

Density of air at 293 K $= (29/22.4)(273/293) = 1.206$ kg/m³

and mass flow of dry air $= (1.206 \times 0.677) = 0.817$ kg/m²s

Slope of operating line: $(L'C_L/G') = 0.26 \times 4.18/0.817 = 1.33$

The coordinates of the bottom of the operating line are:

$$\theta_{L1} = 293 \text{ K}, \quad H_{G1} = 27.67 \text{ kJ/kg}$$

Hence on an enthalpy–temperature diagram, the operating line of slope 1.33 is drawn through the point (293, 27.67) (θ_{L1}, H_{G1}).

The top point of the operating line is given by $\theta_{L2} = 328$ K, and H_{G2} is found to be 76.5 kJ/kg (Fig. 11.18).

From Figs. 11.4 and 11.5 the curve representing the enthalpy of saturated air as a function of temperature is obtained and drawn in. Alternatively, this plot may be calculated from:

$$H_F = C_a(\theta_f - 273) + \mathscr{H}_0[C_w(\theta_f - 273) + \lambda] \text{ kJ/kg}$$

The curve represents the relation between enthalpy and temperature at the interface, that is H_f as a function of θ_f.

It now remains to evaluate the integral $\int dH_G/(H_f - H_G)$ between limits, $H_{G1} = 27.7$ kJ/kg and $H_{G2} = 76.5$ kJ/kg. Various values of H_G between these limits are selected and the value of θ obtained from the operating line. At this value of θ_1 now θ_f, the corresponding value of H_f is obtained from the curve for saturated air. The working is as follows:

H_G	$\theta = \theta_f$	H_f	$(H_f - H_G)$	$1/(H_f - H_G)$
27.7	293	57.7	30	0.0330
30	294.5	65	35	0.0285
40	302	98	58	0.0172
50	309	137	87	0.0114
60	316	190	130	0.0076
70	323	265	195	0.0051
76.5	328	355	279	0.0035

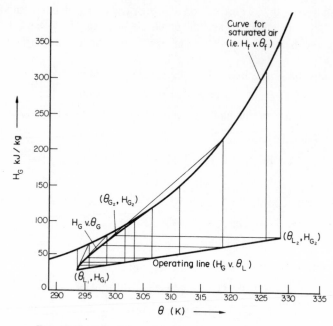

FIG. 11.18. Calculation of height of water-cooling tower.

A plot of $1/(H_f - H_G)$ and H_G is now made as in Fig. 11.19 from which the area under the curve $= 0·648$.

This value may be checked using the approximate solution of Carey and Williamson.[5] At the bottom of the column:

$$H_{G1} = 27·7 \text{ kJ/kg}, \quad H_{f1} = 57·7 \text{ kJ/kg} \quad \therefore \Delta H_1 = 30 \text{ kJ/kg}$$

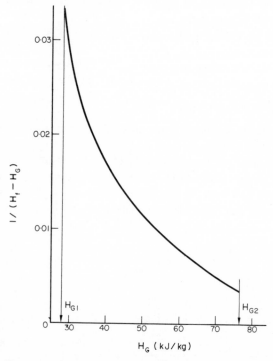

FIG. 11.19. Evaluation of the integral of $dH_G/(H_f - H_G)$.

At the top of the column:

$$H_{G2} = 76 \cdot 5 \text{ kJ/kg}, \quad H_{f2} = 355 \text{ kJ/kg} \quad \therefore \Delta H_2 = 279 \text{ kJ/kg}$$

At the mean water temperature of $0 \cdot 5(328 + 293) = 310 \cdot 5$ K:

$$H_{Gm} = 52 \text{ kJ/kg}, \quad H_f = 145 \text{ kJ/kg} \quad \therefore \Delta H_m = 93 \text{ kJ/kg}$$

$$\Delta H_m / \Delta H_1 = 3 \cdot 10, \quad \Delta H_m / \Delta H_2 = 0 \cdot 333,$$

and from Fig. 11.17: $f = 0 \cdot 79$

Thus $(H_{G2} - H_{G1})/f \Delta H_m = (76 \cdot 5 - 27 \cdot 7)/(0 \cdot 79 \times 93) = 0 \cdot 66$

which agrees with the value obtained by graphical integration.
    Thus in equation 11.51:

$$\text{height of packing, } z = \int_{H_{G1}}^{H_{G2}} [\mathrm{d}H_G/(H_f - H_G)]G'/h_D a \rho$$

$$= 0 \cdot 65 \times 0 \cdot 817/(0 \cdot 2 \times 1 \cdot 206)$$

$$= \underline{\underline{2 \cdot 20 \text{ m}}}$$

Assuming that the resistance to mass transfer lies entirely within the gas phase, the lines connecting θ_L and θ_f are parallel with the enthalpy axis.
    In Fig. 11.18 a plot of H_G and θ_G is obtained using the construction given in Section 11.6.3 and shown in Fig. 11.6. From this curve, the value of θ_{G2} corresponding to $H_{G2} = 76 \cdot 5$ kJ/kg is 300 K. From Fig. 11.5, under these conditions, the exit air has a humidity of $0 \cdot 019$ kg/kg which from Fig. 11.4 corresponds to a relative humidity of 83 per cent.

11.6.5. Humidifying Towers

If the main function of the tower is to produce a stream of humidified air, the final temperature of the liquid will not be specified but the humidity of the gas leaving the top of the tower will be given instead. It is therefore not possible to fix any point on the operating line, though its slope can be calculated from the liquid and gas rates. In designing a humidifier, therefore, it is necessary to calculate the temperature and enthalpy, and hence the humidity, of the gas leaving the tower for a number of assumed water outlet temperatures and thereby determine the outlet water temperature resulting in the air leaving the tower with the required humidity. The operating line for this water-outlet temperature is then used in the calculation of the height of the tower required to effect this degree of humidification. The calculation of the dimensions of a humidifier is therefore rather more tedious than that for the water-cooling tower.
    In a humidifier in which the make-up liquid is only a small proportion of the total liquid circulating, its temperature approaches the adiabatic saturation temperature θ_s, and remains constant, so that there is no temperature gradient in the liquid. The gas in contact with the liquid surface is approximately saturated and has a humidity $\mathcal{H}_s$.

Thus $\mathrm{d}\theta_L = 0$

and $\theta_{L1} = \theta_{L2} = \theta_L = \theta_f = \theta_s$

Hence $-G's\, \mathrm{d}\theta_G = h_G a\, \mathrm{d}z\, (\theta_G - \theta_s)$ (from equation 11.40)

and $-G'\, \mathrm{d}\mathcal{H} = h_D \rho a\, \mathrm{d}z(\mathcal{H} - \mathcal{H}_s)$ (from equation 11.42)

Integration of these equations gives:

$$\ln\left[(\theta_{G1} - \theta_s)/(\theta_{G2} - \theta_s)\right] = h_G a z / G' s \tag{11.58}$$

and

$$\ln\left[(\mathscr{H}_s - \mathscr{H}_1)/(\mathscr{H}_s - \mathscr{H}_2)\right] = h_D \rho a z / G' \tag{11.59}$$

assuming h_G, h_D, and s to remain approximately constant.

From these equations the temperature θ_{G2} and the humidity $\mathscr{H}_2$ of the gas leaving the humidifier can be calculated in terms of the height of the tower. Rearrangement of equation 11.60 gives:

$$\ln\left(1 + \frac{\mathscr{H}_1 - \mathscr{H}_2}{\mathscr{H}_s - \mathscr{H}_1}\right) = (h_D \rho a z)/G'$$

i.e.

$$(\mathscr{H}_2 - \mathscr{H}_1)/(\mathscr{H}_s - \mathscr{H}_1) = 1 - e^{-h_D \rho a z / G'} \tag{11.60}$$

Thus the ratio of the actual increase in humidity produced in the saturator to the maximum possible increase in humidity (i.e. the production of saturated gas) is equal to $(1 - e^{-h_D \rho a z / G'})$, and complete saturation of the gas is reached exponentially. A similar relation exists for the change in the temperature of the gas stream, namely:

$$(\theta_{G1} - \theta_{G2})/(\theta_{G2} - \theta_s) = 1 - e^{-h_G a z / G' s} \tag{11.61}$$

Further, the relation between the temperature and the humidity of the gas at any stage in the adiabatic humidifier is given by:

$$d\mathscr{H}/d\theta_G = (\mathscr{H} - \mathscr{H}_s)/(\theta_G - \theta_s) \quad \text{(from equation 11.50)}$$

On integration

$$\ln\frac{(\mathscr{H}_s - \mathscr{H}_2)}{(\mathscr{H}_s - \mathscr{H}_1)} = \ln\frac{(\theta_{G2} - \theta_s)}{(\theta_{G1} - \theta_s)} \tag{11.62}$$

or

$$\frac{(\mathscr{H}_s - \mathscr{H}_2)}{(\mathscr{H}_s - \mathscr{H}_1)} = \frac{(\theta_{G2} - \theta_s)}{(\theta_{G1} - \theta_s)} \tag{11.63}$$

11.7. SYSTEMS OTHER THAN AIR–WATER

Calculations relating to systems where the Lewis relation is not applicable are very much more complicated because the adiabatic saturation temperature and the wet-bulb temperature do not coincide. Thus the significance of the adiabatic cooling lines on the temperature–humidity chart is very much restricted. They no longer represent the changes which take place in a gas as it is humidified by contact with liquid initially at the adiabatic saturation temperature of the gas, but simply give the compositions of all gases with the same adiabatic saturation temperature.

Calculation of the change in the condition of the liquid and the gas in a humidification tower is rendered more difficult since equation 11.47, which was derived for the air–water system, is no longer applicable. Lewis and White[13] have developed a method of calculation based on the use of a *modified enthalpy* in place of the true enthalpy of the system.

For the air–water system, from equation 11.11:

$$h_G = h_D \rho s \tag{11.64}$$

This relationship applies quite closely for the conditions normally encountered in practice. For other systems, the relation between the heat and mass transfer coefficients in the gas phase is given by:

$$h_G = b h_D \rho s \tag{11.65}$$

where b is approximately constant and generally has a value greater than unity.

For these systems, therefore, equation 11.44 becomes:

$$G's \, d\theta_G = bh_D\rho a \, dz(s\theta_f - s\theta_G) \tag{11.66}$$

Adding equations 11.66 and 11.45 to obtain the relationship corresponding to equation 11.46:

$$G'(s \, d\theta_G + \lambda \, d\mathcal{H}) = h_D\rho a \, dz[(bs\theta_f + \lambda\mathcal{H}_f) - (bs\theta_G + \lambda\mathcal{H})] \tag{11.67}$$

Lewis and White use a *modified latent heat of vaporisation* λ' defined by:

$$b = \lambda/\lambda' \tag{11.68}$$

and a *modified enthalpy* per unit mass of dry gas defined by:

$$H'_G = s(\theta_G - \theta_0) + \lambda'\mathcal{H} \tag{11.69}$$

Substituting in equation 11.65; from equations 11.36, 11.68, and 11.69:

$$G' \, dH_G = bh_D\rho a \, dz(H'_f - H'_G) \tag{11.70}$$

and

$$\frac{dH_G}{H'_f - H'_G} = bh_D\rho a \, dz/G' \tag{11.71}$$

Combining equations 11.35, 11.38, and 11.70:

$$\frac{H'_G - H'_f}{\theta_L - \theta_f} = -\frac{h_L}{h_D\rho b} \quad \text{(cf. equation 11.48)} \tag{11.72}$$

From equations 11.64 and 11.70:

$$\frac{H'_G - H'_f}{\theta_G - \theta_f} = \frac{dH_G}{d\theta_G} \quad \text{(cf. equation 11.49)} \tag{11.73}$$

From equations 11.42 and 11.65:

$$\frac{\mathcal{H} - \mathcal{H}_f}{\theta_G - \theta_f} = b\frac{d\mathcal{H}}{d\theta_G} \quad \text{(cf. equation 11.50)} \tag{11.74}$$

The calculation of conditions within a countercurrent column operating with a system other than air–water is carried out in a similar manner to that already described by applying equations 11.71, 11.72, and 11.73 in conjunction with equation 11.37:

$$G'(H_{G2} - H_{G1}) = L'C_L(\theta_{L2} - \theta_{L1}) \quad \text{(equation 11.37)}$$

On an enthalpy–temperature diagram (Fig. 11.20) the enthalpy of saturated gas is plotted against its temperature. If equilibrium between the liquid and gas exists at the interface, this curve PQ represents the relation between gas enthalpy and temperature at the interface (H_f v. θ_f). The modified enthalpy of saturated gas is then plotted against temperature (curve RS)' to give the relation between H'_f and θ_f. Since b is greater than unity, RS will lie below PQ. By combining equations 11.33, 11.68, and 11.70, H'_G is obtained in terms of H_G:

$$H'_G = \frac{1}{b}[H_G + (b - 1)s(\theta_G - \theta_0)] \tag{11.75}$$

H'_G can conveniently be plotted against H_G for a number of constant temperatures. If b and s are constant, a series of straight lines is obtained. The operating line AB given by equation 11.37 is drawn in Fig. 11.20. Point A has coordinates (θ_{L1}, H_{G1}) corresponding to the bottom of the column. Point a has coordinates (θ_{L1}, H'_{G1}), H'_{G1} being obtained from equation 11.75.

From equation 11.70, a line through a, of slope $-(h_L/h_D\rho b)$, will intersect curve RS at c, (θ_{f1}, H'_{f1}) to give the interface conditions at the bottom of the column. The corresponding air enthalpy is given by C, (θ_{f1}, H_{f1}). The difference between the ordinates of c and a then gives

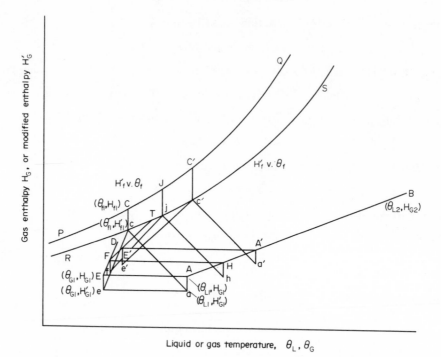

FIG. 11.20. Construction for height of humidifying column for vapour other than water.

the driving force in terms of modified enthalpy at the bottom of the column $(H'_{f1} - H'_{G1})$. A similar construction at other points, such as A', enables the driving force to be calculated at any other point. Hence $(H'_f - H'_G)$ is obtained as a function of H_G throughout the column. The height of column corresponding to a given change in air enthalpy can be obtained from equation 11.69 since the left-hand side can now be evaluated.

Thus

$$\int_{H_{G1}}^{H_{G2}} \frac{dH_G}{H'_f - H'_G} = \frac{b h_D \rho a z}{G'} \tag{11.76}$$

The change in the condition of the gas stream is obtained as follows: E, with coordinates (θ_{G1}, H_{G1}), represents the condition of the inlet gas. The modified enthalpy of this gas is given by $e(\theta_{G1}, H'_{G1})$. From equation 11.73 it is seen that ec gives the rate of change of gas enthalpy with temperature $(dH_G/d\theta_G)$ at the bottom of the column. Thus ED, parallel to ec, describes the way in which gas enthalpy changes at the bottom of the column. At some arbitrary small distance from the bottom, F represents the condition of the gas and H gives the corresponding liquid temperature. In exactly the same way the next small change is obtained by drawing through h a line hj parallel to ac. The slope of fj gives the new value of $(dH_G/d\theta_G)$ and therefore the gas condition at a higher point in the column is obtained by drawing FT parallel to fj. In this way the change in the condition of the gas through the column can be followed by continuing the procedure until the gas enthalpy reaches the value H_{G2} corresponding to the top of the column.

11.8.   FURTHER READING

ECKERT, E. R. G. and DRAKE, R. M.: *Analysis of Heat and Mass Transfer* (McGraw-Hill, New York, 1972).
NORMAN, W. S.: *Absorption, Distillation and Cooling Towers* (Longmans, London, 1961).
BACKHURST, J. R., HARKER, J. H., and PORTER, J. E.: *Problems in Heat and Mass Transfer* (Edward Arnold, London, 1974).

JACKSON, J.: *Cooling Towers* (Butterworths Scientific Publications, London, 1951).
STANFORD, W.: *Cooling Towers* (Carter Thermal Engineering Ltd., 1970).

11.9.   REFERENCES

1. SHERWOOD, T. K. and COMINGS, E. W.: *Trans. Am. Inst. Chem. Eng.* **28** (1932) 88. An experimental study of the wet bulb hygrometer.
2. WOOD, B. and BETTS, P.: *Proc. Inst. Mech. Eng.* (*Steam Group*) **163** (1950) 54. A contribution to the theory of natural draught cooling towers.
3. MICKLEY, H. S.: *Chem. Eng. Prog.* **45** (1949) 739. Design of forced draught air conditioning equipment.
4. MERKEL, F.: *Ver. deut. Ing. Forschungsarb.* No. 275 (1925). Verdunstungs-Kühlung.
5. CAREY, W. F. and WILLIAMSON, G. J.: *Proc. Inst. Mech. Eng.* (*Steam Group*) **163** (1950) 41. Gas cooling and humidification: design of packed towers from small scale tests.
6. THOMAS, W. J. and HOUSTON, P.: *Brit. Chem. Eng.* **4** (1959) 160, 217. Simultaneous heat and mass transfer in cooling towers.
7. CRIBB, G.: *Brit. Chem. Eng.* **4** (1959) 264. Liquid phase resistance in water cooling.
8. LOWE, H. J. and CHRISTIE, D. G.: *Inst. Mech. Eng. Symposium on Heat Transfer* (1962), Paper 113, 933. Heat transfer and pressure drop data on cooling tower packings, and model studies of the resistance of natural-draught towers to airflow.
9. WOOD, B. and BETTS, P.: *Engineer* **189** (4912) 337, (4913) 349 (1950). A total heat–temperature diagram for cooling tower calculations.
10. ZIVI, S. M. and BRAND, B. B.: *Refrig. Eng.* **64** (1956) 8, 31, 90. Analysis of Cross-Flow cooling towers.
11. *Counter-flow Cooling Tower Performance* (J. F. Pritchard & Co., Kansas City, 1957).
12. COOLING TOWER INSTITUTE: *Performance Curves* (Houston, 1967).

11.10.   NOMENCLATURE

		Units in SI System	Dimensions in M, L, T, θ
A	Interfacial area	m²	L^2
a	Interfacial area per unit volume of column	m²/m³	L^{-1}
b	Psychrometric ratio ($h/h_D\rho_A s$)	—	—
C_a	Specific heat of gas at constant pressure	J/kg K	$L^2T^{-2}\theta^{-1}$ or —
C_L	Specific heat of liquid	J/kg K	$L^2T^{-2}\theta^{-1}$ or —
C_p	Specific heat of gas and vapour mixture at constant pressure	J/kg K	$L^2T^{-2}\theta^{-1}$
C_s	Specific heat of solid	J/kg K	$L^2T^{-2}\theta^{-1}$ or —
C_w	Specific heat of vapour at constant pressure	J/kg K	$L^2T^{-2}\theta^{-1}$ or —
c	Mass concentration of vapour	kg/m³	ML^{-3}
c_0	Mass concentration of vapour in saturated gas	kg/m³	ML^{-3}
f	Correction factor for mean driving force	—	—
G'	Mass rate of flow of gas per unit area	kg/m²s	$ML^{-2}T^{-1}$
H	Enthalpy of humid gas per unit mass of dry gas	J/kg	L^2T^{-2}
H_a	Enthalpy per unit mass, of dry gas	J/kg	L^2T^{-2}
H_w	Enthalpy per unit mass, of vapour	J/kg	L^2T^{-2}
H_1	Enthalpy of stream of gas per unit mass of dry gas	J/kg	L^2T^{-2}
H_2	Enthalpy of another stream of gas, per unit mass of dry gas	J/kg	L^2T^{-2}
H_3	Enthalpy per unit mass of liquid or vapour	J/kg	L^2T^{-2}
H'	Modified enthalpy of humid gas defined by (11.67)	J/kg	L^2T^{-2}
ΔH	Enthalpy driving force ($H_f - H_G$)	J/kg	L^2T^{-2}
h	Heat transfer coefficient	W/m² K	$MT^{-3}\theta^{-1}$
h_D	Mass transfer coefficient	kmol/(kmol/m³)m²s	LT^{-1}
h_G	Heat transfer coefficient for gas phase	W/m² K	$MT^{-3}\theta^{-1}$
h_L	Heat transfer coefficient for liquid phase	W/m² K	$MT^{-3}\theta^{-1}$
$\mathcal{H}$	Humidity	kg/kg	—
$\mathcal{H}_s$	Humidity of gas saturated at the adiabatic saturation temperature	kg/kg	—
$\mathcal{H}_w$	Humidity of gas saturated at the wet bulb temperature	kg/kg	—
$\mathcal{H}_0$	Humidity of saturated gas	kg/kg	—

		Units in SI system	Dimensions in M, L, T, θ
$\mathcal{H}_1$	Humidity of a gas stream	kg/kg	—
$\mathcal{H}_2$	Humidity of second gas stream	kg/kg	—
K	Mass transfer coefficient	kg/m²s	$\mathbf{ML^{-2}T^{-1}}$
L'	Mass rate of flow of liquid per unit area	kg/m²s	$\mathbf{ML^{-2}T^{-1}}$
M_A	Molecular weight of gas	kg/kmol	—
M_w	Molecular weight of vapour	kg/kmol	—
m, m_1, m_2	Masses of dry gas	kg	$\mathbf{M}$
m_3	Mass of liquid or vapour	kg	$\mathbf{M}$
P	Total pressure	N/m²	$\mathbf{ML^{-1}T^{-2}}$
P_A	Mean partial pressure of gas	N/m²	$\mathbf{ML^{-1}T^{-2}}$
P_w	Partial pressure of vapour	N/m²	$\mathbf{ML^{-1}T^{-2}}$
P_{w0}	Partial pressure of vapour in saturated gas	N/m²	$\mathbf{ML^{-1}T^{-2}}$
Q	Rate of transfer of heat to liquid surface	W	$\mathbf{ML^2T^{-3}}$
$\mathbf{R}$	Universal gas constant	8·314 kJ/kmol K	$\mathbf{L^2T^{-2}\theta^{-1}}$
s	Humid heat of gas	J/kg K	$\mathbf{L^2T^{-2}\theta^{-1}}$
T	Absolute temperature	K	$\mathbf{\theta}$
V	Active volume per plan area of column	m³/m²	$\mathbf{L}$
W	Rate of evaporation	kg/s	$\mathbf{MT^{-1}}$
Z	Percentage relative humidity	—	—
z	Height from bottom of tower	m	$\mathbf{L}$
θ	Temperature of gas stream	K	$\mathbf{\theta}$
θ_0	Reference temperature, taken as the melting point of the material	K	$\mathbf{\theta}$
θ_s	Adiabatic saturation temperature	K	$\mathbf{\theta}$
θ_w	Wet bulb temperature	K	$\mathbf{\theta}$
λ	Latent heat of vaporisation per unit mass, at datum temperature	J/kg	$\mathbf{L^2T^{-2}}$
λ_f	Latent heat of freezing per unit mass, at datum temperature	J/kg	$\mathbf{L^2T^{-2}}$
λ'	Modified latent heat of vaporisation per unit mass defined by (11.66)	J/kg	$\mathbf{L^2T^{-2}}$
ρ	Mean density of gas and vapour	kg/m³	$\mathbf{ML^{-3}}$
ρ_A	Mean density of gas at partial pressure P_A	kg/m³	$\mathbf{ML^{-3}}$
Pr	Prandtl number	—	—
Sc	Schmidt number	—	—

Suffixes 1, 2, f, L, G denote conditions at the bottom of the tower, the top of the tower, the interface, the liquid, and the gas, respectively.

Suffix m refers to the mean water temperature.

Tables of Physical Properties

Steam Tables

TABLE 1. Temperature conversion chart

General formula: $°F = (°C \times \tfrac{9}{5}) + 32$; $°C = (°F - 32) \times \tfrac{5}{9}$; $K = °C + 273.15$;

C.		F.	C.		F.	C.		F.	C.		F.	C.		F.	C.		F.	C.		F.	C.		F.	C.		F.
-273.1	-459.7		-17.8	0	32	10.0	50	122.0	38	100	212	260	500	932	538	1000	1832	816	1500	2732	1093	2000	3632	1371	2500	4532
-268	-450		-17.2	1	33.8	10.6	51	123.8	43	110	230	266	510	950	543	1010	1850	821	1510	2750	1099	2010	3650	1377	2510	4550
-262	-440		-16.7	2	35.6	11.1	52	125.6	49	120	248	271	520	968	549	1020	1868	827	1520	2768	1104	2020	3668	1382	2520	4568
-257	-430		-16.1	3	37.4	11.7	53	127.4	54	130	266	277	530	986	554	1030	1886	832	1530	2786	1110	2030	3686	1388	2530	4586
-251	-420		-15.6	4	39.2	12.2	54	129.2	60	140	284	282	540	1004	560	1040	1904	838	1540	2804	1116	2040	3704	1393	2540	4604
-246	-410		-15.0	5	41.0	12.8	55	131.0	66	150	302	288	550	1022	566	1050	1922	843	1550	2822	1121	2050	3722	1399	2550	4622
-240	-400		-14.4	6	42.8	13.3	56	132.8	71	160	320	293	560	1040	571	1060	1940	849	1560	2840	1127	2060	3740	1404	2560	4640
-234	-390		-13.9	7	44.6	13.9	57	134.6	77	170	338	299	570	1058	577	1070	1958	854	1570	2858	1132	2070	3758	1410	2570	4658
-229	-380		-13.3	8	46.4	14.4	58	136.4	82	180	356	304	580	1076	582	1080	1976	860	1580	2876	1138	2080	3776	1416	2580	4676
-223	-370		-12.8	9	48.2	15.0	59	138.2	88	190	374	310	590	1094	588	1090	1994	866	1590	2894	1143	2090	3794	1421	2590	4694
-218	-360		-12.2	10	50.0	15.6	60	140.0	93	200	392	316	600	1112	593	1100	2012	871	1600	2912	1149	2100	3812	1427	2600	4712
-212	-350		-11.7	11	51.8	16.1	61	141.8	99	210	410	321	610	1130	599	1110	2030	877	1610	2930	1154	2110	3830	1432	2610	4730
-207	-340		-11.1	12	53.6	16.7	62	143.6	100	212	413	327	620	1148	604	1120	2048	882	1620	2948	1160	2120	3848	1438	2620	4748
-201	-330		-10.6	13	55.4	17.2	63	145.4	104	220	428	332	630	1166	610	1130	2066	888	1630	2966	1166	2130	3866	1443	2630	4766
-196	-320		-10.0	14	57.2	17.8	64	147.2	110	230	446	338	640	1184	616	1140	2084	893	1640	2984	1171	2140	3884	1449	2640	4784
-190	-310		-9.44	15	59.0	18.3	65	149.0	116	240	464	343	650	1202	621	1150	2102	899	1650	3002	1177	2150	3902	1454	2650	4802
-184	-300		-8.89	16	60.8	18.9	66	150.8	121	250	482	349	660	1220	627	1160	2120	904	1660	3020	1182	2160	3920	1460	2660	4820
-179	-290		-8.33	17	62.6	19.4	67	152.6	127	260	500	354	670	1238	632	1170	2138	910	1670	3038	1188	2170	3938	1466	2670	4838
-173	-280		-7.78	18	64.4	20.0	68	154.4	132	270	518	360	680	1256	638	1180	2156	916	1680	3056	1193	2180	3956	1471	2680	4856
-169.5	-273.1	-459.7	-7.22	19	66.2	20.6	69	156.2	138	280	536	366	690	1274	643	1190	2174	921	1690	3074	1199	2190	3974	1477	2690	4874
-168	-270	-454	-6.67	20	68.0	21.1	70	158.0	143	290	554	371	700	1292	649	1200	2192	927	1700	3092	1204	2200	3992	1482	2700	4892
-162	-260	-436	-6.11	21	69.8	21.7	71	159.8	149	300	572	377	710	1310	654	1210	2210	932	1710	3110	1210	2210	4010	1488	2710	4910
-157	-250	-418	-5.56	22	71.6	22.2	72	161.6	154	310	590	382	720	1328	660	1220	2228	938	1720	3128	1216	2220	4028	1493	2720	4928
-151	-240	-400	-5.00	23	73.4	22.8	73	163.4	160	320	608	388	730	1346	666	1230	2246	943	1730	3146	1221	2230	4046	1499	2730	4946
-146	-230	-382	-4.44	24	75.2	23.3	74	165.2	166	330	626	393	740	1364	671	1240	2264	949	1740	3164	1227	2240	4064	1504	2740	4964
-140	-220	-364	-3.89	25	77.0	23.9	75	167.0	171	340	644	399	750	1382	677	1250	2282	954	1750	3182	1232	2250	4082	1510	2750	4982
-134	-210	-346	-3.33	26	78.8	24.4	76	168.8	177	350	662	404	760	1400	682	1260	2300	960	1760	3200	1238	2260	4100	1516	2760	5000
-129	-200	-328	-2.78	27	80.6	25.0	77	170.6	182	360	680	410	770	1418	688	1270	2318	966	1770	3218	1243	2270	4118	1521	2770	5018
-123	-190	-310	-2.22	28	82.4	25.6	78	172.4	188	370	698	416	780	1436	693	1280	2336	971	1780	3236	1249	2280	4136	1527	2780	5036
-118	-180	-292	-1.67	29	84.2	26.1	79	174.2	193	380	716	421	790	1454	699	1290	2354	977	1790	3254	1254	2290	4154	1532	2790	5054

Interpolation factors

C.		F.	C.		F.
0·56	1	1·8	3·33	6	10·8
1·11	2	3·6	3·89	7	12·6
1·67	3	5·4	4·44	8	14·4
2·22	4	7·2	5·00	9	16·2
2·78	5	9·0	5·56	10	18·0

C	°	F
– 112	– 170	– 274
– 107	– 160	– 256
– 101	– 150	– 238
– 95·6	– 140	– 220
– 90·0	– 130	– 202
– 84·4	– 120	– 184
– 78·9	– 110	– 166
– 73·3	– 100	– 148
– 67·8	– 90	– 130
– 62·2	– 80	– 112
– 56·7	– 70	– 94
– 51·1	– 60	– 76
– 45·6	– 50	– 58
– 40·0	– 40	– 40
– 34·4	– 30	– 22
– 28·9	– 20	– 4
– 23·3	– 10	14
– 17·8	0	32

C	°	F
– 1·11	30	86·0
– 0·56	31	87·8
0	32	89·6
0·56	33	91·4
1·11	34	93·2
1·67	35	95·0
2·22	36	96·8
2·78	37	98·6
3·33	38	100·4
3·89	39	102·2
4·44	40	104·0
5·00	41	105·8
5·56	42	107·6
6·11	43	109·4
6·67	44	111·2
7·22	45	113·0
7·78	46	114·8
8·33	47	116·6
8·89	48	118·4
9·44	49	120·2

C	°	F
26·7	80	176·0
27·2	81	177·8
27·8	82	179·6
28·3	83	181·4
28·9	84	183·2
29·4	85	185·0
30·0	86	186·8
30·6	87	188·6
31·1	88	190·4
31·7	89	192·2
32·2	90	194·0
32·8	91	195·8
33·3	92	197·6
33·9	93	199·4
34·4	94	201·2
35·0	95	203·0
35·6	96	204·8
36·1	97	206·6
36·7	98	208·4
37·2	99	210·2
37·8	100	212·0

C	°	F
199	390	734
204	400	752
210	410	770
216	420	788
221	430	806
227	440	824
232	450	842
238	460	860
243	470	878
249	480	896
254	490	914

C	°	F
427	800	1472
432	810	1490
438	820	1508
443	830	1526
449	840	1544
454	850	1562
460	860	1580
466	870	1598
471	880	1616
477	890	1634
482	900	1652
488	910	1670
493	920	1688
499	930	1706
504	940	1724
510	950	1742
516	960	1760
521	970	1778
527	980	1796
532	990	1814

C	°	F
704	1300	2372
710	1310	2390
716	1320	2408
721	1330	2426
727	1340	2444
732	1350	2462
738	1360	2480
743	1370	2498
749	1380	2516
754	1390	2534
760	1400	2552
766	1410	2570
771	1420	2588
777	1430	2606
782	1440	2624
788	1450	2642
793	1460	2660
799	1470	2678
804	1480	2696
810	1490	2714

C	°	F
982	1800	3272
988	1810	3290
993	1820	3308
999	1830	3326
1004	1840	3344
1010	1850	3362
1016	1860	3380
1021	1870	3398
1027	1880	3416
1032	1890	3434
1038	1900	3452
1043	1910	3470
1049	1920	3488
1054	1930	3506
1060	1940	3524
1066	1950	3542
1071	1960	3560
1077	1970	3578
1082	1980	3596
1088	1990	3614
1093	2000	3632

C	°	F
1260	2300	4172
1266	2310	4190
1271	2320	4208
1277	2330	4226
1282	2340	4244
1288	2350	4262
1293	2360	4280
1299	2370	4298
1304	2380	4316
1310	2390	4334
1316	2400	4352
1321	2410	4370
1327	2420	4388
1332	2430	4406
1338	2440	4424
1343	2450	4442
1349	2460	4460
1354	2470	4478
1360	2480	4496
1366	2490	4514

C	°	F
1538	2800	5072
1543	2810	5090
1549	2820	5108
1554	2830	5126
1560	2840	5144
1566	2850	5162
1571	2860	5180
1577	2870	5198
1582	2880	5216
1588	2890	5234
1593	2900	5252
1599	2910	5270
1604	2920	5288
1610	2930	5306
1616	2940	5324
1621	2950	5342
1627	2960	5360
1632	2970	5378
1638	2980	5396
1643	2990	5414
1649	3000	5432

Note.—The numbers in bold-face type refer to the temperature (in either Centigrade or Fahrenheit degrees) which it is desired to convert into the other scale. If converting from Fahrenheit degrees to Centigrade degrees the equivalent temperature is in the left column, while if converting from degrees Centigrade to degrees Fahrenheit, the equivalent temperature is in the column on the right. This table, made by Albert Sauveur, is published by permission of Mrs. Albert Sauveur.

TABLE 2. Thermal conductivities of liquids*

A linear variation with temperature may be assumed. The extreme values given constitute also the temperature limits over which the data are recommended.

Liquid	k (W/m K)	K	k (Btu/h ft °F)	Liquid	k (W/m K)	K	k (Btu/h ft °F)
Acetic acid 100 per cent	0·171	293	0·099	Hexane (n-)	0·138	303	0·080
50 per cent	0·35	293	0·20		0·135	333	0·078
Acetone	0·177	303	0·102	Heptyl alcohol (n-)	0·163	303	0·094
	0·164	348	0·095		0·157	348	0·091
Allyl alcohol	0·180	298 to 303	0·104	Hexyl alcohol (n-)	0·161	303	0·093
Ammonia	0·50	258 to 303	0·29		0·156	348	0·090
Ammonia, aqueous	0·45	293	0·261	Kerosene	0·149	293	0·086
	0·50	333	0·29		0·140	348	0·081
Amyl acetate	0·144	283	0·083	Mercury	8·36	301	4·83
Amyl alcohol (n-)	0·163	303	0·094	Methyl alcohol 100 per cent	0·215	293	0·124
	0·154	373	0·089	80 per cent	0·267	293	0·154
Amyl alcohol (iso-)	0·152	303	0·088	60 per cent	0·329	293	0·190
	0·151	348	0·087	40 per cent	0·405	293	0·234
Aniline	0·173	273 to 293	0·100	20 per cent	0·492	293	0·284
Benzene	0·159	303	0·092	100 per cent	0·197	323	0·114
	0·151	333	0·087	Methyl Chloride	0·192	258	0·111
Bromobenzene	0·128	303	0·074		0·154	303	0·089
	0·121	373	0·070				
Butyl acetate (n-)	0·147	298 to 303	0·085	Nitrobenzene	0·164	303	0·095
Butyl alcohol (n-)	0·168	303	0·097		0·152	373	0·088
	0·164	348	0·095	Nitromethane	0·216	303	0·125
Butyl alcohol (iso-)	0·157	283	0·091		0·208	333	0·120
Calcium chloride brine							
30 per cent	0·55	303	0·32	Nonane (n-)	0·145	303	0·084
15 per cent	0·59	303	0·34		0·142	333	0·082
Carbon disulphide	0·161	303	0·093	Octane (n-)	0·144	303	0·083
	0·152	348	0·088		0·140	333	0·081
Carbon tetrachloride	0·185	273	0·107	Oils, Petroleum	0·138–0·156	273	0·08–0·09
	0·163	341	0·094	Oil, Castor	0·180	293	0·104
Chlorobenzene	0·144	283	0·083		0·173	373	0·100
Chloroform	0·138	303	0·080	Oil, Olive	0·168	293	0·097
Cymene (para)	0·135	303	0·078		0·164	373	0·095
	0·137	333	0·079				
Decane (n-)	0·147	303	0·085	Paraldehyde	0·145	303	0·084
	0·144	333	0·083		0·135	373	0·078
Dichlorodifluoromethane	0·099	266	0·057	Pentane (n-)	0·135	303	0·078
	0·092	289	0·053		0·128	348	0·074

Substance		t	
Dichloroethane	0·083	311	0·048
	0·074	333	0·043
	0·066	355	0·038
Dichloromethane	0·142	323	0·082
Ethyl acetate	0·192	258	0·111
	0·166	303	0·096
Ethyl alcohol 100 per cent	0·175	293	0·101
80 per cent	0·182	293	0·105
60 per cent	0·237	293	0·137
40 per cent	0·305	293	0·176
20 per cent	0·388	293	0·224
100 per cent	0·486		0·281
Ethyl benzene	0·151	323	0·087
	0·149	303	0·086
Ethyl bromide	0·142	333	0·082
	0·121	293	0·070
Ethyl ether	0·138	303	0·080
	0·135	348	0·078
Ethyl iodide	0·111	313	0·064
	0·109	348	0·063
Ethylene glycol	0·265	273	0·153
Gasoline	0·135	303	0·078
Glycerol 100 per cent	0·284	293	0·164
80 per cent	0·327	293	0·189
60 per cent	0·381	293	0·220
40 per cent	0·448	293	0·259
20 per cent	0·481	293	0·278
100 per cent	0·284	373	0·164
Heptane (n-)	0·140	303	0·081
	0·137	333	0·079
Perchloroethylene	0·159	323	0·092
Petroleum ether	0·130	303	0·075
	0·126	348	0·073
Propyl alcohol (n-)	0·171	303	0·099
	0·164	348	0·095
Propyl alcohol (iso-)	0·157	303	0·091
	0·155	333	0·090
Sodium	85	373	49
	80	483	46
Sodium chloride brine 25·0 per cent	0·57	303	0·33
12·5 per cent	0·59	303	0·34
Sulphuric acid 90 per cent	0·36	303	0·21
60 per cent	0·43	303	0·25
30 per cent	0·52	303	0·30
Sulphur dioxide	0·22	258	0·128
	0·192	303	0·111
Toluene	0·149	303	0·086
β-trichloroethane	0·145	348	0·084
Trichloroethylene	0·133	323	0·077
Turpentine	0·138	323	0·080
Vaseline	0·128	288	0·074
	0·184	288	0·106
Water	0·57	273	0·330
	0·615	303	0·356
	0·658	333	0·381
	0·688	353	0·398
Xylene (ortho-)	0·155	293	0·090
(meta-)	0·155	293	0·090

*By permission from *Heat Transmission*, by W. H. McAdams, Copyright 1942, McGraw-Hill.

TABLE 3. Latent heats of vaporisation*

For water at 373 K, $\theta_c - \theta = 647 - 373 = 274$, and the latent heat of vaporisation is 2257 kJ/kg

No.	Compound	Range $\theta_c - \theta$ (K)	θ_c (K)	No.	Compound	Range $\theta_c - \theta$ (K)	θ_c (K)
18	Acetic acid	100–225	594	2	Freon-12 (CCl_2F_2)	40–200	384
22	Acetone	120–210	508	5	Freon-21 ($CHCl_2F$)	70–250	451
29	Ammonia	50–200	406	6	Freon-22 ($CHClF_2$)	50–170	369
13	Benzene	10–400	562	1	Freon-113 ($CCl_2F\text{-}CClF_2$)	90–250	487
16	Butane	90–200	426	10	Heptane	20–300	540
21	Carbon dioxide	10–100	304	11	Hexane	50–225	508
4	Carbon disulphide	140–275	546	15	Isobutane	80–200	407
2	Carbon tetrachloride	30–250	556	27	Methanol	40–250	513
7	Chloroform	140–275	536	20	Methyl chloride	70–250	416
8	Dichloromethane	150–250	489	19	Nitrous oxide	25–150	309
3	Diphenyl	175–400	800	9	Octane	30–300	569
25	Ethane	25–150	305	12	Pentane	20–200	470
26	Ethyl alcohol	20–140	516	23	Propane	40–200	369
28	Ethyl alcohol	140–300	516	24	Propyl alcohol	20–200	537
17	Ethyl chloride	100–250	460	14	Sulphur dioxide	90–160	430
13	Ethyl ether	10–400	467	30	Water	100–500	647
2	Freon-11 (CCl_3F)	70–250	471				

Latent heats of vaporisation

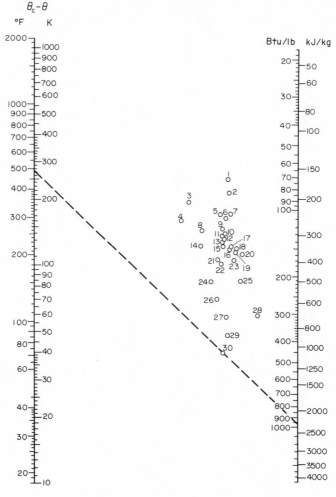

TABLE 4. Specific heats of liquids*

No.	Liquid	Range (K)
29	Acetic acid, 100 per cent	273–353
32	Acetone	293–323
52	Ammonia	203–323
37	Amyl alcohol	223–298
26	Amyl acetate	273–373
30	Aniline	273–403
23	Benzene	283–353
27	Benzyl alcohol	253–303
10	Benzyl chloride	243–303
49	Brine, 25 per cent CaCl₂	233–293
51	Brine, 25 per cent NaCl	233–293
44	Butyl alcohol	273–373
2	Carbon disulphide	173–298
3	Carbon tetrachloride	283–333
8	Chlorobenzene	273–373
4	Chloroform	273–323
21	Decane	193–298
6A	Dichloroethane	243–333
5	Dichloromethane	233–323
15	Diphenyl	353–393
22	Diphenylmethane	303–373
16	Diphenyl oxide	273–473
16	Dowtherm A	273–473
24	Ethyl acetate	223–298
42	Ethyl alcohol, 100 per cent	303–353
46	Ethyl alcohol, 95 per cent	293–353
50	Ethyl alcohol, 50%	293–353
25	Ethyl benzene	273–373
1	Ethyl bromide	278–298
13	Ethyl chloride	243–313
36	Ethyl ether	173–298
7	Ethyl iodide	273–373
39	Ethylene glycol	233–473
2A	Freon-11 (CCl₃F)	253–343
6	Freon-12 (CCl₂F₂)	233–288
4A	Freon-21 (CHCl₂F)	253–343
7A	Freon-22 (CHClF₂)	253–333
3A	Freon-113 (CCl₂F-CClF₂)	253–343
38	Glycerol	233–293
28	Heptane	273–333
35	Hexane	193–293
48	Hydrochloric acid, 30 per cent	293–373
41	Isoamyl alcohol	283–373
43	Isobutyl alcohol	273–373
47	Isopropyl alcohol	253–323
31	Isopropyl ether	193–293
40	Methyl alcohol	233–293
13A	Methyl chloride	193–293
14	Naphthalene	363–473
12	Nitrobenzene	273–373
34	Nonane	223–298
33	Octane	223–298
3	Perchlorethylene	243–413
45	Propyl alcohol	253–373
20	Pyridine	223–298
9	Sulphuric acid, 98 per cent	283–318
11	Sulphur dioxide	253–373
23	Toluene	273–333
53	Water	283–473
19	Xylene *ortho*	273–373
18	Xylene *meta*	273–373
17	Xylene *para*	273–373

Specific heats of liquids

Temperature °F / K axis; Specific heat axis (kJ/kg K) (Btu/lb °F)

TABLE 5. Specific heats C_p of gases and vapours at 101·3 kN/m²*

No.	Gas	Range (K)	No.	Gas	Range (K)
10	Acetylene	273–473	1	Hydrogen	273–873
15	Acetylene	473–673	2	Hydrogen	873–1673
16	Acetylene	673–1673	35	Hydrogen bromide	273–1673
27	Air	273–1673	30	Hydrogen chloride	273–1673
12	Ammonia	273–873	20	Hydrogen fluoride	273–1673
14	Ammonia	873–1673	36	Hydrogen iodide	273–1673
18	Carbon dioxide	273–673	19	Hydrogen sulphide	273–973
24	Carbon dioxide	673–1673	21	Hydrogen sulphide	973–1673
26	Carbon monoxide	273–1673	5	Methane	273–573
32	Chlorine	273–473	6	Methane	573–973
34	Chlorine	473–1673	7	Methane	973–1673
3	Ethane	273–473	25	Nitric oxide	273–973
9	Ethane	473–873	28	Nitric oxide	973–1673
8	Ethane	873–1673	26	Nitrogen	273–1673
4	Ethylene	273–473	23	Oxygen	273–773
11	Ethylene	473–873	29	Oxygen	773–1673
13	Ethylene	873–1673	33	Sulphur	573–1673
17B	Freon-11 (CCl₃F)	273–423	22	Sulphur dioxide	273–673
17C	Freon-21 (CHCl₂F)	273–423	31	Sulphur dioxide	673–1673
17A	Freon-22 (CHClF₂)	273–423	17	Water	273–1673
17D	Freon-113 (CCl₂F-CClF₂)	273–423			

*By permission from *Heat Transmission*, by W. H. McAdams, copyright, 1942, McGraw-Hill.

Specific heats of gases at constant pressures

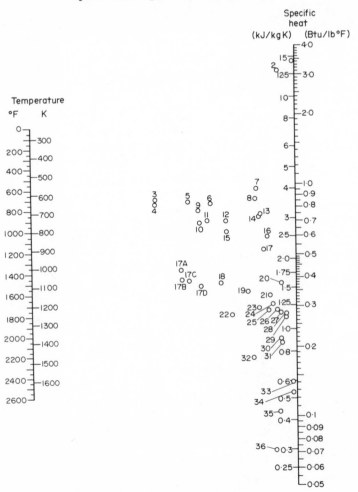

TABLE 6. Viscosity of water*

Temp. (K)	Viscosity (mN s/m²)	Temp. (K)	Viscosity (mN s/m²)	Temp. (K)	Viscosity (mN s/m²)
273	1·7921	306	0·7523	340	0·4233
274	1·7313	307	0·7371	341	0·4174
275	1·6728	308	0·7225	342	0·4117
276	1·6191	309	0·7085	343	0·4061
277	1·5674	310	0·6947	344	0·4006
278	1·5188	311	0·6814	345	0·3952
279	1·4728	312	0·6685	346	0·3900
280	1·4284	313	0·6560	347	0·3849
281	1·3860	314	0·6439	348	0·3799
282	1·3462	315	0·6321	349	0·3750
283	1·3077	316	0·6207	350	0·3702
284	1·2713	317	0·6097	351	0·3655
285	1·2363	318	0·5988	352	0·3610
286	1·2028	319	0·5883	353	0·3565
287	1·1709	320	0·5782	354	0·3521
288	1·1404	321	0·5683	355	0·3478
289	1·1111	322	0·5588	356	0·3436
290	1·0828	323	0·5494	357	0·3395
291	1·0559	324	0·5404	358	0·3355
292	1·0299	325	0·5315	359	0·3315
293	1·0050	326	0·5229	360	0·3276
293·2	1·0000	327	0·5146	361	0·3239
294	0·9810	328	0·5064	362	0·3202
295	0·9579	329	0·4985	363	0·3165
296	0·9358	330	0·4907	364	0·3130
297	0·9142	331	0·4832	365	0·3095
298	0·8937	332	0·4759	366	0·3060
299	0·8737	333	0·4688	367	0·3027
300	0·8545	334	0·4618	368	0·2994
301	0·8360	335	0·4550	369	0·2962
302	0·8180	336	0·4483	370	0·2930
303	0·8007	337	0·4418	371	0·2899
304	0·7840	338	0·4355	372	0·2868
305	0·7679	339	0·4293	373	0·2838

*Calculated by the formula:

$$1/\mu = 2\cdot1482[(\theta - 281\cdot435) + \sqrt{(8078\cdot4 + (\theta - 281\cdot435)^2)}] - 120$$

(By permission from *Fluidity and Plasticity*, by E. C. Bingham. Copyright 1922, McGraw-Hill Book Company Inc.)

TABLE 7. Thermal conductivities of gases and vapours*

The extreme temperature values given constitute the experimental range. For extrapolation to other temperatures, it is suggested that the data given be plotted as log k vs. log T, or that use be made of the assumption that the ratio $C_p\mu/k$ is practically independent of temperature (or of pressure, within moderate limits).

Substance	k (W/m K)	K	k (Btu/h ft °F)	Substance	k (W/m K)	K	k (Btu/h ft °F)
Acetone	0·0098	273	0·0057	Dichlorodifluoromethane	0·0083	273	0·0048
	0·0128	319	0·0074		0·0111	323	0·0064
	0·0171	373	0·0099		0·0139	373	0·0080
	0·0254	457	0·0147		0·0168	423	0·0097
Acetylene	0·0118	198	0·0068				
	0·0187	273	0·0108	Ethane	0·0114	203	0·0066
	0·0242	323	0·0140		0·0149	239	0·0086
	0·0298	373	0·0172		0·0183	273	0·0106
	0·0164	173	0·0095		0·0303	373	0·0175
Air	0·0242	273	0·0140	Ethyl acetate	0·0125	319	0·0072
	0·0317	373	0·0183		0·0166	373	0·0096
	0·0391	473	0·0226		0·0244	457	0·0141
	0·0459	573	0·0265	Alcohol	0·0154	293	0·0089

403

TABLE 7. Thermal conductivities of gases and vapours—cont'd

Substance	k (W/m K)	K	k (Btu/h ft °F)	Substance	k (W/m K)	K	k (Btu/h ft °F)
Ammonia	0·0164	213	0·0095		0·0215	373	0·0124
	0·0222	273	0·0128	Chloride	0·0095	273	0·0055
	0·0272	323	0·0157		0·0164	373	0·0095
	0·0320	373	0·0185		0·0234	457	0·0135
					0·0263	485	0·0152
Benzene	0·0090	273	0·0052	Ether	0·0133	273	0·0077
	0·0126	319	0·0073		0·0171	319	0·0099
	0·0178	373	0·0103		0·0227	373	0·0131
	0·0263	457	0·0152		0·0327	457	0·0189
	0·0305	485	0·0176		0·0362	485	0·0209
Butane (n-)	0·0135	273	0·0078	Ethylene	0·0111	202	0·0064
	0·0234	373	0·0135		0·0175	273	0·0101
(iso-)	0·0138	273	0·0080		0·0267	323	0·0131
	0·0241	373	0·0139		0·0279	373	0·0161
Carbon dioxide	0·0118	223	0·0068	Heptane (n-)	0·0194	473	0·0112
	0·0147	273	0·0085		0·0178	373	0·0103
	0·0230	373	0·0133	Hexane (n-)	0·0125	273	0·0072
	0·0313	473	0·0181		0·0138	293	0·0080
	0·0396	573	0·0228	Hexene	0·0106	273	0·0061
disulphide	0·0069	273	0·0040		0·0189	373	0·0109
	0·0073	280	0·0042	Hydrogen	0·0113	173	0·065
monoxide	0·0071	84	0·0041		0·0144	223	0·083
	0·0080	94	0·0046		0·0173	273	0·100
	0·0234	213	0·0135		0·0199	323	0·115
tetrachloride	0·0071	319	0·0041		0·0223	373	0·129
	0·0090	373	0·0052		0·0308	573	0·178
	0·0112	457	0·0065	Hydrogen and carbon dioxide		273	
Chlorine	0·0074	273	0·0043	0 per cent H_2	0·0144		0·0083
Chloroform	0·0066	273	0·0038	20 per cent	0·0286		0·0165
	0·0080	319	0·0046	40 per cent	0·0467		0·0270
	0·0100	373	0·0058	60 per cent	0·0709		0·0410
	0·0133	457	0·0077	80 per cent	0·1070		0·0620
Cyclohexane	0·0164	375	0·0095	100 per cent	0·173		0·10
Hydrogen and nitrogen		273		Nitric oxide	0·0178	203	0·0103
0 per cent H_2	0·0230		0·0133		0·0239	273	0·0138
20 per cent	0·0367		0·0212	Nitrogen	0·0164	173	0·0095
40 per cent	0·0542		0·0313		0·0242	273	0·0140
60 per cent	0·0758		0·0438		0·0277	323	0·0160
80 per cent	0·1098		0·0635		0·0312	373	0·0180
Hydrogen and nitrous oxide		273		Nitrous oxide	0·0116	201	0·0067
0 per cent H_2	0·0159		0·0092		0·0157	273	0·0087
20 per cent	0·0294		0·0170		0·0222	373	0·0128
40 per cent	0·0467		0·0270				
60 per cent	0·0709		0·0410	Oxygen	0·0164	173	0·0095
80 per cent	0·112		0·0650		0·0206	223	0·0119
Hydrogen sulphide	0·0132	273	0·0076		0·0246	273	0·0142
					0·0284	323	0·0164
Mercury	0·0341	473	0·0197		0·0321	373	0·0185
Methane	0·0173	173	0·0100				
	0·0251	223	0·0145	Pentane (n-)	0·0128	273	0·0074
	0·0302	273	0·0175		0·0144	293	0·0083
	0·0372	323	0·0215	(iso-)	0·0125	273	0·0072
Methyl alcohol	0·0144	273	0·0083		0·0220	373	0·0127
	0·0222	373	0·0128	Propane	0·0151	273	0·0087
Acetate	0·0102	273	0·0059		0·0261	373	0·0151
	0·0118	293	0·0068				
Chloride	0·0092	273	0·0053	Sulphur dioxide	0·0087	273	0·0050
	0·0125	319	0·0072		0·0119	373	0·0069
	0·0163	373	0·0094				
	0·0225	457	0·0130	Water vapour	0·0208	319	0·0120
	0·0256	485	0·0148		0·0237	373	0·0137
Methylene chloride	0·0067	273	0·0039		0·0324	473	0·0187
	0·0085	319	0·0049		0·0429	573	0·0248
	0·0109	373	0·0063		0·0545	673	0·0315
	0·0164	485	0·0095		0·0763	773	0·0441

*By permission from *Heat Transmission*, by W. H. McAdams, copyright 1942, McGraw-Hill.

TABLE 8. Viscosities of gases*

Co-ordinates for use with graph on opposite page

No.	Gas	X	Y
1	Acetic acid	7·7	14·3
2	Acetone	8·9	13·0
3	Acetylene	9·8	14·9
4	Air	11·0	20·0
5	Ammonia	8·4	16·0
6	Argon	10·5	22·4
7	Benzene	8·5	13·2
8	Bromine	8·9	19·2
9	Butene	9·2	13·7
10	Butylene	8·9	13·0
11	Carbon dioxide	9·5	18·7
12	Carbon disulphide	8·0	16·0
13	Carbon monoxide	11·0	20·0
14	Chlorine	9·0	18·4
15	Chloroform	8·9	15·7
16	Cyanogen	9·2	15·2
17	Cyclohexane	9·2	12·0
18	Ethane	9·1	14·5
19	Ethyl acetate	8·5	13·2
20	Ethyl alcohol	9·2	14·2
21	Ethyl chloride	8·5	15·6
22	Ethyl ether	8·9	13·0
23	Ethylene	9·5	15·1
24	Fluorine	7·3	23·8
25	Freon-11 (CCl_3F)	10·6	15·1
26	Freon-12 (CCl_2F_2)	11·1	16·0
27	Freon-21 ($CHCl_2F$)	10·8	15·3
28	Freon-22 ($CHClF_2$)	10·1	17·0
29	Freon-113 (CCl_2F-$CClF_2$)	11·3	14·0
30	Helium	10·9	20·5
31	Hexane	8·6	11·8
32	Hydrogen	11·2	12·4
33	$3H_2 + 1N_2$	11·2	17·2
34	Hydrogen bromide	8·8	20·9
35	Hydrogen chloride	8·8	18·7
36	Hydrogen cyanide	9·8	14·9
37	Hydrogen iodide	9·0	21·3
38	Hydrogen sulphide	8·6	18·0
39	Iodine	9·0	18·4
40	Mercury	5·3	22·9
41	Methane	9·9	15·5
42	Methyl alcohol	8·5	15·6
43	Nitric oxide	10·9	20·5
44	Nitrogen	10·6	20·0
45	Nitrosyl chloride	8·0	17·6
46	Nitrous oxide	8·8	19·0
47	Oxygen	11·0	21·3
48	Pentane	7·0	12·8
49	Propane	9·7	12·9
50	Propyl alcohol	8·4	13·4
51	Propylene	9·0	13·8
52	Sulphur dioxide	9·6	17·0
53	Toluene	8·6	12·4
54	2, 3, 3-trimethylbutane	9·5	10·5
55	Water	8·0	16·0
56	Xenon	9·3	23·0

(By permission from *Chemical Engineers' Handbook*, by J. H. PERRY. Copyright 1975, McGraw-Hill Book Company Inc.)

To convert to lb/ft-hr multiply by 2·42

Viscosities of gases

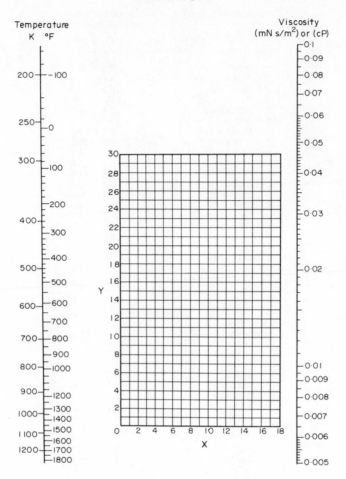

TABLE 9. Viscosities and densities of liquids*

Co-ordinates for graph on page **409**

No.	Liquid	X	Y	Density at 293 K (kg/m³)
1	Acetaldehyde	15·2	4·8	783 (291 K)
2	Acetic acid, 100 per cent	12·1	14·2	1049
3	Acetic acid, 70 per cent	9·5	17·0	1069
4	Acetic anhydride	12·7	12·8	1083
5	Acetone, 100 per cent	14·5	7·2	792
6	Acetone, 35 per cent	7·9	15·0	948
7	Allyl alcohol	10·2	14·3	854
8	Ammonia, 100 per cent	12·6	2·0	817 (194 K)
9	Ammonia, 26 per cent	10·1	13·9	904
10	Amyl acetate	11·8	12·5	879
11	Amyl alcohol	7·5	18·4	817
12	Aniline	8·1	18·7	1022
13	Anisole	12·3	13·5	990
14	Arsenic trichloride	13·9	14·5	2163
15	Benzene	12·5	10·9	880
16	Brine, CaCl₂, 25 per cent	6·6	15·9	1228
17	Brine, NaCl, 25 per cent	10·2	16·6	1186 (298 K)
18	Bromine	14·2	13·2	3119
19	Bromotoluene	20·0	15·9	1410
20	Butyl acetate	12·3	11·0	882
21	Butyl alcohol	8·6	17·2	810
22	Butyric acid	12·1	15·3	964
23	Carbon dioxide	11·6	0·3	1101 (236 K)
24	Carbon disulphide	16·1	7·5	1263
25	Carbon tetrachloride	12·7	13·1	1595
26	Chlorobenzene	12·3	12·4	1107
27	Chloroform	14·4	10·2	1489
28	Chlorosulphonic acid	11·2	18·1	1787 (298 K)
29	Chlorotoluene, *ortho*	13·0	13·3	1082
30	Chlorotoluene, *meta*	13·3	12·5	1072
31	Chlorotoluene, *para*	13·3	12·5	1070
32	Cresol, *meta*	2·5	20·8	1034
33	Cyclohexanol	2·9	24·3	962
34	Dibromoethane	12·7	15·8	2495
35	Dichloroethane	13·2	12·2	1256
36	Dichloromethane	14·6	8·9	1336
37	Diethyl oxalate	11·0	16·4	1079
38	Dimethyl oxalate	12·3	15·8	1148 (327 K)
39	Diphenyl	12·0	18·3	992 (346 K)
40	Dipropyl oxalate	10·3	17·7	1038 (273 K)
41	Ethyl acetate	13·7	9·1	901
42	Ethyl alcohol, 100 per cent	10·5	13·8	789
43	Ethyl alcohol, 95 per cent	9·8	14·3	804
44	Ethyl alcohol, 40 per cent	6·5	16·6	935
45	Ethyl benzene	13·2	11·5	867
46	Ethyl bromide	14·5	8·1	1431
47	Ethyl chloride	14·8	6·0	917 (279 K)
48	Ethyl ether	14·5	5·3	708 (298 K)
49	Ethyl formate	14·2	8·4	923
50	Ethyl iodide	14·7	10·3	1933
51	Ethylene glycol	6·0	23·6	1113
52	Formic acid	10·7	15·8	1220
53	Freon-11 (CCl₃F)	14·4	9·0	1494 (290 K)
54	Freon-12 (CCl₂F₂)	16·8	5·6	1486 (293 K)
55	Freon-21 (CHCl₂F)	15·7	7·5	1426 (273 K)
56	Freon-22 (CHClF₂)	17·2	4·7	3870 (273 K)
57	Freon-113 (CCl₂F-CClF₂)	12·5	11·4	1576
58	Glycerol, 100 per cent	2·0	30·0	1261
59	Glycerol, 50 per cent	6·9	19·6	1126
60	Heptane	14·1	8·4	684
61	Hexane	14·7	7·0	659

TABLE 9. Viscosities and densities of liquids—*cont'd*

No.	Liquid	X	Y	Density at 293 K (kg/m^3)
62	Hydrochloric acid, 31·5 per cent	13·0	16·6	1157
63	Isobutyl alcohol	7·1	18·0	779 (299 K)
64	Isobutyric acid	12·2	14·4	949
65	Isopropyl alcohol	8·2	16·0	789
66	Kerosene	10·2	16·9	780–820
67	Linseed oil, raw	7·5	27·2	930–938 (288 K)
68	Mercury	18·4	16·4	13546
69	Methanol, 100 per cent	12·4	10·5	792
70	Methanol, 90 per cent	12·3	11·8	820
71	Methanol, 40 per cent	7·8	15·5	935
72	Methyl acetate	14·2	8·2	924
73	Methyl chloride	15·0	3·8	952 (273K)
74	Methyl ethyl ketone	13·9	8·6	805
75	Naphthalene	7·9	18·1	1145
76	Nitric acid, 95 per cent	12·8	13·8	1493
77	Nitric acid, 60 per cent	10·8	17·0	1367
78	Nitrobenzene	10·6	16·2	1205 (291 K)
79	Nitrotoluene	11·0	17·0	1160
80	Octane	13·7	10·0	703
81	Octyl alcohol	6·6	21·1	827
82	Pentachloroethane	10·9	17·3	1671 (298 K)
83	Pentane	14·9	5·2	630 (291 K)
84	Phenol	6·9	20·8	1071 (298 K)
85	Phosphorus tribromide	13·8	16·7	2852 (288 K)
86	Phosphorus trichloride	16·2	10·9	1574
87	Propionic acid	12·8	13·8	992
88	Propyl alcohol	9·1	16·5	804
89	Propyl bromide	14·5	9·6	1353
90	Propyl chloride	14·4	7·5	890
91	Propyl iodide	14·1	11·6	1749
92	Sodium	16·4	13·9	970
93	Sodium hydroxide, 50 per cent	3·2	25·8	1525
94	Stannic chloride	13·5	12·8	2226
95	Sulphur dioxide	15·2	7·1	1434 (273 K)
96	Sulphuric acid, 110 per cent	7·2	27·4	1980
97	Sulphuric acid, 98 per cent	7·0	24·8	1836
98	Sulphuric acid, 60 per cent	10·2	21·3	1498
99	Sulphuryl chloride	15·2	12·4	1667
100	Tetrachloroethane	11·9	15·7	1600
101	Tetrachloroethylene	14·2	12·7	1624 (288 K)
102	Titanium tetrachloride	14·4	12·3	1726
103	Toluene	13·7	10·4	866
104	Trichloroethylene	14·8	10·5	1466
105	Turpentine	11·5	14·9	861–867
106	Vinyl acetate	14·0	8·8	932
107	Water	10·2	13·0	998
108	Xylene, *ortho*	13·5	12·1	881
109	Xylene, *meta*	13·9	10·6	867
110	Xylene, *para*	13·9	10·9	861

*By permission from *Chemical Engineers' Handbook*, by J. H. Perry, copyright 1975, McGraw-Hill.

Viscosities of liquids

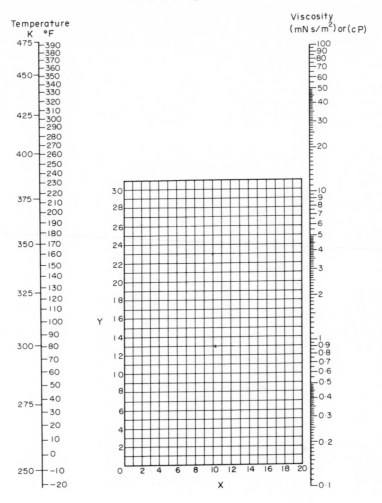

Temperature
K °F

Viscosity
$(mN \, s/m^2)$ or (cP)

TABLE 10. Critical constants of gases*

	Critical Temperature T_c (K)	Critical Pressure P_c, (MN/m^2)	Compressibility Constant in critical state Z_c
Paraffins			
Methane	191	4·64	0·290
Ethane	306	4·88	0·284
Propane	370	4·25	0·276
n-Butane	425	3·80	0·274
Isobutane	408	3·65	0·282
n-Pentane	470	3·37	0·268
Isopentane	461	3·33	0·268
Neopentane	434	3·20	0·268
n-Hexane	508	3·03	0·264
n-Heptane	540	2·74	0·260
n-Octane	569	2·49	0·258
Monoolefins			
Ethylene	282	5·07	0·268
Propylene	365	4·62	0·276
1-Butene	420	4·02	0·276
1-Pentene	474	4·05	
Miscellaneous organic compounds			
Acetic acid	595	5·78	0·200
Acetone	509	4·72	0·237
Acetylene	309	6·24	0·274
Benzene	562	4·92	0·274
1,3-Butadiene	425	4·33	0·270
Cyclohexane	553	4·05	0·271
Dichlorodifluoromethane (Freon-12)	385	4·01	0·273
Diethyl ether	467	3·61	0·261
Ethyl alcohol	516	6·38	0·249
Ethylene oxide	468	7·19	0·25
Methyl alcohol	513	7·95	0·220
Methyl chloride	416	6·68	0·276
Methyl ethyl ketone	533	4·00	0·26
Toluene	594	4·21	0·27
Trichlorofluoromethane (Freon-11)	471	4·38	0·277
Trichlorotrifluoroethane (Freon-113)	487	3·41	0·274
Elementary gases			
Bromine	584	10·33	0·307
Chlorine	417	7·71	0·276
Helium	5·3	0·23	0·300
Hydrogen	33·3	1·30	0·304
Neon	44·5	2·72	0·307
Nitrogen	126	3·39	0·291
Oxygen	155	5·08	0·29
Miscellaneous inorganic compounds			
Ammonia	406	11·24	0·242
Carbon dioxide	304	7·39	0·276
Carbon monoxide	133	3·50	0·294
Hydrogen chloride	325	8·26	0·266
Hydrogen sulfide	374	9·01	0·284
Nitric oxide (NO)	180	6·48	0·25
Nitrous oxide (N_2O)	310	7·26	0·271
Sulphur	1313	11·75	
Sulphur dioxide	431	7·88	0·268
Sulphur trioxide	491	8·49	0·262
Water	647	22·1	0·23

*Selected values from K. A. Kobe and R. E. Lynn, Jr., *Chem. Rev.*, **52**, 117 (1953). By permission.

Steam Tables

Tables 11A, 11B, 11C, 11D and 11E are adapted from the
Abridged Callendar Steam Tables
by permission of Messrs Edward Arnold (Publishers) Ltd.

TABLE 11A. Properties of saturated steam

Pressure			Temperature		Heat per unit mass						Entropy per unit mass (kJ/kg K)		Steam	
					(kJ/kg)			(Btu/lb)					Specific volume	
(lb/in.²)	(kN/m²)	Vacuum (in.)	(K)	(°F)	Water	Latent	Steam	Water	Latent	Steam	Water	Steam	(ft³/lb)	(m³/kg)
0·5	3·45	28·99	299·58	79·6	110·7	2438·8	2549·5	47·6	1048·5	1096·1	0·387	8·529	643·0	40·14
0·6	4·14	28·79	302·73	85·3	123·7	2431·6	2555·3	53·2	1045·4	1098·6	0·430	8·463	540·6	33·75
0·7	4·83	28·58	305·44	90·1	135·1	2425·3	2560·4	58·1	1042·7	1100·8	0·468	8·407	466·6	29·13
0·8	5·52	28·38	307·83	94·4	145·1	2419·8	2564·9	62·4	1040·3	1102·7	0·501	8·361	411·7	25·70
0·9	6·21	28·17	309·96	98·2	154·0	2414·6	2568·6	66·2	1038·1	1104·3	0·529	8·319	368·7	23·02
1·0	6·89	27·97	311·90	101·7	162·1	2410·0	2572·1	69·7	1036·1	1105·8	0·555	8·282	334·0	20·85
1·1	7·58	27·76	313·68	104·9	169·6	2405·8	2575·7	72·9	1034·3	1107·2	0·578	8·248	305·2	19·05
1·2	8·27	27·56	315·33	107·9	176·5	2401·6	2578·1	75·9	1032·5	1108·4	0·600	8·219	281·1	17·55
1·3	8·96	27·35	316·86	110·7	183·1	2397·9	2581·0	78·7	1030·9	1109·6	0·621	8·190	260·5	16·26
1·4	9·65	27·15	318·30	113·3	189·1	2394·6	2583·7	81·3	1029·5	1110·8	0·639	8·165	243·0	15·17
1·5	10·34	26·95	319·65	115·7	194·7	2391·4	2586·1	83·7	1028·1	1111·8	0·657	8·140	228·0	14·23
1·6	11·03	26·74	320·93	118·0	200·0	2388·3	2588·3	86·0	1026·8	1112·8	0·674	8·117	214·3	13·38
1·7	11·72	26·54	322·14	120·2	205·2	2385·3	2590·5	88·2	1025·5	1113·7	0·691	8·096	202·5	12·64
1·8	12·41	26·33	323·29	122·2	209·8	2382·8	2592·6	90·2	1024·4	1114·6	0·704	8·076	191·8	11·97
1·9	13·10	26·13	324·38	124·2	214·2	2380·2	2594·4	92·1	1023·3	1115·4	0·718	8·057	182·3	11·38
2·0	13·79	25·92	325·43	126·1	218·6	2377·6	2596·2	94·0	1022·2	1116·2	0·732	8·039	173·7	10·84
3·0	20·68	23·88	333·99	141·5	254·5	2356·7	2611·2	109·4	1013·2	1122·6	0·841	7·900	118·7	7·41
4·0	27·58	21·84	340·39	153·0	281·5	2341·6	2623·1	121·0	1006·7	1127·7	0·921	7·801	90·63	5·66
5·0	34·47	19·80	345·54	162·3	302·9	2329·7	2632·6	130·2	1001·6	1131·8	0·983	7·724	73·52	4·59
6·0	41·37	17·76	349·88	170·1	321·2	2318·1	2639·3	138·1	996·6	1134·7	1·034	7·661	61·98	3·87
7·0	48·26	15·71	353·65	176·9	337·0	2307·9	2644·9	144·9	992·2	1137·1	1·081	7·610	53·64	3·35
8·0	55·16	13·67	357·00	182·9	351·2	2299·3	2650·5	151·0	988·5	1139·5	1·120	7·563	47·35	2·96
9·0	62·05	11·63	360·00	188·3	364·0	2291·6	2655·6	156·5	985·2	1141·7	1·156	7·523	42·40	2·65
10·0	68·95	9·59	362·74	193·2	375·2	2285·3	2660·5	161·3	982·5	1143·8	1·187	7·488	38·42	2·40
11·0	75·84	7·55	365·26	197·8	385·9	2278·6	2664·5	165·9	979·6	1145·5	1·217	7·455	35·14	2·19
12·0	82·74	5·50	367·60	202·0	395·7	2272·3	2668·0	170·1	976·9	1147·0	1·243	7·429	32·40	2·02
13·0	89·63	3·46	369·78	205·9	404·5	2266·9	2671·4	173·9	974·6	1148·5	1·268	7·399	30·05	1·88
14·0	96·53	1·42	371·81	209·6	413·3	2261·3	2674·6	177·7	972·2	1149·9	1·292	7·374	28·03	1·75
14·696	101·33	Gauge lb/in.²	373·16	212·0	418·9	2257·6	2676·5	180·1	970·6	1150·7	1·307	7·358	26·80	1·673
15	103·42	0·3	373·73	213·0	421·5	2256·2	2677·7	181·2	970·0	1151·2	1·313	7·350	26·28	1·641
16	110·32	1·3	375·56	216·3	429·2	2257·3	2680·5	184·5	967·9	1152·4	1·334	7·329	24·74	1·544
17	117·21	2·3	377·29	219·5	436·4	2246·7	2683·1	187·6	965·9	1153·5	1·353	7·308	23·38	1·460
18	124·11	3·3	378·94	222·4	443·3	2242·3	2685·6	190·6	964·0	1154·6	1·372	7·289	22·17	1·384
19	131·00	4·3	380·52	225·2	450·1	2238·1	2688·2	193·5	962·2	1155·7	1·389	7·272	21·07	1·315
20	137·90	5·3	382·03	228·0	456·6	2233·9	2690·5	196·3	960·4	1156·7	1·406	7·255	20·09	1·254
21	144·79	6·3	383·48	230·6	462·6	2230·2	2692·8	198·9	958·8	1157·7	1·422	7·237	19·19	1·198
22	151·69	7·3	384·87	233·1	468·5	2226·5	2695·0	201·4	957·2	1158·6	1·437	7·222	18·38	1·148
23	158·58	8·3	386·21	235·5	474·3	2222·7	2697·0	203·9	955·6	1159·5	1·452	7·208	17·63	1·101
24	165·48	9·3	387·50	237·8	479·9	2219·0	2698·9	206·3	954·0	1160·3	1·466	7·193	16·94	1·057
25	172·37	10·3	388·75	240·1	485·2	2215·5	2700·7	208·6	952·5	1161·1	1·480	7·180	16·30	1·018
26	179·27	11·3	389·96	242·2	490·3	2212·3	2702·6	210·8	951·1	1161·9	1·493	7·167	15·72	0·981
27	186·16	12·3	391·13	244·4	495·2	2209·0	2704·2	212·9	949·7	1162·6	1·505	7·155	15·17	0·947
28	193·05	13·3	392·27	246·4	500·1	2205·8	2705·9	215·0	948·3	1163·3	1·518	7·143	14·67	0·916
29	199·95	14·3	393·37	248·4	504·7	2202·7	2707·4	217·0	947·0	1164·0	1·530	7·131	14·19	0·886
30	206·85	15·3	394·46	250·3	509·4	2199·5	2708·9	219·0	945·6	1164·6	1·542	7·119	13·73	0·857

TABLE 11A. Properties of saturated steam—*cont'd*

Pressure			Temperature		Heat per unit mass (kJ/kg)			Heat per unit mass (Btu/lb)			Entropy per unit mass (kJ/kg K)		Steam Specific volume	
(lb/in.²)	(kN/m²)	Gauge (lb/in.²)	(K)	(°F)	Water	Latent	Steam	Water	Latent	Steam	Water	Steam	(ft³/lb)	(m³/kg)
32	220·6	17·3	396·5	254·0	518·0	2193·7	2711·7	222·7	943·1	1165·8	1·564	7·097	12·93	0·807
34	234·4	19·3	398·5	257·6	526·4	2188·1	2714·5	226·3	940·7	1167·0	1·585	7·078	12·21	0·762
36	248·2	21·3	400·4	260·9	534·3	2182·9	2717·3	229·7	938·5	1168·2	1·605	7·059	11·58	0·723
38	262·0	23·3	402·1	264·1	541·9	2178·1	2720·0	233·0	936·4	1169·4	1·624	7·041	11·02	0·688
40	275·8	25·3	403·9	267·2	549·2	2173·4	2722·6	236·1	934·4	1170·5	1·643	7·024	10·50	0·655
42	289·6	27·2	405·5	270·3	556·1	2168·5	2724·6	239·1	932·3	1171·4	1·660	7·007	10·30	0·643
44	303·4	29·3	407·1	273·1	562·9	2163·9	2726·8	242·0	930·3	1172·3	1·676	6·992	9·600	0·599
46	317·2	31·3	408·6	275·8	569·6	2159·2	2728·8	244·9	928·3	1173·2	1·692	6·977	9·209	0·575
48	330·9	33·3	410·1	278·5	575·9	2154·8	2730·7	247·6	926·4	1174·0	1·707	6·963	8·848	0·552
50	344·7	35·3	411·5	281·0	581·9	2150·6	2732·5	250·2	924·6	1174·8	1·722	6·949	8·516	0·532
52	358·5	37·3	412·9	283·5	587·8	2146·7	2734·5	252·7	922·9	1175·6	1·736	6·936	8·208	0·512
54	372·3	39·3	414·2	285·9	593·6	2142·5	2736·1	255·2	921·1	1176·3	1·750	6·923	7·922	0·495
56	386·1	41·3	415·5	288·3	599·1	2138·5	2737·6	257·6	919·4	1177·0	1·763	6·911	7·656	0·478
58	399·9	43·3	416·8	290·5	604·5	2134·8	2739·3	259·9	917·8	1177·7	1·776	6·899	7·407	0·462
60	413·7	45·3	418·1	292·7	609·9	2131·1	2741·0	262·2	916·2	1178·4	1·789	6·887	7·175	0·448
62	427·5	47·3	419·3	294·9	615·0	2127·4	2742·4	264·4	914·6	1179·0	1·801	6·876	6·957	0·434
64	441·3	49·3	420·5	296·9	619·9	2123·8	2743·8	266·5	913·1	1179·6	1·813	6·866	6·752	
66	455·1	51·3	421·6	299·0	624·8	2120·4	2745·2	268·6	911·6	1180·2	1·825	6·855	6·560	0·410
68	468·8	53·3	422·7	301·0	629·7	2116·9	2746·6	270·7	910·1	1180·8	1·836	6·845	6·378	0·398
70	482·6	55·3	423·8	302·9	634·3	2113·6	2747·9	272·7	908·7	1181·4	1·847	6·836	6·206	0·387
72	496·4	57·3	424·8	304·8	638·7	2110·6	2749·3	274·6	907·4	1182·0	1·858	6·826	6·044	0·377
74	510·2	59·3	425·8	306·7	643·1	2107·4	2750·5	276·5	906·0	1182·5	1·868	6·817	5·890	0·368
76	524·0	61·3	426·8	308·5	647·6	2104·1	2751·7	278·4	904·6	1183·0	1·878	6·808	5·743	0·359
78	537·8	63·3	427·8	310·3	651·9	2100·8	2752·7	280·3	903·2	1183·0	1·888	6·799	5·604	0·350
80	551·6	65·3	428·8	312·0	656·2	2097·8	2754·0	282·1	901·9	1184·0	1·898	6·791	5·472	0·342
82	565·4	67·3	429·7	313·7	660·4	2094·8	2755·2	283·9	900·6	1184·5	1·908	6·782	5·346	0·334
84	579·2	69·3	430·7	315·4	664·3	2092·0	2756·3	285·6	899·4	1185·0	1·917	6·775	5·226	0·326
86	592·9	71·3	431·6	317·1	668·3	2089·0	2757·3	287·3	898·1	1185·4	1·926	6·766	5·110	0·319
88	606·7	73·3	432·6	318·7	672·2	2085·9	2758·1	289·0	896·8	1185·8	1·935	6·758	5·000	0·312
90	620·5	75·3	433·5	320·3	676·2	2082·9	2759·1	290·7	895·5	1186·2	1·944	6·751	4·896	0·306
92	634·3	77·3	434·4	321·9	679·9	2080·1	2760·0	292·3	894·3	1186·6	1·953	6·743	4·796	0·299
94	648·1	79·3	435·2	323·3	683·6	2077·4	2761·0	293·9	893·1	1187·0	1·961	6·736	4·699	0·293
96	661·9	81·3	436·0	324·8	687·3	2074·5	2761·9	295·5	891·9	1187·4	1·969	6·729	4·607	0·288
98	675·7	83·3	436·8	326·6	690·8	2072·0	2762·8	297·0	890·8	1187·8	1·977	6·722	4·519	0·282
100	689·5	85·3	437·6	327·8	694·3	2069·4	2763·7	298·5	889·7	1188·2	1·985	6·715	4·434	0·278
105	723·9	90·3	439·6	331·3	702·9	2062·9	2765·8	302·2	886·9	1189·1	2·005	6·699	4·230	0·264
110	758·4	95·3	441·4	334·8	711·1	2056·6	2767·7	305·7	884·2	1189·9	2·023	6·683	4·046	0·252
115	792·9	100·3	443·2	338·1	719·2	2050·4	2769·6	309·2	881·5	1190·7	2·041	6·668	3·880	0·292
120	827·4	105·3	445·0	341·3	726·9	2044·3	2771·2	312·5	878·9	1191·4	2·059	6·653	3·729	0·233
125	861·9	110·3	446·7	344·4	734·3	2038·5	2772·8	315·7	876·4	1192·1	2·076	6·639	3·587	0·224
130	896·3	115·3	448·4	347·3	741·5	2032·9	2774·4	318·8	874·0	1192·8	2·092	6·625	3·456	0·216
135	930·8	120·3	450·0	350·2	748·7	2027·1	2775·8	321·9	871·5	1193·4	2·108	6·612	3·335	0·208
140	965·3	125·3	451·5	353·0	755·7	2021·5	2777·2	324·9	869·1	1194·0	2·123	6·600	3·222	0·201
145	999·8	130·3	453·0	355·8	762·5	2016·2	2778·7	327·8	866·8	1194·6	2·138	6·587	3·116	0·195
150	1034·2	135·3	454·5	358·4	769·0	2010·8	2779·8	330·6	864·5	1195·1	2·152	6·575	3·015	0·188

TABLE 11B. Total heat H of dry

P (lb/in.² abs)	(kN/m²)	θ (K)	H_s	10	20	30	40	50	60	70	P (kN/m²)
		Saturation				Degrees of superheat (K)					
15	103·4	373·8	2677	2703	2719	2739	2759	2779	2799	2819	103·4
20	137·9	382·1	2690	2712	2732	2752	2772	2793	2813	2833	137·9
30	206·8	394·5	2709	2732	2752	2772	2792	2813	2833	2854	206·8
40	275·8	403·9	2723	2745	2765	2786	2807	2828	2848	2869	275·8
50	344·7	411·5	2733	2755	2777	2797	2818	2839	2860	2881	344·7
60	413·7	418·1	2742	2763	2785	2806	2828	2849	2870	2891	413·7
70	482·6	423·8	2749	2770	2792	2814	2836	2857	2878	2899	482·6
80	551·6	428·8	2754	2777	2799	2821	2843	2864	2886	2907	551·6
90	620·5	433·5	2760	2782	2806	2827	2850	2871	2893	2914	620·5
100	689·5	437·6	2764	2787	2813	2832	2857	2876	2900	2920	689·5
120	827·3	445·0	2771	2795	2822	2842	2867	2887	2911	2931	827·3
140	965·3	451·5	2778	2802	2829	2849	2874	2895	2919	2940	965·3
160	1103	457·4	2783	2808	2835	2856	2881	2902	2926	2947	1103
180	1241	462·7	2787	2813	2840	2862	2887	2908	2932	2954	1241
200	1378	467·5	2790	2817	2844	2866	2892	2914	2938	2960	1378
250	1724	478·1	2798	2824	2850	2876	2902	2925	2950	2973	1724
300	2068	487·3	2801	2829	2857	2884	2910	2934	2959	2983	2068
400	2758	502·4	2803	2835	2863	2893	2920	2946	2972	2997	2758
500	3447	514·9	2804	2835	2865	2897	2926	2954	2981	3007	3447
600	4137	525·5	2801	2836	2867	2900	2930	2960	2988	3015	4137
700	4827	534·9	2796	2835	2868	2902	2932	2963	2992	3020	4827
800	5516	543·3	2790	2831	2866	2902	2933	2966	2996	3025	5516
1000	6895	558·0	2775	2820	2861	2899	2935	2968	3001	3031	6895
2000	13790	608·6	2643	2728	2794	2850	2900	2946	2988	3027	13790

TABLE 11C. Total heat H of dry

P (lb/in.² abs.)	θ (°F)	H_s	20°	40°	60°	80°	100°	120°	140°	P (lb/in.² abs)
	Saturation				Degrees of superheat (°F)					
15	213·0	1151·2	1161·2	1170·9	1180·5	1190·0	1199·5	1208·9	1218·4	15
20	228·0	1156·7	1167·0	1176·7	1186·3	1195·8	1205·3	1214·8	1224·3	20
30	250·3	1164·6	1175·1	1184·9	1194·7	1204·4	1214·0	1223·6	1233·2	30
40	267·2	1170·5	1181·0	1190·9	1200·8	1210·7	1220·6	1230·3	1240·0	40
50	281·0	1174·8	1185·5	1195·7	1205·7	1215·6	1225·5	1235·3	1245·1	50
60	292·7	1178·4	1189·3	1199·5	1209·7	1219·8	1229·7	1239·6	1249·5	60
70	302·9	1181·4	1192·5	1202·8	1213·0	1223·2	1233·3	1243·3	1253·4	70
80	312·0	1184·0	1195·2	1205·6	1215·9	1226·2	1236·5	1246·7	1256·8	80
90	320·3	1186·2	1197·5	1208·0	1218·5	1228·9	1239·3	1249·6	1259·8	90
100	327·8	1188·2	1199·3	1210·2	1220·9	1231·4	1241·8	1252·2	1262·5	100
120	341·3	1191·4	1202·8	1214·0	1224·9	1235·7	1246·3	1256·8	1267·2	120
140	353·0	1194·0	1205·8	1217·1	1228·2	1239·1	1249·9	1260·6	1271·2	140
160	363·6	1196·1	1208·3	1219·7	1231·0	1242·1	1253·1	1263·9	1274·6	160
180	373·1	1198·0	1210·3	1222·0	1233·5	1244·7	1255·9	1266·8	1277·7	180
200	381·8	1199·5	1212·1	1224·0	1235·6	1247·0	1258·2	1269·3	1280·2	200
250	401·0	1202·1	1215·3	1227·6	1239·7	1251·7	1263·4	1274·9	1286·2	250
300	417·3	1203·8	1217·3	1230·3	1242·9	1255·1	1267·2	1278·9	1290·3	300
400	444·6	1205·5	1219·9	1234·1	1247·5	1260·5	1273·1	1285·3	1297·2	400
500	467·0	1205·4	1220·8	1235·7	1250·0	1263·6	1277·0	1289·7	1302·1	500
600	486·2	1204·2	1220·7	1236·5	1251·7	1266·0	1279·7	1292·7	1305·5	600
700	503·1	1202·2	1220·0	1236·6	1252·2	1267·0	1281·3	1295·0	1308·1	700
800	518·2	1199·6	1218·7	1236·1	1252·2	1267·4	1282·1	1296·3	1309·9	800
1000	544·6	1192·8	1214·5	1233·9	1251·6	1268·4	1283·9	1298·9	1313·2	1000
2000	635·8	1136·1	1176·1	1206·6	1232·9	1256·0	1276·7	1295·6	1313·5	2000

steam (superheated) (kJ/kg).

P (lb/in.² abs)	(kN/m²)	Degrees of superheat (K)								P (kN/m²)
		80	90	100	120	140	160	180	200	
15	103·4	2838	2858	2877	2917	2957	2997	3036	3076	103·4
20	137·9	2853	2873	2892	2932	2972	3013	3052	3092	137·9
30	206·8	2874	2894	2913	2953	2994	3035	3075	3115	206·8
40	275·8	2889	2909	2929	2969	3010	3051	3092	3133	275·8
50	344·7	2901	2921	2942	2983	3023	3064	3105	3147	344·7
60	413·7	2911	2931	2952	2993	3033	3075	3117	3158	413·7
70	482·6	2920	2940	2961	3002	3042	3085	3127	3168	482·6
80	551·6	2928	2948	2970	3011	3051	3094	3136	3177	551·6
90	620·5	2936	2955	2977	3019	3059	3102	3143	3185	620·5
100	689·5	2942	2961	2984	3026	3066	3110	3151	3193	689·5
120	827·3	2952	2971	2995	3038	3078	3122	3164	3206	827·3
140	965·3	2962	2981	3005	3048	3088	3133	3175	3217	965·3
160	1103	2969	2988	3014	3057	3097	3142	3185	3228	1103
180	1241	2976	2995	3021	3065	3105	3151	3194	3237	1241
200	1378	2983	3002	3028	3071	3112	3159	3202	3245	1378
250	1724	2996	3015	3042	3087	3128	3175	3219	3262	1724
300	2068	3006	3025	3053	3099	3140	3188	3233	3277	2068
400	2758	3023	3046	3071	3118	3161	3210	3255	3301	2758
500	3447	3034	3058	3084	3132	3176	3227	3273	3319	3447
600	4137	3042	3067	3094	3143	3187	3239	3287	3334	4137
700	4827	3048	3074	3102	3153	3198	3251	3299	3347	4827
800	5516	3053	3081	3108	3161	3207	3262	3311	3359	5516
1000	6895	3061	3089	3118	3171	3219	3275	3325	3374	6895
2000	13790	3063	3097	3131	3194	3254	3311	3366	3419	13790

steam (superheated), (Btu/lb).

P (lb/in.² abs.)	Degrees of superheat (°F)								P (lb/in.² abs.)
	160°	180°	200°	240°	280°	320°	360°	400°	
15	1227·7	1237·0	1246·2	1265·0	1284·0	1303·1	1322·3	1341·5	15
20	1233·7	1243·1	1252·5	1271·5	1290·6	1309·8	1329·1	1348·5	20
30	1242·8	1252·3	1261·8	1281·1	1300·5	1319·9	1339·3	1358·7	30
40	1249·7	1259·3	1268·9	1288·3	1307·7	1327·2	1346·7	1366·3	40
50	1254·9	1264·7	1274·4	1293·9	1313·4	1332·9	1352·6	1372·4	50
60	1259·4	1269·3	1279·1	1298·7	1318·3	1338·0	1357·8	1377·6	60
70	1263·4	1273·3	1283·1	1302·8	1322·5	1342·3	1362·1	1382·0	70
80	1266·9	1276·9	1286·8	1306·5	1326·2	1346·0	1365·8	1385·7	80
90	1270·0	1280·1	1290·1	1309·9	1329·7	1349·5	1369·3	1389·2	90
100	1272·7	1282·9	1293·0	1313·0	1332·9	1352·7	1372·6	1392·5	100
120	1277·5	1287·8	1298·0	1319·2	1338·2	1358·1	1378·1	1398·0	120
140	1281·7	1292·0	1302·3	1322·6	1342·8	1362·9	1383·0	1403·0	140
160	1285·2	1295·7	1306·0	1326·3	1346·6	1366·9	1387·2	1407·5	160
180	1288·3	1298·8	1309·2	1329·7	1350·2	1370·7	1391·1	1411·5	180
200	1291·0	1301·6	1312·1	1322·9	1353·7	1374·3	1394·9	1415·3	200
250	1297·1	1307·7	1318·5	1340·0	1361·2	1382·0	1402·6	1423·1	250
300	1301·6	1312·6	1323·4	1345·1	1366·5	1387·6	1408·6	1429·4	300
400	1308·9	1320·3	1331·6	1353·9	1375·9	1397·5	1418·7	1439·8	400
500	1314·0	1325·8	1337·5	1360·5	1383·4	1405·3	1427·0	1448·3	500
600	1317·9	1330·1	1342·1	1365·5	1388·6	1411·2	1433·5	1455·3	600
700	1320·9	1333·5	1345·9	1369·9	1393·4	1416·3	1438·8	1460·9	700
800	1323·2	1336·2	1349·0	1373·6	1397·5	1420·9	1443·7	1465·8	800
1000	1326·8	1340·0	1353·2	1378·5	1402·8	1426·9	1450·2	1473·1	1000
2000	1330·1	1346·0	1361·4	1390·4	1417·9	1444·4	1469·7	1494·0	2000

TABLE 11D. Entropy S of dry

P (lb/in.² abs)	(kN/m²)	Saturation θ (K)	Saturation S	Degrees of superheat (K) 10	20	30	40	50	60	70	P (kN/m²)
15	103·4	373·8	7·350	7·407	7·457	7·507	7·557	7·606	7·653	7·699	103·4
20	137·9	382·1	7·255	7·310	7·360	7·410	7·459	7·507	7·553	7·599	137·9
30	206·8	394·5	7·119	7·173	7·223	7·272	7·320	7·367	7·412	7·456	206·8
40	275·8	403·9	7·024	7·076	7·126	7·174	7·222	7·268	7·312	7·355	275·8
50	344·7	411·5	6·949	6·999	7·048	7·097	7·144	7·190	7·234	7·276	344·7
60	413·7	418·1	6·887	6·938	6·987	7·035	7·082	7·128	7·172	7·215	413·7
70	482·6	423·8	6·836	6·886	6·936	6·984	7·031	7·078	7·121	7·164	482·6
80	551·6	428·8	6·791	6·842	6·892	6·940	6·987	7·033	7·077	7·120	551·6
90	620·5	433·5	6·751	6·803	6·853	6·902	6·949	6·994	7·038	7·081	620·5
100	689·5	437·6	6·715	6·767	6·818	6·867	6·914	6·959	7·004	7·046	689·5
120	827·3	445·0	6·653	6·705	6·756	6·805	6·852	6·897	6·941	6·984	827·3
140	965·3	451·5	6·600	6·651	6·703	6·752	6·800	6·845	6·889	6·931	965·3
160	1103	457·4	6·553	6·604	6·656	6·705	6·754	6·799	6·843	6·886	1103
180	1241	462·7	6·512	6·562	6·615	6·664	6·713	6·758	6·802	6·845	1241
200	1378	467·5	6·475	6·525	6·578	6·627	6·675	6·721	6·765	6·808	1378
250	1724	478·1	6·396	6·449	6·503	6·553	6·601	6·648	6·693	6·737	1724
300	2068	487·3	6·329	6·385	6·440	6·491	6·541	6·589	6·634	6·678	2068
400	2758	502·4	6·220	6·281	6·337	6·390	6·440	6·489	6·535	6·579	2758
500	3447	514·9	6·132	6·195	6·253	6·308	6·359	6·410	6·954	6·501	3447
600	4137	525·5	6·057	6·119	6·181	6·238	6·293	6·343	6·391	6·438	4137
700	4827	534·9	5·991	6·058	6·122	6·181	6·236	6·288	6·337	6·384	4827
800	5516	543·3	5·931	6·003	6·068	6·130	6·186	6·240	6·289	6·337	5516
1000	6895	558·0	5·823	5·903	5·976	6·041	6·101	6·156	6·208	6·258	6895
2000	13790	608·6	5·383	5·516	5·623	5·713	5·792	5·862	5·924	5·981	13790

TABLE 11E. Entropy S of dry

P (lb/in.² abs)	Saturation θ	Saturation S	Degrees of superheat (°F) 20°	40°	60°	80°	100°	120°	140°	P (lb/in.² abs)
15	213·0	1·7556	1·7706	1·7844	1·7969	1·8101	1·8230	1·8353	1·8470	15
20	228·0	1·7327	1·7475	1·7610	1·7742	1·7867	1·7991	1·8110	1·8227	20
30	250·3	1·7004	1·7148	1·7281	1·7410	1·7534	1·7655	1·7773	1·7887	30
40	267·2	1·6776	1·6912	1·7045	1·7172	1·7295	1·7416	1·7531	1·7644	40
50	280·9	1·6597	1·6729	1·6861	1·6987	1·7109	1·7228	1·7344	1·7457	50
60	292·7	1·6450	1·6580	1·6713	1·6839	1·6961	1·7084	1·7200	1·7312	60
70	302·9	1·6327	1·6459	1·6593	1·6719	1·6840	1·6962	1·7078	1·7189	70
80	312·0	1·6219	1·6353	1·6488	1·6615	1·6736	1·6855	1·6971	1·7082	80
90	320·3	1·6124	1·6261	1·6396	1·6524	1·6645	1·6763	1·6878	1·6989	90
100	327·8	1·6038	1·6176	1·6311	1·6440	1·6561	1·6680	1·6795	1·6906	100
120	341·3	1·5891	1·6031	1·6166	1·6296	1·6417	1·6537	1·6652	1·6762	120
140	353·0	1·5762	1·5903	1·6038	1·6169	1·6290	1·6411	1·6526	1·6635	140
160	363·6	1·5652	1·5791	1·5926	1·6058	1·6179	1·6301	1·6416	1·6527	160
180	373·1	1·5554	1·5690	1·5825	1·5958	1·6080	1·6203	1·6318	1·6429	180
200	381·8	1·5466	1·5598	1·5735	1·5865	1·5990	1·6111	1·6228	1·6339	200
250	401·0	1·5276	1·5417	1·5557	1·5690	1·5818	1·5939	1·6057	1·6169	250
300	417·3	1·5117	1·5264	1·5408	1·5545	1·5675	1·5798	1·5918	1·6031	300
400	444·6	1·4857	1·5015	1·5164	1·5302	1·5434	1·5560	1·5680	1·5794	400
500	467·0	1·4646	1·4810	1·4959	1·5107	1·5240	1·5372	1·5594	1·5612	500
600	486·2	1·4466	1·4632	1·4793	1·4943	1·5084	1·5216	1·5341	1·5459	600
700	503·1	1·4308	1·4486	1·4654	1·4807	1·4950	1·5085	1·5211	1·5333	700
800	518·2	1·4165	1·4353	1·4529	1·4677	1·4832	1·4969	1·5099	1·5221	800
1000	544·6	1·3909	1·4119	1·4307	1·4477	1·4630	1·4772	1·4906	1·5031	1000
2000	635·8	1·2857	1·3206	1·3481	1·3710	1·3908	1·4082	1·4241	1·4387	2000

steam (superheated) (kJ/kg K)

P (lb/in.² abs)	(kN/m²)	Degrees of superheat (K)								P (kN/m²)
		80	90	100	120	140	160	180	200	
15	103·4	7·743	7·786	7·827	7·909	7·988	8·064	8·137	8·208	103·4
20	137·9	7·642	7·684	7·725	7·807	7·885	7·960	8·032	8·102	137·9
30	206·8	7·499	7·541	7·582	7·662	7·738	7·812	7·884	7·953	206·8
40	275·8	7·398	7·439	7·480	7·558	7·634	7·707	7·778	7·847	275·8
50	344·7	7·318	7·360	7·400	7·478	7·553	7·626	7·697	7·765	344·7
60	413·7	7·257	7·298	7·339	7·416	7·490	7·563	7·633	7·701	413·7
70	482·6	7·207	7·247	7·287	7·363	7·437	7·509	7·579	7·647	482·6
80	551·6	7·162	7·202	7·242	7·319	7·392	7·463	7·532	7·600	551·6
90	620·5	7·123	7·163	7·203	7·278	7·352	7·423	7·492	7·560	620·5
100	689·5	7·088	7·128	7·168	7·244	7·317	7·388	7·456	7·523	689·5
120	827·3	7·025	7·066	7·105	7·182	7·255	7·325	7·394	7·460	827·3
140	965·3	6·973	7·013	7·053	7·129	7·202	7·272	7·341	7·406	965·3
160	1103	6·928	6·969	7·008	7·085	7·157	7·227	7·295	7·360	1103
180	1241	6·887	6·928	6·968	7·044	7·118	7·187	7·255	7·319	1241
200	1378	6·850	6·892	6·931	7·007	7·081	7·151	7·219	7·283	1378
250	1724	6·779	6·820	6·860	6·937	7·010	7·080	7·147	7·211	1724
300	2068	6·721	6·762	6·802	6·878	6·951	7·021	7·089	7·153	2068
400	2758	6·621	6·664	6·705	6·781	6·854	6·294	6·990	7·055	2758
500	3447	6·545	6·588	6·629	6·706	6·779	6·849	6·917	6·982	3447
600	4137	6·483	6·525	6·566	6·644	6·717	6·788	6·856	6·922	4137
700	4827	6·429	6·472	6·522	6·592	6·666	6·738	6·804	6·872	4827
800	5516	6·383	6·426	6·469	6·547	6·622	6·698	6·761	6·828	5516
1000	6895	6·305	6·349	6·392	6·472	6·547	6·619	6·687	6·753	6895
2000	13790	6·036	6·086	6·132	6·217	6·297	6·372	6·443	6·510	13790

steam (superheated) (Btu/lb °F).

P (lb/in.² abs)	Degrees of superheat (°F)								P (lb/in.² abs)
	160°	180°	200°	240°	280°	320°	360°	400°	
15	1·8584	1·8694	1·8802	1·9015	1·9220	1·9415	1·9604	1·9786	15
20	1·8340	1·8450	1·8556	1·8770	1·8973	1·9166	1·9352	1·9529	20
30	1·8000	1·8108	1·8213	1·8422	1·8621	1·8812	1·8996	1·9173	30
40	1·7755	1·7865	1·7969	1·8176	1·8374	1·8563	1·8745	1·8920	40
50	1·7567	1·7675	1·7778	1·7980	1·8175	1·8363	1·8546	1·8724	50
60	1·7422	1·7529	1·7628	1·7829	1·8023	1·8210	1·8392	1·8569	60
70	1·7299	1·7406	1·7503	1·7702	1·7895	1·8080	1·8261	1·8437	70
80	1·7192	1·7298	1·7397	1·7595	1·7787	1·7970	1·8150	1·8325	80
90	1·7099	1·7204	1·7302	1·7500	1·7691	1·7874	1·8054	1·8228	90
100	1·7015	1·7120	1·7220	1·7418	1·7607	1·7789	1·7969	1·8142	100
120	1·6870	1·6974	1·7073	1·7270	1·7460	1·7639	1·7818	1·7990	120
140	1·6744	1·6848	1·6947	1·7144	1·7333	1·7512	1·7690	1·7861	140
160	1·6636	1·6740	1·6840	1·7036	1·7224	1·7402	1·7579	1·7749	160
180	1·6539	1·6643	1·6743	1·6939	1·7128	1·7305	1·7482	1·7652	180
200	1·6449	1·6554	1·6655	1·6854	1·7042	1·7220	1·7395	1·7563	200
250	1·6280	1·6384	1·6485	1·6684	1·6872	1·7050	1·7225	1·7390	250
300	1·6142	1·6246	1·6349	1·6547	1·6734	1·6914	1·7086	1·7250	300
400	1·5905	1·6011	1·6114	1·6313	1·6501	1·6679	1·6851	1·7016	400
500	1·5724	1·5831	1·5935	1·6133	1·6320	1·6502	1·6675	1·6842	500
600	1·5574	1·5682	1·5785	1·5986	1·6176	1·6358	1·6532	1·6698	600
700	1·5448	1·5558	1·5664	1·5864	1·6056	1·6236	1·6410	1·6578	700
800	1·5338	1·5449	1·5557	1·5759	1·5951	1·6133	1·6306	1·6472	800
1000	1·5151	1·5264	1·5374	1·5579	1·5771	1·5954	1·6129	1·6296	1000
2000	1·4522	1·4648	1·4768	1·4988	1·5191	1·5380	1·5557	1·5720	2000

Problems

1.1. 98% sulphuric acid of viscosity 0.025 N s/m^2 and density 1840 kg/m^3 is pumped at 685 cm^3/s through a 25 mm line. Calculate the value of the Reynolds number.

1.2. Compare the costs of electricity at 1p per kWh and town gas at 15p per 100 MJ.

1.3. A boiler plant raises 5.2 kg/s of steam at 1825 kN/m^2 pressure, using coal of calorific value 27.2 MJ/kg. If the boiler efficiency is 75%, how much coal is consumed per day? If the steam is used to generate electricity, what is the power generation in kilowatts, assuming a 20% conversion efficiency of the turbines and generators?

1.4. The power required by an agitator in a tank is a function of the following four variables:

(a) Diameter of impeller.
(b) Number of rotations of impeller per unit time.
(c) Viscosity of liquid.
(d) Density of liquid.

From a dimensional analysis, obtain a relation between the power and the four variables. The power consumption is found, experimentally, to be proportional to the square of the speed of rotation. By what factor would the power be expected to increase if the impeller diameter were doubled?

1.5. It is found experimentally that the terminal settling velocity u_0 of a spherical particle in a fluid is a function of the following quantities:

> particle diameter d,
> buoyant weight of particle (weight of particle–weight of displaced fluid) W,
> fluid density ρ,
> fluid viscosity μ.

Obtain a relationship for u_0 using dimensional analysis.
Stokes established, from theoretical considerations, that for small particles which settle at very low velocities, the settling velocity is independent of the density of the fluid except in so far as this affects the buoyancy. Show that the settling velocity *must* then be inversely proportional to the viscosity of the fluid.

2.1. Calculate the ideal available energy produced by the discharge to atmosphere through a nozzle of air stored in a cylinder of capacity 0.1 m^3 at a pressure of 5 MN/m^2. The initial temperature of the air is 290 K and the ratio of the specific heats is 1.4.

2.2. Obtain expressions for the variation of:

(a) internal energy with change of volume,
(b) internal energy with change of pressure, and
(c) enthalpy with change of pressure,

all at constant temperature, for a gas whose equation of state is given by van der Waals' law.

2.3. Calculate the energy stored in 1000 cm^3 of gas at 80 MN/m^2 and 290 K using a datum of STP.

2.4. Compressed gas is distributed from a works in cylinders which are filled to a pressure P by connecting them to a large reservoir of gas which remains at a steady pressure P and temperature T. If the small cylinders are initially at a temperature T and pressure P_0, what is the final temperature of the gas in the cylinders if heat losses can be neglected and if the compression can be regarded as reversible? Assume that the ideal gas laws are applicable.

3.1. 1250 cm^3/s of water is to be pumped through a steel pipe, 25 mm diameter and 30 m long, to a tank 12 m higher than its reservoir. Calculate approximately the power required. What type of pump would you install for the purpose and what power motor (in kW) would you provide?

> Viscosity of water $= 1.30$ mN s/m^2.
> Density of water $\quad = 1000$ kg/m^3.

3.2. Calculate the pressure drop in, and the power required to operate, a condenser consisting of 400 tubes 4.5 m long and 10 mm internal diameter. The coefficient of contraction at the entrance of the tubes is 0.6, and 0.04 m^3/s of water is to be pumped through the condenser.

3.3. 75% sulphuric acid, of density 1650 kg/m^3 and viscosity 8.6 mN s/m^2, is to be pumped for 0.8 km along a 50 mm internal diameter pipe at the rate of 3.0 kg/s, and then raised vertically 15 m by the pump. If the pump is electrically driven and has an efficiency of 50%, what power will be required? What type of pump would you use and of what material would you construct the pump and pipe?

3.4. 60% sulphuric acid is to be pumped at the rate of 4000 cm³/s through a lead pipe 25 mm diameter and raised to a height of 25 m. The pipe is 30 m long and includes two right-angled bends. Calculate the theoretical power required.

The specific gravity of the acid is 1·531 and its kinematic viscosity is 0·425 cm²/s. The density of water may be taken as 1000 kg/m³.

3.5. 1·3 kg/s of 98% sulphuric acid is to be pumped through a 25 mm diameter pipe, 30 m long, to a tank 12 m higher than its reservoir. Calculate the power required and indicate the type of pump and material of construction of the line that you would choose.

Viscosity of acid = 0·025 N s/m².
Specific gravity = 1·84.

3.6. Calculate the hydraulic mean diameter of the annular space between a 40 mm and a 50 mm tube.

3.7. 0·015 m³/s of acetic acid is pumped through a 75 mm diameter horizontal pipe 70 m long. What is the pressure drop in the pipe?

Viscosity of acid   = 2·5 mN s/m².
Density of acid   = 1060 kg/m³.
Roughness of pipe surface = 6 × 10⁻⁵ m.

3.8. A cylindrical tank, 5 m in diameter, discharges through a mild steel pipe 90 m long and 230 mm diameter connected to the base of the tank. Find the time taken for the water level in the tank to drop from 3 m to 1 m above the bottom. Take the viscosity of water as 1 mN s/m².

3.9. Two storage tanks A and B containing a petroleum product discharge through pipes each 0·3 m in diameter and 1·5 km long to a junction at D. From D the product is carried by a 0·5 m diameter pipe to a third storage tank C, 0·8 km away. The surface of the liquid in A is initially 10 m above that in C and the liquid level in B is 7 m higher than that in A. Calculate the initial rate of discharge of the liquid if the pipes are of mild steel. Take the density of the petroleum product as 870 kg/m³ and the viscosity as 0·7 mN s/m².

3.10. Find the drop in pressure due to friction in a pipe 300 m long and 150 mm diameter when water is flowing at the rate of 0·05 m³/s. The pipe is of glazed porcelain.

3.11. Two tanks, the bottoms of which are at the same level, are connected with one another by a horizontal pipe 75 mm diameter and 300 m long. The pipe is bell-mouthed at each end so that losses on entry and exit are negligible. One tank is 7 m diameter and contains water to a depth of 7 m. The other tank is 5 m diameter and contains water to a depth of 3 m. If the tanks are connected to each other by means of the pipe, how long will it take before the water level in the larger tank has fallen to 6 m? Assume the pipe to be an old mild steel pipe.

3.12. Two immiscible fluids A and B, of viscosities μ_A and μ_B, flow under streamline conditions between two horizontal parallel planes of width b, situated a distance 2a apart (where a is much less than b), as two distinct parallel layers one above the other, each of depth a. Show that the volumetric rate of flow of A is

$$\frac{\Delta P a^3 b}{12 \mu_A l} \times \frac{7\mu_A + \mu_B}{\mu_A + \mu_B},$$

where ΔP is the pressure drop over a length l in the direction of flow.

3.13. A petroleum fraction is pumped 2 km from a distillation plant to storage tanks through a mild steel pipeline, 150 mm in diameter, at the rate of 0·04 m³/s. What is the pressure drop along the pipe and the power supplied to the pumping unit if it has an efficiency of 50%?

The pump impeller is eroded and the pressure at its delivery falls to one half. By how much is the flowrate reduced?

Specific gravity of the liquid = 0·705.
Viscosity of the liquid   = 0·5 mN s/m².
Roughness of pipe surface   = 0·004 mm.

3.14. Glycerol is pumped from storage tanks to rail cars through a single 50 mm diameter main 10 m long, which must be used for all grades of glycerol. After the line has been used for commercial material, how much pure glycerol must be pumped before the issuing liquid contains not more than 1% of the commercial material? The flow in the pipeline is streamline and the two grades of glycerol have identical densities and viscosities.

3.15. A viscous fluid flows through a pipe with slightly porous walls so that there is a leakage of kP m³/m² s, where P is the local pressure measured above the discharge pressure and k is a constant. After a length L the liquid is discharged into a tank. If the internal diameter of the pipe is D m and the volumetric rate of flow at the inlet is Q m³/s, show that the pressure drop in the pipe is given by

$$P = \frac{Q}{\pi k D} a \tanh aL,$$

where $a = (128 \, k\mu/D^3)^{0·5}$

Assume a fully developed flow with $(R/\rho\mu^2) = 8Re^{-1}$.

3.16. A petroleum product of viscosity $0 \cdot 5$ mN s/m^2 and specific gravity $0 \cdot 7$ is pumped through a pipe of $0 \cdot 15$ m diameter to storage tanks situated 100 m away. The pressure drop along the pipe is 70 kN/m^2. The pipeline has to be repaired and it is necessary to pump the liquid by an alternative route consisting of 70 m of 20 cm pipe followed by 50 m of 10 cm pipe. If the existing pump is capable of developing a pressure of 300 kN/m^2, will it be suitable for use during the period required for the repairs? Take the roughness of the pipe surface as $0 \cdot 00005$ m.

3.17. Explain the phenomenon of hydraulic jump which occurs during the flow of a liquid in an open channel.

A liquid discharges from a tank into an open channel under a gate so that the liquid is initially travelling at a velocity of $1 \cdot 5$ m/s and a depth of 75 mm. Calculate, from first principles, the corresponding velocity and depth after the jump.

3.18. What is a non-Newtonian fluid? Describe the principal types of behaviour exhibited by these fluids. The viscosity of a non-Newtonian fluid changes with the rate of shear according to the approximate relationship

$$\mu = \mu_0(du/dr),$$

where μ is the viscosity, and du/dr is the velocity gradient normal to the direction of motion.

Show that the volumetric rate of streamline flow through a horizontal tube of radius a is

$$0 \cdot 25 \pi a^{3 \cdot 5} \sqrt{\frac{\Delta P}{\mu_0 l}},$$

where ΔP is the pressure drop over a length l of the tube.

3.19. Calculate the pressure drop when 3 kg/s of sulphuric acid flows through 60 m of 25 mm pipe ($\rho = 1840$ kg/m^3, $\mu = 0 \cdot 025$ N s/m^2).

3.20. Calculate the power required to pump oil of specific gravity $0 \cdot 85$ and viscosity 3 mN s/m^2 at 4000 cm^3/s through a 50 mm pipeline 100 m long, the outlet of which is 15 m higher than the inlet. The efficiency of the pump is 50%. What effect does the nature of the surface of the pipe have on the resistance?

3.21. 600 cm^3/s of water at 320 K is pumped in a 40 mm i.d. pipe through a length of 150 m in a horizontal direction and up through a vertical height of 10 m. In the pipe there is a control valve which may be taken as equivalent to 200 pipe diameters and other pipe fittings equivalent to 60 pipe diameters. Also in the line there is a heat exchanger across which there is a loss in head of $1 \cdot 5$ m of water. If the main pipe has a roughness of $0 \cdot 0002$ m, what power must be delivered to the pump if the unit is 60% efficient?

3.22. A pump developing a pressure of 800 kN/m^2 is used to pump water through a 150 mm pipe 300 m long to a reservoir 60 m higher. With the valves fully open, the flowrate obtained is $0 \cdot 05$ m^3/s. As a result of corrosion and scaling the effective absolute roughness of the pipe surface increases by a factor of 10. By what percentage is the flowrate reduced?

Viscosity of water $= 1$ mN s/m^2.

3.23. The relation between cost per unit length C of a pipeline installation and its diameter d is given by

$$C = a + bd,$$

where a and b are independent of pipe size. Annual charges are a fraction β of the capital cost. Obtain an expression for the optimum pipe diameter on a minimum cost basis for a fluid of density ρ and viscosity μ flowing at a mass rate of G. Assume that the fluid is in turbulent flow and that the Blasius equation is applicable, i.e. the friction factor is proportional to the Reynolds number to the power of minus one quarter. Indicate clearly how the optimum diameter depends on flowrate and fluid properties.

3.24. A heat exchanger is to consist of a number of tubes each 25 mm diameter and 5 m long arranged in parallel. The exchanger is to be used as a cooler with a rating of 4 MW and the temperature rise in the water feed to the tubes is to be 20 K.

If the pressure drop over the tubes is not to exceed 2 kN/m^2, calculate the minimum number of tubes that are required. Assume that the tube walls are smooth and that entrance and exit effects can be neglected.

Viscosity of water $= 1$ mN s/m^2.

3.25. Sulphuric acid is pumped at 3 kg/s through a 60 m length of smooth 25 mm pipe. Calculate the drop in pressure. If the pressure drop falls by one half, what will the new flowrate be?

Density of acid $= 1840$ kg/m^3.
Viscosity of acid $= 25$ mN s/m^2.

3.26. A Bingham plastic material is flowing under streamline conditions in a pipe of circular cross-section. What are the conditions for one half of the total flow to be within the central core across which the velocity profile is flat? The shear stress acting within the fluid R_y varies with velocity gradient du_x/dy according to the relation

$$R_y - R_c = -k(du_x/dy),$$

where R_c and k are constants for the material.

4.1. Town gas, having a molecular weight of 13 kg/kmol and a kinematic viscosity of $0 \cdot 25$ cm^2/s, is flowing through a pipe $0 \cdot 25$ m internal diameter and 5 km long at the rate of $0 \cdot 4$ m^3/s and is delivered at atmospheric pressure. Calculate the pressure required to maintain this rate of flow.

The volume occupied by 1 kmol at 289 K and $101 \cdot 3 \, kN/m^2$ may be taken as $24 \cdot 0 \, m^3$.

What effect on the pressure required would result if the gas was delivered at a height of 150 m (i) above and (ii) below its point of entry into the pipe?

4.2. Nitrogen at $12 \, MN/m^2$ pressure is fed through a 25 mm diameter mild steel pipe to a synthetic ammonia plant at the rate of $1 \cdot 25 \, kg/s$. What will be the drop in pressure over a 30 m length of pipe for isothermal flow of the gas at 298 K?

> Absolute roughness of the pipe surface = 0·005 mm.
> Kilogram molecular volume   = 22·4 m³.
> Viscosity of nitrogen   = 0·02 mN s/m².

4.3. Hydrogen is pumped from a reservoir at $2 \, MN/m^2$ pressure through a clean horizontal mild steel pipe 50 mm diameter and 500 m long. The downstream pressure is also $2 \, MN/m^2$ and the pressure of this gas is raised to $2 \cdot 6 \, MN/m^2$ by a pump at the upstream end of the pipe. The conditions of flow are isothermal and the temperature of the gas is 293 K. What is the flowrate and what is the effective rate of working of the pump?

> Viscosity of hydrogen = $0 \cdot 009 \, mN \, s/m^2$ at 293 K.

4.4. In a synthetic ammonia plant the hydrogen is fed through a 50 mm steel pipe to the converters. The pressure drop over the 30 m length of pipe is $500 \, kN/m^2$, the pressure at the downstream end being $7 \cdot 5 \, MN/m^2$. What power is required in order to overcome friction losses in the pipe? Assume isothermal expansion of the gas at 298 K. What error is introduced by assuming the gas to be an incompressible fluid of density equal to that at the mean pressure in the pipe? $\mu = 0 \cdot 02 \, mN \, s/m^2$.

4.5. A vacuum distillation plant operating at $7 \, kN/m^2$ pressure at the top has a boil-up rate of $0 \cdot 125 \, kg/s$ of xylene. Calculate the pressure drop along a 150 mm bore vapour pipe used to connect the column to the condenser. The pipe length may be taken as equivalent to 6 m, $e/d = 0 \cdot 002$ and $\mu = 0 \cdot 01 \, mN \, s/m^2$.

4.6. Nitrogen at $12 \, MN/m^2$ pressure is fed through a 25 mm diameter mild steel pipe to a synthetic ammonia plant at the rate of $0 \cdot 4 \, kg/s$. What will be the drop in pressure over a 30 m length of pipe assuming isothermal expansion of the gas at 300 K? What is the average quantity of heat per unit area of pipe surface that must pass through the walls in order to maintain isothermal conditions? What would be the pressure drop in the pipe if it were perfectly lagged?

4.7. Air, at a pressure of $10 \, MN/m^2$ and a temperature of 290 K, flows from a reservoir through a mild steel pipe of 10 mm diameter and 30 m long into a second reservoir at a pressure P_2. Plot the mass rate of flow of the air as a function of the pressure P_2. Neglect any effects attributable to differences in level and assume an adiabatic expansion of the air. $\mu = 0 \cdot 018 \, mN \, s/m^2$, $\gamma = 1 \cdot 36$.

4.8. In order to reduce transport costs it has been suggested that gasworks should be located in the coal-mining areas and the gas be distributed to the centres of population through high-pressure mains. Discuss the engineering and economic problems which are likely to be involved in such a project.

If a trunk pipeline is 200 mm in diameter and 80 km long, derive an expression for the rate of flow of coal gas in terms of the upstream and downstream pressures P_1 and P_2 respectively, on the assumption that the gas temperature remains at 290 K and that the ideal gas law is followed. Calculate the maximum rate of flow when the upstream pressure is $10 \, MN/m^2$. What is the pressure at the downstream end of the pipe under these conditions?

> Viscosity of coal gas   = 0·02 mN s/m².
> Mean molecular weight = 13 kg/kmol.
> Molecular volume   = 22·4 m³/kmol.
> Pipe roughness   = 0·05 mm.

4.9. Over a 30 m length of 150 mm vacuum line carrying air at 293 K the pressure falls from $1 \, kN/m^2$ to $0 \cdot 1 \, kN/m^2$. If the relative roughness e/d is $0 \cdot 002$, what is the approximate flowrate?

4.10. A vacuum system is required to handle 10 g/s of vapour (molecular weight 56 kg/kmol) so as to maintain a pressure of $1 \cdot 5 \, kN/m^2$ in a vessel situated 30 m from the vacuum pump. If the pump is able to maintain a pressure of $0 \cdot 15 \, kN/m^2$ at its suction point, what diameter pipe is required? The temperature is 290 K, and isothermal conditions may be assumed in the pipe, whose surface can be taken as smooth. The ideal gas law is followed.

> Gas viscosity = $0 \cdot 01 \, mN \, s/m^2$.

4.11. In a vacuum system, air is flowing isothermally at 290 K through a 150 mm diameter pipeline 30 m long. If the relative roughness of the pipewall e/d is $0 \cdot 002$ and the downstream pressure is $130 \, N/m^2$, what will the upstream pressure be if the flow rate of air is $0 \cdot 025 \, kg/s$?

Assume that the ideal gas law applies and that the viscosity of air is constant at $0 \cdot 018 \, mN \, s/m^2$.

What error would be introduced if the change in kinetic energy of the gas as a result of expansion were neglected?

4.12. Air is flowing at a rate of $30 \, kg/m^2 \, s$ through a smooth pipe of 50 mm diameter and 300 m long. If the upstream pressure is $800 \, kN/m^2$, what will the downstream pressure be if the flow is isothermal at 273 K? Take the viscosity of air as $0 \cdot 015 \, mN \, s/m^2$ and the kg molecular volume as $22 \cdot 4 \, m^3$. What is the significance of the change in kinetic energy of the fluid?

4.13. If temperature does not change with height, estimate the boiling point of water at a height of 3000 m above sea-level. The barometer reading at sea-level is $98 \cdot 4 \, kN/m^2$ and the temperature is $288 \cdot 7 \, K$. The vapour pressure of water at $288 \cdot 7 \, K$ is $1 \cdot 77 \, kN/m^2$. The molecular weight of air is 29 kg/kmol.

4.14. A 150 mm gas main is used for transferring coal gas (molecular weight 13 kg/kmol and kinematic viscosity $0 \cdot 25 \, cm^2/s$) at 295 K from a plant to a storage station 100 m away, at a rate of $1 \, m^3/s$. Calculate the pressure drop, if the pipe can be considered to be smooth.

If the maximum permissible pressure drop is 10 kN/m^2, is it possible to increase the flowrate by 25%?

5.1. Sulphuric acid of specific gravity 1·3 is flowing through a pipe of 50 mm internal diameter. A thin-lipped orifice, 10 mm diameter, is fitted in the pipe and the differential pressure shown by a mercury manometer is 10 cm. Assuming that the leads to the manometer are filled with the acid, calculate (a) the weight of acid flowing per second, and (b) the approximate loss of pressure (in kN/m^2) caused by the orifice.

The coefficient of discharge of the orifice may be taken as 0·61, the specific gravity of mercury as 13·55, and the density of water as 1000 kg/m^3.

5.2. The rate of discharge of water from a tank is measured by means of a notch for which the flowrate is directly proportional to the height of liquid above the bottom of the notch. Calculate and plot the profile of the notch if the flowrate is $0·01 \text{ m}^3/\text{s}$ when the liquid level is 150 mm above the bottom of the notch.

5.3. Water flows at between 3000 and 4000 cm^3/s through a 50 mm pipe and is metered by means of an orifice. Suggest a suitable size of orifice if the pressure difference is to be measured with a simple water manometer. What approximately is the pressure difference recorded at the maximum flowrate?

5.4. The rate of flow of water in a 150 mm diameter pipe is measured with a venturi meter with a 50 mm diameter throat. When the pressure drop over the converging section is 100 mm of water, the flowrate is 2·7 kg/s. What is the coefficient for the converging cone of the meter at that flowrate and what is the head lost due to friction? If the total loss of head over the meter is 15 mm water, what is the coefficient for the diverging cone?

5.5. A venturi meter with a 50 mm throat is used to measure a flow of slightly salt water in a pipe of inside diameter 100 mm. The meter is checked by adding 20 cm^3/s of normal sodium chloride solution above the meter and analysing a sample of water downstream from the meter. Before addition of the salt, 1000 cm^3 of water requires 10 cm^3 of 0·1 M silver nitrate solution in a titration. 1000 cm^3 of the downstream sample required 23·5 cm^3 of 0·1 M silver nitrate. If a mercury underwater manometer connected to the meter gives a reading of 205 mm, what is the discharge coefficient of the meter? Assume that the density of the liquid is not appreciably affected by the salt.

5.6. A gas cylinder containing 30 m^3 of air at 6 MN/m^2 pressure discharges to the atmosphere through a valve which may be taken as equivalent to a sharp-edged orifice of 6 mm diameter (coefficient of discharge = 0·6). Plot the rate of discharge against the pressure in the cylinder. How long will it take for the pressure in the cylinder to fall to (a) 1 MN/m^2, and (b) 150 kN/m^2?

Assume an adiabatic expansion of the gas through the valve and that the contents of the cylinder remain at 273 K.

5.7. Air at a pressure of 1500 kN/m^2 and a temperature of 370 K flows through an orifice of 30 mm^2 to atmospheric pressure. If the coefficient of discharge is 0·65, the critical pressure ratio is 0·527, and the ratio of the specific heats is 1·4, calculate the weight flowing per second.

5.8. Water flows through an orifice of 25 mm diameter situated in a 75 mm pipe at the rate of 300 cm^3/s. What will be the difference in level on a water manometer connected across the meter? Take the viscosity of water as 1 mN s/m^2.

5.9. Water flowing at $1500 \text{ cm}^3/\text{s}$ in a 50 mm diameter pipe is metered by means of a simple orifice of diameter 25 mm. If the coefficient of discharge of the meter is 0·62, what will be the reading on a mercury-under-water manometer connected to the meter?

What is the Reynolds number for the flow in the pipe?

$$\text{Density of water} = 1000 \text{ kg/m}^3.$$
$$\text{Viscosity of water} = 1 \text{ mN s/m}^2.$$

5.10. What size of orifice would give a pressure difference of 0·3 m water gauge for the flow of a petroleum product of specific gravity 0·9 at $0·05 \text{ m}^3/\text{s}$ in a 150 mm diameter pipe?

5.11. The flow of water through a 50 mm pipe is measured by means of an orifice meter with a 40 mm aperture. The pressure drop recorded is 150 mm on a mercury-under-water manometer and the coefficient of discharge of the meter is 0·6. What is the Reynolds number in the pipe and what would you expect the pressure drop over a 30 m length of the pipe to be?

$$\text{Friction factor, } \phi = R/\rho u^2 = 0·0025.$$
$$\text{Specific gravity of mercury} = 13·6.$$
$$\text{Viscosity of water} \qquad = 1 \text{ mN s/m}^2.$$

What type of pump would you use, how would you drive it, and what material of construction would be suitable?

5.12. A rotameter has a tube 0·3 m long which has an internal diameter of 25 mm at the top and 20 mm at the bottom. The diameter of the float is 20 mm, its effective specific gravity is 4·80, and its volume 6·6 cm^3. If the coefficient of discharge is 0·72, at what height will the float be when metering water at 100 cm^3/s?

5.13. Explain why there is a critical pressure ratio across a nozzle at which, for a given upstream pressure, the flowrate is a maximum.

Obtain an expression for the maximum flow for a given upstream pressure for isentropic flow through a horizontal nozzle. Show that for air (ratio of specific heats $\gamma = 1·4$) the critical pressure ratio is 0·53 and calculate the maximum flow through an orifice of area 30 mm^2 and coefficient of discharge 0·65 when the upstream pressure is $1·5 \text{ MN/m}^2$ and the upstream temperature 293 K.

$$\text{Kilogram molecular volume} = 22·4 \text{ m}^3.$$

5.14. A gas cylinder containing air discharges to atmosphere through a valve whose characteristics may be considered similar to those of a sharp-edged orifice. If the pressure in the cylinder is initially 350 kN/m^2, by how much will the pressure have fallen when the flowrate has decreased to one-quarter of its initial value?

The flow through the valve may be taken as isentropic and the expansion in the cylinder as isothermal. The ratio of the specific heats at constant pressure and constant volume is 1·4.

5.15. Water discharges from the bottom outlet of an open tank 1·5 m by 1 m in cross-section. The outlet is equivalent to an orifice 40 mm diameter with a coefficient of discharge of 0·6. The water level in the tank is regulated by a float valve on the feed supply which shuts off completely when the height of water above the bottom of the tank is 1 m and which gives a flowrate which is directly proportional to the distance of the water surface below this maximum level. When the depth of water in the tank is 0·5 m the inflow and outflow are directly balanced.

As a result of a short interruption in the supply, the water level in the tank falls to 0·25 m above the bottom but is then restored again. How long will it take the level to rise to 0·45 m above the bottom?

5.16. The flowrate of air at 298 K in a 0·3 m diameter duct is measured with a pitot tube which is used to traverse the cross-section. Readings of the differential pressure recorded on a water manometer are taken with the pitot tube at ten different positions in the cross-section. These positions are so chosen as to be the mid-points of ten concentric annuli each of the same cross-sectional area. The readings are as follows:

Position	1	2	3	4	5
Manometer reading (mm water)	18·5	18·0	17·5	16·8	15·7
Position	6	7	8	9	10
Manometer reading (mm water)	14·7	13·7	12·7	11·4	10·2

The flow is also metered using a 15 cm orifice plate across which the pressure differential is 50 mm on a mercury-under-water manometer. What is the coefficient of discharge of the orifice meter?

5.17. Explain the principle of operation of the pitot tube and indicate how it can be used in order to measure the total flowrate of fluid in a duct.

If a pitot tube is inserted in a circular cross-section pipe in which a fluid is in streamline flow, calculate at what point in the cross-section it should be situated so as to give a direct reading representative of the mean velocity of flow of the fluid.

6.1. A three-stage compressor is required to compress air from 140 kN/m^2 and 283 K to 4000 kN/m^2. Calculate the ideal intermediate pressures, the work required per kilogram of gas, and the isothermal efficiency of the process. ASsume the compression to be adiabatic and the interstage cooling to cool the air to the initial temperature. Show qualitatively, by means of temperature–entropy diagrams, the effect of unequal work distribution and imperfect intercooling, on the performance of the compressor.

6.2. A twin-cylinder, single-acting compressor, working at 5 Hz, delivers air at 515 kN/m^2 pressure, at the rate of $0·2 \text{ m}^3/\text{s}$. If the diameter of the cylinder is 20 cm, the cylinder clearance ratio 5% and the temperature of the inlet air 283 K, calculate the length of stroke of the piston and the delivery temperature.

6.3. A single-stage double-acting compressor running at 3 Hz is used to compress air from 110 kN/m^2 and 282 K to 1150 kN/m^2. If the internal diameter of the cylinder is 20 cm, the length of stroke 25 cm and the piston clearance 5%, calculate (a) the maximum capacity of the machine, referred to air at the initial temperature and pressure, and (b) the theoretical power requirements under isentropic conditions.

6.4. Methane is to be compressed from atmospheric pressure to 30 MN/m^2 in four stages.

Calculate the ideal intermediate pressures and the work required per kilogram of gas. Assume compression to be isentropic and the gas to behave as an ideal gas. Indicate on a temperature–entropy diagram the effect of imperfect intercooling on the work done at each stage.

6.5. An air-lift raises $0·01 \text{ m}^3/\text{s}$ of water from a well 100 m deep through a 100 mm diameter pipe. The level of the water is 40 m below the surface. The air consumed is $0·1 \text{ m}^3/\text{s}$ of free air compressed to 800 kN/m^2.

Calculate the efficiency of the pump and the mean velocity of the mixture in the pipe.

6.6. In a single-stage compressor:

$$\text{Suction pressure} \quad = 101·3 \text{ kN/m}^2.$$
$$\text{Suction temperature} = 283 \text{ K.}$$
$$\text{Final pressure} \quad = 380 \text{ kN/m}^2.$$
$$\text{Compression is adiabatic.}$$

If each new charge is heated 18 K by contact with the clearance gases, calculate the maximum temperature attained in the cylinder.

6.7. A single-acting reciprocating pump has a cylinder diameter of 115 mm and a stroke of 230 mm. The suction line is 6 m long and 50 mm in diameter, and the level of the water in the suction tank is 3 m below the cylinder of the pump. What is the maximum speed at which the pump can run without an air vessel if separation is not to occur in the suction line? The piston undergoes approximately simple harmonic motion. Atmospheric pressure is equivalent to a head of 10·4 m of water and separation occurs at a pressure corresponding to a head of 1·22 m of water.

6.8. An air-lift pump is used for raising 800 cm³/s of a liquid of specific gravity 1·2 to a height of 20 m. Air is available at 450 kN/m². If the efficiency of the pump is 30%, calculate the power requirement, assuming isentropic compression of the air ($\gamma = 1\cdot4$).

6.9. A single-acting air compressor supplies 0·1 m³/s of air (at STP) compressed to 380 kN/m² from 101·3 kN/m² pressure. If the suction temperature is 288·5 K, the stroke is 250 mm, and the speed is 4 Hz, find the cylinder diameter. Assume the cylinder clearance is 4% and compression and re-expansion are isentropic ($\gamma = 1\cdot4$). What is the theoretical power required for the compression?

6.10. Air at 290 K is compressed from 101·3 to 2000 kN/m² pressure in a two-stage compressor operating with a mechanical efficiency of 85%. The relation between pressure and volume during the compression stroke and expansion of the clearance gas is $PV^{1\cdot25} = $ constant. The compression ratio in each of the two cylinders is the same and the interstage cooler may be taken as perfectly efficient. If the clearances in the two cylinders are 4% and 5% respectively, calculate:

(a) the work of compression per unit mass of gas compressed;
(b) the isothermal efficiency;
(c) the isentropic efficiency ($\gamma = 1\cdot4$);
(d) the ratio of the swept volumes in the two cylinders.

6.11. Explain briefly the significance of the "specific speed" of a centrifugal or axial-flow pump.
A pump is designed to be driven at 10 Hz and to operate at a maximum efficiency when delivering 0·4 m³/s of water against a head of 20 m. Calculate the specific speed. What type of pump does this value suggest?
A pump, built for these operating conditions, has a measured maximum overall efficiency of 70%. The same pump is now required to deliver water at 30 m head. At what speed should the pump be driven if it is to operate at maximum efficiency? What will be the new rate of delivery and the power required?

6.12. A centrifugal pump is to be used to extract water from a condenser in which the vacuum is 640 mm of mercury. At the rated discharge the net positive suction head must be at least 3 m above the cavitation vapour pressure of 710 mm mercury vacuum. If losses in the suction pipe account for a head of 1·5 m, what must be the least height of the liquid level in the condenser above the pump inlet?

6.13. What is meant by the Net Positive Suction Head (NPSH) required by a pump? Explain why it exists and how it can be made as low as possible. What happens if the necessary NPSH is not provided?
A centrifugal pump is to be used to circulate liquid (sp. gr. 0·80 and viscosity 0·5 mN s/m²) from the reboiler of a distillation column through a vaporiser at the rate of 400 cm³/s, and to introduce the superheated liquid above the vapour space in the reboiler which contains liquid to a depth of 0·7 m. Suggest a suitable layout if a smooth bore 25 mm pipe is to be used. The pressure of the vapour in the reboiler is 1 kN/m² and the NPSH required by the pump is 2 m of liquid.

7.1. Calculate the time taken for the distant face of a brick wall, of thermal diffusivity, $D_H = 0\cdot0042$ cm²/s and thickness $l = 0\cdot45$ m, initially at 290 K, to rise to 470 K if the near face is suddenly raised to a temperature of $\theta' = 870$ K and maintained at that temperature. Assume that all the heat flow is perpendicular to the faces of the wall and that the distant face is perfectly insulated.

7.2. Calculate the time for the distant face to reach 470 K under the same conditions, except that the distant face is not perfectly lagged. Instead, a very large thickness of material of the same thermal properties as the brickwork is stacked against it.
Note: $p^{-1} e^{-k\sqrt{p}}$ is the Laplace transform of erfc $k/(2\sqrt{t})$.

Tables for erfc x for various values of x are given on p. 373 of *Conduction of Heat in Solids* by Carslaw and Jaeger.

7.3. Benzene vapour, at atmospheric pressure, condenses on a plane surface 2 m long and 1 m wide, maintained at 300 K and inclined at an angle of 45° to the horizontal. Plot the thickness of the condensate film and the point heat transfer coefficient against distance from the top of the surface.

7.4. It is desired to warm 0·9 kg/s of air from 283 to 366 K by passing it through the pipes of a bank consisting of 20 rows with 20 pipes in each row. The arrangement is in-line with centre to centre spacing, in both directions, equal to twice the pipe diameter. Flue gas, entering at 700 K and leaving at 366 K with a free flow mass velocity of 10 kg/m² s, is passed across the outside of the pipes.
Neglecting gas radiation, how long should the pipes be?
For simplicity, outer or inner pipe diameter may be taken as 12 mm.
Values of k and μ, which may be used for both air and flue gases, are given below. The specific heat of air and flue gases is 1·0 kJ/kg K.

Temperature (K)	Thermal conductivity k (W/m K)	Viscosity μ (mN s/m^2)
250	0·022	0·0165
500	0·040	0·0276
800	0·055	0·0367

7.5. A cooling coil, consisting of a single length of tubing through which water is circulated, is provided in a reaction vessel, the contents of which are kept uniformly at 360 K by means of a stirrer. The inlet and outlet temperatures of the cooling water are 280 and 320 K respectively. What would the outlet water temperature become if the length of the cooling coil were increased 5 times? Assume the overall heat transfer coefficient to be constant over the length of the tube and independent of the water temperature.

7.6. In an oil cooler 60 g/s of hot oil enters a thin metal pipe of diameter 25 mm. An equal mass of cooling water flows through the annular space between the pipe and a larger concentric pipe, the oil and water moving in opposite directions. The oil enters at 420 K and is to be cooled to 320 K. If the water enters at 290 K, what length of pipe will be required? Take coefficients of 1·6 kW/m^2 K on the oil side and 3·6 kW/m^2 K on the water side and 2·0 kJ/kg K for the specific heat of the oil.

7.7. The walls of a furnace are built up to 150 mm thickness of a refractory of thermal conductivity 1·5 W/m K. The surface temperatures of the inner and outer faces of the refractory are 1400 and 540 K respectively.

If a layer of insulating material 25 mm thick, of thermal conductivity 0·3 W/m K, is added, what temperatures will its surfaces attain assuming the inner surface of the furnace to remain at 1400 K? The coefficient of heat transfer from the outer surface of the insulation to the surroundings, which are at 290 K, may be taken as 4·2, 5·0, 6·1, and 7·1 W/m K, for surface temperatures of 370, 420, 470, and 520 K respectively. What will be the reduction in heat loss?

7.8. A pipe of outer diameter 50 mm, maintained at 1100 K, is covered with 50 mm of insulation of thermal conductivity 0·17 W/m K.

Would it be feasible to use a magnesia insulation which will not stand temperatures above 615 K and has a thermal conductivity 0·09 W/m K for an additional layer thick enough to reduce the outer surface temperature to 370 K in surroundings at 280 K? Take the surface coefficient of heat transfer by radiation and convection as 10 W/m^2 K.

7.9 In order to warm 0·5 kg/s of a heavy oil from 311 to 327 K, it is passed through tubes of inside diameter 19 mm and length 1·5 m, forming a bank, on the outside of which steam is condensing at 373 K. How many tubes will be needed?

In calculating Nu, Pr, and Re, the thermal conductivity of the oil may be taken as 0·14 W/m K and the specific heat as 2·1 kJ/kg K, irrespective of temperature. The viscosity is to be taken at the mean oil temperature. Viscosity of the oil at 319 and 373 K is 154 and 19·2 mN s/m^2 respectively.

7.10. A metal pipe of 12 mm outer diameter is maintained at 420 K. Calculate the rate of heat loss per metre run in surroundings uniformly at 290 K, (a) when the pipe is covered with 12 mm thickness of a material of thermal conductivity 0·35 W/m K and surface emissivity 0·95, and (b) when the thickness of the covering material is reduced to 6 mm, but the outer surface is treated so as to reduce its emissivity to 0·10.

The coefficients of radiation from a perfectly black surface in surroundings at 290 K are 6·25, 8·18, and 10·68 W/m^2 K at 310, 370, and 420 K respectively.

The coefficients of convection may be taken as $1·22 (\theta/d)^{0·25}$ W/m^2 K, where θ (K) is the temperature difference between the surface and the surrounding air, and d (m) is the outer diameter.

7.11. A condenser consists of 30 rows of parallel pipes of outer diameter 230 mm and thickness 1·3 mm, with 40 pipes, each 2 m long, per row. Water, inlet temperature 283 K, flows through the pipes at 1 m/s, and steam at 372 K condenses on the outside of the pipes. There is a layer of scale 0·25 mm thick, of thermal conductivity 2·1 W/m K, on the inside of the pipes.

Taking the coefficients of heat transfer on the water side as 4·0, and on the steam side as 8·5 kW/m^2 K, calculate the outlet water temperature and the total weight of steam condensed per second. The latent heat of steam at 372 K is 2250 kJ/kg. 1 m^3 water weighs 1000 kg.

7.12. In an oil cooler, water flows at the rate of 100 g/s per tube through metal tubes of outer diameter 19 mm and thickness 1·3 mm, along the outside of which oil flows in the opposite direction at the rate of 75 g/s per tube.

If the tubes are 2 m long, and the inlet temperatures of the oil and water are respectively 370 and 280 K, what will be the outlet oil temperature? The coefficient of heat transfer on the oil side is 1·7 and on the water side 2·5 kW/m^2 K, and the specific heat of the oil is 1·9 kJ/kg K.

7.13. Waste gases flowing across the outside of a bank of pipes are being used to warm air which flows through the pipes. The bank consists of 12 rows of pipes with 20 pipes, each 0·7 m long, per row. They are arranged in-line, with centre-to-centre spacing equal in both directions to one-and-a-half times the pipe diameter. Both inner and outer diameter may be taken as 12 mm. Air, with mass velocity 8 kg/m^{2}s, enters the pipes at 290 K. The initial gas temperature is 480 K and the total weight of the gases crossing the pipes per second is the same as the total weight of the air flowing through them.

Neglecting gas radiation, estimate the outlet temperature of the air. The physical constants for the waste gases may be assumed the same as for air, given on the next page:

Temperature (K)	Thermal conductivity (W/m K)	Viscosity (mN s/m^2)
250	0·022	0·0165
310	0·027	0·0189
370	0·030	0·0214
420	0·033	0·0239
480	0·037	0·0260

Specific heat = 1·00 kJ/kg K.

7.14. Oil is to be warmed from 300 to 344 K by passing it at 1 m/s through the pipes of a shell-and-tube heat exchanger. Steam at 377 K condenses on the outside of the pipes, which have outer and inner diameters of 48 and 41 mm respectively; but, owing to fouling, the inside diameter has been reduced to 38 mm, and the resistance to heat transfer of the pipe wall and dirt together, based on this diameter, is 0·0009 m^2 K/W.

It is known from previous measurements under similar conditions that the oil side coefficients of heat transfer for a velocity of 1 m/s, based on a diameter of 38 mm, vary with the temperature of the oil according to the table below:

Oil temperature (K)	300	311	322	333	344
Oil side coefficient of heat transfer (W/m^2 K)	74	80	97	136	244

The specific heat and density of the oil may be assumed constant at 1·9 kJ/kg K and 900 kg/m^3 respectively, and any resistance to heat transfer on the steam side neglected.

Find the length of tube bundle required.

7.15. It is proposed to construct a heat exchanger to condense 7·5 kg/s of n-hexane at a pressure of 150 kN/m^2, involving a heat load of 4·5 MW. The hexane is to reach the condenser from the top of a fractionating column at its condensing temperature of 356 K.

From experience it is anticipated that the overall heat transfer coefficient will be 450 W/m^2 K. The available cooling water is at 289 K.

Outline the proposals that you would make for the type and size of the exchanger and explain the details of the mechanical construction that you consider require special attention.

7.16. A heat exchanger is to be mounted at the top of a fractionating column about 15 m high to condense 4 kg/s of n-pentane at 205 kN/m^2, corresponding to a condensing temperature of 333 K. Give an outline of the calculations you would make to obtain an approximate idea of the size and construction of the exchanger required.

For purposes of standardisation, the company will use 19 mm outer diameter tubes of 1·65 mm wall thickness, and these may be 2·5, 3·6, or 5 m in length. The film coefficient for condensing pentane on the outside of a horizontal tube bundle may be taken as 1·1 kW/m^2 K. The condensation is effected by pumping water through the tubes, the initial water temperature being 288 K.

The latent heat of condensation of pentane is 335 kJ/kg.

For these 19 mm tubes, a water velocity of 1 m/s corresponds to a flowrate of 200 g/s of water.

7.17. An organic liquid is boiling at 340 K on the inside of a metal surface of thermal conductivity 42 W/m K and thickness 3 mm. The outside of the surface is heated by condensing steam. Assuming that the heat transfer coefficient from steam to the outer metal surface is constant at 11 kW/m^2 K, irrespective of the steam temperature, find what value of the steam temperature would give a maximum rate of evaporation.

The coefficients of heat transfer from the inner metal surface to the boiling liquid depend upon the temperature difference as shown below:

Temperature difference metal surface to boiling liquid (K)	Heat transfer coefficient metal surface to boiling liquid (kW/m^2 K)
22·2	4·43
27·8	5·91
33·3	7·38
36·1	7·30
38·9	6·81
41·7	6·36
44·4	5·73
50·0	4·54

7.18. It is desired to warm an oil of specific heat 2·0 kJ/kg K from 300 to 325 K by passing it through a tubular heat exchanger with metal tubes of inner diameter 10 mm. Along the outside of the tubes flows water, inlet temperature 372 K and outlet temperature 361 K.

The overall heat transfer coefficient from water to oil, reckoned on the inside area of the tubes, may be assumed constant at 230 W/m² K, and 75 g/s of oil is to be passed through each tube.

The oil is to make two passes through the heater. The water makes one pass along the outside of the tubes. Calculate the length of the tubes required.

7.19. A condenser consists of a number of metal pipes of outer diameter 25 mm and thickness 2·5 mm. Water, flowing at 0·6 m/s, enters the pipes at 290 K, and it is not permissible that it should be discharged at a temperature in excess of 310 K.

If 1·25 kg/s of a hydrocarbon vapour is to be condensed at 345 K on the outside of the pipes, how long should each pipe be and how many pipes should be needed?

Take the coefficient of heat transfer on the water side as 2·5 and on the vapour side as 0·8 kW/m² K and assume that the overall coefficient of heat transfer from vapour to water, based upon these figures, is reduced 20% by the effects of the pipe walls, dirt, and scale.

The latent heat of the hydrocarbon vapour at 345 K is 315 kJ/kg.

7.20. An organic vapour is being condensed at 350 K on the outside of a nest of pipes through which water flows at 0·6 m/s, its inlet temperature being 290 K. The outer and inner diameters of the pipes are 19 mm and 15 mm respectively, but a layer of scale 0·25 mm thick and thermal conductivity 2·0 W/m K has formed on the inside of the pipes.

If the coefficients of heat transfer on the vapour and water sides respectively are 1·7 and 3·2 kW/m² K and it is required to condense 25 g/s of vapour on each of the pipes, how long should these be and what will be the outlet temperature of the water?

The latent heat of condensation is 330 kJ/kg.

Neglect any small resistance to heat transfer in the pipe walls.

7.21. A heat exchanger is required to cool continuously 20 kg/s of warm water from 360 to 335 K by means of 25 kg/s of cold water, inlet temperature 300 K.

Assuming that the water velocities are such as to give an overall coefficient of heat transfer of 2 kW/m² K, assumed constant, calculate the total area of surface required (a) in a counterflow heat exchanger, i.e. one in which the hot and cold fluids flow in opposite directions, and (b) in a multipass heat exchanger, with the cold water making two passes through the tubes and the hot water making one pass along the outside of the tubes. In case (b) assume that the hot water flows in the same direction as the inlet cold water and that its temperature over any cross-section is uniform.

7.22. Find the heat loss per m² of surface through a brick wall 0·5 m thick when the inner surface is at 400 K and the outside at 310 K: the thermal conductivity of the brick may be taken as 0·7 W/m K.

7.23. A furnace is constructed with 225 mm of firebrick, 120 mm of insulating brick, and 225 mm of building brick. The inside temperature is 1200 K and the outside temperature 330 K. If the thermal conductivities are 1·4, 0·2, and 0·7 W/m K, find the heat loss per unit area and the temperature at the junction of the firebrick and insulating brick.

7.24. Calculate the total heat loss by radiation and convection from an unlagged horizontal steam pipe of 50 mm outside diameter at 415 K to air at 290 K.

7.25. Toluene is continuously nitrated to mononitrotoluene in a cast-iron vessel of 1 m diameter fitted with a propeller agitator of 0·3 m diameter driven at 2 Hz. The temperature is maintained at 310 K by circulating cooling water at 0·5 kg/s through a stainless steel coil of 25 mm outside diameter and 22 mm inside diameter wound in the form of a helix of 0·81 m diameter. The conditions are such that the reacting material may be considered to have the same physical properties as 75% sulphuric acid. If the mean water temperature is 290 K, what is the overall heat transfer coefficient?

7.26. 7·5 kg/s of pure iso-butane is to be condensed at a temperature of 331·7 K in a horizontal tubular exchanger using a water inlet temperature of 301 K. It is proposed to use 19 mm outside diameter tubes of 1·6 mm wall arranged on a 25 mm triangular pitch. Under these conditions the resistance of the scale may be taken as 0·0005 m² K/W. It is required to determine the number and arrangement of the tubes in the shell.

7.27. 37·5 kg/s of crude oil is to be heated from 295 to 330 K by heat exchange with the bottom product from a distillation column. The bottom product, flowing at 29·6 kg/s, is to be cooled from 420 to 380 K. There is available a tubular exchanger with an inside shell diameter of 0·60 m having one pass on the shell side and two passes on the tube side. It has 324 tubes, 19 mm outside diameter with 2·1 mm wall and 3·65 m long, arranged on a 25 mm square pitch and supported by baffles with a 25% cut, spaced at 230 mm intervals. Would this exchanger be suitable?

7.28. A 150 mm internal diameter steam pipe is carrying steam at 444 K and is lagged with 50 mm of 85% magnesia. What will be the heat loss to the air at 294 K?

7.29. A refractory material which has an emissivity of 0·40 at 1500 K and 0·43 at 1420 K is at a temperature of 1420 K and is exposed to black furnace walls at a temperature of 1500 K. What is the rate of gain of heat by radiation per unit area?

7.30. The total emissivity of clean chromium as a function of surface-temperature T K is given approximately by the empirical expression

$$e = 0·38[1 - (263/T)].$$

Obtain an expression for the absorptivity of solar radiation as a function of surface temperature and compare the absorptivity and emissivity at 300, 400, and 1000 K.

Assume that the sun behaves as a black body at 5500 K.

7.31. Repeat question 7.30 for the case of aluminium, assuming the emissivity to be $1 \cdot 25e$.

7.32. Calculate the heating due to solar radiation on the flat concrete roof of a building, 8 m by 9 m, in Africa, if the surface temperature of the roof is 330 K. What would be the effect of covering the roof with a highly reflecting surface such as polished aluminium?

The total emissivity of concrete at 330 K is 0·89, whilst the total absorptivity of solar radiation (sun temperature = 5500 K) at this temperature is 0·60. Use the data of question 7.31 for aluminium.

7.33. A rectangular iron ingot 15 cm $\times$ 15 cm $\times$ 30 cm is supported at the centre of a reheating furnace. The furnace has walls of silica-brick at 1400 K, and the initial temperature of the ingot is 290 K. How long will it take to heat the ingot to 600 K?

It may be assumed that the furnace is large compared with the ingot-size, and that the ingot remains at uniform temperature throughout its volume; convection effects are negligible.

The total emissivity of the oxidised iron surface is 0·78 and both emissivity and absorptivity are independent of the surface temperature.

Density of iron = $7 \cdot 2$ Mg/m^3.
Specific heat of iron = $0 \cdot 50$ kJ/kg K.

7.34. A wall is made of brick, of thermal conductivity 1·0 W/m K, 230 mm thick, lined on the inner face with plaster of thermal conductivity 0·4 W/m K and of thickness 10 mm. If a temperature difference of 30 K is maintained between the two outer faces, what is the heat flow per unit area of wall?

7.35. A 50 mm diameter pipe of circular cross-section and with walls 3 mm thick is covered with two concentric layers of lagging, the inner layer having a thickness of 25 mm and a thermal conductivity of 0·08 W/m K, and the outer layer has a thickness of 40 mm and a thermal conductivity of 0·04 W/m K. What is the rate of heat loss per metre length of pipe if the temperature inside the pipe is 550 K and the outside surface temperature is 330 K?

7.36. The temperature of oil leaving a co-current flow cooler is to be reduced from 370 to 350 K by lengthening the cooler. The oil and water flow rates and inlet temperatures, and the other dimensions of the cooler, will remain constant. The water enters at 285 K and the oil at 420 K. The water leaves the original cooler at 310 K. If the original length is 1 m, what must be the new length?

7.37. In a countercurrent-flow heat exchanger, 1·25 kg/s of benzene (specific heat 1·9 kJ/kg K and specific gravity 0·88) is to be cooled from 350 to 300 K with water which is available at 290 K. In the heat exchanger, tubes of 25 mm external and 22 mm internal diameter are employed and the water passes through the tubes. If the film coefficients for the water and benzene are 0·85 and 1·70 kW/m^2 K respectively and the scale resistance can be neglected, what total length of tube will be required if the minimum quantity of water is to be used and its temperature is not to be allowed to rise above 320 K?

7.38. Calculate the rate of loss of heat from a 6 m long horizontal steam pipe of 50 mm internal diameter and 60 mm external diameter when carrying steam at 800 kN/m^2. The temperature of the atmosphere and surroundings is 290 K.

What would be the cost of steam saved by coating the pipe with a 50 mm thickness of 85% magnesia lagging of thermal conductivity 0·07 W/m K, if steam costs £0·5 per 100 kg? The emissivity of the surface of the bare pipe and of the lagging may be taken as 0·85, and the coefficient h for heat loss by natural convection can be calculated from the expression

$$h = 1 \cdot 65 \, (\Delta T) 0 \cdot 25 \, (\text{W/m}^2 \text{ K}),$$

where ΔT is the temperature difference in K.

Take the Stefan–Boltzmann constant as $5 \cdot 67 \times 10^{-8}$ W/m^2 K^4.

7.39. A stirred reactor contains a batch of 700 kg reactants of specific heat 3·8 kJ/kg K initially at 290 K, which is heated by dry saturated steam at 170 kN/m^2 fed to a helical coil. During the heating period the steam supply rate is constant at 0·1 kg/s and condensate leaves at the temperature of the steam. If heat losses are neglected, calculate the true temperature of the reactants when a thermometer immersed in the material reads 360 K. The bulb of the thermometer is approximately cylindrical and is 100 mm long by 10 mm diameter with a water equivalent of 15 g, and the overall heat transfer coefficient to the thermometer is 300 W/m^2 K. What would a thermometer with a similar bulb of half the length and half the heat capacity indicate under these conditions?

7.40. Derive an expression relating the pressure drop for the turbulent flow of a fluid in a pipe to the heat transfer coefficient at the walls on the basis of the simple Reynolds analogy. Indicate the assumptions which are made and the conditions under which you would expect it to apply closely. Air at 320 K and atmospheric pressure is flowing through a smooth pipe of 50 mm internal diameter, and the pressure drop over a 4 m length is found to be 150 mm water gauge. By how much would you expect the air temperature to fall over the first metre if the wall temperature there is 290 K?

Viscosity of air   = 0·018 mN s/m^2.
Specific heat (C_p)   = 1·05 kJ/kg K.
Kilogram molecular volume = 22·4 m^3 at STP.

7.41. The radiation received by the earth's surface on a clear day with the sun overhead is 1 kW/m^2 and an additional 0.3 kW/m^2 is absorbed by the earth's atmosphere. Calculate approximately the temperature of the sun, assuming its radius to be 700,000 km and the distance between the sun and the earth to be 150,000,000 km. The sun may be assumed to behave as a black body.

7.42. A thermometer is immersed in a liquid which is heated at the rate of 0.05 K/s. If the thermometer and the liquid are both initially at 290 K, what rate of passage of liquid over the bulb of the thermometer is required if the error in the thermometer reading after 600 s is to be no more than 1 K? Take the water equivalent of the thermometer as 30 g and the heat transfer coefficient to the bulb to be given by $U = 735\,u^{0.8}$ W/m^2 K, and the area of the bulb as 0.01 m^2, where u is the velocity in m/s.

7.43. In a shell and tube type of heat exchanger with horizontal tubes 25 mm external diameter and 22 mm internal diameter, benzene is condensed on the outside by means of water flowing through the tubes at the rate of 0.03 m^3/s. If the water enters at 290 K and leaves at 300 K and the heat transfer coefficient on the water side is 850 W/m^2 K, what total length of tubing will be required?

7.44. In a contact sulphuric acid plant, the gases leaving the first convertor are to be cooled from 845 to 675 K by means of the air required for the combustion of the sulphur. The air enters the heat exchanger at 495 K. If the flow of each of the streams is 2 m^3/s at NTP, suggest a suitable design for a shell and tube type of heat exchanger employing tubes of 25 mm internal diameter.

(a) Assume parallel co-current flow of the gas streams.
(b) Assume parallel countercurrent flow.
(c) Assume that the heat exchanger is fitted with baffles giving cross-flow outside the tubes.

7.45. A large block of material of thermal diffusivity $D_H = 0.0042$ cm^2/s is initially at a uniform temperature of 290 K and one face is raised suddenly to 875 K and maintained at that temperature. Calculate the time taken for the material at a depth of 0.45 m to reach a temperature of 475 K on the assumption of unidirectional heat transfer and that the material can be considered to be infinite in extent in the direction of transfer.

7.46. A 50% glycerol–water mixture is flowing at a Reynolds number of 1500 through a 25 mm diameter pipe. Plot the mean value of the heat transfer coefficient as a function of pipe length assuming that

$$Nu = 1.62(Re\,Pr\,d/L)^{0.33}.$$

Indicate the conditions under which this is consistent with the predicted value $Nu = 4.1$ for fully developed flow.

7.47. A liquid is boiled at a temperature of 360 K using steam fed at a temperature of 380 K to a coil heater. Initially the heat transfer surfaces are clean and an evaporation rate of 0.08 kg/s is obtained from each square metre of heating surface. After a period, a layer of scale of resistance 0.0003 m^2 K/W is deposited by the boiling liquid on the heat transfer surface. On the assumption that the coefficient on the steam side remains unaltered and that the coefficient for the boiling liquid is proportional to its temperature difference raised to the power of 2.5, calculate the new rate of boiling.

7.48. A batch of reactants of specific heat 3.8 kJ/kg K and weighing 1000 kg is heated by means of a submerged seam coil of area 1 m^2 fed with steam at 390 K. If the overall heat transfer coefficient is 600 W/m^2 K, calculate the time taken to heat the material from 290 to 360 K, if heat losses to the surroundings are neglected.

If the external area of the vessel is 10 m^2 and the heat transfer coefficient to the surroundings at 290 K is 8.5 W/m^2 K, what will be the time taken to heat the reactants over the same temperature range and what is the maximum temperature to which the reactants can be raised?

What methods would you suggest for improving the rate of heat transfer?

7.49. What do you understand by the terms "black body" and "grey body" when applied to radiant heat transfer?

Two large, parallel plates with grey surfaces are situated 75 mm apart; one has an emissivity of 0.8 and is at a temperature of 350 K and the other has an emissivity of 0.4 and is at a temperature of 300 K. Calculate the net rate of heat exchange by radiation per square metre taking the Stefan–Boltzmann constant as 5.67×10^{-8} W/m^2 K^4. Any formula (other than Stefan's law) which you use must be proved.

7.50. A longitudinal fin on the outside of a circular pipe is 75 mm deep and 3 mm thick. If the pipe surface is at 400 K, calculate the heat dissipated per metre length from the fin to the atmosphere at 290 K if the coefficient of heat transfer from its surface may be assumed constant at 5 W/m^2 K. The thermal conductivity of the material of the fin is 50 W/m K and the heat loss from the extreme edge of the fin may be neglected. It should be assumed that the temperature is uniformly 400 K at the base of the fin.

7.51. Liquid oxygen is distributed by road in large spherical insulated vessels, 2 m internal diameter, well lagged on the outside. What thickness of magnesia lagging, of thermal conductivity 0.07 W/m K must be used so that not more than 1% of the liquid oxygen evaporates during a journey of 10 ks if the vessel is initially 80% full?

Latent heat of vaporisation of oxygen = 215 kJ/kg.
Boiling point of oxygen   = 90 K.
Density of liquid oxygen   = 1140 kg/m^3.
Atmospheric temperature   = 288 K.
Head transfer coefficient from outside
    lagging to atmosphere   = 4.5 W/m^2 K.

7.52. Benzene is to be condensed at the rate of 1.25 kg/s in a vertical shell and tube type of heat exchanger fitted with tubes of 25 mm outside diameter and 2.5 m long. The vapour condenses on the outside of the tubes and the cooling water enters at 295 K and passes through the tubes at 1·05 m/s. Calculate the number of tubes required if the heat exchanger is arranged for a single pass of the cooling water. The tube wall thickness is 1·6 mm.

7.53. One end of a metal bar 25 mm in diameter and 0·3 m long is maintained at 375 K and heat is dissipated from the whole length of the bar to surroundings at 295 K. If the coefficient of heat transfer from the surface is 10 W/m² K, what is the rate of loss of heat? Take the thermal conductivity of the metal as 85 W/m K.

7.54. A shell and tube heat exchanger consists of 120 tubes of internal diameter 22 mm and length 2·5 m. It is operated as a single pass condenser with benzene condensing at a temperature of 350 K on the outside of the tubes and water of inlet temperature 290 K passing through the tubes. Initially there is no scale on the walls and a rate of condensation of 4 kg/s is obtained with a water velocity of 0·7 m/s through the tubes. After prolonged operation, a scale of resistance 0·0002 m² K/W is formed on the inner surface of the tubes. To what value must the water velocity be increased in order to maintain the same rate of condensation on the assumption that the transfer coefficient on the water side is proportional to the velocity raised to the 0·8 power, and that the coefficient for the condensing vapour is 2·25 kW/m² K based on the inside area? The latent heat of vaporisation of benzene is 400 kJ/kg.

7.55. Derive an expression for the radiant heat transfer rate per unit area between two large parallel planes of emissivities e_1 and e_2 and at absolute temperatures T_1 and T_2 respectively.
Two such planes are situated 2·5 mm apart in air; one has an emissivity of 0·1 and is at a temperature of 350 K and the other has an emissivity of 0·05 and is at a temperature of 300 K. Calculate the percentage change in the total heat transfer rate by coating the first surface so as to reduce its emissivity to 0·025.

Stefan–Boltzmann constant $= 5·67 \times 10^{-8}$ W/m² K⁴.
Thermal conductivity of air $= 0·026$ W/m K.

7.56. Water flows at 2 m/s through a 2·5 m length of a 25 mm diameter tube. If the tube is at 320 K and the water enters and leaves at 293 and 295 K respectively, what is the value of the heat transfer coefficient? How would the outlet temperature change if the velocity were increased by 50%?

7.57. A liquid hydrocarbon is fed at 295 K to a heat exchanger consisting of a 25 mm diameter tube heated on the outside by condensing steam at atmospheric pressure. The flow rate of the hydrocarbon is measured by means of a 19 mm orifice fitted to the 25 mm feed pipe. The reading on a differential manometer containing the hydrocarbon-over-water is 450 mm and the coefficient of discharge of the meter is 0·6.
Calculate the initial rate of rise of temperature (K/s) of the hydrocarbon as it enters the heat exchanger.
Outside film coefficient = 6·0 kW/m² K. Inside film coefficient h is given by:

$$hd/k = 0·023(ud\rho/\mu)^{0·8}(C\mu/k)^{0·4}.$$
u = linear velocity of hydrocarbon (m/s).
d = tube diameter (m).
ρ = liquid density (800 kg/m³).
μ = liquid viscosity (9×10^{-4} N s/m²).
C = specific heat of liquid ($1·7 \times 10^3$ J/kg K).
k = thermal conductivity of liquid (0·17 W/m K).

7.58. Water passes at 1·2 m/s through a series of 25 mm diameter tubes 5 m long maintained at 320 K. If the inlet temperature is 290 K, at what temperature would you expect it to leave?

7.59. Heat is transferred from one fluid stream to a second fluid across a heat transfer surface. If the film coefficients for the two fluids are, respectively, 1·0 and 1·5 kW/m² K, the metal is 6 mm thick (thermal conductivity 20 W/m K) and the scale coefficient is equivalent to 850 W/m² K, what is the overall heat transfer coefficient?

7.60. A pipe of outer diameter 50 mm carries hot fluid at 1100 K. It is covered with a 50 mm layer of insulation of thermal conductivity 0·17 W/m K. Would it be feasible to use magnesia insulation, which will not stand temperatures above 615 K and has a thermal conductivity of 0·09 W/m K for an additional layer thick enough to reduce the outer surface temperature to 370 K in surroundings at 280 K? Take the surface coefficient of transfer by radiation and convection as 10 W/m² K.

7.61. A jacketed reaction vessel containing 0·25 m³ of liquid of specific gravity 0·9 and specific heat 3·3 kJ/kg K is heated by means of steam fed to a jacket on the walls. The contents of the tank are agitated by a stirrer running at 3 Hz. The heat transfer area is 2·5 m² and the steam temperature is 380 K. The outside film heat transfer coefficient is 1·7 kW/m² K and the 10 mm thick wall of the tank has a thermal conductivity of 6·0 W/m K. The inside film coefficient was found to be 1·1 kW/m² K for a stirrer speed of 1·5 Hz and to be proportional to the two-thirds power of the speed of rotation.
Neglecting heat losses and the heat capacity of the tank, how long will it take to raise the temperature of the liquid from 295 to 375 K?

7.62. By dimensional analysis, derive a relationship for the heat transfer coefficient h for natural convection between a surface and a fluid on the assumption that the coefficient is a function of the following variables:
k = thermal conductivity of the fluid.
C = specific heat of the fluid.
ρ = density of the fluid.
μ = viscosity of the fluid.
βg = the product of the acceleration due to gravity and the coefficient of cubical expansion of the fluid.
l = a characteristic dimension of the surface.
T = the temperature difference between the fluid and the surface.

Indicate why each of the above quantities would be expected to influence the heat transfer coefficient and explain how the orientation of the surface affects the process.

Under what conditions is heat transfer by convection important in chemical engineering?

7.63. A shell and tube heat exchanger is used for preheating the feed to an evaporator. The liquid of specific heat 4·0 kJ/kg K and specific gravity 1·1 passes through the inside of the tubes and is heated by steam condensing at 395 K on the outside. The exchanger heats liquid at 295 K to an outlet temperature of 375 K when the flow rate is 175 cm³/s and to 370 K when the flowrate is 325 cm³/s. What is the heat transfer area and the value of the overall heat transfer coefficient when the flow is 175 cm³/s?

Assume that the film heat transfer coefficient for the liquid in the tubes is proportional to the 0·8 power of the velocity that the transfer coefficient for the condensing steam remains constant at 3·4 kW/m² K and that the resistance of the tube wall and scale can be neglected.

8.1. Ammonia gas is diffusing at a constant rate through a layer of stagnant air 1 mm thick. Conditions are fixed so that the gas contains 50% by volume of ammonia at one boundary of the stagnant layer. The ammonia diffusing to the other boundary is quickly absorbed and the concentration is negligible at that plane. The temperature is 295 K and the pressure atmospheric, and under these conditions the diffusivity of ammonia in air is 0·18 cm²/s. Calculate the rate of diffusion of ammonia through the layer.

8.2. A simple rectifying column consists of a tube arranged vertically and supplied at the bottom with a mixture of benzene and toluene as vapour. At the top a condenser returns some of the product as a reflux which flows in a thin film down the inner wall of the tube. The tube is insulated and heat losses can be neglected. At one point in the column the vapour contains 70 mol% benzene and the adjacent liquid reflux contains 59 mol% benzene. The temperature at this point is 365 K. Assuming the diffusional resistance to vapour transfer to be equivalent to the diffusional resistance of a stagnant vapour layer 0·2 mm thick, calculate the rate of interchange of benzene and toluene between vapour and liquid. The molar latent heats of the two materials can be taken as equal. The vapour pressure of toluene at 365 K is 54·0 kN/m² and the diffusivity of the vapours is 0·051 cm²/s.

8.3. By what percentage would the rate of absorption be increased or decreased by increasing the total pressure from 100 to 200 kN/m² in the following cases?

(a) The absorption of ammonia from a mixture of ammonia and air containing 10% of ammonia by volume, using pure water as solvent. Assume that all the resistance to mass transfer lies within the gas phase.
(b) The same conditions as (a) but the absorbing solution exerts a partial vapour pressure of ammonia of 5 kN/m².

The diffusivity can be assumed to be inversely proportional to the absolute pressure.

8.4. In the Danckwerts model of mass transfer it is assumed that the fractional rate of surface renewal s is constant and independent of surface age. Under such conditions the expression for the surface age distribution function ϕ is $\phi = se^{-st}$. If the fractional rate of surface renewal were proportional to surface age (say $s = bt$, where b is a constant), show that the surface age distribution function would then assume the form

$$\phi = (2b/\pi)^{1/2}e^{-bt^2/2}.$$

8.5. By consideration of the appropriate element of a sphere show that the general equation for molecular diffusion in a stationary medium and in the absence of a chemical reaction is

$$\frac{\partial C}{\partial t} = D\left(\frac{\partial^2 C}{\partial r^2} + \frac{1}{r^2}\frac{\partial^2 C}{\partial \beta^2} + \frac{1}{r^2 \sin^2 \beta}\frac{\partial^2 C}{\partial \phi^2} + \frac{2}{r}\frac{\partial C}{\partial r} + \frac{\cot \beta}{r^2}\frac{\partial C}{\partial \beta}\right),$$

where C is the concentration of the diffusing substance, D the molecular diffusivity, t the time, and r, β, and ϕ are spherical polar coordinates, β being the latitude angle.

8.6. Prove that for equimolecular counterdiffusion from a sphere to a surrounding stationary, infinite medium, the Sherwood number based on the diameter of the sphere is equal to 2.

8.7. Show that the concentration profile for unsteady-state diffusion into a bounded medium of thickness L, when the concentration at the interface is suddenly raised to a constant value C_i and kept constant at the initial value of C_o at the other boundary is:

$$C = C_o + (C_i - C_o)\left[1 - \frac{z}{L} - \frac{2}{\pi}\sum_{n=1}^{n=\infty}\frac{1}{n}\exp(-n^2\pi^2 Dt/L^2)\sin nz\pi/L\right].$$

N.B. Assume the solution to be the sum of the solution for infinite time (steady-state part) and the solution of a second unsteady-state part; this simplifies the boundary conditions for the second part.

8.8. Show that under the conditions specified in question 8.7 and assuming the Higbie model of surface renewal, the average mass flux at the interface is given by

$$N_A = (C_i - C_o)D/L\left\{1 + (2L^2/\pi^2 Dt)\sum_{n=1}^{n=\infty}\left[\frac{\pi^2}{6} - \frac{1}{n^2}\exp(-n^2\pi^2 Dt/L^2)\right]\right\}.$$

N.B. Use the relation $\sum_{n=1}^{n=\infty}\frac{1}{n^2} = \pi^2/6$.

8.9. According to the simple penetration theory the instantaneous mass flux, N_A° is

$$N_A^\circ = (C_i - C_o)\left(\frac{D}{\pi t}\right)^{0.5}.$$

What is the equivalent expression for the instantaneous heat flux under analogous conditions?

Pure SO_2 is absorbed at 295 K and atmospheric pressure into a laminar water jet. The solubility of SO_2, assumed constant over a small temperature range, is $1.54 \, kmol/m^3$ under these conditions and the heat of solution is 28 kJ/kmol.

Calculate the resulting jet surface temperature if the Lewis number is 90. Neglect heat transfer between the water and the gas.

8.10. In a packed column, operating at approximately atmospheric pressure and 295 K, a 10% ammonia–air mixture is scrubbed with water and the concentration is reduced to 0·1%. If the whole of the resistance to mass transfer may be regarded as lying within a thin laminar film on the gas side of the gas–liquid interface, derive from first principles an expression for the rate of absorption at any position in the column. At some intermediate point where the ammonia concentration in the gas phase has been reduced to 5%, the partial pressure of ammonia in equilibrium with the aqueous solution is $660 \, N/m^2$ and the transfer rate is $10^{-3} \, kmol/m^2 \, s$. What is the thickness of the hypothetical gas film if the diffusivity of ammonia in air is $0.24 \, cm^2/s$?

8.11. An open bowl, 0·3 m in diameter, contains water at 350 K evaporating into the atmosphere. If the air currents are sufficiently strong to remove the water vapour as it is formed and if the resistance to its mass transfer in air is equivalent to that of a 1 mm layer for conditions of molecular diffusion, what will be the rate of cooling due to evaporation? The water can be considered as well mixed and the water equivalent of the system is equal to 10 kg. The diffusivity of water vapour in air may be taken as $0.20 \, cm^2/s$ and the kilogram molecular volume at NTP as $22.4 \, m^3$.

8.12. Show by substitution that when a gas of solubility C^+ is absorbed into a stagnant liquid of infinite depth, the concentration at time t and depth x is

$$C^+ \, \text{erfc} \, \frac{x}{2\sqrt{Dt}}.$$

Hence, on the basis of the simple penetration theory, show that the rate of absorption in a packed column will be proportional to the square root of the diffusivity.

8.13. Show that in steady-state diffusion through a film of liquid, accompanied by a first-order irreversible reaction, the concentration of solute in the film at depth z below the interface is given by:

$$C = \sinh \frac{\sqrt{(\alpha/D)}(z_L - z)}{\sinh \sqrt{(\alpha/D)}z_L} \, C_i$$

if $C = 0$ at $z = z_L$ and $C = C_i$ at $z = 0$, corresponding to the interface. Hence show that according to the "film theory" of gas-absorption, the rate of absorption per unit area of interface N_A is given by

$$N_A = K_L C_i \frac{\beta}{\tanh \beta},$$

where $\beta = \sqrt{(D\alpha)}/K_L$, D is the diffusivity of the solute, α the rate constant of the reaction, K_L the liquid film mass transfer coefficient for physical absorption, C_i the concentration of solute at the interface, z the distance normal to the interface, and z_L the liquid film thickness.

8.14. The diffusivity of the vapour of a volatile liquid in air can be conveniently determined by Winkelmann's method, in which liquid is contained in a narrow diameter vertical tube maintained at a constant temperature, and an air stream is passed over the top of the tube sufficiently rapidly to ensure that the partial pressure of the vapour there remains approximately zero. On the assumption that the vapour is transferred from the surface of the liquid to the air stream by molecular diffusion, calculate the diffusivity of carbon tetrachloride vapour in air at 321 K and atmospheric pressure from the following experimentally obtained data:

Time from commencement of experiment (ks)	Liquid level (cm)
0	0·00
1·6	0·25
11·1	1·29
27·4	2·32
80·2	4·39
117·5	5·47
168·6	6·70
199·7	7·38
289·3	9·03
383·1	10·48

The vapour pressure of carbon tetrachloride at 321 K is 37·6 kN/m² and the density of the liquid is 1540 kg/m³. Take the kilogram molecular volume as 22·4 m³.

8.15. Ammonia is absorbed in water from a mixture with air using a column operating at atmospheric pressure and 295 K. The resistance to transfer can be regarded as lying entirely within the gas phase. At a point in the column the partial pressure of the ammonia is 6·6 kN/m². The back pressure at the water interface is negligible and the resistance to transfer can be regarded as lying in a stationary gas film 1 mm thick. If the diffusivity of ammonia in air is 0·236 cm²/s, what is the transfer rate per unit area at that point in the column? If the gas were compressed to 200 kN/m² pressure, how would the transfer rate be altered?

8.16. What are the general principles underlying the two-film, penetration and film-penetration theories for mass transfer across a phase boundary? Give the basic differential equations which have to be solved for these theories with the appropriate boundary conditions.

According to the penetration theory, the instantaneous rate of mass transfer per unit area (N_A) at some time t after the commencement of transfer is given:

$$N_A = \Delta C \sqrt{\frac{D}{\pi t}},$$

where ΔC is the concentration force and D is the diffusivity.

Obtain expressions for the average rates of transfer on the basis of the Higbie and Danckwerts assumptions.

8.17. A solute diffuses from a liquid surface at which its molar concentration is C_i into a liquid with which it reacts. The mass transfer rate is given by Fick's law and the reaction is first order with respect to the solute. In a steady-state process the diffusion rate falls at a depth L to one half the value at the interface. Obtain an expression for the concentration C of solute at a depth z from the surface in terms of the molecular diffusivity D and the reaction rate constant α. What is the molar flux at the surface?

8.18. 4 cm³ of mixture formed by adding 2 cm³ of acetone to 2 cm³ of dibutyl phthalate is contained in a 6 mm diameter vertical glass tube immersed in a thermostat maintained at 315 K. A stream of air at 315 K and atmospheric pressure is passed over the open top of the tube to maintain a zero partial pressure of acetone vapour at that point. The liquid level is initially 1·15 cm below the top of the tube and the acetone vapour is transferred to the air stream by molecular diffusion alone. The dibutyl phthalate can be regarded as completely non-volatile and the partial pressure of acetone vapour may be calculated from Raoult's law on the assumption that the density of dibutyl phthalate is sufficiently greater than that of acetone for the liquid to be completely mixed.

Calculate the time taken for the liquid level to fall to 5 cm below the top of the tube, neglecting the effects of bulk flow in the vapour.

Kilogram molecular volume = 22·4 m³.
Molecular weights of acetone, dibutyl phthalate = 58 and 278 kg/kmol respectively.
Liquid densities of acetone, dibutyl phthalate = 764 and 1048 kg/m³ respectively.
Vapour pressure of acetone at 315 K = 60·5 kN/m².
Diffusivity of acetone vapour in air at 315 K = 0·123 cm²/s.

9.1. Calculate the thickness of the boundary layer at a distance of 75 mm from the leading edge of a plane surface over which water is flowing at a rate of 3 m/s. Assume that the flow in the boundary layer is streamline and that the velocity u of the fluid at a distance y from the surface can be represented by the relation $u = a + by + cy^2 + dy^3$ (where the coefficients a, b, c, and d are independent of y). Take the viscosity of water as 1 mN s/m².

9.2. Water flows at a velocity of 1 m/s over a plane surface 0·6 m wide and 1 m long. Calculate the total drag force acting on the surface if the transition from streamline to turbulent flow in the boundary layer occurs when the Reynolds group Re_x equals 10^5.

9.3. Calculate the thickness of the boundary layer at a distance of 150 mm from the leading edge of a surface over which oil, of viscosity 50 mN s/m² and density 990 kg/m³, flows with a velocity of 0·3 m/s. What is the displacement thickness of the boundary layer?

9.4. Calculate the thickness of the laminar sub-layer when benzene flows through a pipe of 50 mm diameter at 3000 cm³/s. What is the velocity of the benzene at the edge of the laminar sub-layer? Assume fully developed flow exists within the pipe.

9.5. Calculate the rise in temperature of water passed at 4 m/s through a smooth 25 mm diameter pipe, 6 m long. The water enters at 300 K and the temperature of the wall of the tube can be taken as approximately constant at 330 K. Use:

(a) The simple Reynolds analogy.
(b) The Taylor–Prandtl modification.
(c) The universal velocity profile.
(d) $Nu = 0·023\, Re^{0·8}\, Pr^{0·33}$.

Comment on the differences in the results so obtained.

9.6. Calculate the rise in temperature of a stream of air, entering at 290 K and passing at 4 m/s through the tube maintained at 350 K, other conditions remaining the same as in the previous question.

9.7. Air is flowing at a velocity of 5 m/s over a plane surface. Derive an expression for the thickness of the laminar sub-layer and calculate its value at a distance of 1 m from the leading edge of the surface.

Assume that within the boundary layer outside the laminar sub-layer the velocity of flow is proportional to the one-seventh power of the distance from the surface and that the shear stress R at the surface is given by

$$\frac{R}{\rho u_s^2} = 0.03 \left(\frac{u_s \rho x}{\mu}\right)^{-0.2},$$

where ρ is the density of the fluid (1.3 kg/m^3 for air), μ the viscosity of the fluid ($17 \times 10^{-6} \text{ N s/m}^2$ for air), u_s the stream velocity (m/s), and x the distance from the leading edge (m).

9.8. Air flows through a smooth circular duct of internal diameter 0.25 m at an average velocity of 15 m/s. Calculate the fluid velocity at points 50 mm and 5 mm from the wall. What will be the thickness of the laminar sub-layer if this extends to $u^+ = y^+ = 5$? The density of the air may be taken as 1.12 kg/m^3 and its viscosity as 0.02 mN s/m^2.

9.9. Obtain the momentum equation for an element of boundary layer. If the velocity profile in the laminar region can be represented approximately by a sine function, calculate the boundary-layer thickness in terms of distance from the leading edge of the surface.

10.1. If the temperature rise per metre length along a pipe carrying air at 12·2 m/s is 66 K, what will be the corresponding pressure drop for a pipe temperature of 420 K and an air temperature of 310 K?
The density of air at 310 K is 1.14 kg/m^3.

10.2. It is required to warm a quantity of air from 289 to 313 K by passing it through a number of parallel metal tubes of inner diameter 50 mm maintained at 373 K. The pressure drop must not exceed 250 N/m^2.
How long should the individual tubes be?
The density of air at 301 K is 1.19 kg/m^3.
The coefficients of heat transfer by convection from tube to air are 45, 62, and 77 W/m^2 K for velocities of 20, 24, and 30 m/s at 301 K respectively.

10.3. Air at 330 K, flowing at 10 m/s, enters a pipe of inner diameter 25 mm, maintained at 415 K. The drop of static pressure along the pipe is 80 N/m^2 per metre length. Using the Reynolds analogy between heat transfer and friction, estimate the temperature of the air 0·6 m along the pipe.

10.4. Air flows at 12 m/s through a pipe of inside diameter 25 mm. The rate of heat transfer by convection between the pipe and the air is 60 W/m^2 K. Neglecting the effects of temperature variation, estimate the pressure drop per metre length of pipe.

10.5. Apply Reynolds analogy to solve the following problem. Air at 320 K and atmospheric pressure is flowing through a smooth pipe of 50 mm internal diameter and the pressure drop over a 4 m length is found to be 1.5 kN/m^2. By how much would you expect the air temperature to fall over the first metre of pipe length if the wall temperature there is kept constant at 295 K?

$$\text{Viscosity of air} \quad = 0.018 \text{ mN s/m}^2.$$
$$\text{Specific heat of air} = 1.05 \text{ kJ/kg K.}$$

10.6. Obtain an expression for the simple Reynolds analogy between heat transfer and friction. Indicate the assumptions which are made in the derivation and the conditions under which you would expect the relation to be applicable.
The Reynolds number of a gas flowing at 2.5 kg/m^2 s through a smooth pipe is 20000. If the specific heat of the gas at constant pressure is 1.67 kJ/kg K, what will the heat transfer coefficient be?

11.1. In a process in which benzene is used as a solvent, it is evaporated into dry nitrogen. The resulting mixture at a temperature of 297 K and a pressure of 101.3 kN/m^2 has a relative humidity of 60%. It is desired to recover 80% of the benzene present by cooling to 283 K and compressing to a suitable pressure. What must this pressure be?

$$\text{Vapour pressures of benzene: at } 297 \text{ K} = 12.2 \text{ kN/m}^2; \text{ at } 283 \text{ K} = 6.0 \text{ kN/m}^2.$$

11.2. $0.6 \text{ m}^3/\text{s}$ of gas is to be dried from a dew point of 294 K to a dew point of 277·5 K. How much water must be removed and what will be the volume of the gas after drying?

$$\text{Vapour pressure of water at } 294 \text{ K} \quad = 2.5 \text{ kN/m}^2.$$
$$\text{Vapour pressure of water at } 277.5 \text{ K} = 0.85 \text{ kN/m}^2.$$

11.3. Wet material, containing 70% moisture, is to be dried at the rate of 0·15 kg/s in a countercurrent dryer to give a product containing 5% moisture (both on the wet basis). The drying medium consists of air heated to 373 K and containing water vapour equivalent to a partial pressure of 1.0 kN/m^2. The air leaves the dryer at 313 K and 70% saturated. Calculate how much air will be required to remove the moisture. The vapour pressure of water at 313 K may be taken as 7.4 kN/m^2.

11.4. 30000 m^3 of coal gas (measured at 289 K and 101.3 kN/m^2 saturated with water vapour) are compressed to 340 kN/m^2 pressure, cooled to 289 K and the condensed water is drained off. Subsequently the pressure is reduced to 170 kN/m^2 and the gas is distributed at this pressure and 289 K. What is the percentage humidity of the gas after this treatment?
The vapour pressure of water at 289 K is 1.8 kN/m^2.

11.5. A rotary countercurrent dryer is fed with ammonium nitrate containing 5% moisture at the rate of 1·5 kg/s, and discharges the nitrate with 0·2% moisture. The air enters at 405 K and leaves at 355 K, the humidity of the entering air being 0·007 kg of moisture per kg of dry air. The nitrate enters at 294 K and leaves at 339 K.

Neglecting radiation losses, calculate the weight of dry air passing through the dryer and the humidity of the air leaving the dryer.

Latent heat of water at 294 K   = 2450 kJ/kg.
Specific heat of ammonium nitrate = 1·88 kJ/kg K.
Specific heat of dry air   = 0·99 kJ/kg K.
Specific heat of water vapour   = 2·01 kJ/kg K.

11.6. Material is fed to a dryer at the rate of 0·3 kg/s and the moisture removed is 35% of the wet charge. The stock enters and leaves the dryer at 324 K. The air temperature falls from 341 to 310 K, its humidity rising from 0·01 to 0·02 kg/kg.
Calculate the heat loss to the surroundings.

Latent heat of water at 324 K = 2430 kJ/kg.
Specific heat of dry air   = 0·99 kJ/kg K.
Specific heat of water vapour = 2·01 kJ/kg K.

11.7. A rotary dryer is fed with sand at a rate of 1 kg/s. The feed is 50% wet and the sand is discharged with 3% moisture. The entering air is at 380 K and has an absolute humidity of 0·007 kg/kg. The wet sand enters at 294 K and leaves at 309 K and the air leaves at 310 K.
Calculate the mass of air passing through the dryer and the humidity of the air leaving the dryer. Allow a radiation loss of 25 kJ/kg of dry air.

Latent heat of water at 294 K = 2450 kJ/kg.
Specific heat of sand   = 0·88 kJ/kg K.
Specific heat of dry air   = 0·99 kJ/kg K.
Specific heat of vapour   = 2·01 kJ/kg K.

11.8. Water is to be cooled in a packed tower from 330 to 295 K by means of air flowing countercurrently. The liquid flows at a rate of 275 cm³/m² s and the air at 0·7 m³/m² s. The entering air has a temperature of 295 K and a relative humidity of 20%. Calculate the required height of tower and the condition of the air leaving at the top.
The whole of the resistance to heat and mass transfer can be considered as being within the gas phase and the product of the mass transfer coefficient and the transfer surface per unit volume of column ($h_D a$) can be taken as 0·2 per s.

11.9. Water is to be cooled in a small packed column from a temperature of 330 to 285 K by means of air flowing countercurrently. The rate of flow of liquid is 1400 cm³/m² s and the flow rate of the air, which enters at a temperature of 295 K and a relative humidity of 60%, is 3·0 m³/m² s. Calculate the required height of tower if the whole of the resistance to heat and mass transfer can be considered as being in the gas phase and the product of the mass transfer coefficient and the transfer surface per unit volume of column is 2 per s.
What is the condition of the air which leaves at the top?

11.10. Air containing 0·005 kg of water vapour per kg of dry air is heated to 325 K in a dryer and passed to the lower shelves. It leaves these shelves at 60% relative humidity and is reheated to 325 K and passed over another set of shelves, again leaving at 60% relative humidity. This is again reheated for the third and fourth sets of shelves, after which the air leaves the dryer. On the assumption that the material in each shelf has reached the wet bulb temperature and that heat losses from the dryer can be neglected, determine:

(a) the temperature of the material on each tray;
(b) the rate of water removal if 5 m³/s of moist air leaves the dryer;
(c) the temperature to which the inlet air would have to be raised to carry out the drying in a single stage.

11.11. 0·08 m³/s of air at 305 K and 60% relative humidity is to be cooled to 275 K. Calculate, by use of a psychrometric chart, the amount of heat to be removed for each 10 K interval of the cooling process. What total weight of moisture will be deposited? What is the humid heat of the air at the beginning and end of the process?

11.12. A hydrogen stream at 300 K and atmospheric pressure has a dew point of 275 K. It is to be further humidified by adding to it (through a nozzle) saturated steam at 240 kN/m² at the rate of 1 kg steam to 30 kg of the hydrogen feed. What will be the temperature and humidity of the resultant stream?

11.13. In a countercurrent packed column n-butanol flows down at a rate of 0·25 kg/m² s and is cooled from 330 to 295 K. Air at 290 K, initially free of n-butanol vapour, is passed up the column at the rate of 0·7 m³/m² s. Calculate the required height of tower and the condition of the exit air.

Data:

Mass transfer coefficient per unit volume: $h_D a = 0.1$ per s.

Psychrometric ratio: $\dfrac{h}{h_D C_A s} = 2.34$

Heat transfer coefficients: $h_L = 3h_G$ kg/m²s

Latent heat of vaporisation of n-butanol, $\lambda = 590$ kJ/kg.

Specific heat of liquid: $C_L = 2.5$ kJ/kg K.

Humid heat of gas: $s = 1.05$ kJ/kg K.

Temperature (K)	Vapour pressure (kN/m²)
295	0·59
300	0·86
305	1·27
310	1·75
315	2·48
320	3·32
325	4·49
330	5·99
335	7·89
340	10·36
345	14·97
350	17·50

Index